The Nature of the
Lower Continental Crust

GEOLOGICAL SOCIETY SPECIAL PUBLICATION NO 24

The Nature of the Lower Continental Crust

Papers read at a Joint Meeting of the
Metamorphic Studies Group of the Geological Society and the
Mineralogical Society, the Joint Association for Geophysics
and the 3rd Alfred Wegener Conference at
Burlington House, London, 24–26 October, 1984

edited by
J.B. Dawson and D.A. Carswell
Department of Geology, University of Sheffield

J. Hall
Department of Geology, University of Glasgow

K.H. Wedepohl
Geochemisches Institut der Universität, Göttingen

1986

Published for
The Geological Society by
Blackwell Scientific Publications
Oxford   London   Edinburgh
Boston   Palo Alto   Melbourne

Published for
The Geological Society by
Blackwell Scientific Publications
Osney Mead, Oxford OX2 0EL
8 John Street, London WC1N 2ES
23 Ainslie Place, Edinburgh EH3 6AJ
52 Beacon Street, Boston, Massachusetts 02108, USA
667 Lytton Avenue, Palo Alto, California 94301, USA
107 Barry Street, Carlton, Victoria 3053, Australia

First published 1986

DISTRIBUTORS

USA and Canada
   Blackwell Scientific Publications Inc.
   PO Box 50009, Palo Alto
   California 94303

Australia
   Blackwell Scientific Publications
   (Australia) Pty Ltd,
   107 Barry Street, Carlton,
   Victoria 3053

Printed in Great Britain
by Alden Press Ltd, Oxford

British Library Cataloguing in Publication Data

The nature of the lower continental crust.—
   (Geological Society special publication,
   ISSN 0305–8719; 24)
   1. Earth—Crust—Case studies
   2. Geophysics—Case studies
   I. Geological Society of London. *Metamorphic
   Studies Group* II. Alfred Wegener Conference
   (3rd : 1984 : Burlington House)
   III. Dawson, J.B. IV. Series
   551.1′4′0722   QE511

ISBN 0–632–01561–6

Library of Congress Cataloging-in-Publication Data

The nature of the lower continental crust.
   (Geological Society special publication; no. 24)
   "Papers read at a joint meeting of the Metamorphic
Studies Group of the Geological Society and 3rd
Alfred Wegener Conference at Burlington House,
London, 24–26 October, 1984"—
   Bibliography: p.
   Includes index.
   1. Earth—Crust—Congresses. I. Geological
Society of London. II. Series.
QE511.N28 1986   551.1′3   86–8309
ISBN 0–632–01561–6

Contents

Introduction

Continental crust was first defined from earthquake seismology by Mohorovicic as a surface layer with a predominant P-wave velocity of 6 km s^{-1} overlying a mantle with velocity $\geqslant 8$ km s^{-1}. Over sixty years ago Conrad showed the existence of a deep layer within the crust with a higher velocity (6.4 km s^{-1}) than most surface rocks. In the early models this crustal layer with higher velocity was assumed to be of gabbroic composition. The nature of this voluminous deeper layer is poorly known compared with that of the more accessible upper crust. Samples of what has once been the lower crust are available from deeply-eroded Precambrian terrains, in upthrust tectonic slices in orogenic belts and as xenoliths in volcanic rocks. Samples from these different sources complement each other, in that those from the more extensive terrains provide the best data on field-relationships and structure, whereas the xenoliths, having suffered neither the tectonic degradation nor the long exposures to lower temperatures and pressures suffered by the Precambrian terrains, probably provide us with the least equivocal chemical and isotope characteristics of the lower crust.

Earth scientists have been studying various aspects of the lower crust over the past three decades, but surprisingly, it had never formed the focus of a major conference. In 1983, the Metamorphic Studies Group (a Joint Group of the Geological Society and the Mineralogical Society) proposed a multidisciplinary conference to review our knowledge of this major layer of the earth, this proposal being strongly supported by the Joint Association for Geophysics of the Geological Society and the Royal Astronomical Society. Quite independently the same subject was being adopted in Germany by the Alfred Wegener Foundation as the theme of the 3rd Alfred Wegener Conference. Liaison within Working Group 5 of the International Lithosphere Programme happily resulted in a decision to hold a collaborative meeting in London, at the Geological Society in October 1984. As a result of this collaboration we were able to attract and support speakers from many parts of the world, both to address the conference and contribute to the present volume.

It is perhaps relevant to briefly review the type of lower crustal models that prevailed prior to the conference. Permissible models for the deep part of the continental crust must have seismic velocity and density greater than the average of upper-crustal rocks, together with distinctly lower heat production than typical upper-crustal rocks. For these reasons, and for indications from mineral stabilities, the granulite terrains have attracted attention as possible lower-crustal models. The most abundant rock type in some Precambrian high-grade terrains is a quartzofeldspathic granulite containing orthopyroxene and, often, garnet; this can be the granulite-facies equivalent of the familiar tonalite to granite-gneiss suite of the Archaean cratons. Intermediate quartzofeldspathic granulites have seismic velocities (at about 6 kbar) in the range 6.4–6.8 km s^{-1} and densities around 2.8 Mg m^{-3}, these ranges being suitable for some of the lower crust, though the velocities fall somewhat short of the 7 km s^{-1} characteristic of the deepest crustal regions in areas as diverse as the southern Yilgarn block of SW Australia and the Caledonides of northern Britain. The low surface heat flow of continental regions requires that the radiogenic heat production of the lower half of the crust be no greater than one quarter that of normal granite and the moderate to extreme depletion of certain acid granulites in the heat-producers U, Th and K, together with other large-ion lithophile (LIL) elements, qualifies them as a major deep-crustal component. Mafic rocks of normal radioactivity can be equally satisfactory in this respect whilst also satisfying the higher seismic velocities sometimes observed in deepest crust. This has led to more complex models of deep crust in which the identity of a mid-crustal Conrad discontinuity becomes confused. While aware of such complications, many authors have advanced specific granulite terrains as plausible deep-crustal models, including Enderby Land, Antarctica; the Adirondack Highlands of New York; the Fraser Range of Western Australia; the Ivrea Zone of northern Italy and the Scourie Complex of north-west Scotland, to name only a few. An integral part of most models envisages that the lower crustal granulites represent a water-poor restite after anatectic melting of pre-existing rock types, with migration of volatile- and LILE-enriched granitic melt fractions to the upper crust.

This type of lower-crustal model has become so prevalent that it was timely to pause and reconsider additional observations on the lithologies, structures, chemistry and physical properties of granulite-facies terrains that are at variance with, or require modification of, the granulite models. For example, many granulite terrains contain supracrustal lithologies including marbles, metapelites, calc-silicates, graphitic paragneisses and para-amphibolites; these exist alongside the dominant pyroxene-bearing orthogneisses in some of the highest-grade, most LILE-depleted terrains such as Scourie, NW Scotland and the Bamble area of SW Norway; this paradox of survival of supracrustal lithologies in a depleted (?restite) terrain is compounded by observations on the granulites of Jequie (Brazil) that are not LILE-depleted. Clearly tectonic mechanisms must exist for the transport of supracrustal lithologies to lower crustal depths as well as for the exhumation of relatively unmodified lower crustal segments. It was also time to examine the geotectonic setting of deep crustal terrains, together with their broader physical make-up. The COCORP and BIRPS deep seismic reflection data had revealed a highly heterogeneous deep crust with numerous reflectors, and also revealed the presence of major, low-angle thrusts and lamellar structures of alternating seismic velocities. Furthermore, in most granulite terrains, the largest-scale and earliest deformation style had been observed to be sub-horizontal, with large recumbent folds, flat-lying foliation and over-thrusts; commonly, the high-grade metamorphism began during the early deformation and outlasted it, and granulite-facies metamorphism gave rise to regional zonation with mappable isograds, such as the sillimanite-K-feldspar and orthopyroxene isograds in the Bamble area. Together these lithologic and structural features are reminiscent of those in younger orogenic terrains which might imply, pursuing the granulite deep-crustal model, that the continental crust has been built up by successive orogenic events operating on continental shelves, back-arc basins and island arcs.

Examination of these and other aspects of the lower continental crust, such as ductility, chemistry and regional studies, forms the subject of this volume. In addition, several important problems are identified. For example; does the Moho represent a lithologic or structural break at the base of the crust, or both? Why is the Conrad discontinuity sometimes absent? What is the reason for the presence of the layer of high electrical conductivity in the lower crust which, if logically interpreted as being due to the presence of water or free carbon, conflicts with the observed 'dry' granulite facies mineralogy? What are the wider results and implications of the carbonic metasomatism observed in many granulites? And, finally, why it is, in contrast to the granulites in exposed terrains, that the granulite xenoliths in volcanic pipes are dominantly basic in composition; is this the result of non-random volcanic sampling? These, and other problems, remain to be fully resolved. Nonetheless, the abundance of new data presented here and elsewhere show the existence of significant lateral and vertical heterogeneity, with the consequence that a simple model for the petrological make-up of the lower continental crust is redundant.

ACKNOWLEDGEMENTS: Financial assistance for the conference was provided by the Alfred Wegener Foundation, the Geological Society, the Royal Astronomical Society and the International Lithosphere Project. During both the preparation period and the conference itself we received the unstinting assistance of the Executive Secretary of the Geological Society, Mr R.M. Bateman, and his staff, of whom Miss C. Symonds deserves special mention. Finally, we express our deepest appreciation to our numerous colleagues who undertook the task of reviewing the papers for this volume.

J.B. Dawson
D.A. Carswell
J. Hall
K.H. Wedepohl.

Twenty years of deep seismic reflection profiling in Germany— a contribution to our knowledge of the nature of the lower Variscan crust

R. Meissner

SUMMARY: Near-vertical reflection studies in Germany have concentrated on the following objectives:

(1) The nature of crustal reflectors
(2) A detailed velocity structure
(3) Zones of increased tectonic activity, e.g. faults.

Approximately 20 years ago, the identity between reflectors in the near-vertical and the wide-angle range was established, wide-angle events being strongly influenced by velocity gradients. A pronounced, but limited low-velocity body below the Urach geothermal anomaly was detected by using an especially long spread length. An alternating polarity of deep crustal reflectors could be demonstrated. A deep fault zone, cutting the whole crust at steep angles, and shallow-angle overthrust structures were detected in the Variscan area. There is generally a strong reflectivity in the Variscan lower crust between the Conrad level at 6 s two-way travel time, i.e., 18 km, and the Moho at 9–10 s, i.e., 27–30 km depth. The lower crust seems to consist of thin lenses or lamellae with alternating velocity. They were probably generated by an extensive syn- and post-orogenic melting process leaving behind mafic and ultramafic 'splinters' in the lower crust and a compact ultramafic residue in the uppermost mantle, the Moho being formed as a new and uniform boundary. The poorly reflective upper crust is probably dominated by vertically oriented plutons in a rigid environment.

The early seismic data

Extensive seismic refraction and wide-angle observations have been made in West Germany since 1947, providing tight constraints on the velocity structure in, and adjacent to, the strongly differentiated Variscan crust (Giese *et al.* 1976). In addition, since 1964, specific seismic reflection studies across geological structures of specific interest were performed. Fig. 1 shows a location map of these profiles. The first two crustal reflection lines in the Bavarian Molasse trough and in the Rhenish Massif were combined with a very wide-angle common midpoint (CMP) configuration of 180 km width in an attempt to collect velocity data as accurately as possible although restricted to the area below the CMP. Based on these two experiments two major conclusions were obtained:

(i) the reality of the near-vertical reflectors was demonstrated by their coincidence with jumps or gradient zones in velocities as obtained from the CMP-observations

(ii) the concept of lamellae for the lowermost crust was born, based on a joint interpretation of near-vertical and wide angle data and on studies of continuity, frequency and amplitude response of reflections (Meissner 1967, 1976; Glocke & Meissner 1976).

Based on these early data the next two near-vertical reflection experiments in 1973 and 1975 were directed towards two steeply dipping fault zones. Here, the continuous profiling technique of exploration seismics was used in a slightly modified version. The objective of these experiments was to locate the dip and depth range of the fault zones. A large listric fault could be observed at the transition from the Devonian rocks of the Rhenish Massif into the Permian Saar Nahe trough, where up to 4 km of sediments were accumulated during the post-Variscan break-up, at a time when extensional tectonics dominated. Steeply dipping reflectors in the half-graben-shaped Saar-Nahe trough are clearly separated from the sub-horizontal layering in the Rhenish Massif. It is interesting to note that the fault zone itself in the lower crust acts as a reflecting boundary (Meissner *et al.* 1980), although seismicity ends at about 15 km. The fault zone cuts the whole crust and reaches the mantle at 30 to 35 km.

A completely different picture of reflectivity was obtained along a short profile crossing one of the western branches of the Tertiary Rhinegraben fault. Only short segments of reflecting boundaries were observed, but these extended well into the upper mantle. No continuous reflections, no separation of dips, and only some small amplitude maxima could be mapped. The whole pattern of reflectivity seems to indicate that vertically oriented mixing processes are still going on

From: DAWSON, J.B., CARSWELL, D.A., HALL, J. & WEDEPOHL, K.H. (eds) 1986, *The Nature of the Lower Continental Crust*, Geological Society Special Publication No. 24, pp. 1–10.

I

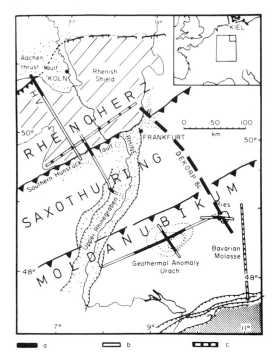

FIG. 1. Location map of main tectonic units and seismic reflection profiles in Western Germany. Rhenoherz = Rhenohercynian, Saxothuring = Saxothuringian, Moldanubikum = Moldanubian. a = reflection survey using the continuous profiling technique; b = wide-angle-refraction observations in connection with reflection surveys; c = near-vertical and wide-angle observations with a common depth point (CMP) configuration. DEKORP = first DEKORP profile, observed in 1984

between mantle and crust, and a final separation of crustal material has not yet taken place.

Another short profile through the Ries-impact crater in the Moldanubian (Angenheister & Pohl 1976) showed that the lower crust is highly reflective, but effects of the crater walls obscure some of the deeper information.

The Aachen and Urach data of 1978

A 35 km long reflection line was observed across the Aachen–Nohes Venn thrust fault, known in France and Belgium as the Faille due Midi fault system which marks the latest and northernmost tectonic effect of the Variscan orogenesis (Ziegler 1978, 1982). The overthrust fault appears as a strong sub-horizontal reflector of negative reflection polarity between 1 and 1.2 s travel time. In Fig. 2 it is the uppermost reflecting interface. As

usually seen in Variscan areas the lower crust is full of reflectors; some of them might also be thrust zones as indicated by the rather continuous bands of reflecting interfaces between 4 and 7 s two way travel time which are especially clear in the migrated version of Fig. 2. The Moho itself shows up as a rather weak and sporadic accumulation of sub-horizontal reflectors around 10 s (Meissner et al. 1981, 1983).

The two Urach profiles were directed toward the investigation of the Urach geothermal anomaly, and especially long spreads of 23 km were used in order to be able to calculate interval velocities as accurately as possible. In addition the structural resolution obtained by 8-fold coverage was quite good. Fig. 3a shows a long line U_1 in the strike direction of the Variscides. An extremely reflective lower crust with only minor possible faults is seen. In contrast, the short U_2-line perpendicular to the Variscan strike shows a more discontinuous appearance with at least two centres of diffraction patterns (Fig. 3b). Such a difference between seismic profiles along and perpendicular to the geological strike direction seems to be a more general phenomenon and is observed also in the BIRPS network (Brewer et al. 1983). The strong dips of reflectors of the lower crust along U_2 is a combined effect from the final uplift of the Rhinegraben shoulders and the general deepening of the crust towards the Alps.

A major feature of the investigation was the detection of a big, low velocity body (LVB) below the centre of the geothermal anomaly. Using an iterative method of Krey (1976, 1978) the stacking velocities were transformed into interval velocities. Fig. 4a shows the LVB with lateral velocity anomalies up to 10%. As demonstrated by Bartelsen et al. (1982) and Meissner et al. (1982) a 10% decrease in velocity would correspond to about 400°C increase in temperature if laboratory relationships are used. Such a strong temperature anomaly is neither compatible with the small heat flow anomaly of only 20–40 mW/m² compared to the surrounding of the Urach anomaly, nor with the rather smooth gravity picture which even exhibits a small positive Bouguer anomaly. Probably, intruding basalting magma has modified, e.g. cracked, the area of the LVB by thermal effects of heating and subsequent cooling.

Another feature which seems to be characteristic of the lower crust is the polarity of reflections. As seen in Fig. 4b the polarity is positive for two strong reflectors inside the lower crust, one of them belonging to the Moho proper. Both these boundaries are also observed in the over-critical angle range in the form of strong and continuous wide-angle events proving their positive polarity,

Aachen Profile

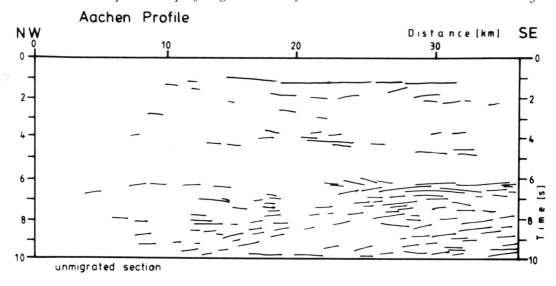

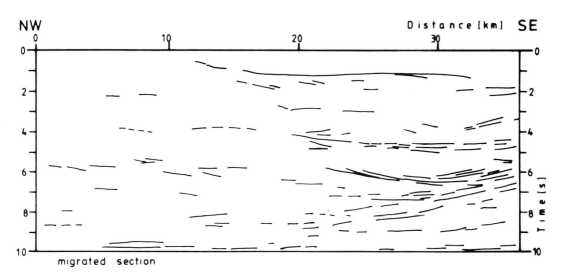

FIG. 2. Line drawing of reflecting boundaries of the Aachen–Hohes Venn profile; unmigrated and migrated section.

caused by a velocity jump (or strong gradient) to higher values. But in between these jumps to higher velocities (strictly, higher impedance $\rho \cdot V_p$), there are clear reflectors with inverse polarity. This picture is proof of the lamellar nature of the lower crust. Reflections are generated by alternating layers of higher and lower velocities. Dense layering results in an interference with a high probability for discontinuous reflectors of various frequencies. Lamellae with thicknesses slightly larger than the dominant wave-length produce individual reflectors with both kinds of polarities.

Histograms of reflections

In order to approach the problem of the nature of reflections and to compare reflection data from different tectonic provinces some quantitative

R. Meissner

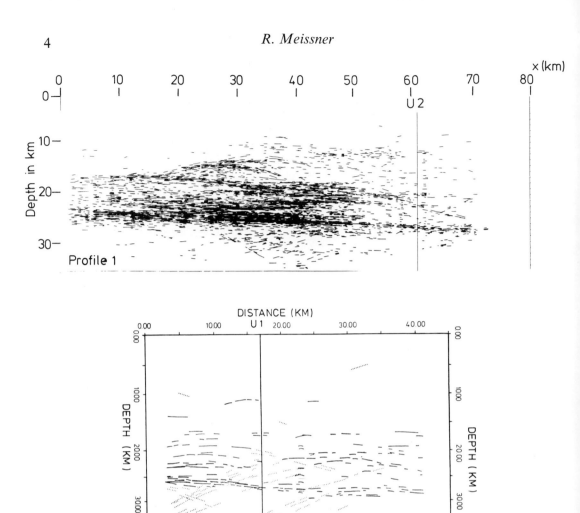

FIG. 3. Line drawings from the two Urach profiles: 3a: Line U1, composition of single records, 8-fold. 3b: Line U2, stacked version.

measure for the number of correlatable reflectors had to be introduced. In a study by Wever (1984) the number of reflections per constant travel time intervals of 1 s was normalized with regard to the number of CMP points per length and the total length of the reflection profile. Six histograms from various areas of the North American and European continent are presented in Fig. 5. The data were obtained with different field techniques: COCORP data on land by vibrators, BIRPS data by marine air-guns and streamers and the German data by explosives in boreholes. But looking at similar geologic provinces e.g. the Palaeozoic areas of BIRPS' WINCH lines (Brewer *et al.*

1983) and the Urach line in the strike direction (Bartelsen *et al.* 1982), histograms show the same features: a strong concentration of reflections in the lower crust. Also the COCORP data near the (Variscan) coastal plains (Cook *et al.* 1979, 1981) and the (Variscan) Hohes Venn data (Meissner *et al.* 1983), two areas where compressional tectonics had created large sub-horizontal overthrusts, certainly give similar histograms. This means that the different field techniques do not have a significant influence on the histograms; the geological age of the area seems to play a much stronger role. The old shield areas in the North American continent as demonstrated by the

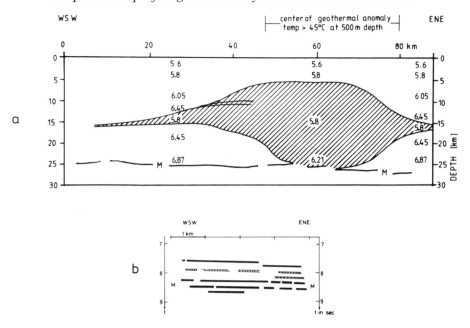

FIG. 4. The Urach low velocity body (a), as obtained from the calculated interval velocity (values are in km s^{-1}) and an example of reflectors with different polarity in the lower crust. Black bars: positive polarity, i.e. $\rho \cdot V$ increasing; hatched bars: negative polarity, i.e. $\rho \cdot V$ decreasing.

profiles through the Wind River area and Minnesota (Gibbs *et al.* 1984) do not show prominent reflectors in the lower crust, and by and large even the crust–mantle boundary cannot be detected. Instead, reflections concentrate in the upper crust. They are especially strong in those areas where later tectonic events have modified the old shield area.

In general, reflectivity seems to correlate with the latest tectonic event. The extensional tectonics which followed the Caledonian and Variscan orogeneses apparently produced a thorough heating, melting and a resettling of the remaining crustal material. This process will be discussed again in the final section.

Seismic velocities as indicators of crustal maturation processes

It has long been known that the old shield and platform areas have thicker crusts with higher seismic velocities than have the later accumulated younger crusts of Caledonian and Variscan age (Vetter & Meissner 1970). The old and the younger areas are generally not far away from a perfect isostatic balance. Because of the proportionality between density ρ and the seismic velo-

city V_p one may state that the younger, thinner crusts are also less dense (and possibly more sialic) than the old shield areas.

Fig. 6 shows a simple picture of this situation: the shield curve with a 40 km thick crust of higher velocity causes the same gravity effect as does a 'Variscan–Caledonian' crust with only about 30 km crustal thickness. But this relationship may have a genetic explanation. A melting of a high-velocity, i.e. mafic, lower crust results in an increase of the upper (low velocity) crustal part by the addition of light melt components, while the mafic–ultramafic residues will become a part of the upper mantle, thereby reducing the total crustal thickness. Certainly the generation of younger crusts at the periphery of old cratons has many more genetic aspects, but a thorough syn- and post-orogenic melting seems to be a general phenomenon. This is indicated by (i) the modified crustal composition, (ii) numerous Variscan granitic plutons and (iii) a surprisingly uniform and small Moho depth in the whole area of the different Variscan collision zones (Meissner 1985). Melting and subsequent cooling of mafic material of the lower crust may result in crystallization seams which seem to be responsible for the high reflectivity of the lower Variscan and Caledonian crust, as will be discussed later.

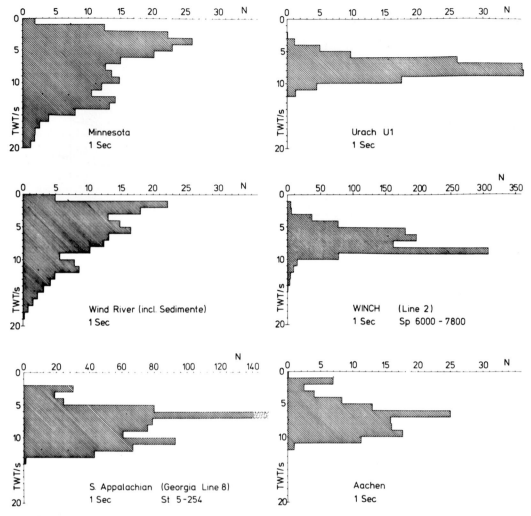

FIG. 5. Histograms of normalized reflection density. N $(\Delta t) = (n/s) \sum l_{\Delta t}$; $\Delta t = 1s$, $l_{\Delta t} =$ length of correlatable reflection in cm, n = number of CMP-points per cm, s = length of profile in km.

Relationships between crustal viscosities and seismicity

Based on Byerlee's relationship between maximum stresses (= frictional strength) and shear stress in combination with stresses obtained from ductile behaviour of rocks under elevated temperatures, Meissner and Strehlau (1982) obtained theoretical curves of maximum stresses (STRESSMAX) as a function of temperature and water content. Fig. 7 shows two such diagrams, the upper one for dry conditions, the lower one for rocks containing water under hydrostatic pressure. The surface heat flow density is shown

as a parameter in both sets of curves. Different values of strain rate have only a small influence on the shape of the curves. Different creep laws also result in the same shape, always showing a strong peak at the transition from the upper brittle to the lower ductile part of the crust. Curves of the lower diagram of Fig. 7 having been calculated for wet rocks show a striking similarity with curves of earthquake hypocentres. The depth–frequency distribution of earthquakes is influenced by the temperature (or the surface heat flow) in a similar way as are the curves of maximum stresses, as shown in the example of Fig. 8. Both sets of data, the results of the stress calculation and the

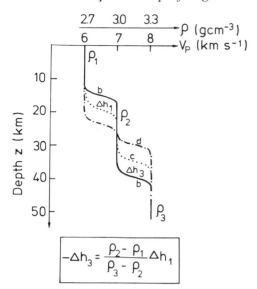

$$-\Delta h_3 = \frac{\rho_2 - \rho_1}{\rho_3 - \rho_2} \Delta h_1$$

FIG. 6. Varieties of isostatic crust showing the same gravity values: b = shield and platform, c = transitional–Caledonian, d = Caledonian–Variscan.
The sequence from b to d might be understood as a modification of density ρ, velocity V_p, and composition by melting of the crust's mafic parts with $\rho = \rho_2$.
For the example: $\rho_2 = 3000$ kg/m³, $\rho_1 = 2700$ kg/m³, $\rho_3 = 3300$ kg/m³, Δh = depth intervals; ρ–V_p relationship after Woollard (1959).

observation of the limited depth of continental earthquakes, strongly indicate that the lower continental crust cannot sustain significant shear stresses. It must be rheologically weak, the later accumulated and warmer crusts being weaker than the cold crusts of old shields and platforms.

For the ductile regime similar formulae exist for the shear stress τ and the viscosity η. Hence, the two diagrams of Fig. 9 are labelled both with τ and η. The diagrams show that the lower continental crust is a zone of low viscosity embedded between the brittle upper crust and the transitional-to-brittle uppermost mantle. The reason for this sandwich-like behaviour, with the lower crust as the 'jelly', lies in the inverse relationship between viscosity and temperature, and in the higher activation energy for diffusional processes (or the higher melting point) of mantle rocks. A change of the composition of crustal rocks or a change in their water content—of course—result in secondary, minor, viscosity maxima or minima inside the crust. This would not affect the general behaviour of the lower crust as a sandwich-like zone of low viscosity. It is interesting to note that Chen and Molnar (1983) found some

areas where the upper mantle and the upper crust show earthquake activity while the lower crust is free of earthquakes. It is of course tempting to correlate the 'jelly-like' lower crust, and especially that of the younger and warmer areas, with its sub-horizontal reflecting lamellae.

About the nature of crustal reflectors

Before coming to a conclusion about the nature of crustal reflectors the more important results from the different fields of research will be summarized. From the reflection work *sensu stricto* it follows that:

1   The lower crust in Variscan (and Caledonian) areas is highly reflective, showing a surprisingly uniform lower limit, the Moho, at about 30 km.
2   The maximum length of phase correlation of deep crustal reflectors is about 35 km (Urach 1). A more typical length is 5 km. Most reflectors are sub-horizontal or at least plane.
3   The length of correlatable phases and the continuity of reflectors in general are larger along profiles *in* the direction of tectonic strike than perpendicular to it.
4   In addition to the highly reflective lower crust in the Variscides and Caledonides, fault zones are also good reflectors.
5   Lateral variations in seismic velocities may reach values up to 10% (Urach 1).
6   Reflectivity generally *in*creases toward the base of the Variscan (and Caledonian) crust. In old shield and platform areas reflectivity *de*creases downwards.

Additional conclusions evolve from the comparison of reflection data with those of wide-angle observations and of viscosity studies, as follows.

7   Strong near-vertical reflections originate at jumps or strong-gradient zones of the velocity obtained from wide-angle data.
8   Reflections in the lower crust are created from lamellae of alternating low and high velocity.
9   These lamellae probably consist of crystallization seams, caused by a general melting process of a primary pre-Variscan gabbroic lower crust, leaving behind ultramafic residues in the form of gravity dominated subhorizontal 'splinters'.
10   The upper crust of gneissic–granitic composition is generally poorly reflective, probably because of a more vertically dominated pattern of previous melt intrusions and plutonic activity within its rigid and highly viscous material.
11   The uppermost mantle is also poorly reflective. In the Variscides it may consist of compact ultramafic residues from a general syn- and postorogenic melting in zones of extension, the Moho

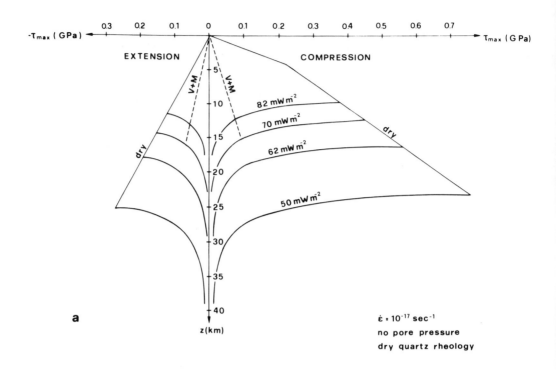

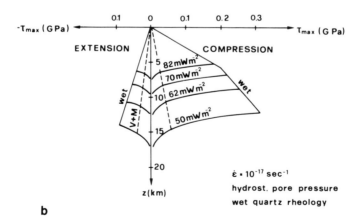

FIG. 7. Shear strength = frictional shear stress (curves with negative gradients) and ductile rheologic stress (curves with positive gradients) combined with curves of maximum stresses as a function of depth for dry quartz rheology (a) and wet quartz rheology (b). Only those values are shown as solid lines where frictional stress τ_{fr} < ductile stress τ_{du} (for the negative-gradient curves) and where $\tau_{du} < \tau_{fr}$ (for the positive-gradient curves), i.e. the maximum possible stress a rock can bear without breaking (negative gradient curves) or without creeping (positive gradient curves). Parameters are values of surface heat flow density q_0. V + M = Vermiculite and montmorillonite. $\dot{\varepsilon}$ = strain rate.

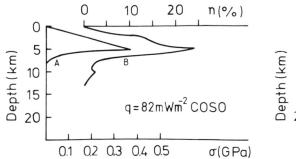
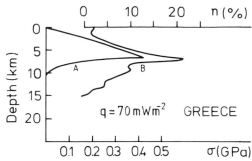

FIG. 8. Comparison of selected curves (A) of maximum stresses from Fig. 7. with the depth-frequency distribution of earthquakes (B), in two areas with different heat flow q. n = number of earthquakes per 1 km depth interval, smoothed.

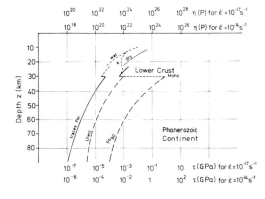

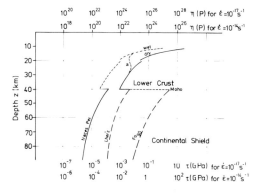

FIG. 9. Maximum stress and viscosity as a function of depth for the lower crust and uppermost mantle. A suggested transition from a wet upper crust to a dry lower crust is indicated by dotted line. Upper part: Phanerozoic lithosphere; Lower part: Shield lithosphere. Per = Peridotite; Lherz = Lherzolite; Fo$_{90}$ = Forsterite 90.

being created as a new crust–mantle boundary in a similar way as in the oceanic (and continental) rifts.

There are other candidates for the reflecting horizons like laminated fault zones, metasediments guided down along sutures, and lenses of melt or water. While the general appearance of melt in present Variscan crusts is not compatible with heat flow studies, fault zones and metasediments certainly may act as strong reflectors, although they are supposed to decrease in number toward the lower crust. The assumption of abundant free water from dehydration processes or metasomatism is controversial. It would be needed in huge ponds in order to explain the many and strong reflectors of the Variscan crust. It would be compatible with the high conductivity found in many parts of lower crust but not with a granulitic composition.

Like metamorphic ages, crustal reflectivity patterns indicate the latest tectonic event. This may be a general melting and extension, creating a highly reflective lower crust, it may be a compression, creating shallow angle thrust faults, or a recent crust–mantle exchange of magmas and residues, resulting in a chaotic appearance of short reflection segments, or it may be a zone of vertical intrusions, creating transparent zones without any seismic reflectors.

ACKNOWLEDGEMENTS. Thanks are due to the Federal Ministry of Research and Technology (BMFT) and the German Research Association (DFG) for their financial support of the investigations. Critical discussions and help from various colleagues of the Institute of Geophysics in Kiel are gratefully acknowledged.

References

ANGENHEISTER, G. & POHL, J. 1976. Results of seismic investigation in the Ries Crater (Southern Germany). *In*: GIESE, P., PRODEHL, C. & STEIN, A. (eds) *Explosion Seismology in Central Europe*, Springer, Berlin, 290–302.

BARTELSEN, H., LUESCHEN, E., KREY, T., MEISSNER, R., SCHMOLL, H. & WALTHER, C. 1982. The combined seismic reflection–refraction investigation of the Urach geothermal anomaly. *In: The Urach Geothermal Project*, Schweitzerbart'sche Verlagsbuchhandlung, 247–262.

BREWER, J., MATTHEWS, D., WARNER, M., HALL, J., SMYTHE, D. & WHITTINGTON, R. 1983. BIRPS deep seismic reflection studies of the British Caledonides, *Nature*, **305**, 206–210.

CHEN, W-P. & MOLNAR, P. 1983. Focal depth of intercontinental and intra-plate earthquakes and their implications for the thermal and mechanical properties of the lithosphere, *J. Geophys. Res.* **88**, 4183–4214.

COOK, F., ALBAUGH, D., BROWN, L.D., KAUFMAN, S., OLIVER, J. & HATCHER, R.D. 1979. Thin-skinned tectonics in the crystalline southern Appalachians: COCORP seismic reflection profiling of the Blue Ridge and Piedmont, *Geology*, **7**, 563–567.

—— BROWN, L.D., KAUFMAN, S., OLIVERS, J. & PETERSEN, T. 1981. COCORP seismic profiling of the Appalachian orogen beneath the coastal plain of Georgia, *Bull. geol. Soc. Am.*, Part I, **92**, 738–748.

GIBBS, A., PAYNE, B., SETZER, T., BROWN, O.D., OLIVER, J. & KAUFMAN, S. 1984. Seismic reflection studies of the Precambrian crust of central Minnesota, *Bull. geol. Soc. Am.*, **95**, 280–294.

GIESE, P., PRODEHL, C. & STEIN, A. 1976. *Explosion Seismology in Central Europe*, Springer-Verlag, Berlin, 418 pp.

GLOCKE, A. & MEISSNER, R. 1976. Near-vertical reflections recorded at the wide-angle profile in the Rhenish Massif. *In*: GIESE, P., PRODEHL, C. & STEIN, A. (eds) *Explosion Seismology in Central Europe*, 252–256, Springer Verlag, Heidelberg.

KREY, T. 1976. Computation of interval velocities from common reflection point moveout times for layers with arbitrary dips and curvatures in three dimensions when assuming small shot-geophone distances. *Geophys. Prospect.*, **14**, 91–111.

—— 1978. Seismic stripping helps unravel deep relections, *Geophysics*, **43**, 499–911.

MEISSNER, R. 1967. Zum Aufbau der Erdkruste, Ergebnisse der Weitwinkelmessungen im bayerischen Molassebecken, *Geol. Beitr. Geoph.*, **76**, 211–254.

—— 1976. Comparison of wide-angle measurements in the USSR and the Fed. Rep. of Germany. *In*: GIESE, P., PRODEHL, C. & STEIN, A. (eds) *Explosion Seismology in Central Europe*, 380–384.

—— 1986. *The Continental Crust*, Academic Press, New York, 462 pp.

—— BARTELSEN, H., KREY, T. & SCHMOLL, J. 1982. Detecting velocity anomalies in the region of the Urach geothermal anomaly by means of new seismic field arrangements. *In*: CERMAK, V. & HAENEL, R. (eds) *Geothermics and Geothermal Energy*, Schweizerbarth, Stuttgart, 285–292.

—— BARTELSEN, H. & MURAWSKI, H. 1980. Seismic reflection and refraction studies for investigating fault zones along the Geotraverse Rhenohercynicum, *Tectonophysics*, **64**, 59–84.

—— BARTELSEN, H. & MURAWSKI, H. 1981. Thin-skinned tectonics in the northern Rhenish Massif, Germany, *Nature*, **290**, 399–401.

—— SPRINGER, M., MURAWSKI, H., BARTELSEN, H., FLÜH, E. & DÜRSCHNER, H. 1983. Combined seismic reflection–refraction investigations in the Rhenish Massif and their relation to recent tectonic movement. *In*: FUCHS, K., V. GEHLEN, K., MURAWSKI, H. & SEMMEL, A. (eds) *Plateau Uplift*, 276–287.

—— & STREHLAU, J. 1982. Limits of stresses in continental crusts and their relation to the depth–frequency distribution of shallow earthquakes, *Tectonics*, **1**, 73–89.

SMITHSON, S.B., BREWER, J., KAUFMAN, S., OLIVER, J. & HURICH, C. 1979. Structure of the Laramide Wind River Uplift, Wyoming, from COCORP deep reflection data and from gravity data, *J. Geophys. Res.*, **84**, 5955–5972.

VETTER, U. & MEISSNER, R. 1970. Überprüfung der Isostasie durch tiefenseismische Sondierungen, *Zeitschr. Geoph.*, **36**, 225–228.

WEBER, K. 1981. The structural development of the Rheinische Schiefergabirge, *Geologie en Mijnbouw*, 149–159.

WEVER, T. 1984. *Häufigkeitsstatistik im Steilwinkelreflexionen aus Gebieten der Kontinentalen Kruste und ihre Anwendung auf die Analyse der kontinentalen Krustenstrukturen*, Diploma Thesis, 163.

WOOLLARD, G.P. 1959. Crustal structure from gravity and seismic measurements, *J. Geophys. Res.*, **64**, 1521–1544.

ZIEGLER, P. 1978. North-western Europe: tectonics and basin development. *Geol. Mijbouw*, **57**, 589–626.

—— 1982. Geological atlas of Western and Central Europe. Elsevier, Amsterdam, 130.

R. MEISSNER, Institut für Geophysik der Christian-Albrechts-Universität D-2300 Kiel, West Germany

Seismic reflections from the lower crust around Britain

D.H. Matthews

SUMMARY: A cartoon is presented which represents a time-section typical of the first 3000 km of BIRPS profiling over the continental shelf north, west and south of Britain. It shows a Mesozoic sedimentary basin in the hanging wall of a low-angle normal fault zone. The dipping fault reflection merges with layering in the lower crust which appears as several strong but short (c. 4 km) horizontal reflections between about 6 s two-way travel time and the base of the crust at about 10 s. The uppermost mantle is devoid of reflections except where dipping events extend from the layered lower crust towards the base of the section at 15 s two-way time. The cartoon is justified by reference to examples from profiles.

Discussion states present hypotheses that are being tested. The layering in the lower crust is real and due to lenses of flat-lying foliation about 4 km across and tens of metres in thickness in which the seismic velocity has been reduced by the presence of fluids. Brine has been suggested to account for the electrical conductivity of the lower crust, and carbon dioxide may be associated with granulites. The geometry implies that the lower crust was ductile at the time of faulting, and that the foliation became horizontal during Mesozoic stretching of West Europe.

The British Institutions Reflection Profiling Syndicate—BIRPS—obtained 3000 km of multichannel seismic reflection profile over the continental shelf around Britain during 1981–1983 (Fig. 1). Data were acquired and processed by the seismic prospecting companies acknowledged at the end of the paper. Experience in the UK has shown that it is twenty times cheaper to profile at sea than on land, and that the records of the deep crust obtained using arrays of airguns at sea are substantially better than those that have been obtained by work on land. On practically all of the BIRPS lines it is possible to determine the echo-time to the Moho, and in many places coherent reflections can be seen from structures in the upper mantle.

This paper is based on the results of lines MOIST, WINCH, and SWAT, all of which lie within 500 km of the edge of the continental shelf (Fig. 1). The area contains some thirty elongated post-Carboniferous sedimentary basins, ranging in size from the Central and Viking grabens of the North Sea, about 150 km across, down to quite small basins, like the Minch at sea or the Vale of Eden on land, about 35 km and 10 km across respectively. The existence of so many Mesozoic basins implies that stretching of the brittle upper crust took place prior to the opening of the North Atlantic. The oldest dated oceanic crust southwest of Ireland was formed in anomaly 32 time (72 m.y., late Cretaceous) and west of Scotland in anomaly 34 time (56 m.y., early Eocene).

Seismic refraction information from the same area indicates that the crust is thinner than normal in this part of north-western Europe. Eleven seismic refraction lines cross the BIRPS lines (Fig. 2): all of them indicate a crustal thickness of 27–30 km (Blundell 1981; Edwards & Blundell 1984). I have not been able to locate a recent map of the area west of the Urals contoured in crustal thickness, but a thickness of 27–30 km may be contrasted with 35–40 km generally recorded in Europe north and east of the Tesseyre–Tornquist line (Beloussov & Pavlenkova 1984; Bungum et al. 1980). For this reason I suggest that the crust of the British area may have been thinned by stretching during Mesozoic time prior to the opening of the North Atlantic.

The paper is in two parts: the first presents a cartoon of a line drawing of a typical BIRPS record and attempts to establish its validity by describing a few key records. The second part presents a snapshot of developing views on the origin of reflections from dipping faults and from horizontal reflectors in the lower crust.

It is notoriously difficult to reduce large seismic reflection records to the size of a journal illustration. Instead, this article is illustrated with line drawings showing correlations observed on the records. Copies of the original records of all BIRPS data that is more than one year from its date of acquisition are available at the cost of reproduction by writing to Programme Manager, Marine Geophysics Research Programme, British Geological Survey, Murchison House, West Mains Road, Edinburgh EH9 3LA. Telex 727343.

A typical record

Fig. 3 is a cartoon, a sketch of an imaginary typical BIRPS record. Like the line drawings of Figs 4 and 5, it shows those correlations on the

From: DAWSON, J.B., CARSWELL, D.A., HALL, J. & WEDEPOHL, K.H. (eds) 1986, *The Nature of the Lower Continental Crust*, Geological Society Special Publication No. 24, pp. 11–22.

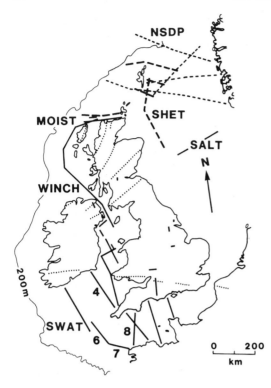

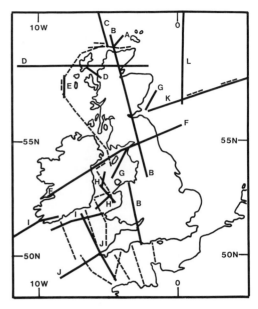

FIG. 2. Map showing the reflection profiles at sea on which this paper is based (broken lines) and wide-angle seismic lines, A–L, which give the velocity structure down to the Moho (thick lines). For a comparison of echo-times to the Moho calculated from refraction and measured from reflection profiles see Table 1.

FIG. 1. The continental shelf of NW Europe showing BIRPS profiles completed prior to June 1984 (solid lines): MOIST (February 1981), WINCH (June 1982), SWAT (November 1984), SALT (February 1983). Also shown, by broken lines, profiles SHET (August 1984) and NSDP (GECO, October 1984). Lines on land have been shot by British Geological Survey (Whittaker & Chadwick 1983). The 200 m contour marks the top of the continental slope. SWAT line numbers increase anticlockwise from 0 to 11. Dotted lines mark the approximate position of the principal faults on land. From NW to SE across Scotland and England they are the Outer Hebrides Fault, Moine Thrust, Naver Slide (short segment), Great Glen Fault, Highland Boundary Fault, Southern Uplands Fault and Variscan Front.

equalized record sections which we believe to be primary reflections from underground. We have eliminated what we believe to be noise, diffractions, and multiples, although because we are looking deep into a 3-dimensional earth this involves some subjective judgements. There is no geological interpretation other than that implied by the naming of the features on the drawings. Faults from which there are no reflections are not shown, despite the fact that they can sometimes be inferred with confidence from the offset of sub-horizontal reflections from sedimentary rocks.

The reflections sketched in Fig. 3 are interpreted as Mesozoic (post-Carboniferous) sedimentary basins and as sub-horizontal layering in the lower crust extending from about 6 s to a relatively abrupt cut-off at about 10 s, the echo time to the Moho. Reflectors dipping at about 30° from the near-surface in the time section have been interpreted as low-angle normal faults which are reactivated Caledonian and Variscan thrust faults. Dipping events seen in the upper mantle have been similarly interpreted. In these unmigrated time sections both sets of dipping reflectors appear to merge with and be lost in the layering in the lower crust. Neither set appear to be associated with substantial shifts in Moho reflection time which can generally be agreed to within ±0.5 second.

Selected line drawings

It seemed best in 1980 to start with a structure that could be traced down to depth from its outcrop at the surface, and the most notorious example in Britain was the Moine Thrust, the eastward dipping Caledonian Front in Scotland. After being persuaded to work at sea we decided

A TYPICAL BIRP

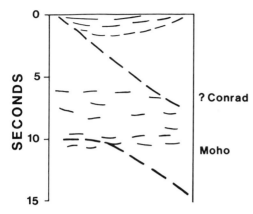

FIG. 3. Cartoon of a typical BIRP. Line drawing of a processed (but not migrated) time section. Assuming 6 km s^{-1} mean velocity, 15 seconds two-way travel time is 45 km depth and horizontal and vertical scales are about equal. Shows Mesozoic sedimentary basin as half-graben above deeply penetrating low angle normal fault which may be reactivated older thrust, no other reflections in the upper crust, short and sub-horizontal reflections in lower crust, and dipping reflections in the upper mantle which terminate in the layered lower crust. The Moho is at the base of the layering in the lower crust.

on MOIST, the Moine and Outer Isles Seismic Traverse, as a trial line 180 km long running as close as practicable to the north coast of Scotland (Fig. 1). MOIST crosses, from east to west, the southern edge of the West Orkney Basin, the Minch Basin and an unnamed basin north of the island of Lewis (Naylor & Shannon 1982). These basins are filled with Permo-Triassic, and in some cases Mesozoic, rocks (Evans *et al.* 1982). A parallel line on land, 15 km farther south, would have crossed the Devonian of the Caithness Basin and the Moine Schists. The Moine Schists were last metamorphosed during the Caledonian orogeny 450 Ma ago but have older, Lewisian-like inliers caught up in them. They are cut by easterly dipping shear zones like the Naver Slide (Butler & Coward 1984). Farther west, it would have passed over the Moine Thrust and on to relatively undeformed Cambro–Ordovician rocks and late Precambrian Torridonian sandstone resting on the Proterozoic Lewisian gneiss, which is the North American foreland of the Caledonides. After crossing Mesozoic and Tertiary sediments, exposed in the Minch Basin between Scotland and the Hebrides it would have ended on Lewisian gneiss exposed on the island of Lewis. In

Lewis it would have crossed exposures of the Outer Hebrides fault zone which cuts the Lewisian gneiss, dips eastwards, and has a north northeast (Caledonian) trend. Mylonites are developed in the fault zone which has been regarded as a thrust of Caledonian age (Sibson 1977) until the iconoclastic suggestions of Wernicke (1984) who argues that it may be a low angle normal shear zone unrelated to Caledonian orogenesis.

Fig. 4a is a line drawing of MOIST (Brewer & Smythe 1984). The profile crosses the line of the Outer Hebrides fault zone some 20 km offshore. A strong eastward dipping reflection has been interpreted as coming from the fault zone. The relations between this reflection and the westward dipping sediments in the basin above it leave no room for doubt that, offshore, the most recent movement of the Outer Hebrides fault zone is as a down-to-the-east low angle normal fault. Farther to the west along MOIST, several eastward dipping reflectors can be seen. All of them terminate half-grabens containing westward dipping sediments. All these eastward dipping reflectors have so far been accepted as low angle normal faults that are reactivated Caledonian thrusts and akin to the Moine Thrust (which is of shallow origin) or the Naver Slide (which is of mid-crustal origin).

East of the Outer Hebrides fault zone a set of reflections from 8–10 seconds were originally interpreted as coming from the vicinity of the Moho (Smythe *et al.* 1982). MOIST crosses two wide-angle seismic lines, A and B in Fig. 2 and Table 1, which yield structures from which the echo-time (twt) to the Moho can be calculated. Comparisons between the normal incidence and the wide-angle Moho times are shown in Table 1 and were the basis for the identification of the Moho on MOIST.

The reflection from the Outer Hebrides fault zone can be traced down on the time section with confidence to a point about 1 s above the Moho. Modelling has been used by Blundell *et al.* (1985) to suggest that the Outer Hebrides fault zone extends right through the crust and by Peddy (1984) to suggest that it displaces the Moho by about 0.5 seconds down to the east, that is, as a normal fault. There is still some controversy about this result. The other eastward dipping reflectors seen at the eastern end of MOIST are lost in reflections from the lower crust and no one has yet claimed that they extend into the upper mantle, nor would it be easy to make a case that other prominent dipping reflectors mentioned below, the South Irish Sea lineament (Brewer *et al.* 1983) or the Variscan Front (Fig. 5, mid.) do not sole out in the layered lower crust.

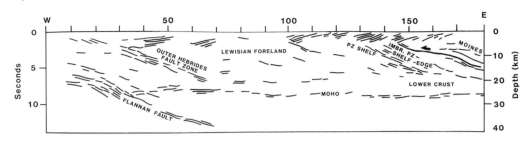

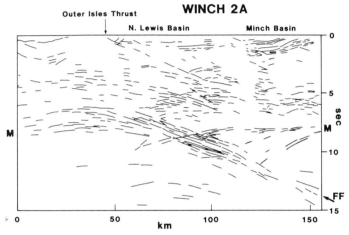

FIG. 4. Line drawings of two reflection profiles off NW Scotland. Both show dipping reflectors that have been interpreted as low angle normal faults which may be reactivated Caledonian thrusts. Time sections but approximately true to scale. a) MOIST. The heavy lines and arrow mark one possible position of the Moine Thrust. Horizontal scale in kilometres. b) WINCH 2a. Moho, MM. Flannan Fault, FF.

The most unexpected result of the MOIST profile was the discovery of an apparently strong and continuous reflector in the upper mantle subparallel to the Outer Hebrides fault zone and extending from near the bottom of the section at 15 seconds echo-time through the Moho and into the short reflections in the lower crust which appear west of the Outer Hebrides fault zone. If this reflector were to be continued in a straight line to the surface it would outcrop close to the Flannan Islets near the continental margin, so it was named the 'Flannan Thrust (?)'. In this paper it will be called the Flannan Fault. No reflector like this had been seen before in the upper mantle and we were naturally dubious of its reality, so the northernmost lines of the next year's programme, WINCH, were arranged to embrace the western end of MOIST.

WINCH, the Western Isles and North Channel traverse, was planned to cross the major faults of the Caledonides in Britain (Fig. 1). On it we first became aware of the ubiquitous reflections that

characterize the lower crust west and south-west of Britain (Brewer *et al.* 1983). The two northernmost segments did confirm the reality of the Flannan Fault. The longer of the two, WINCH 2a running west southwest off the Butt of Lewis, is shown in Fig. 4b. At the west end of this section the Flannan Fault appears to be turning over to merge with the layering in the lower crust. It does not appear to displace the Moho.

A complete line drawing of the WINCH time section has been published (Brewer *et al.* 1983). Another reflector in the upper mantle reappears south of the Hebrides, approximately where it would be expected from extrapolation of the strike of the Flannan Fault determined from the three crossings north of the Hebrides. The reflection appears to be repeated by the Great Glen Fault, implying 100–150 km of left lateral displacement along the Great Glen. Other examples of dipping reflections in the upper mantle are seen on the SWAT lines southwest of Britain, and a particularly unequivocal example which, being

TABLE 1. *Comparisons between two-way travel times (twt) in seconds to the Moho calculated from structures obtained from wide-angle (refraction) experiments and twt to the base of the lower crust layering observed on BIRPS reflection profiles. For positions of crossover points see Fig. 2.*

	Refraction			Reflection	
Fig. 2	Reference	twt	twt	profile	
A	Jones *et al.* (1984)	8.9	8.5	MOIST	
B	Bamford *et al.* (1978)	8.6	8.5	MOIST	
C	Smith & Bott (1975)	8.5	8.5	MOIST	
D	Bott *et al.* (1979)	8.5	9.0	WINCH2	
E	Jones *et al.* (1984)	8.7	9.0	WINCH2	
F	Jacob *et al.* (in press, 1985)	10.2	10.0	WINCH2	
			10.0	WINCH3	
G	Agger & Carpenter (1965)				
H	Blundell & Parks (1969)	10.1	10.5	WINCH4	
			11.0	SWAT1	
			9.0	SWAT2	
I	Bamford & Blundell (1970)	9.7	9.0	SWAT2	
			10.7	SWAT4	
			8.5	SWAT5	
J	Holder & Bott (1971)	8.8	10.0	SWAT4	
			10.5	SWAT8	
			9.7	SWAT6	
K	Barton *et al.* (1984)	11.0	11.0	SALT1	
	Barton (1985)	10.5–9.5	10.5–9.5	SALT2	
L	Christie (1982)				

concave upwards will not be closed up by 2-D migration, can be seen under the English Channel south of Start Point on SWAT line 8 (Fig. 5, bot.).

Further examples of reflections from the upper crust that appear to come from major low angle normal faults that have post Permo-Triassic sedimentary basins in their hanging walls can be seen in the BIRPS data: on WINCH is the northward dipping South Irish Sea Lineament with the St George's Channel basin in its hanging wall (Brewer *et al.* 1983). The SWAT lines show reflectors dipping south from the Cornish coast which have been described by Day and Edwards (1983) as Variscan thrusts. The dipping reflectors under the Plymouth Bay basin seen on SWAT 9 (not illustrated) and SWAT 8 (Fig. 5, bot.) could be low angle normal faults following the line of Variscan thrusts mapped on land. Fig. 5, mid. shows a crossing of the North Celtic Sea Basin which, here at least, is situated in the hanging wall of a fault reflection identified as the Variscan Front. Inspection of industrial reflection survey data shows clearly that the reflector has an east–west (Variscan) strike. The North Celtic Sea basin is complexly broken up by transverse faulting, but on Swat 4 it is clear that the underlying fault, which may have been an Upper Palaeozoic thrust, has been reactivated or formed as a low angle normal fault. Line-drawing migrations of this record made by Dr Sue McGeary (Fig. 6) confirm the ramp and flat shape of the surface.

All these low angle faults, except possibly the Outer Hebrides Fault, terminate in the layered lower crust. So do the dipping events in the upper mantle. This would be readily explained if the lower crust was ductile when they were formed.

Reflections from the lower crust

Three varieties of the almost ubiquitous reflections from the lower crust are illustrated in Fig. 5. The top drawing shows the part of the WINCH line west of the Hebrides. Here, west of the Outer Hebrides Fault, the layering becomes more prominent just above the Moho. The echo-time (twt) to the Moho is particularly well established by a wide-angle reflection line carried out along the WINCH line, E in Fig. 2 and Table 1. This part of the line runs over an area where the Lewisian gneiss is exposed or buried at shallow depth, but the upper crust is conspicuously devoid of correlatable reflections. Southward

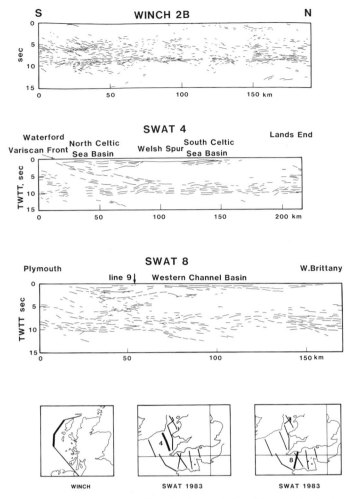

FIG. 5. Simplified line drawing of reflection profiles showing varieties of layering in the lower crust. True scale time sections. (top) WINCH 2b, west of the Hebrides; (mid.) SWAT 4 (provisional data), in the Celtic Sea; (bot.) SWAT 8 (provisional data), across the W. English Channel; location maps.

along WINCH (Brewer *et al.* 1983) reflections become more widespread throughout the lower crust between about 5 s echo-time and 10 s (the Moho). No reflections occur under the Southern Uplands of Scotland but they return abruptly south of a northward dipping feature in the lower crust tentatively equated with the Iapetus Suture.

SWAT 4 (Fig. 5, mid.) shows particularly clear reflections from the lower crust. This section may be regarded as typical. Fig. 5 (bot.) shows the line SWAT 8. Off the coast of Brittany the reflections from the lower crust appear to be concentrated in two bands. This feature is also seen on the brute stacks of SWAT 9, 10 and 11 where those lines approach the coast of France. It is not yet clear

whether this pattern of reflections is a mappable characteristic of the Variscides in Europe.

Photographs of part of record sections showing the reflective lower crust will be published elsewhere (Matthews & Cheadle 1985; Blundell & Raynaud 1985). Correlations extend over distances that range from 2 km (smaller than the Fresnel Zone) up to 25 km but they are most commonly about 4 km in length. They are subhorizontal. Attribute analyses show that the lower crust reflections are comparable in amplitude and frequency with those from the sediments.

The base of the layering is believed to be the Moho. The Moho is defined in terms of the

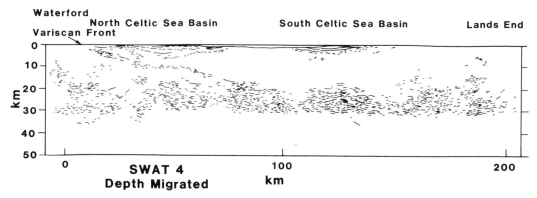

FIG. 6. Depth migration of a line drawing of SWAT 4, cf. Fig. 5.

velocity of refracted waves received at wide-angle. Comparisons between normal incidence echo-times (twt) from reflection records and two-way travel times calculated from published velocity structures obtained by wide-angle ('refraction') experiments are shown in Table 1. Interpretation methods have improved over the years and the more recent lines have more shots and more receivers. There is good agreement with the newer data, particularly off NW Scotland. Probably the best comparison published so far was obtained along the SALT line in the North Sea (Fig. 1). A wide-angle line was interpreted by 2-D amplitude modelling using a model in which velocity varied both vertically and horizontally (Barton & Wood 1984). Two-way travel times calculated from the wide-angle results were superimposed on line drawings of the reflection data. The results suggested that the base of the layering is the wide-angle Moho within the limits of error (Barton *et al.* 1984; Barton 1985). Errors on the best wide-angle determinations are thought to be about ±1 km but most are far less reliable; the base of the layering can usually be determined with an uncertainty of ±0.5 seconds, about ±1.5 km. It has been argued that the Conrad discontinuity, where the velocity increases from *c.* 6.0 to *c.* 6.7 km s^{-1}, is at the top of the layering in Germany (Dohr & Fuchs 1967), but the Conrad discontinuity is seldom reported from Britain.

The Moho appears to be flat on the time-sections where they cross sedimentary basins of moderate size north, west and south of Britain. This is an artefact, brought about by the opposed effects of velocity pull-down and actual crustal thinning (McKenzie 1978). However, a depth migration of a line drawing of SWAT 4 (Fig. 6) shows that the Moho rises only 2–3 km under the North Celtic Sea basin which is 7 km deep. The crystalline crust beneath the sediments in the basin is thinned from nearly 31 km to 23 km.

Discussion: what are the reflectors?

The dipping reflectors

Several examples of dipping reflectors that traverse the crystalline upper crust have been cited and illustrated. Such dipping events have been seen wherever the reflection profiles have crossed the line of a major Palaeozoic thrust fault or shear zone extrapolated from land geological mapping (Fig. 1). But the evidence of the profiles taken at sea is unequivocal: they have moved as low angle normal faults during Mesozoic time. Their dip does not exceed 30° in the time section, although it increases by a few degrees after depth migration—compare the Variscan Front in Fig. 4 (mid.) and in Fig. 6. An event dipping at 40–45° in the time section would become near vertical in the depth section after migration, and the geometry of reflection profiling is not well adapted to imaging near-vertical reflectors. Steep faults can be traced through undeformed sediments in spite of there being no reflection from the fault zone because they displace the reflections from the flat-lying strata, but they cannot be traced thus through crystalline rocks because, in the upper crust, sub-horizontal reflections are absent and, in the lower crust the short sub-horizontal reflections are insufficiently persistent.

Reflection events may be due to constructive interference of seismic waves reflected from layers as little as 10 m apart (Blundell & Raynaud 1985). We do not yet understand the physics of reflections from shear zones deep in the earth, in crust or mantle. Jones & Nur (1984) and others have suggested that mylonites may be the source of the acoustic impedance contrasts needed to explain reflections from fault zones in metamorphic rocks; fluids, reducing the acoustic impedance, may also be important. From the BIRPS data alone one might assert that we obtain reflections only when thrusts have been reactivated as normal faults. However no one has yet claimed that the Wind River Thrust (Brewer *et al.* 1980), the thrusts beneath the Southern Appalachians (Cook *et al.* 1983), or the Faille du Midi, the Variscan Front in west Europe, have been reactivated as normal faults so that a local explanation will not serve.

Horizontal reflections from the lower crust

For BIRPS, the lower crust defines itself as the reflective lower crust. How many reflectors are there? We are acutely aware that we are profiling over a three-dimensional structure. Blundell & Raynaud (1985) have modelled an eggbox-shaped reflector at about 20 km depth and obtained several reflections from it that extend over one or two seconds of the record. So reflections from out of the plane of the section can increase the number seen but cannot account for the relatively sharp cut-off at the crust–mantle boundary. BIRPS records show eight or nine reflections from the lower crust in a typical vertical column, between 6 and 10 s twt. To account for them we need at least 4 reflecting surfaces suitably disposed in 3-D, and probably more.

How thick are the reflecting layers? Blundell & Raynaud have applied the ideas of earlier authors to the case of the lower crust. Reflections come from constructive interference of waves returned from closely spaced interfaces, for example from the top and bottom surfaces of layers within the foliation in which the velocity is reduced. With a frequency of 20 Hz and an interval velocity of 6.5 km s^{-1} the seismic wavelength is 325 m. Blundell & Raynaud showed that wavelets with amplitudes comparable with the incident wavelet are returned from a single layer one quarter of a wavelength in thickness (81 m) in which the velocity is reduced by 0·1 km s^{-1}, or a layer 1/32 of a wavelength (10 m) in which the velocity is reduced by 2.5 km s^{-1}. So to account for our typical reflection we need a patch of foliation perpendicular to the ray path in which the acoustic impedance is adequately reduced or increased and which is about 3–4 km in diameter and several tens of metres in thickness. Acoustic impedance is defined as the product of seismic velocity and density of the rock. It can be spectacularly reduced by the presence of fluid-filled cracks, particularly if the fluid is overpressured. Difference in acoustic impedance controls the reflected amplitude.

Reflections from the lower crust were first reported from Germany during the 1950s (Dohr & Fuchs 1967). More recently they have been described from the USA (Oliver *et al.* 1983), from Australia (Mathur 1983) and from Germany (Meissner & Lueschen 1983). They are particularly clearly imaged by ECORS work in France and in BIRPS profiles from the stretched margin of Europe. In this context it is relevant to point out that COCORP profiling has often not shown reflections from the lower crust in the USA but it has shown them abundantly in profiles across the stretched Basin and Range province (Hauge *et al.* 1984).

The author knows of four published explanations. When lower crust reflectors were first identified they were seen near the Moho, so that earlier papers deal with the nature of that boundary, on which of course there was already information from wide-angle modelling (Helmburger 1968; Fuchs 1969; Davydova 1972). Meissner (1973) advocated ultrabasic layering to account for the complex reflections from the Moho. Phinney & Jurdy (1979) discussed deep reflections on the early COCORP lines in terms of flowage whose effect is to draw even crosscutting structures into parallelism with the foliation (Ramsay 1982). Hale & Thompson (1982) collected specimens from the exposure of the base of the continental crust at Ivrea in Italy, measured their acoustic impedance in the laboratory under appropriate conditions and used the results to calculate a synthetic seismogram which resembles those obtained from the Moho by COCORP. McKenzie (1984) advocated basic sills as the origin of the lower crust reflections. An additional, fifth explanation, is the presence of fluid-filled cracks. I envisage the possibility of lenses of sub-horizontally foliated rocks containing fluids in intercrystal pores and in shear fractures. The fluid might be overpressured.

These explanations are not mutually exclusive. They fall into two groups, those that explain the necessary acoustic impedance contrasts in terms of changes in rock type in essentially dry rocks and those that appeal to changes in bulk physical properties due to the presence of fluid-filled cracks. Fluids that have been suggested are water, brine, carbon dioxide and methane. The presence of aqueous fluids concentrated into thin intercommunicating lenses, particularly brine and carbonic acid which at the temperature and pressure of the lower crust is a strong acid (Fyfe, pers. comm.), would account for the fact that the electrical conductivity of the lower continental crust appears to be one to two orders of magnitude higher than the conductivity of dry samples of lower crustal rocks, amphibolites or granulites, when they are measured in the laboratory (Shankland & Ander 1983; Lee *et al.* 1983). At the meeting from which this volume stems, Vine mentioned the dramatic effect on conductivity of as little as 0.1% of saline pore fluid in gneisses of basic composition. He was thinking of essentially uniformly distributed pore fluids. In order to account for a reduction of seismic velocity from 6.5 to 6.4 km s^{-1} in a layer within the foliation, a rough calculation using the time-average equation (which may not be applicable) and assuming that the velocity of sound in the pore fluid and under lower crustal conditions is still 1.5 km s^{-1} indicates a local crack-filling pore fluid concentration of about 0.4%. The numerical feasibility of explaining horizontal reflectors by fluid concentrations in flat-lying foliation is being investigated. If cracks remain open, experience measuring the velocity of sound in rocks in the laboratory (Birch 1961) or in the field (Lort & Matthews 1972) indicates that porosity is more important than mineralogy in controlling the seismic velocity.

The consensus at the meeting appeared to favour explanations involving basic or ultrabasic sills within the foliation, but I do not concur. The Lewisian gneiss, exposed in the Hebrides, was metamorphosed to amphibolite and granulite grade at pressures that correspond to those of the upper part of the lower crust today. Unfortunately we cannot tell whether it exhibited layered reflections then. We have profiled west of the Hebrides now (Fig. 5, top), and we see no reflections from the upper part of the crust under the exposed gneiss. Where have all the sills gone?

It is possible that the lower continental crust is never raised to the surface without a cataclysm like that which took place at Vredefort or Ivrea. There is one single example of a reflection profile which revealed layering that looks very similar to that described from the lower crust and was within reach of the drill. It crosses a metamorphic core complex in the Basin and Range province of Arizona. The results of drilling, logging, and 1-D synthetic seismograms suggested to Reif & Robinson (1981) that the reflections were due to compositional changes in the dry foliation. Their conclusion does not agree with the view expressed here.

Rheology

The cartoon of Fig. 3 suggests that faults or shear zones originating in the brittle uppermost crust and uppermost mantle (Chen & Molnar 1983) vanish into the layered lower crust. This implies that the lower crust is ductile, or was so at the time of faulting.

The rheology of the crust was discussed at the London meeting by Kusznir and Park, Murrell, and Weber (Kusznir & Park 1984). Several years

ago Meissner pointed out that the ductility of rocks depends upon their temperature, upon the strain rate, and upon their mineralogy and water content (Meissner & Strehlau 1982). At the temperature of the lower crust quartzo-felspathic rocks approach their melting point and soften. Although hotter, the upper mantle is more refractory and therefore harder, being olivine. Our ability to image thrusts and low-angle normal faults and trace them deep into the crust, coupled with developing techniques in rheological modelling offers the prospect of reconstructing the temperature distribution during periods both of orogenesis and of tensional deformation.

The relation of deep reflection data to the rheology of the lithosphere is discussed by Matthews & Cheadle (1985) and briefly by Matthews & Hirn (1984).

Conclusions

The cartoon of Fig. 3 gives a fair summary of the first 3000 km of BIRPS profiles. The continuous dipping reflections in the crust and in the upper mantle are from Mesozoic low angle normal faults or slide zones which in most if not all cases follow the line of Palaeozoic (Caledonian or Variscan) thrusts. The lower crust west of Britain is layered, and despite the effects of offline reflectors (the eggbox effect), 6 or 7 separate reflectors are crossed by most lines. These reflections are from lenses, typically 4 km across and several tens of metres in thickness, within the sub-horizontal foliation. The necessary impedance contrasts could be provided from basic and ultrabasic sills intruded and incorporated into the foliation. However, we intend to examine the effects of fluids contained within near-horizontal broken-up foliation, to confirm that they could also be responsible for the reflections. The foliation was dragged into a near-horizontal attitude during Mesozoic extension of the ductile lower crust. The same tensional regime, preceding the opening of the North Atlantic, was responsible for faulting and the formation of half grabens in the cool brittle upper crust, much like the crevasses in the surface of a flowing polar ice sheet which is cooled from the top. It seems to me to be significant that by far the clearest lower crustal layering in the continental US has been seen by COCORP on profiles across the strongly extended Basin and Range province.

Each new data set that comes in provides surprises liable to cause a volte-face in such speculative conclusions.

ACKNOWLEDGEMENTS: The contractors who

have worked for us are Western Geophysical (MOIST and SALT), GECO (WINCH and SHET), Merlin Profilers and Seismograph Services (England) Ltd (SWAT). We thank all of these for generosity exceeding the calls of a commercial relation. Funding is from the Natural Environment Research Council through the Deep Geology Committee. Shell UK have migrated data and funded Core Group associates from other British universities. BGS has seconded staff to work as Associates. We are indebted to most of the oil companies in Britain for access to well logs and records and to the contracting companies for making their Non-Exclusive Data (Spec Surveys) freely available to BIRPS. I am indebted to members of the Core Group and to Prof. D.J. Blundell for innumerable ideas, to Nicky White for Fig. 2 and Table 1, and to Dr Sue McGeary and Bernard Raynaud for Fig. 6.

References

AGGER, H.E. & CARPENTER, E.W. 1965. A crustal study in the vicinity of the Eskdalemuir seismological array station. *Geophys. J.R. astr. Soc.* **9,** 69–83.

BAMFORD, S.A.D. & BLUNDELL, D.J. 1970. Southwest Basin Continental Margin Experiment. *Inst. Geol. Sci. Rep.* 70/14, 143–156. HMSO, London.

BAMFORD, D., NUNN, K., PRODEHL, C. & JACOB, B. 1978. LISPB-IV Crustal structure of Northern Britain. *Geophys. J.R. astr. Soc.* **54,** 43–60.

BARTON, P.J. (in press) 1985. Comparison of deep reflection and refraction structures in the North Sea. *In:* BARAZANGI, M. (ed.), *Deep Structure of the Continental Crust,* Amer. Geophys. Un., Geodynamics Series.

—— MATTHEWS, D.H., HALL, J. & WARNER, M. 1984. Moho beneath the North Sea compared on normal incidence and wide-angle seismic records. *Nature,* **308,** 55–56.

—— & WOOD, R.J. 1984. Tectonic evolution of the North Sea basic: crustal stretching and subsidence. *Geophys. J. R. astr. Soc.* **79,** 987–1022.

BELOUSSOV, V.V. & PAVLENKOVA, N.I. 1984. The types of the earth's crust. *J. Geodynamics,* **1,** 167–183.

BIRCH, F. 1961. The velocity of compressional waves in rocks to 10 kilobars, Part 2. *J. Geophys. Res.* **66,** 2199–2224.

BLUNDELL, D.J. 1981. The Nature of the Continental Crust beneath Britain. *In:* ILLING, L.V. & HOBSON, J.B. (eds), *Petroleum Geology of the Continental Shelf of North-West Europe.* Inst. of Petroleum, London, 58–64.

—— & PARKS, R. 1969. A study of the crustal structure beneath the Irish Sea. *Geophys. J. R. astr. Soc.* **17,** 45–62.

—— HURICH, C.A. & SMITHSON, S.B. 1985. A model for the MOIST seismic reflection profile, N. Scotland. *J. Geol. Soc. London,* **142,** 245–258.

—— & RAYNAUD, B. (in press). 1985. Modelling lower crust reflections observed on BIRPS profiles. *In:* BARAZANGI, M. (ed.), *Deep Structure of the Continental Crust,* Amer. Geophys. Un., Geodynamics Series.

BOTT, M.H.P., ARMOUR, A.R., HIMSWORTH, E.M., MURPHY, T. & WYLIE, G. 1979. An explosion seismology investigation of the continental margin west of the Hebrides, Scotland, at 58°N. *Tectonophys.* **59,** 217–231.

BREWER, J.A., SMITHSON, S.B., OLIVER, J.E., KAUFMAN, S. & BROWN, L.D. 1980. The Laramide Orogeny: evidence from COCORP deep crustal seismic profiles in the Wind River Mountains, Wyoming. *Tectonophys.* **62,** 165–189.

—— MATTHEWS, D.H., WARNER, M.R., HALL, J.R., SMYTHE, D.K. & WHITTINGTON, R.J. 1983. BIRPS deep seismic reflection studies of the British Caledonides. *Nature,* **305,** 206–210.

—— & SMYTHE, D.K. 1984. MOIST and the continuity of crustal reflector geometry along the Caledonian–Appalachian orogeny. *J. geol. Soc. London,* **141,** 105–120.

BUNGUM, H., PIRHONEN, S.E. & HUSEBYE, E.S. 1980. Crustal thicknesses in Fennoscandia. *Geophys. J. R. astr. Soc.* **63,** 759–774.

BUTLER, R.W.H. & COWARD, M.P. 1984. Geological constraints, structural evolution and deep geology of the Northwest Scottish Caledonides. *Tectonics,* **3,** 317–408.

CHEN, W-P. & MOLNAR, P. 1983. Focal depths of intracontinental and intraplate earthquakes and their implications for the thermal and mechanical properties of the lithosphere. *J. Geophys. Res.* **88,** 4183–4214.

CHRISTIE, P.A.F. 1982. Interpretation of refraction experiments in the North Sea. *Phil Trans. Roy. Soc. A.* **305,** 101–112.

COOK, F.A., BROWN, L.D., KAUFMAN, S. & OLIVER, J.E. 1983. The COCORP seismic reflection traverse across the southern Appalachians. *AAPG Studies in Geology,* No. 14. Amer. Assoc. Petr. Geol., Tulsa, Oklahoma, U.S.A.

DAVYDOVA, N.I. (ed.). 1972. *Seismic Properties of the Mohorovicic Discontinuity.* Transl. from Russian by R. Teteruka 1975, U.S. Dept. of Commerce, Nat. Tech. Inf. Service, Springfield, Va. 22151.

DAY, G.A. & EDWARDS, J.W.H. 1983. Variscan thrusting in the basement of the English Channel and Southwest Approaches. *Proc. Ussher Soc.* **5,** 432–436.

DOHR, G. & FUCHS, K. 1967. Statistical evaluation of deep crustal reflections in Germany. *Geophysics,* **32,** 951–967.

EDWARDS, J.W.F. & BLUNDELL, D.J. 1984. *Summary of Seismic Refraction Experiments in the English Channel, Celtic Sea and St. George's Channel.* Report No. 144, British Geological Survey, Marine Sci. Directorate, Murchison House, West Mains Road, Edinburgh.

EVANS, D., CHESHER, J.A., DEEGAN, C.E. & FANNIN, N.G.T. 1982. The offshore geology of Scotland in relation to the I.G.S. Shallow Drilling Programme,

1970–78. *Institute of Geological Sciences Rep. 81/12*, London, HMSO.

FUCHS, K. 1969. On the properties of deep crustal reflectors. *Zeitschrift fur Geophysik*, **35**, 133–149.

HALE, L.D. & THOMPSON, G.D. 1982. The seismic reflection character of the continental Mohorovicic discontinuity. *J. Geophys. Res.* **87**, 4625–4635.

HAUGE *et al.* 1984. COCORP Basin and Range Profile. (abstr. with programs). GSA.

HELMBERGER, D.V. 1968. The Crust–Mantle transition in the Bering Sea. *Bull. Seismol. Soc. Amer.* **58**, 179–214.

HOLDER, A.P. & BOTT, M.H.P. 1971. Crustal structure in the vicinity of southwest England. *Geophys. J. R. astr. Soc.* **23**, 465–489.

JACOB, A.W.B., KAMINISKI, W., MURPHY, T., PHILLIPS, W.E.A. & PRODEHL, C. (in press). 1985. A crustal model for a Northeast–Southwest profile through Ireland, *Tectonophysics*.

JONES, E.J.W., WHITE, R.S., HUGHES, V.J., MATTHEWS, D.H. & CLAYTON, B.R. 1984. Crustal structure of the continental shelf off northwest Britain from two-ship seismic experiments. *Geophys.* **49**, 1605–1621.

JONES, T. & NUR, A. 1984. The nature of seismic reflections from deep crustal fault zones. *J. Geophys. Res.* **89**, 3153–3171.

KUSZNIR, M.J. & PARK, R.G. 1984. Intraplate lithosphere deformation and the strength of the lithosphere. *Geophys. J. R. astr. Soc.* **79**, 513–538.

LEE, C.D., VINE, F.J. & ROSS, R.G. 1983. Electrical conductivity models for the continental crust based on laboratory measurements on high-grade metamorphic rocks. *Geophys. J. R. astr. Soc.* **72**, 353–371.

LORT, J.M. & MATTHEWS, D.H. 1972. Seismic velocities measured in rocks of the Troodos Igneous Complex. *Geophys. J. R. astr. Soc.* **27**, 383–392.

MATHUR, S.P. 1983. Deep reflection probes in eastern Australia reveal differences in the nature of the crust. *First Break*, July 1983, 9–16.

MATTHEWS, D.H. & CHEADLE, M.J. (in press). 1985. Deep reflections from the Caledonides and Variscides west of Britain and comparison with the Himalayas. *In:* BARAZANGI, M. (ed.), *Deep Structure of the Continental Crust*, Amer. Geophys. Un., Geodynamics Series.

—— & HIRN, A. 1984. Crustal thickening in Himalayas and Caledonides. *Nature*, **308**, 497–498.

MCKENZIE, D.P. 1978. Some remarks on the development of sedimentary basins. *Earth Planet. Sci. Lett.* **40**, 25–32.

—— 1984. A possible mechanism for epeirogenic uplift. *Nature*, **307**, 616–618.

MEISSNER, R. 1973. The Moho as a transition zone. *Geophysical Surveys*, **1**, 195–216.

—— & STREHLAU, J. 1982. Limits of stresses in continental crusts and their relation to the depth-frequency distribution of shallow earthquakes. *Tectonics*, **1**, 73–89.

—— & LUESCHEN, E. 1983. Seismic near-vertical reflection studies of the Earth's crust in the Federal Republic of Germany. *First Break*, February 1983, 19–24.

NAYLOR, D. & SHANNON, P.M. 1982. *Geology of Offshore Ireland and West Britain.* Graham & Trotman, London.

OLIVER, J., COOK, F. & BROWN, L. 1983. COCORP and the continental crust. *J. Geophys. Res.*, **88**, 3329–3347.

PEDDY, C.P. 1984. Displacement of the Moho by the Outer Isles thrust shown by seismic modelling. *Nature*, **312**, 628–630.

PHINNEY, R.A. & JURDY, D.M. 1979. Seismic imaging of deep crust. *Geophysics*, **44**, 1637–1660.

RAMSAY, J.G. 1982. Rock ductility and its influence on the development of tectonic structures in mountain belts. *In:* HSU, K.J. (ed.), *Mountain Building Processes.* Academic Press, London and New York, 111–127.

REIF, D.M. & ROBINSON, J.P. 1981. Geophysical, geochemical and petrographic data and regional correlation from the Arizona State A-1 Well, Pinal Country, Arizona. *Arizona Geol. Soc. Digest.* **13**, 99–109.

SHANKLAND, T.J. & ANDER, M.E. 1983. Electrical conductivity, temperatures, and fluids in the lower crust. *J. geophys. Res.* **88**, 9475–9484.

SIBSON, R.H. 1977. Fault rocks and fault mechanisms. *J. geol. Soc. London*, **133**, 191–213.

SMITH, P.J. & BOTT, M.H.P. 1975. Structure of the crust beneath the Californian foreland and Caledonian belt of the North Scottish Shelf region. *Geophys. J. R. astr. Soc.* **40**, 185–205.

SMYTHE, D.K. *et al.* 1982. Deep structure of the Scottish Caledonides revealed by the MOIST reflection profile. *Nature*, **299**, 338 340.

WERNICKE, B. 1984. Structural discordance between thrusts and low-angle normal faults in southern Nevada. Implications for deep reflection profiles in Utah and Scotland. (abstr.) *EOS*, **16**, 281.

WHITTAKER, A. & CHADWICK, R.A. 1983. Deep seismic reflection profiling onshore United Kingdom. *First Break*, September 1983, 9–13.

D.H. MATTHEWS, BIRPS Core Group, Department of Earth Sciences, Bullard Labs., Madingley Road, Cambridge, CB3 0EZ, England.

A physical model of the lower crust from North America based on seismic reflection data

S.B. Smithson

SUMMARY: In speculations about crustal composition, structure, and genesis, the nature of the deep crust is always the major unknown. Seismic reflection data indicate a complex, finely (about 30–100 m) layered lower crust in some areas and an apparently transparent lower crust in other areas. Lower crustal layering is folded to sub-horizontal. The number of seismic events may be highly influenced by recording conditions. Variations in lower crustal structure appear extensive. Best lower crustal reflections are found in areas of young extension in the western US, ancient gneiss in Minnesota, and in offshore Long Island. The lower crust may be affected by a wide range of processes after its formation. Variability and complexity of the lower crust is consistent with postulated exposed deep crustal sections.

Recent geological and geophysical research provides an increasingly sophisticated view of the continental crust, yet the composition, structure and genesis of the crust are uncertainties. Most of these uncertainties stem from our lack of knowledge about the lower continental crust, and these uncertainties severely limit our hypotheses about crustal evolution and genesis. This problem becomes particularly acute in estimation of composition which is a central parameter in any hypothesis on crustal genesis and evolution. Geophysical methods offer the best approach to sample the lower crust in any given location. Reflection methods reveal geometry with a resolution far better than other geophysical techniques and also resolve abrupt velocity contrasts in the crust. The method has the potential to determine interval velocities over relatively small increments (about 5 km) and thus provide estimates of average rock composition (Meissner 1967).

Results from reflection seismology in the United States

Seismic reflection investigations of the deep continental crust in the United States have accelerated with expansion from the initial COCORP efforts by researchers at the US Geological Survey, Virginia Polytechnic Institute and University of Wyoming. Numerous lines have been run to image a variety of targets but only some of these lines provide useful data about the lower crust. Profiles reviewed in this section include data from Minnesota, Laramie Mountains, Wind River Mountains, Adirondacks, Kansas, Long Island, Utah, Rio Grande Rift and Nevada including the Ruby Mountains. Particular attention is drawn to data quality, abundance of and geometry of reflectors, and interpretations of lithology, structure and tectonics of the lower crust.

Several reflection profiles have been conducted over high-grade Precambrian crystalline terrains which include granulite-facies rocks in Minnesota, the Wind River Mountains, Wyoming and the Adirondacks; all of these areas represent deep levels of exposure, i.e., areas that were in the lower crust at some past time. COCORP data from Minnesota (Fig. 1) contain abundant reflections beneath gneisses probably equivalent to the 3.6 Ba gneiss terrain further south (Gibb et al. 1984; Pierson 1984), some of the oldest rocks in North America. The gneiss includes pelitic and amphibolitic members but much of it is tonalitic, and dips are generally moderate (Grant 1972). Reflections are multicyclic spanning 1–3 seconds of travel time so that they must come from layered sequences 3–10 km thick. The layered zone defined by different 'packets' of reflections is over 20 km thick extending to a depth of about 30 km. Packets of curved reflections occur between 2 and 8 s, and straight horizontal reflections occur at 9 s (Fig. 1).

The deep crust beneath the Wind River Mountains, Wyoming shows complex arcuate events at 8–11 s on Line 1A (Smithson et al. 1980, Fig. 3, p. 300). Interpretation of these events on the basis of fold structures exposed up-plunge in the hanging-wall block of the Wind River uplift suggests that the lower crust is not significantly different from the folded high-grade metamorphic rocks intruded by granites exposed in the core of the range. The reflection data provides little information on the lower 10 km of the crust, but an unreversed refraction profile (Braile et al. 1974) indicates that it is a high velocity material, perhaps relatively mafic. Xenolith data suggest that the lowermost crust could consist of gabbros,

From: DAWSON, J.B., CARSWELL, D.A., HALL, J. & WEDEPOHL, K.H. (eds) 1986, *The Nature of the Lower Continental Crust*, Geological Society Special Publication No. 24, pp. 23–34.

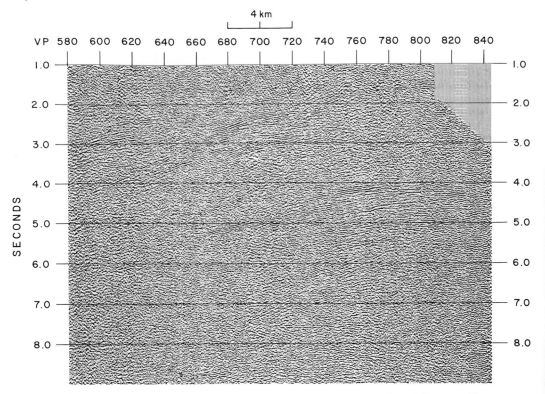

F<small>IG</small>. 1. COCORP crustal reflection profile from the 3.6 Ba gneiss terrain in southern Minnesota. Upper-amphibolite to granulite facies gneisses at surface represent a deep level of exposure. Arcuate reflections from 3 to 8 s beneath the gneiss and flat-lying reflections at 9–10 s. Seismic data reprocessed at the University of Wyoming.

troctolites, two-pyroxene granulites, and anorth-osites. The lower crust may consist of deformed quartzofeldspathic granulites interlayered with mafic granulites intermixed with and possibly underlain by differentiated mafic rocks.

Reflection response is highly variable in the Proterozoic granulite-facies terrain of the Adir-ondacks which must have been at lower crustal levels in the past. COCORP Lines 8 and 10 show very few reflections from the deep crust (Brown *et al.* 1983), however Line 7 located over anorthosite shows some of the better deep crustal reflections yet to be recorded in the US. These events extend from 5 to 10 s as straight, gently dipping reflec-tions suggesting a layered sequence of rocks at least 16 km thick (Fig. 2). Wide-angle reflections are found from 9–24 km depth near Line 8 (Kubichek *et al.* 1984) where none were found at vertical incidence; it is important to know whether this difference in seismic sections is due to actual structural variations within a short distance or to recording technique. This deep

crust beneath an exposed granulite terrain could be strongly layered to within a few kilometres of the base of the crust; other parts appear homoge-neous in standard reflection profiles.

Seismic sections from COCORP data in Kan-sas (Serpa *et al.* 1984) exhibit a complicated wavefield at travel-times corresponding to lower crustal depths (Fig. 3); the wavefield consists of arcuate crossing events which cannot successfully be migrated to separate domains. This suggests that at least some of these events are energy arriving out of the plane of the section, but whether these events represent deep signal or shallow side-arriving energy (noise) is uncertain. This is a fundamental problem in the interpreta-tion of many such seismic sections. If these events are coming from depth, then this seismic section represents a classic, complicated (not plane layered) deep crustal section consisting of some combination of folded metamorphic rocks and intrusions.

COCORP data from Hardeman County,

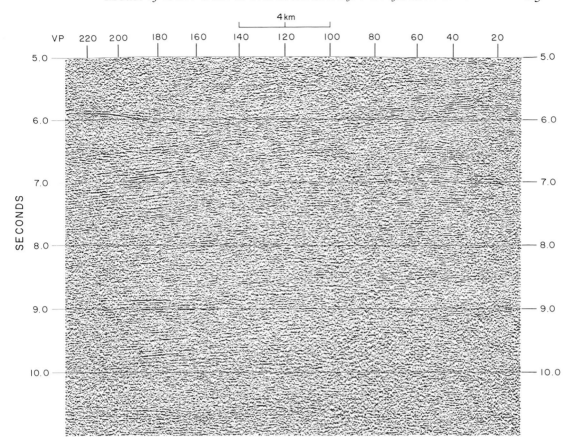

FIG. 2. COCORP crustal reflection profile from the granulite-facies terrain in the Adirondacks, N.Y., on the flank of an anorthosite complex. Note the abundant sub-horizontal reflections.

Texas (Oliver *et al.* 1976) and adjacent areas in Oklahoma (Brewer *et al.* 1981) show some of the strongest, most continuous crustal reflectors yet found (Fig. 4). These come from a layered sequence in the Hardeman basin believed to be a 10–15 km thick sequence of interlayered volcanic and sedimentary rocks. Here the deep crust generates many arcuate events (Schilt *et al.* 1981) of the type that would come from any complex structure (Smithson 1979; Smithson *et al.* 1980). These have been interpreted as originating from point diffractors between depths of 10 and 22 km (Schilt *et al.* 1981), but reflections from synformal features might be a more likely explanation because of the relatively high amplitude of these events. The deep crust slightly to the north in Oklahoma, however, seems to be transparent, while the Hardeman area has highly heterogeneous deep crust.

Seismic profiles from offshore parallel to Long Island (Hutchinson *et al.* 1984) in New York are of great interest because they cross the strike of the Northern Appalachian orogen and because they show numerous events down to 10 s below a transparent zone from 1.5–4 s (Fig. 5). They also show complex arcuate and crossing events that appear to come from the deep crust. These events may be related to the complex structure of the orogen and indicate strong lithologic heterogeneity if we presume at least some of these events are from the deep crust and not sideswipe. A dramatic change in the number of events occurs between the reflection-rich offshore data and the reflection-poor COCORP profiles on land (Brown *et al.* 1983, Fig. 2) which lie along strike of the Northern Appalachians. If the change in data quality is caused by recording conditions, then great care must be exercised in evaluating deep crustal structure on land and comparing it with offshore continental seismic profiles.

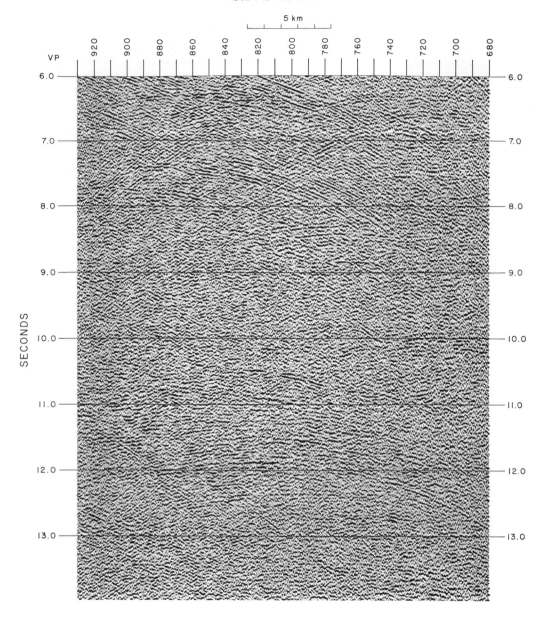

F<small>IG</small>. 3. COCORP crustal reflection profile from Kansas. Note the arcuate dipping and crossing events, which ·
if actually coming from the lower crust indicate folded layers and possibly intrusions.

In general, the Moho does not stand out in crustal reflection profiles within the US and, at best, is hard to identify. Segments of possible Moho reflections are found between 10.5 and 11.5 s in COCORP data from west-central Utah (Allmendinger *et al.* 1983, Fig. 2, p. 533). A multicyclic reflection at 11 s in the Rio Grande Rift data (Brown *et al.* 1980, Fig. 3a, p. 4779) may be from the Moho. Several possibilities are that the Moho is accentuated as layering developed by ductile flow under simple shear or that the 'Moho zone' is underplated by basaltic intrusions. In the COCORP Appalachian data, events are rare at travel times greater than 6–8 s (Brown *et al.* 1983)

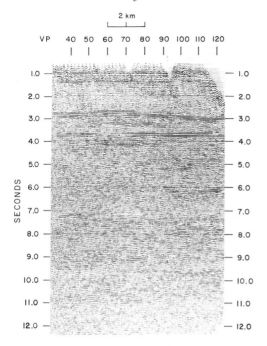

FIG. 4. COCORP crustal reflection profile from Hardeman County, Texas. Note strong, flat reflections at 3 and 4 s are from probable supracrustal rocks in a Proterozoic basin. Arcuate events between 5 and 9 s are from deep crust and represent seismic reflection response of a geologically complex deep crust.

but a short event attributed to the Moho is found at 11 s in the Southern Appalachians (Cook *et al.* 1979). A strong event that could represent Moho appears at 10 s (Fig. 5) on the Long Island profile (Hutchinson *et al.* 1984). In the Laramie Range in Wyoming, an event at 16 s interpreted as a Moho reflection (Allmendinger *et al.* 1982) only occurs under the sediments of the Denver–Julesberg basin where the field records are extremely ringy so it is probably a multiple reflection (Johnson 1984). The best potential Moho events in the US are found in the new COCORP line across Nevada where a discontinuous multicyclic event is found sporadically at about 10 s (Hauser *et al.* 1984).

Deep crustal reflectors

Special high-resolution seismic reflection profiles have been conducted in the US to image upper crustal targets because those targets may be actualistic models of structures in the deep crust. To test a recent proposal that mylonite zones should be good crustal reflectors due to their

layered structure, relatively planar nature and seismic anisotropy (Fountain *et al.* 1984) the University of Wyoming seismograph crew recorded new seismic profiles across the Kettle dome in Washington (Hurich *et al.* in press) and the Ruby Mountains in Nevada (Hurich *et al.* 1984) in order to obtain reflections from exposed mylonite zones. The Ruby Mountains, Nevada exposes a metamorphic core complex (Snoke 1980) with crystalline nappes underlying a thick mylonite zone and detachment fault. This area represents a window into the middle crust due to tectonic denudation. The present-day upper crust here shows abundant sub-horizontal reflections that image the mylonite zone, and a wide-angle line recorded from an offset of 25 km has many reflections between 6 and 9 s corresponding to depths of 15–25 km (Hurich *et al.* 1984). Of great interest (and possibly concern) is the fact that this data set shows events throughout the crust whereas the COCORP line across the same range 70 km to the south shows a transparent seismic section to Moho depth (Hauser *et al.* 1984). Reflection characteristics similar to the Ruby Mountains data are observed in the crust beneath the Kettle dome core complex (Fig. 6) in north-eastern Washington (Hurich *et al.* in press), which formed at depths (~ 20 km) much greater than its present level of exposure as evidenced by exposed sillimanite-grade rocks. Both these areas experienced shortening and isoclinal folding followed by crustal-scale extension. The present fabric of the crust apparently consists of sub-horizontal layering, and the reflection quality is so good that reflections throughout the crystalline crust resemble events from sedimentary sections.

These data suggest that we can follow fault zones through the crust and determine how the crystalline crust has deformed. With depth and increasing ductility, movement passes into discrete mylonite zones at intermediate depth (Sibson 1977). In deep zones, the mylonite may become broader until it eventually passes into semi-homogeneous strain in the lower crust, or it may continue as a discrete zone (Smythe *et al.* 1982; Brewer & Smythe 1984). If so, earlier igneous and metamorphic structures should get pulled out into the movement plane and form a new 'layering'. Such layered zones would be excellent reflectors if dips were not too steep. This may be an explanation for some of the planar layered reflectors in the lower crust. Such features would be hard to distinguish seismically from a metasedimentary rock sequence but could, in fact, contain rocks of any origin and any earlier geometry. Mechanisms for generating multicyclic reflections such as this would be promoted by higher ductility in the lower crust.

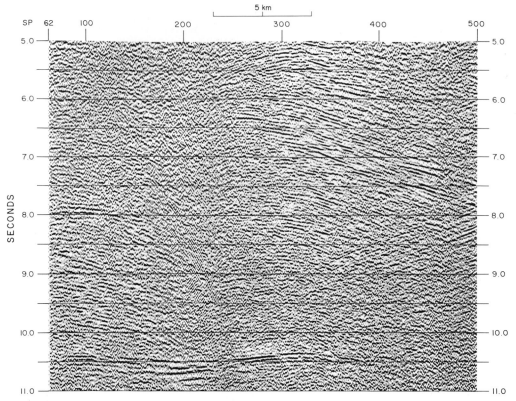

FIG. 5. U.S. Geological Survey reflection profile from offshore parallel to Long Island, N.Y. Crossing arcuate events at travel times corresponding to lower crust typify this seismic profile. Strong event at 10.5 s could be a Moho reflection.

Structure and composition of the lower crust

Modelling

Seismic reflection results provide a growing body of data and constraints on the nature of the lower crust, and this data will become vastly more useful as our understanding of it increases. Models of some typical structures that might be found in the deep crust give important insight into interpretation of crustal reflection data. Early efforts in modelling were primarily directed toward understanding reflections from the Moho and lower crust (Clowes *et al.* 1968; Fuchs 1969; Clowes & Kanasewich 1970; Dohr & Meissner 1975). Theoretical considerations indicated that high amplitude reflections from deep crustal levels are likely interference effects from alternating lamellae with high and low velocities. Layer thickness necessary to produce this interference phenomenon can be as small as 30 m (Fuchs 1969). Recently, Wong *et al.* (1982), Fountain *et*

al. (1984) and Jones & Nur (1984) applied this same approach to explain complex, multicyclic events which apparently emanate from deep fault zones. In this case, lamellae consist of mylonitic rocks which vary in velocity because of compositional variation or seismic anisotropy due to fabric or both. Similar multicyclic reflections could also originate from layered sequences such as layered gneisses in general and possibly a layered mafic intrusion (Johnson & Smithson, in press). It is, therefore, likely that strong, complex, multicyclic events from the lower crust result from layered sequences of rocks which implies fine-scale lithological variation at depth. This conclusion has been confirmed by seismic reflection profiling over mylonite zones formed at depth (about 20 km) and now exposed at the surface (Hurich *et al.* 1984; Hurich *et al.* in press). The layering can be composed of various lithologies including igneous, meta-igneous and metasedimentary rocks. The lithologic diversity observed in xenolith suites (Kay & Kay, 1981) and in lower crustal

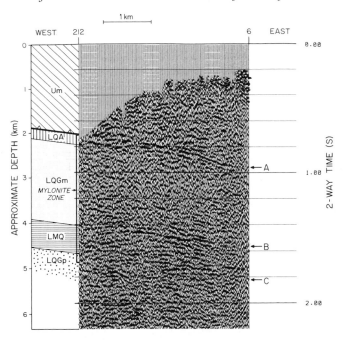

FIG. 6. Seismic reflection profile recorded by the University of Wyoming seismograph crew on the east flank of the Kettle dome, Washington, a metamorphic core complex. Mylonitic gneisses of the lower plate are variable in composition and contain sillimanite and K-feldspar indicating a depth of exposure of about 20 km. Lithologies in the gneiss dome are projected down dip and correlated with the reflection response in the seismic section indicating that mylonites formed deep in the crust are highly reflective. Um = upper plate, LQA = mylonitized supracrustal rocks, LQG = quartzo-feldspathic mylonite zone, LMQ = lower mylonitized supracrustal rocks.

cross-sections (Fountain & Salisbury, 1981) are consistent with this view. This does not mean that all reflections come from mylonites but rather represents a field calibration of the suggestion that strongest reflections and multicyclic reflections come from layering. Such layering could be found in mylonites, gneisses in general, or possibly in layered mafic intrusions. In general, the most planar layering may be represented by mylonites, which may also have gentler dips than layering in the surrounding crystalline rocks and thus form the best reflectors.

Models of actual geological structures can aid in understanding other features of lower crustal reflections. A migmatite terrain, represented by folded discontinuous layers cut by a granitic body, is modelled in Fig. 7. The seismic response of this model is a series of short discontinuous reflections that pass into diffraction tails. The diffractions would probably be too weak to appear in reflection sections so that the expected response would be short, horizontal events. This is a fairly common occurrence in crustal reflection

data and such a migmatite terrain is one explanation for such reflections. A model of a layered mafic intrusion presented by Wong *et al.* (1982, Fig. 4, p. 96) shows a dipping sequence of reflections that become arcuate where the curvature of the layers becomes greater. If the mafic intrusion in the model had been buried deeper, some of the reflections would have become more arcuate. Another plausible geometrical model for deep crust is one of folded recumbent folds (Smithson *et al.* 1980, Fig. 5, p. 302). The reflection response consists of a series of convex-upward arcuate events and is, in fact, the reflection response for almost any kind of typical nonplanar structural feature found in crystalline rocks (Smithson 1979; Wong *et al.* 1982). Thus, folds and intrusions, if buried at any depth, will generate a series of arcuate reflections and diffractions. These characteristics should be the common reflection response of the deep crust and may be illustrated by the wavefield observed in the offshore Long Island data (Fig. 5), Hardeman County COCORP data (Fig. 4) and possibly the

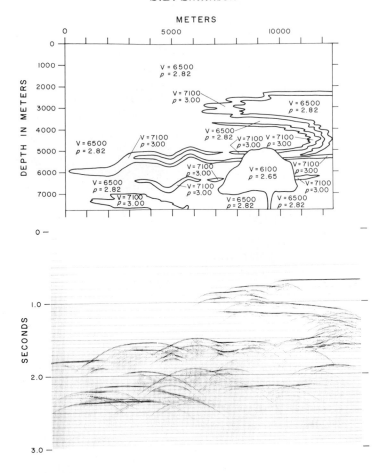

FIG. 7. Seismic model of migmatitic crust cut by granite intrusion (upper) and its computer-generated seismic response (lower). Short discontinuous reflections pass into weaker arcuate diffraction tails that might not be present in an actual seismic section. This model of deep crust represents one explanation for the short discontinuous reflections commonly found in deep crustal seismic sections.

Kansas COCORP data set (Fig. 3). Fountain (in press) modelled the Ivrea zone in 2-D and showed that its seismic response is an interfering sequence of arcuate events. This could be a standard response for lower crust. That such arcuate events are not particularly common in crustal reflection data raises the possibility that 'tails' of arcuate events are not recorded so that crests of events preferentially appear. If these events emanate from deep structures, the crests would appear as flat, discontinuous reflection segments. Three-dimensional lower-crustal structure will only complicate the wavefield in a 2-D seismic section (Blundell, in press) and render 2-D migration ineffective.

Classification of lower crustal types

Lower crust may be classified on the basis of its reflection response, but such a classification is extremely limited at present by our incomplete understanding of reflection results. These classes are: 1) non-reflective lower crust, 2) planar, near horizontal reflections, 3) broadly arched reflections, 4) arcuate, criss-crossing events. Like all classifications this one is arbitrary in its divisions. Of all these, the category and significance of non-reflective crust is least satisfactory. Firstly, lack of reflections may simply indicate homogeneity, and if this were the only (or even most likely) explanation this response would be extremely important.

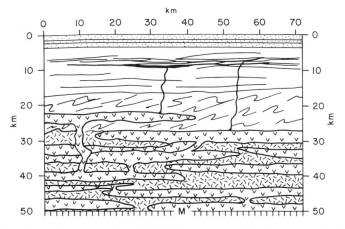

FIG. 8. Interpretation of crustal structure from SW Oklahoma beneath the Proterozoic basin. Basin subsidence that allowed the accumulation of 12 km of supracrustal rocks may be related to underplating of lower crust by basaltic magma. Interpretation based on combined reflection and refraction results.

Numerous effects may cause a given volume of crust to appear non-reflective (Smithson 1979), but the first reason that must be considered is recording and/or processing effects. Results from several areas mentioned earlier suggest that different recording techniques may produce different results regarding reflection density, and recent reprocessing of the Sevier Desert COCORP data sets (Branch *et al.* 1985) demonstrates how near-surface velocity variations degrade deep reflections. In fact, the effect of recording conditions is one of the burning issues demanding immediate attention in crustal reflection seismology. Other factors that might produce no reflections are steep dips, strong folding, gradational changes in composition and very small scale heterogeneity.

Significance of lower-crustal reflection response

If a volume of the lower crust is non-reflective because it is truly homogeneous, then it is almost certainly an igneous intrusion, for sedimentary and metamorphic rocks are rarely homogeneous over large volumes. Planar near-horizontal reflections could be caused by intrusions of gabbroic sills in more felsic rocks, discrete mylonite zones, homogeneous ductile deformation of a heterogeneous lower crust, or possibly a layered mafic intrusion. If underplating occurs by intrusion of gabbro as discrete sills into lower-velocity material, the reflection response would probably consist of subparallel, sub-horizontal reflections, i.e. class (2). If, however, the sills coalesce, probably no visible reflections would be generated and the lower (or lowermost crust) would appear transparent but give a higher (gabbroic) lower crustal

velocity (Meissner 1967), i.e. class (1). Another possibility which is likely in a lower crustal environment, is that a sill differentiates into mafic layers like the Palisades sill and generates weak sub-horizontal reflections. Whether they would be visible would depend on the signal to noise ratio.

Supracrustal rocks would certainly produce multicyclic and probably broadly arched reflections, i.e. class (3) or if late folding was stronger would form crossing arcuate events, i.e. class (4). Xenolith suites (Kay & Kay 1981) and exposed-crustal sections (Fountain & Salisbury 1981; Percival & Card 1983) suggest that a metasedimentary component is common and widespread in the lower crust although the actual amount is hard to estimate and probably varies greatly from area to area if we accept the proportions implied by exposed crustal sections. The metasedimentary rocks are typically 'dry' granulite-facies rocks and do not explain the rather common suggestion of high electrical conductivity in the lower crust (Shankland & Ander 1983), a contradictory observation that remains one of the major questions concerning the nature of the lower crust. The presence of metasedimentary rocks in the lower crust is extremely important and must be explained by hypotheses on crustal genesis.

Recent studies (Fountain *et al.* 1984; Hurich *et al.* in press; Smithson *et al.* in press) demonstrate that mylonites are excellent reflectors, and this reflectivity is probably largely due to layering on all scales. Two of the mylonite zones studied were exposed at what was formerly mid-crustal level (sillimanite grade) and must thus extend into the lower crust; or in some areas mylonites probably extend through it in order to explain the reflection from the Outer Isles thrust in the MOIST data (Brewer & Smythe, 1984). While mylonites could be abundant in the lower crust, we would expect them to form relatively discrete zones from 1–3 km thick (0.3–1 s). If they were anastomosing, they would have a discontinuous irregularly dipping reflection pattern but not the criss-crossing arcuate events typical of Kansas and offshore Long Island. Both mylonites and supra-crustal sequences have been deformed so much that layering would not have any primary signifi-cance, and large amounts of intrusive igneous components could be incorporated into what is now a strongly layered sequence. Rather similarly if the lower crust were subjected to roughly homogeneous strain, mixed rocks that had highly variable and irregular geometries could be drawn into semicontinuous layers parallel to the slip direction (the old 'deck-of-cards' trick) (Sander 1936). Thus lower crustal layering may have several origins, but in its best development is probably related to strain or extreme strain.

Along this line, a reasonably consistent picture seems to be emerging from the metamorphic core complexes of the western US. Here, where extreme extension has been the more recent tectonic event (Crittenden *et al.* 1980), the atte-nuated crust seems to be highly reflective, both in discrete mylonite zones and over greater areas. Although compressional nappe structures undoubtedly contribute to the reflection wave field in areas like the Ruby Mountains, most crystalline terrains throughout the US commonly record an early isoclinal folding event but do not show the numerous reflections of the extended terrains, probably because later superposed fold-ing has complicated the reflection response of the isoclinal structures, e.g. see Fig. 7. Thus the mylonites formed during extension may combine with more homogeneous strain in the ductile deep crust to produce very reflective zones of high strain (mylonites) and intervening zones of moderate strain in which earlier features are rotated toward the horizontal as the crust extends. This same strain pattern may have created multilayered reflectors for the Moho in the Basin and Range–Rio Grande rift.

The best lower crustal reflections in the US are found in the Great Basin affected by late Ceno-zoic extension, paradoxically also in the oldest Archaean crust of Minnesota and the Northern Appalachians in offshore Long Island. This latter data set may appear to have more reflective lower crust simply because it is recorded offshore; we are not sure how much recording conditions are affecting our view of lower crustal reflections. The deep crust from old Archaean of Minnesota is heterogeneous and complexly layered. The gneisses, which contain a distinct metasedimen-tary component, may represent a thick stack of nappes (Matt Walton, pers. comm.). This possibi-lity is entirely consistent with the reflection data which shows broadly arched packets of layering. in this case, much of the lower crust would consist of packets of nappes (the individual reflection 'packages' of Fig. 1) overlying a lowermost crust with some horizontal reflectors. The rocks have had an extended Archaean history of folding intrusion and remobilization which have, how-ever, left a highly reflective moderately dipping layered sequence.

A major question is whether ancient Archaean crust is different, i.e. was it thin and then thick-ened later by crustal underplating (Fyfe 1974)? If so, in Minnesota the mafic magma must have been intruded into more felsic rock as discrete sills tens of metres thick and then broadly folded later during one of the younger events to affect the area. The similarity of the reflection packets to those within the near-surface, however, suggests that they are more likely a layered sequence of rocks forming nappes. The deeper part of the layered sequence could consist of rather mafic migmatites that contain restite layers from partial melting to produce the granitic melts that were emplaced in the exposed gneisses.

Underplating by either thick (> 100 m) sills or coalescing sills of mafic rocks would not produce good reflections from the lower crust so areas with no reflections or weak sub-horizontal reflec-tions at lower crustal depths could be candidates for crustal underplating. One such area is south-ern Oklahoma where refraction data indicate a thick, high-velocity lower crust (Mitchell & Landisman, 1970). Although similar crustal refraction interpretations are found in other parts of the mid-continent, I suggest that this area which contains a large excess mass in the lower crust is a prime candidate for crustal underplating (Fig. 8), which accompanied the subsidence that allowed the thick Proterozoic sedimentary basin (Fig. 4) to form.

The weak near-horizontal reflections in the Adirondacks (Fig. 2) could be caused by layered mafic intrusions, supracrustal gneisses or mylo-nites. The great vertical extent, general sub-horizontal parallelism, and relatively weak reflec-tions suggest that these reflections could be a series of layered mafic intrusions.

The arcuate crossing events found deep in the reflection sections from Kansas are typical of complex geology (Smithson 1979; Smithson *et al.* 1980; Wong *et al.* 1982) such as folded (relatively open folds as opposed to isoclinal folds where the hinges would not generally be visible in reflection data) gneisses cut by igneous intrusions. They thus indicate heterogeneity and complicated geology in the lower crust like some exposed crustal sections (Fountain, in press).

In an earlier paper (Smithson 1978), the presence of lower crustal reflections was used to suggest that the lower crust is more felsic than gabbro, i.e., because of the presence of more felsic material to cause reflections. This proposal needs qualification because we can imagine a situation in which reflections could be generated from thin granitic zones in gabbro yet the amount of gabbro is so dominant that the overall composition is essentially gabbroic. Whether this situation is geologically probable is another question.

Processes that may modify deep crust are thrusting, extension, underplating, anatexis, and metasomatism. Thrusting may locally alter lower crustal geometry by thickening or by crustal doubling when it occurs on a continental scale. Extension may alter geometry by movement along discrete listric faults or by pure shear which might explain some of the good sub-horizontal lower crustal reflectors. Underplating by intrusion of basaltic magma could create a lower crust totally different from the earlier lower crust for a particular area. Anatexis would differentiate lower crust into felsic and mafic zones, and separation of the melt would drive the lower crust toward more mafic compositions. Metasomatism caused by mantle degassing would presumably drive the lower crust toward more felsic compositions. Of these, strain may be the most important process affecting the lower crust.

Conclusions

Seismic reflection sections show a wide range of lower crustal response ranging from seemingly transparent (homogeneous?) to highly folded, layered, structurally- and lithologically-complex lower crust. The significance of the varying reflection response is not yet well understood, and particularly the effect of different recording conditions and processing parameters must be understood in order to progress in determining the significance of variations in reflection response. Some of the best lower crustal reflections come from areas of crustal extension in the western US and may be related to ductile extension, but good reflections indicating lower crustal heterogeneity and complexity are found under the ancient (~ 3.6 Ba) gneisses in Minnesota. Variations in apparent lower crustal structure appear extensive and resemble the variations found in postulated exposed lower crustal sections (Fountain & Salisbury 1981). Lower crust may be affected by and altered by a wide range of processes after initial formation so that in many areas the lower crust may be multigenetic.

ACKNOWLEDGEMENTS: Financial support was received from US National Science Foundation grants EAR-8306542 and EAR-8300659. M.C. Humphreys, Allen Tanner and Robert Tweed carried out the seismic recording. Barbara Cox receives thanks for help with many aspects of the project. Processing was carried out on the DISCO VAX 11/780 computer system in the Program for Crustal Studies.

References

ALLMENDINGER, R.W., BREWER, J.A., BROWN, L.D., KAUFMAN, S., OLIVER, J.E. & HOUSTON, R.S. 1982. COCORP profiling across the Rocky Mountain Front in southern Wyoming, part 2: Precambrian basement structure and its influence on Laramide deformation, *Bull. Geol. Soc. Am.* **93**, 1253–1263.
—— *et al.* 1983. Cenozoic and Mesozoic structure of the eastern Basin and Range Province, Utah from COCORP seismic-reflection data, *Geology*, **11**, 532–536.
BRAILE, L.W., SMITH, R.B., KELLER, G.R., WELCH, R.M. & MEYER, R.P. 1974. Crustal structure across Wasatch front from detailed seismic refraction studies, *J. Geophys. Res.* **79**, 2669–2677.
BRANCH, C., JOHNSON, R.A. & SMITHSON, S.B. 1985. Megastatics applied to Moho imaging on COCORP Utah seismic reflection data (abs.). EOS, *Trans. Amer. Geophys. Un.* **66**, 311.
BREWER, J.A. & SMYTHE, D.K. 1984. MOIST and the continuity of crustal reflector geometry along the Caledonian-Appalachian orogen. *J. Geol. Soc.* **141**, 105–120.
—— BROWN, L.D., STEINER, D., OLIVER, J.E., KAUFMAN, S. & DENISON, R.E. 1981. Proterozoic basin in the southern midcontinent of the United States revealed by COCORP deep seismic reflection profiling, *Geology*, **9**, 569–575.
BROWN, L.D. *et al.* 1983. Adirondack–Appalachian crustal structure: The COCORP Northeast traverse, *Bull. Geol. Soc. Am.* **94**, 1173–1184.
—— CHAPIN, C.E., SANDFORD, A.R., KAUFMAN, S. & OLIVER, J.E. 1980. Deep structure of the Rio Grande rift from seismic reflection profiling, *J. Geophys. Res.* **85**, 4773–4800.
CLOWES, R.M. & KANASEWICH, E.R. 1970. Seismic attenuation and the nature of reflecting horizons within the crust, *J. Geophys. Res.* **75**, 6693–6705.
—— KANASEWICH, E.R. & CUMMING, G.L. 1968. Deep crustal seismic reflections at near-vertical incidence, *Geophysics*, **33**, 441–451.

34

S.B. Smithson

COOK, F., ALBAUGH, D., BROWN, L., KAUFMAN, S., OLIVER, J. & HATCHER, R. 1979. Thin-skinned tectonics in the crystalline southern Appalachians; COCORP seismic reflection profiling of the Blue Ridge and Piedmont, *Geology*, **7**, 563–567.

CRITTENDEN, M.D., CONEY, P.J. & DAVIS, G.H. 1980. Cordilleran metamorphic core complexes, *Geol. Soc. Amer. Memoir* **153**.

DOHR, G.P. & MEISSNER, R. 1975. Deep crustal reflectors in Europe, *Geophysics*, **40**, 25–39.

FOUNTAIN, D.M., HURICH, C.A. & SMITHSON, S.B. 1984. Seismic reflectivity of mylonite zones in the crust, *Geology*, **12**, 195–198.

—— & SALISBURY, M.H. 1981. Exposed cross-sections through the continental crust: Implications for crustal structure, petrology, and evolution, *Earth Planet. Sci. Lett.* **56**, 263–277.

FUCHS, K. 1969. On the properties of deep crustal reflections, *J. Geophys*, **35**, 133–149.

FYFE, W.S. 1974. Archean tectonics, *Nature*, **249**, 338–340.

GIBB, A.K., PAYNE, B., SETZER, T., BROWN, L.D., OLIVER, J.E. & KAUFMAN, S. 1984. Seismic-reflection study of the Precambrian crust of central Minnesota, *Bull. Geol. Soc. Am.* **95**, 280–294.

GRANT, J.A. 1972. Minnesota River Valley, southwestern Minnesota. *In*: SIMS, P.K. & MOREY, G.G. (eds), *Geology of Minnesota—A centennial volume*, Minnesota Geological Survey, 177–198.

HAUSER, E. *et al.* 1984. The COCORP 40° N transect of the North American Cordillera: Part 2 (abs.), Abstracts with Program, *97th Annual Meeting Geol. Soc. Am.* 532.

HURICH, C.A., HUMPHREYS, M.C., SNOKE, A.W., JOHNSON, R.A., FOUNTAIN, D.M. & SMITHSON, S.B. 1984. Crustal reflection profiling in the metamorphic core complex of the Ruby Range, Nevada (abs.), EOS, *Trans. Am. Geophys. Un.* **65**, 985.

—— SMITHSON, S.B., FOUNTAIN, D.M. & HUMPHREYS, M.C. (in press), Mylonite reflectivity: evidence from the Kettle dome, Washington, *Geology*.

HUTCHINSON, D.R., GROW, J.A. & KLITGORD, K.D. 1984. Crustal reflections from the Long Island Platform of the U.S. Atlantic continental margin, *Int'l Symp. on Deep Structure of the Continental Crust* in press.

KAY, R.W. & KAY, S.M. 1981. The nature of the lower continental crust: inferences from geophysics, surface geology, and crustal xenoliths, *Rev. Geophys. Space Phys.* **19**, 271–297.

KUBICHEK, R.F., HUMPHREYS, M.C., JOHNSON, R.A. & SMITHSON, S.B. 1984. Long-range recording of VIBROSEIS data: Simulation and experiment, *Geophys. Res. Lett.* **11**, 809–812.

MEISSNER, R. 1967. Zum Aufbau der Erdkruste. Ergebinisse der Weitwinkelmessungen im bayerischen Molassebecken, Tiel 2, *Ger. Beitr. Z. Geophysik.* **76**, 295–314.

MITCHELL, B.J. & LANDISMAN, M. 1970. Interpretation of a crustal section across Oklahoma. *Bull. Geol. Soc. Am.* **81**, 2647–2656.

OLIVER, J., DOBRIN, M., KAUFMAN, S., MEYER, R. & PHINNEY, R. 1976. Continuous seismic reflection profiling of the deep basement, Hardeman County, Texas, *Geol. Soc. Am. Bull.* **87**, 1537–1546.

PIERSON, W.R. 1984. *A geophysical study of the contact between the greenstone–granite terrain and the gneiss terrain in central Minnesota*. Unpubl. M.S. Thesis, Univ. of Wyoming, Laramie, 84 p.

SANDER, B. 1936. *Einführung in die Gefügekunde der Gesteine*, Springer, Berlin.

SCHILT, F.S., KAUFMAN, S. & LONG, G.H. 1981. A three-dimensional study of seismic diffraction patterns from deep basement sources, *Geophysics*, **46**, 1673–1683.

SERPA, L. *et al.* 1984. Structure of the southern Keweenawan rift from COCORP surveys across the Midcontinent geophysical anomaly in northeastern Kansas, *Tectonics*, **3**, 367–384.

SHANKLAND, T.J. & ANDER, M.E. 1983. Electrical conductivity, temperatures, and fluids in the lower crust, *J. Geophys. Res.* **88**, 9475–9484.

SIBSON, R.H. 1977. Fault rocks and fault mechanisms, *J. Geol. Soc. London*, **133**, 191–213.

SMITHSON, S.B. 1978. Modeling continental crust: structural and chemical constraints, *Geophys. Res. Lett.* **5**, 149–152.

—— 1979. Aspects of continental structure and growth: targets for scientific deep drilling, *Univ. of Wyoming Contributions to Geology*, **17**, 65–75.

—— BREWER, J.A., KAUFMAN, S., OLIVER, J.E. & ZAWISLAK, R.L. 1980. Complex Archean lower crustal structure revealed by COCORP crustal reflection profiling in the Wind River Range, Wyoming, *Earth Planet. Sci. Lett*, **46**, 295–305.

—— JOHNSON, R.A. & HURICH, C.A. in press. Crustal reflections and crustal structure, *Amer. Geophys. Union Mono.*

SMYTHE, D.K. *et al.* 1982. Deep structure of the Scottish Caledonides revealed by the MOIST reflection profile, *Nature*, **299**, 338–340.

SNOKE, A.W. 1980. Transition from infrastructure to suprastructure in the northern Ruby Mountains, Nevada. *In*: CRITTENDEN, M.D. Jr., CONEY, P.J. & DAVIS, G.H. (eds), *Cordilleran metamorphic core complexes*, Geol. Soc. Am. Mem. **153**, 287–334.

WONG, Y.K., SMITHSON, S.B. & ZAWISLAK, R.L. 1982. The role of seismic modeling in deep crustal reflection interpretation, Part I, *Contr. Geol.* **20**, 91–109.

SCOTT B. SMITHSON, Department of Geology and Geophysics, Program for Crustal Studies, University of Wyoming, Laramie, Wyoming 82071, USA.

Electrical resistivity in continental lower crust

V. Haak & R. Hutton

SUMMARY: The application of d.c. resistivity, magnetotelluric—both natural and artificial source—and geomagnetic variation techniques has provided lithospheric electrical resistivity models which generally display a relatively low resistivity region at lower crustal/upper mantle depths. In this paper, model parameters are summarized for the lower crust in tectonically inactive regions and discussed in more detail with respect to specific studies undertaken in Europe, N. America and Africa. Special attention is drawn to the resolution of the model parameters and to the fact that the depth to the low resistivity layers and the ratio of their thickness to their resistivity can normally be well determined.

The petrological interpretation is non-unique. Although electrical resistivity is strongly temperature dependent, the temperature within the lower crust in inactive regions is too low to give an *in situ* decrease in resistivity of the major rock phases. A more acceptable interpretation involves compositional variations—both lateral and vertical—within the crust associated with an increased amount of free fluids (e.g. water) or of free carbon (e.g. graphite). The correlation of the low resistivity structures in the lower crust with sedimentary layers involved in large-scale overthrusting or related tectonic processes is an attractive idea while the generation of low resistivity shear zones during metamorphism could also be an effective explanation of the observations. Some of the elongated anomalies of low resistivity which are discussed may correspond to Proterozoic plate boundaries and subduction zones.

Introduction

The distribution and values of electrical resistivity in the Earth at lower crust and upper mantle depths provide information—in some respects unique—about the nature of these regions for the following two reasons:

(a) electrical resistivity of rocks is a property which is closely related to petrological parameters, and
(b) it can normally be determined for depths ranging from the surface to upper mantle depths from observations made at the Earth's surface with the appropriate frequency sounding range.

Electrical conductivity, σ, which is the reciprocal of resistivity (i.e. $\sigma = 1/\rho$) may also be used to represent the same physical property but in this paper resistivity will normally be used.

For crustal rocks, ρ can vary from 10^{-1}–10^{+5} ohm m, according to their state and to their composition as shown in Fig. 1. Values of ρ are sensitively dependent on temperature and on the presence of specific minerals or even very small amounts of volatiles, e.g. free water distributed in connected channels. Increase in temperature and increase in fluid content and/or its salinity each cause a reduction in rock resistivity and can, in some studies, introduce ambiguity in interpretation. For this reason, in this study of the lower crust the discussion is primarily concerned with the resistivity of *normal* continental crust, for which the temperature is assumed to vary smoothly, i.e. regions in which any vertical or lateral change of resistivity can be primarily attributed to a change in bulk composition, e.g. the amount and/or salinity of fluids. Recent reviews of electromagnetic induction studies of active plate margins, active grabens and explicitly known geothermal areas which all require the additional consideration of strong temperature effects and partial melts have been published elsewhere (e.g. Hutton 1976; Jiracek *et al.* 1979; Hermance 1982).

The normal resistivity of continental lower crust in inactive regions

During the past 15 years, anomalously low resistive zones, e.g. of 10–500 Ω m (ohm metre), within the continental lower crust have been reported in many studies. This classification as 'anomalous' was based on the concept that the normal crust was expected to be highly resistive and to conform to the models proposed by Brace (1971) on the basis of intensive laboratory studies on small rock samples. As in the majority of studies the inferred resistivities of the lower crust are substantially less than for Brace's model, it seems timely to introduce a new classification based on the interpretation of actual field data. For example, Jones (1981) and Shankland & Ander (1983) have recently compiled such data according to resistivity and seismic velocity, and resistivity and regional heat flow respectively and have accepted as the *observed* resistivity-depth

From: DAWSON, J.B., CARSWELL, D.A., HALL, J. & WEDEPOHL, K.H. (eds) 1986, *The Nature of the Lower Continental Crust*, Geological Society Special Publication No. 24, pp. 35–49.

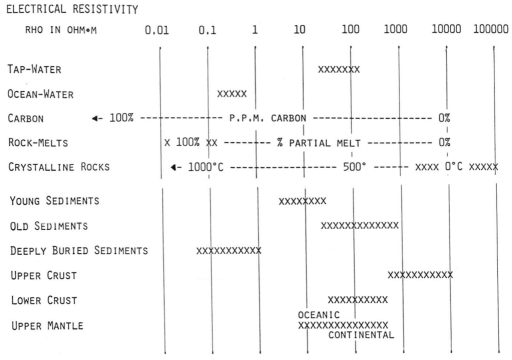

FIG. 1. Range of electrical resistivities of some natural materials (upper portion) and of crustal material from the Earth's surface down to the upper mantle (lower portion).

TABLE 1(a). *Normal lower crustal resistivity (case studies)*

Location	Method	Inversion	Depth range (km)	ρ (ohm m)	Reference
1 Europe/Asia					
(i) Bohemian Massif	GDS/MT	2D	25–45	400–500	Cerv *et al.* (1984)
(ii) N. Scotland	GDS/MT	2D	20–40	300–1000	Hutton *et al.* (1980)
(iii) Sweden (KIR)	HG method	1D	30→	~300	Jones (1982)
(iv) Sweden	MT	1D	20–30→	~300	Rasmussen *et al.* (1984)
(Fennolora Profile)					Roberts *et al.* (1984)
2 N. America					
(i) Wisconsin	Deep dc resistivity	1D	14→	50–1500	Sternberg (1979)
(Canadian Shield	& dipole-dipole				
extension)					
(ii) Canadian Shield	MT/GDS	1 and 2D	9–14	80	Kurtz (1982)
(Grenville Province)	Controlled source EM	1D	10–23	1000	Connerney *et al.* (1980)
(iii) Canadian Shield	Controlled source EM	1D	20→	<270	Duncan *et al.* (1980)
(Superior Province)					
(iv) Prince Edward Is.	MT	2D	24–44	1000	Kurtz & Garland (1976)
3 Africa					
Central Africa	MT	2D	30–40	100–300	Ritz (1983)
(Mobile belt)					

Lower crustal conductance (d/ρ).
The values of the thickness–resistivity ratio—or conductance—from the above studies lie within the range 10–65 Siemens.

TABLE 1(b). *Anomalously low lower crustal resistivity (case studies)*

Location	Method	Inversion	Depth range (km)	ρ (ohm m)	Reference
1 Europe/Asia					
(i) Scotland (S. Uplands)	MT	1 and 2D	20–50	~50	Jones & Hutton (1979) Ingham & Hutton (1982) Sule & Hutton (1985)
(ii) W. Germany	MT	2D	15–19	~10	Jödicke *et al.* (1983)
(iii) Carpathians	MT	2D	<20–>30	1–20	Cerv *et al.* (1984)
	GDS/MT	2D	10→	2–10	Zhdanov *et al.* (1984)
(iv) Fennoscandian Shield (Sau)	MT/HG Method	1D	20→	25	Jones (1983)
(v) Sweden (Central) and Finland	MT	1D	10–20→	~1	Zhang *et al.* (1984)
(vi) USSR (East European platform)	MT	1D	10–30→	(d/ρ=10^2 − 10^3 S)	Vanyan *et al.* (1984)
2 N. America					
(i) Wisconsin	MT	1D	12–40	~20	Dowling (1970)
(ii) Canadian Shield (Grenville Province)	MT/GDS	1 and 2D	24–35	~20	Kurtz (1982)
(iii) Appalachians	HG Method	1D	22–34	25	Connerney *et al.* (1980)
	GDS	2D	20–60	10–20	Mareschal *et al.* (1983)
(iv) Canada (Atlantic)	MT/GDS	2D	10–30	2–20	Cochrane & Hyndman (1974)
3 Africa					
(i) S. Africa (Craton)	DC Schlumberger	1D	12–30	50	Van Zijl (1977) Blohm *et al.* (1977)
(ii) Damara Mobile belt	DC Schlumberger	1D	4–20→	10	De Beer *et al.* (1982b)

Lower crustal conductance (d/ρ).
The values of the thickness–resistivity ratio—or conductance—for the above studies lie within the range 300–1500 Siemens.

TABLE 1(c). *Anomalously high lower crustal resistivity (case studies)*

Location	Method	Inversion	Depth range (km)	ρ (ohm m)	Reference
1 Europe/Asia					
(i) USSR (Karelian Megablock)	MT	1D	10–30	80,000–10,000	Kaikkonnen *et al.* (1983)
2 Africa					
(i) S. Africa (Kapvaal)	DC Schlumberger	1D	12–25	5000	Van Zijl (1977)
(ii) W. Africa	MT	2D	variable	~3000	Ritz (1983)

Lower crustal conductance (d/ρ).
The values of the thickness–resistivity ratio—or conductance—for the above studies is less than 5 Siemens.

profile for old continental crust, one in which the average lower crustal resistivity is about 10^2 to 10^3 Ωm as shown in the lower portion of Fig. 1. It thus seems reasonable to propose that these are normal lower crustal resistivities. Resistivities of less than 10^2 Ωm and greater than 10^3 Ωm can thus be regarded as anomalously low and high respectively. Only a very few examples of anomalously high lower crustal resistivities have been reported in the literature (Kovtun 1976).

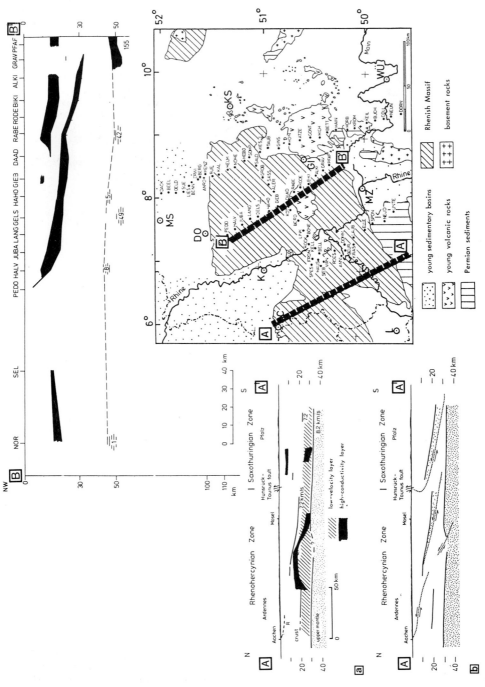

FIG. 2. Electrical resistivity structures in the Rhenish Massif in Germany (Jödicke *et al.* 1983). In the lower right section the high density of observational sites is displayed. Two profiles A–A′ and B–B′ are indicated for which the electrical resistivity structures are presented to the left (A–A′) and above (B–B′). A–A′: Section (a) shows the agreement between the low resistivity structures, which have actually been projected from the western magnetotelluric line, and the seismic low velocity layer (Giese 1983). Section (b) shows a simplified tectonic interpretation (Giese *et al.* 1983). B–B′: This profile shows low resistivity structures dipping southwards.

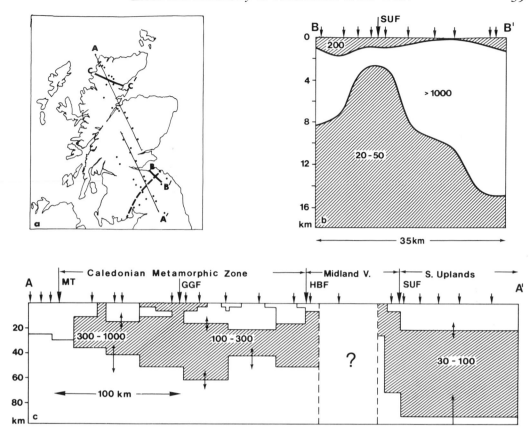

FIG. 3. (a) The locations of magnetotelluric observational sites in Scotland are shown by dots for traverses AA′ and by thick solid lines for multi-station AMT traverses BB′ and CC′. The broken line across S Scotland indicates the axis of the geomagnetic variation anomaly. (b) An electrical resistivity section compiled from a set of 1D inversions of AMT data from traverse BB′ of Fig. 3(a). The site locations are indicated by arrows and the resistivity units are ohm m. (c) A simplified form of Hutton *et al.*'s (1980) electrical resistivity model for the Scottish traverse AA′ of Fig. 3(a). The low resistivity layer is hatched. Resistivity values in ohm m are given for this layer only—all other zones have higher resistivities.

Using this proposed classification some well-established lower-crust resistivity data are presented in Tables 1(a)–(c). In considering the extent to which these data may be used—for example, in comparison with seismic data—their precision is, without doubt, the most important question. For this purpose, it is possible to analyse the data in a manner, called the SVD (Single Value Decomposition) method, which indicates the best determined parameter of a particular model (Edwards *et al.* 1981)—normally the depth to low resistivity layers and their thickness resistivity ratios.

It will be noted in Tables 1(a)–(c) that a variety of geophysical methods have been used to infer a resistivity depth profile in each case and that there

is remarkably close agreement between the results of the different techniques.

Lateral and vertical variations of resistivity

A zone of anomalous resistivity signifies the anomalous state or composition of the rocks in that zone. Since it has become increasingly evident that the physical properties of the lower crust can have significant 'small-scale length' lateral variations, the corresponding structure of anomalously resistive zones can only be detected by use of a dense network of observational sites and a sufficiently broad range of sounding frequencies to probe to upper mantle depths. Unfortunately, in many studies, only single or widely

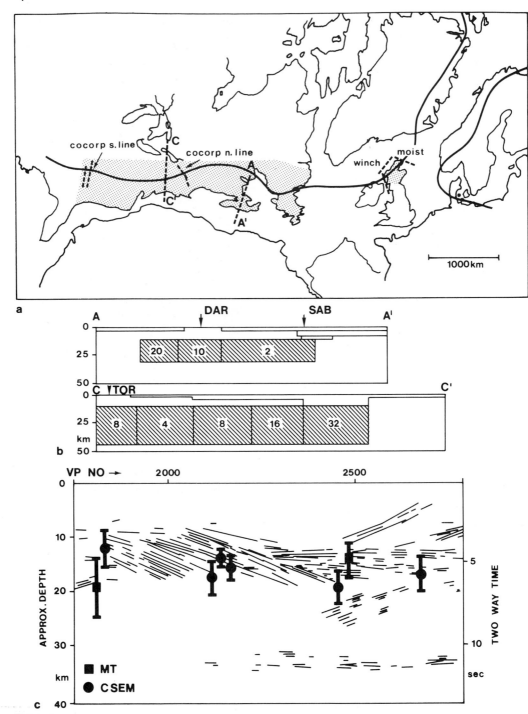

FIG. 4. (a) Regions of intensive magnetotelluric and geomagnetic variation studies in Scotland and eastern N America are indicated by shading on Smythe's map of the pre-drift reconstruction of the Caledonian–Appalachian orogen (re-drawn from Brewer and Smythe 1984). The deep seismic reflection traverses of COCORP and BIRPS (MOIST and WINCH) and two of the geomagnetic variation traverses AA′ and CC′ of Greenhouse and Bailey (1981) are indicated by broken lines. (b) The electrical resistivity models for traverses AA′ and CC′ across the Appalachians (from Greenhouse & Bailey 1981). The low resistivity layer is hatched, with the resistivity values given in ohm m. (c) A line drawing of the COCORP profile in the S Appalachians with the depth range to the top of a low resistivity layer as deduced from magnetotelluric and controlled source electromagnetic studies (from Thompson *et al.* 1983).

spaced observational sites have been used and, with the exception of geothermal fields, only a few studies have been subjected to such intensive study. It is from this latter group that a few specific studies have been selected for further discussion.

Firstly, a systematic multi-station magnetotelluric study has been undertaken on the Rhenish Massif in NW Germany (Fig. 2) by Jödicke *et al.* (1983). The field data from about 100 sites were modelled 1-dimensionally, this being justified on the basis that the electric field was not strongly polarized. Moreover, the 2-dimensional model for one of the profiles is very similar to the section derived from a set of 1-dimensional models. The most prominent feature of the models is that one or sometimes two low resistivity layers (about 10 Ωm) can be observed dipping southwards within the lower crust. Jödicke and his co-authors claim that their mathematical computations on the model resolution have confirmed that the thickness of these layers cannot be increased beyond a few kilometres. This result differs from those of most of the studies listed in Table 1 where it appears that the whole of the lower crust has a low resistivity. For many of these listed studies, however, the data sets were inadequate for such discrimination. This factor should be borne in mind for future work as it is crucial for a petrophysical and tectonic interpretation that the distinction be made between very thin low resistivity 'layers' within the lower crust and a lower crust which has low resistivity—this may require the introduction of constraints from other geophysical techniques.

Another region in which multi-station magnetotelluric observations have been made is Scotland where a progressive series of smaller studies have resulted in data of comparable intensity to those of the Rhenish Massif study. Due to the geological complexity of the region, however, the site density and frequency sounding band width is as yet insufficient to provide adequate resolution of the resistivity structure except in a few locations. Nevertheless, a low resistivity lower crustal layer has been detected along much of the traverse from the Lewisian Foreland to North England (Hutton *et al.* 1980) (Fig. 3(a)–(c)) with an anomalously low resistivity in the Southern Uplands (Fig. 3 (b) & (c)). This anomaly was first detected by magnetic variation studies (Osemeikhian & Everett 1968; Edwards *et al.* 1971) while subsequent similar studies (Hutton & Jones 1980; Banks *et al.* 1983) have confirmed its existence (Fig. 3(a)). Although more intensive systematic observations could improve the resolution of the structure so far derived, these studies are of interest at the present time in view of the current

availability of the off-shore deep seismic reflection profiles of BIRPS (Brewer *et al.* 1983; Brewer & Smythe 1984) and of the comparable electromagnetic and COCORP profiles in the Eastern USA, regions of common tectonic history (Fig. 4(a)).

A petrological interpretation of the low resistivity layers in the lower crust in the Rhenish Massif and some comments on the implications of the electromagnetic and deep seismic studies in Scotland and the Eastern USA follow later in this paper.

Major crustal lineaments of anomalous resistivity

It will be noted in Table 1 that the most frequently used techniques for inferring resistivity-depth profiles are magnetotellurics (MT), vertical electrical sounding (d.c. resistivity) and controlled source electromagnetics and that the studies have been largely undertaken within the past decade. Over the past 25 years, however, simpler techniques have been available for *locating* crustal and upper mantle resistivity anomalies (Alabi 1984). These GDS (Geomagnetic Deep Sounding) studies, based on the observations from single magnetometers or an array are less suited than the former for determining the depth and resistivity of the anomalies. However, since they produce data which can be mapped over large areas, they have revealed the existence of several elongated resistivity anomalies, extending for distances up to more than 1000 km. A comprehensive list, including those associated with active tectonic regions such as rifts and subduction zones where resistivity anomalies are to be expected, has recently been compiled by Haak (1985). In this paper, the locations of some of the major elongated anomalies in the more stable continental regions are given in Table 2. Of major significance are the facts that some of these anomalies have no surface expression and that their presence has been revealed for the first time as a result of these GDS studies. Their association with the major crustal lineaments discussed by Hall (1986) seems worthy of study.

Examination of Table 2 shows that elongated resistivity anomalies have been detected in many countries. Probably the longest of these anomalies known at present is that of the North American Central Plain—almost 2000 km long, striking N–S within the North American craton. The depth, width and E–W extension of the anomaly are unknown but the depth is expected to extend to the lower crust or upper mantle (Alabi *et al.* 1975).

TABLE 2. *Elongated resistivity anomalies*

Location	Length (km)	References
1 Europe/Asia		
(i) Carpathian	$\geqslant 1200$	Ritter (1975)
(ii) Kirovograd	$\geqslant 700$	Rokityansky (1982)
(iii) Ladoga	$\geqslant 500$	Rokityansky (1982)
(iv) Moscow	> 400	Rokityansky (1982)
(v) S. Scotland	> 100	Hutton & Jones (1980)
		Banks *et al.* (1983)
(vi) Pyrenees	> 200	Vasseur *et al.* (1977)
2 America/Africa		
(i) N. American Central Plains	> 1500	Alabi *et al.* (1975)
(ii) S.W. Africa	> 1000	De Beer *et al.* (1982b)
(iii) S. Africa (Cape Fold)	> 1000	De Beer & Gough (1980)

Several low resistivity anomalies of this elongated type have also been discovered in Eastern Europe using intensive magnetic variation studies and the resistivity structure across the anomaly axis has been obtained for a number of traverses for which MT data have also been available. The most prominent is the Carpathian anomaly (Fig. 5(a)) for which the models of Figs. 5(b) and (c) have been obtained by Cerv *et al.* (1984) and Zhdanov *et al.* (1984) respectively. While this anomaly can be associated with a known plate boundary the other three anomalies shown in Fig. 5(a)—Kirovograd, Moscow and Ladoga Lake— are similar to the North American Central Plain anomaly in that they are examples of resistivity anomalies for which there is no surface expression.

The elongated anomalies in S. Africa (Fig. 6) also detected by magnetic variation studies have subsequently been studied using deep Schlumberger soundings.

Several hypotheses about the nature of these elongated anomalies will be discussed in the next section.

Interpretation of lower crustal resistivities

Low electrical resistivities of crustal rocks in continental regions can be caused by the existence of free water, free carbon, some hydrated minerals such as serpentine, by magnetic oxides and by sulphur. However, in recent years following much intensive discussion, it appears that the favoured factor contributing to low electrical resistivity of crustal rocks is the presence of aqueous fluids with a high ionic content (Connerney *et al.* 1980; Shankland & Ander 1983; Lee *et al.* 1983).

Rock-melts at lower crustal depths would of course lower the resistivity considerably, as Schwarz *et al.* (1984) have discussed with reference to their studies in the Andes in Northern Chile. However, in this paper we are excluding

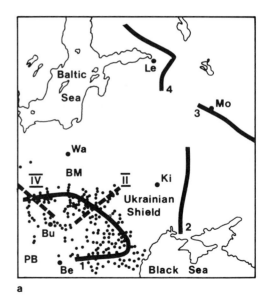

a

FIG. 5 (a) The axes of elongated low resistivity structures in Eastern Europe, 1—Carpathian anomaly, 2—Kirovograd anomaly, 3—Moscow anomaly, 4—Ladoga anomaly, with the locations of the magnetic variation sites used in the determination of the Carpathian anomaly axis and the resistivity models. The traverses II and VI refer to deep seismic profiles along which magnetotelluric and magnetic variation data have provided the electrical resistivity models of (b) and (c).

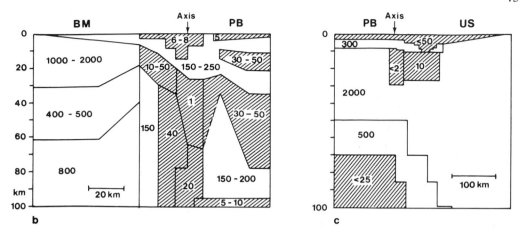

b

c

FIG. 5 (b) The 2-D resistivity model for traverse VI (from Cerv *et al.* 1984). Regions of resistivity less than 50 ohm are stippled. (c) The 2-D resistivity model for traverse II (from Zhdanov *et al.* 1984). Note the very low resistivity—a few ohm m—in the lower crust in both models—(b) and (c)—in the region of the magnetic variation anomaly axis—indicated by an arrowhead—suggesting that low resistivity material has been pulled or pushed into the lower crust by a subduction-like process.

Abbreviations: PB, Pannonian Basin; BM, Bohemian Massif; US, Ukrainian Shield.

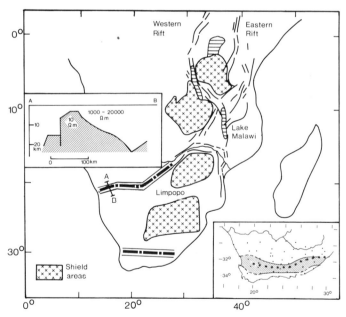

FIG. 6. The locations of elongated low resistivity structures in central and southern Africa. Three distinct types of anomaly are indicated. (a) The East African rift conductors (Hutton 1976) with probably partial melts at crustal and upper mantle depths. (b) The Namibian conductor which may be a prolongation of the E African rifts to the west. Schlumberger soundings across the anomaly axis have provided the resistivity section illustrated—the low resistivity zone probably represents a zone of weakness with increased pore volume for saline fluids (de Beer *et al.* 1982b). (c) The Southern Cape conductive belt (de Beer et al., 1982a). The location of the anomaly—but not the depth or resistivity—is known. Taking account of gravity and magnetic data for the region, it is suggested that all the observations are compatible with an accumulation of highly serpentinized previous oceanic crustal material along the anomaly axis, i.e. it reveals a Proterozoic subduction zone of Andean type.

such tectonically active regions where partial melts may occur at such depths.

Indeed, Shankland & Ander (1983) inferred from their studies that (1) the electrical resistivity in tectonically active regions (with explicit exclusion of regions of partial melting) is due to the same conductivity mechanism as in stable regions and (2) the electrical resistivity is generally lower in the former regions. They therefore concluded that the resistivity is associated with the porosity and hence fluid content and not due to other compositional causes such as the existence of free carbon. It would otherwise be necessary to argue that there is more graphite in the crust in active regions than in shield areas for which there is no petrological argument. The observed normal lower crustal resistivity of 100–1000 Ωm could be due to 0.1–0.01% by volume of fluid content. It has been argued by Shankland & Ander (1983) that the water should be free and not structurally bound. Despite this preferred interpretation, the other possibilities should not be completely ruled out and especially in the case of very low resistivities, the possibilities of an abundance of graphite or amorphous carbon should always be considered.

The existence of free water in particular in the lower crust should play a major role in all geodynamic processes of the continental crust. This has been discussed extensively quite recently by Etheridge & Wall (1983) and Etheridge *et al.* (1984). It seems that the different resistivity depth profiles listed in Tables 1 and 2 may represent different amounts of free water at the corresponding depths within the crust. Sources of free water at depth are now accepted as arising from dehydration reactions although a critical temperature of about 400°C in the lower crust (Ringwood, 1975, Fig. 8.4) is required to initiate abundant dehydration (Walther & Orville 1982; Spear & Selverstone 1983). Existing free water, however, has the tendency by diffusion or by advection to rise to the Earth's surface and thus to leave the lower crust (Walther & Orville 1982; Fyfe *et al.* 1978). Moreover, convective flow of fluids may exist in nearly impermeable crustal rocks of 10^{-18} m^2 according to theoretical studies by Ribando & Torrance (1976), Ribando *et al.* (1976) and Klever (1984). Klever (1984) has demonstrated theoretically that the initiation and geometry of convective flow is strongly influenced by the pattern of anisotropy of permeability. Hence the pattern of low resistivity structures may represent the pattern of such fluid flow as influenced by the anisotropic permeability.

Both the critical temperature and the escape of fluids support the petrological model of a normally dry continental crust—contrary to observation. One possible solution to these problems may be that the source of the free water is in the upper mantle and that this supplies and recharges the water content in the lower crust. One possible explanation for retention of fluids in the lower crust is the existence of an impermeable 'layer' at middle crustal depth created by the precipitation of the fluid mineral content as proposed by Etheridge and Wall (1983). Another is the adsorption of water molecules to the 'inner surface' of the rocks (Olhoeft 1981; Schopper 1982), the adsorption forces causing an effective drop in resistivity due to the surface-conduction-mechanism. In this case, the value of the resistivity would depend on the inner surface area and in turn on the grain size and the time that the crustal rocks have been exposed to the volatiles and fluids rising from the upper mantle. In reality all these solutions for keeping water in the crust may exist—the continuous recharge from the upper mantle, the adsorption of water molecules to the grain surface and the existence of an impermeable middle crust.

Some general comments must also be made about the derivation of petrological–tectonic models from electrical data. In fact it is impossible to deduce a petrological model directly from electrical data. The procedure must be reversed as in the work of Hyndman & Hyndman (1968) who were among the first to present crustal cross-sections at different metamorphic stages and to predict an electrical resistivity–depth function for each of these petrological models. This presentation has been extended quite recently by Greenhouse & Bailey (1981) based on ideas from Feldman (1976). Their sketches illustrating the different stages of a prograde metamorphic process have been redrawn in Fig. 7 a–e in a more simplified manner. For our purpose we need a small but realistic addition to the latter model: this takes account of the mechanical response of the crust to laterally acting tectonic forces during the peak stage of metamorphism. We will distinguish five stages which are demonstrated by the sequence of sketches of Fig. 7(f)–(j). It is open to question whether these rather general comments are applicable to individual electrical resistivity structures. As will be demonstrated for the case studies described in the preceding sections individual discussion of the interpretation is required for each structure.

In the case of the Rhenish Massif, studies of xenoliths appear to preclude the presence of large amounts of free water in the middle and lower crust and no accumulation of hydrated minerals was detected (Voll 1983; Mengel & Wedepohl 1983). However, these studies did indicate thin layers of metamorphic black shales at a depth of

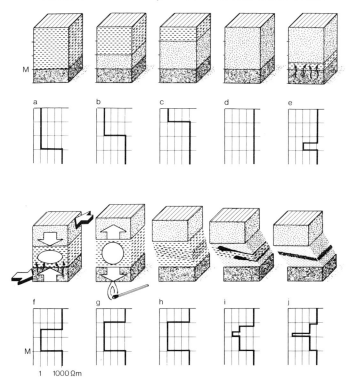

FIG. 7. Estimated electrical resistivity profiles for different stages of development of the continental crust: (a) Thickened but undifferentiated, wet geosynclinal crust, sediments and volcanics in the upper crust, greenschist to amphibolite facies in the lower crust. (b) and (c) Intermediate stages of development: low to high grade metamorphic rocks in the upper part; dehydrated, very high grade metamorphic, granulite rocks in the lower crust. (d) Stable shield: High grade metamorphic rocks and granites in the upper crust, granulites in the lower crust. (e) Intrusion and storage of mantle fluids, in particular water, in the lower crust. (f) The pre-metamorphic stage with lithostatic pressure greater than pore fluid pressure and lateral tectonic forces acting on the crust. (g) Heating of the lower crust during the peak-metamorphic stage: the pore-fluid pressure is now greater than the lithostatic pressure. (h) Lateral tectonic forces decouple the upper from the lower crust. (i) Opening of shear fractures and drainage of the lower crust by the flow of fluids into the fractures. (j) The final cold state with the precipitation of the mineral content of the crustal fluids during the cooling periods and production of well conducting structures throughout the continental crust.

(a)–(d) redrawn with corresponding comments from Hyndman & Hyndman 1968 and (e) from Greenhouse & Bailey 1981.

about 20 km. Moreover, Giese (1983) explains reflection and refraction seismic results by large overthrusting during Variscan folding and there is a striking similarity between his model and the southward dip in the low resistivity layer in the same region (Fig. 2). One hypothesis may be, as has been suggested for the eastern United States, that the low resistivity 'layer' can be identified as overthrusted younger sediments or as graphite bearing black shales of Precambrian age. Another hypothesis (Giese *et al.* 1983) is that the low resistive 'layers' represent the actual shear zones generated during crustal thickening by stacking as explained by the processes in Figs 7(e)–(j).

With respect to the Scottish studies, the corre-spondence between electrical resistivity models in this region and in eastern North America has previously been noted (Jones & Hutton 1979; Hutton *et al.* 1980; Mbipom & Hutton 1983). In this paper, some more recent results from these regions of common tectonic history have been presented—Figs 3(b) and 4(b) and (c). Of special interest in the American studies has been the approximate coincidence of low resistivity zones in the crust with the seismic high reflectivity zones detected in the COCORP northern and southern Appalachian profiles (Thompson *et al.* 1983) with an interpretation of the low resistivities in terms of a water saturated sedimentary layer which would also provide a detachment surface for the tectonic movement. From examination of the correspondence between the seismic reflectivity zones observed in the MOIST and WINCH traverses of BIRPS in the seas to the north and west of Scotland and the mainland electrical resistivity section it can only be stated that the crust is in general both reflecting and relatively conducting along the traverses. In the Southern Uplands region, however, where there is a well-established very low resistivity lower crust with a marked lateral variation in the depth to the low resistivity zone, there is almost a complete absence of lower crustal reflectors in the corresponding off-shore seismic section. It is thought (D. Smythe, pers. comm.) that this may be due to the inability of the seismic signals to penetrate the Upper Palaeozoic sediments in this section of the traverse. Further south in the Northumberland Basin, recent analysis of data from 3 sites in the Northumberland Basin (Beamish 1985) has yielded 1-dimensional models with a thin low resistivity layer which correlates in location and geometry with a deep crustal reflecting wedge detected off-shore by WINCH. The northward dipping slope of this layer is in agreement with the previously published 1- and 2-dimensional models of the region (Jones & Hutton 1979; Hutton et al. 1980). Whether a very thin very low resistivity layer can be regarded as more acceptable than a thicker layer of relatively higher resistivity is open to question (Parker 1983; Edwards et al. 1981).

In the region where more intensive data sets have already been modelled both 1- and 2-dimensionally, and where the axis of an elongated resistivity anomaly has been well determined (Fig. 3(a) and (b)), it is particularly interesting to note the correspondence between these features and those observed along the axis of the Carpathian and other magnetic variation anomalies in regions clearly identified as plate boundaries. This correspondence tends to support the hypothesis that the resistivity anomaly crossing the Southern Uplands (Fig. 3(a)–(c)) is associated with the Iapetus suture. As in the North American electrical studies, the preferred interpretation of the low resistivity layers observed in the UK studies is in general in terms of varying amounts of fluids retained in the crust as a result of the processes described in Fig. 7.

Finally considering the *elongated anomalies* the hypothesis that they represent relics of now extinguished subduction zones and plate margins (Camfield & Gough 1977; Gough 1983), probably applies to all such anomalies in inactive continental regions. This may also be true for the Southern Cape conductive belt in South Africa (Fig. 6) as discussed by De Beer et al. (1982a). A completely opposite explanation has been proposed for the elongated anomaly running approximately E–W through Namibia (Fig. 6). Since at its eastward extremity it approaches the location of the Eastern African Rift systems, they argue that it may be the signature of the very first stage of a new rift (De Beer et al. 1975). The low resistivity in this case might thus be due to saline fluids circulating in an incipiently fractured zone of crustal weakness.

Conclusions on the nature of the continental crust

The electrical resistivity of the continental lower crust in inactive regions—where melting may not occur—is lower than has been predicted from petrological modelling. This may be explained by the existence of free (and partially by adsorbed) water of 0.1% by volume or less. Since free water will escape either by advection or by diffusion even from a nearly impermeable crust there must exist a continuous recharge of fluids from the upper mantle. Furthermore, anomalously low resistivity structures, as thin layers or as elongated anomalies, have been discovered in particular regions within the lower and upper part of the continental crust. They have been generated most probably by major tectonic processes in which water could have played a dominant role. Such low resistivity structures may represent relics of shear fractures which had been generated during crustal stacking processes or relics of Proterozoic plate boundaries.

This study shows that electrical resistivity methods are suited to the detection or reconstruction of tectonic processes of the normal continental crust. It should be emphasized that this can only be accomplished by dense networks of observational sites and appropriate sounding frequencies. In general this is a task for the future.

References

ALABI, A.O. 1984. Magnetometer array studies. *Geophys. Surveys*, **6**, 153–172.
—— CAMFIELD, P.A. & GOUGH, D.I. 1975. The North

American Central Plains conductivity anomaly. *Geophys. J.R. Astr. Soc.*, **43**, 815–833.
BANKS, R.J., BEAMISH, D. & GEAKE, M.J. 1983. Mag-

netic variation anomalies in northern England and southern Scotland. *Nature*, **303**, 316–318.

BEAMISH, D. 1985. Deep crustal conductivity structure beneath the Northumberland Basin. *NERC 3rd Deep Geology Workshop*. Durham (abstract).

BLOHM, E.K., WORZYK, P. & SCRIBA, H. 1977. Geoelectrical deep sounding in Southern Africa using the Cabora Bassa power line. *J. Geophys.*, **43**, 665–679.

BRACE, W.F. 1971. Resistivity of saturated crustal rocks to 40 km based on laboratory measurements. *In:* HEACOCK, J.G. (ed.) *Structure and Physical Properties of the Earth's Crust*. Geophys. Monogr. Ser., **14**, AGU, Washington, D.C., 243–255.

BREWER, J.A., MATTHEWS, D.H., WARNER, M.R., HALL, J., SMYTHE, D.K. & WHITTINGTON, R.J. 1983. BIRPS deep seismic reflection studies of the British Caledonides. *Nature*, **305**, 206–210.

—— & SMYTHE, D. 1984. MOIST and the continuity of crustal reflector geometry along the Caledonian–Appalachian orogen. *J. Geol. Soc., Lond.*, **141**, 105–120.

CAMFIELD, P.A. & GOUGH, D.I. 1977. A possible Proterozoic plate boundary in North America. *Can. J. Earth Sci.*, **14**, 1229–1238.

CERV, V., PEK, J. & PRAUS, O. 1984. Models of induction anomalies in Czechoslovakia. *J. Geophys.*, **55**, 161–168.

COCHRANE, N.A. & HYNDMAN, R.D. 1974. Magnetotelluric and magnetovariational studies in Atlantic Canada. *Geophys. J.R. Astr. Soc.*, **39**, 385–406.

CONNERNEY, J.E.P., NEKUT, A. & KUCKES, A.F. 1980. Deep crustal electrical conductivity in the Adirondacks. *J. Geophys. Res.*, **85**, 2603–2614.

DE BEER, J.H., GOUGH, D.I. & VAN ZIJL, J.S.V. 1975. An electrical conductivity anomaly and rifting in southern Africa. *Nature*, **225**, 678–680.

—— & GOUGH, D.I. 1980. Conductive structures in southernmost Africa: a magnetometer array study. *Geophys. J.R. Astr. Soc.*, **63**, 479–495.

—— VAN ZIJL, J.S.V. & GOUGH, D.I. 1982a. The Southern Cape Conductive Belt (South Africa): its composition, origin and tectonic significance. *Tectonophysics*, **83**, 205–225.

—— HUYSSEN, R.M.J., JOUBERT, S.J. & VAN ZIJL, J.S.V, 1982b. Magnetometer array studies and deep Schlumberger soundings in the Damara orogenic belt, South West Africa. *Geophys. J.R. Astr. Soc.*, **70**, 11–29.

DOWLING, F. 1970. Magnetotelluric measurements across the Winconsin Arch. *J. Geophys. Res.*, **75**, 2683–2698.

DUNCAN, M., HWANG, A., EDWARDS, R.A., BAILEY, R.C. & GARLAND, G.D. 1980. The development and applications of a wide band electromagnetic prospecting system using a pseudo-noise source. *Geophysics*, **45**, 1276–1296.

EDWARDS, R.N., LAW, L.K. & WHITE, A. 1971. Geomagnetic variations in the British Isles and their relation to electrical currents in the oceans and shallow seas. *Phil. Trans. R. Soc. Lond.*, **270**, 289–323.

—— BAILEY, R.C. & GARLAND, G.D. 1981. Conductivity anomalies: lower crust or asthenosphere? *Phys. Earth Planet. Int.*, **25**, 263–272.

ETHERIDGE, M.A. & WALL, V.J. 1983. The role of the fluid phase during regional metamorphism and deformation. *J. metamorph. Geol.*, **1**, 205–226.

—— WALL, V.J., COX, S.J. & VERNON, R.H. 1984. High fluid pressures during regional metamorphism and deformation: implications for mass transport and deformation mechanisms. *J. Geophys. Res.*, **89**, 4344–4358.

FELDMAN, I.S. 1976. On the nature of conductive layers in the Earth's Crust and upper mantle. *In:* ADAM, A. (ed.) *Geoelectric and Geothermal Studies*, Akademiai Kiado, Budapest, 721–730.

FYFE, W.S., PRICE, W.J. & THOMPSON, A.B. 1978. Fluids in the Earth's crust. *In: Developments in Geochemistry*, **1**, Elsevier, Amsterdam.

GIESE, P. 1983. The evolution of the Hercynian crust—some implications to the uplift problem of the Rhenish Massif. *In:* FUCHS, K. *et al.* (eds) *Plateau Uplift*, Springer Verlag, Berlin–Heidelberg, 303–314.

—— JÖDICKE, H., PRODEHL, C., WEBER, K. 1983. The crustal structure of the Hercynian mountain system—a model for crustal thickening by stacking. *In:* MARTIN, H. & EDER, F.W. (eds) *Intracontinental Fold Belts*, Springer Verlag, Berlin, 405–426.

GOUGH, D.I. 1983. Electromagnetic geophysics and global tectonics. *J. Geophys. Res.*, **88**, 3367–3377.

GREENHOUSE, J.P. & BAILEY, R.C. 1981. A review of geomagnetic variation measurements in the eastern United States: implications for continental tectonics. *Can. J. Earth Sci.*, **18**, 1268–1289.

HAAK, V. 1985. Anomalies of the electrical conductivity in the earth's crust and upper mantle, in Landolt-Bornstein: Numerical data and functional relationships. *In:* ANGENHEISTER, G. (ed.) *Science and Technology*. New Series, Springer Verlag, Berlin, V.2b, chapter 2.3.2, 397–436.

HALL, J. 1986. Geophysical lineaments and deep continental structure. *Phil. Trans. R. Soc.* A317, 33–48.

HERMANCE, J. 1982. Magnetotelluric and geomagnetic deep sounding studies in rifts and adjacent regions: constraints on physical processes in the crust and upper mantle. *In: Geodynamics Ser.*, **8**, AGU Washington, D.C., 169–182.

HUTTON, R. 1976. Induction studies in rifts and other active regions. *Acta Geodet. Geophys. et Mont.*, **11**, 347–376.

HUTTON, V.R.S., INGHAM, M.R. & MBIPOM, E.W. 1980. An electrical model of the crust and upper mantle in Scotland. *Nature*, **287**, 30–33.

—— & JONES, A.G. 1980. Magnetovariational and magnetotelluric investigations in Scotland. *J. Geomag. Geoelectr.*, **32**, Suppl. 1, 141–150.

HYNDMAN, R.D. & HYNDMAN, D.W. 1968. Water saturation and high electrical conductivity in the lower continental crust. *Earth Planet. Sci. Lett.*, **4**, 427–432.

INGHAM, M.R. & HUTTON, V.R.S. 1982. The interpretation and tectonic implications of the geo-electric structure of S Scotland. *Geophys. J.R. astr. Soc.* **69**, 579–594.

JIRACEK, G.R., ANDER, M.E. & HOLECOMB, H.T. 1979. Magnetotelluric soundings of crustal conductive zones in major continental rifts. *In:* RUKER, R.E.

(ed.) *Rio Grande Rift: Tectonics and Magmatism*, AGU, Washington, D.C., 209–222.

JÖDICKE, H., UNTIEDT, J., OLGEMANN, W., SCHULTE, L. & WAGENITZ, V. 1983. Electrical conductivity structure of the crust and upper mantle beneath the Rhenish Massif. *In:* FUCHS, K. *et al.* (eds) *Plateau Uplift*, Springer Verlag, Berlin, Heidelberg, 288–362.

JONES, A.G. & HUTTON, V.R.S. 1979. A multi-station magnetotelluric study in southern Scotland, II. Monte Carlo inversion of the data and its geophysical and tectonic implications. *Geophys. J.R. Astr. Soc.*, **56**, 351–368.

—— 1981. On the type classification of lower crustal layers under Precambrian regions. *J. Geophys.*, **49**, 226–233.

—— 1982. On the electrical crust–upper mantle structure in Fennoscandia: no Moho and the Asthenosphere revealed? *Geophys. J.R. astr. Soc.*, **68**, 371–388.

—— 1983. The electrical structure of the lithosphere and asthenosphere beneath the Fennoscandian Shield. *J. Geomag. Geoelect.*, **35**, 811–827.

KAIKKONEN, P., VANYAN, L.L., HJELT, S.C., SHILAVSKY, A.P., PAJUNPÄÄ, K. & SHILOVSKY, P.P. 1983. A preliminary geoelectric model of the Karelian megablock of the Baltic Shield. *Phys. Earth Planet. Int.*, **32**, 301–305.

KLEVER, N. 1984. Stationäre Konvektion in porösen Medien — Numerische Untersuchungen an unterschiedlichen Fragestellungen aus der Hydrothermik und der Schneemetamorphose. *Berliner geowiss. Abh.*, (B), **11**, 114 pp., Verlag von Dietrich Reimer, Berlin.

KOVTUN, A.A. 1976. Induction studies in stable shield and platform areas. *Acta Geodet. Geophys. et Montanist, Acad. Sci. Hung.*, **11** (3–4), 333–346.

KURTZ, R.D. 1982. Magnetotelluric interpretation of crustal and mantle structure in the Grenville Province. *Geophys. J.R. Astr. Soc.*, **70**, 373–397.

—— & GARLAND, G.D. 1976. Magnetotelluric measurements in Eastern Canada. *Geophys. J.R. astr. Soc.*, **45**, 321–347.

LEE, C.D., VINE, F.J. & ROSS, R.G. 1983. Electrical conductivity models for continental crust based on laboratory measurements on high grade metamorphic rocks. *Geophys. J.R. astr. Soc.*, **72**, 353–372.

MARESCHAL, J.-C., MUSSER, J. & BAILEY, R.C. 1983. Geomagnetic variation studies in the southern Appalachians: preliminary results. *Can. J. Earth Sci.*, **20**, 1434–1444.

MBIPOM, E.W. & HUTTON, V.R.S. 1983. Geoelectromagnetic measurements across the Moine Thrust and the Great Glen in northern Scotland. *Geophys. J.R. Astr. Soc.*, **74**, 507–524.

MENGEL, K. & WEDEPOHL, K.H. 1983. Crustal xenoliths in Tertiary volcanics of the Rhenish Massif and the Tertiary basalts of the Northern Hessian depression. *In:* FUCHS, K. *et al.* (eds), *Plateau Uplift*, Springer Verlag, Berlin, Heidelberg, 332–335.

OLHOEFT, G.R. 1981. Electrical properties of granite with implications for the lower crust. *J. Geophys. Res.*, **86**, 931–936.

OSEMEIKHIAN, J.E.A. & EVERETT, J.E. 1968. Anomalous magnetic variations in SW Scotland. *Geophys. J.R. Astr. Soc.* **15**, 361–366.

PARKER, R.L. 1983. The magnetotelluric inverse problem. *Geophys. Surveys*, **6**, 5–25.

RASMUSSEN, T., PEDERSEN, L.B. & ZHANG, P. 1984. Preliminary results from magnetotelluric measurements along the Fennolara Profile. *Proc. Baltic Shield Symposium*, Department of Geophysics, University of Oulu, Finland, Rep. No. **8**, 307–329.

RIBANDO, R.J. & TORRANCE, K.E. 1976. Natural convection in a porous medium: Effect of confinement, variable permeability and thermal boundary conditions. *J. Heat Transfer*, **98**, 42–48.

—— TORRANCE, K.E. & TURCOTTE, D.L. 1976. Numerical models for hydrothermal circulation in the oceanic crust. *J. Geophys. Res.*, **81**, 3007–3012.

RINGWOOD, A.E. 1975. *Composition and Petrology of the Earth's Mantle*. McGraw Hill, New York, 618 pp.

RITTER, E. 1975. Results of geoelectromagnetic deep soundings in Europe. Gerlands Beitr. *Geophysik Leipzig*, **84**, 261–273.

RITZ, M. 1983. Use of the magnetotelluric method for a better understanding of the West African Shield. *J. Geophys. Res.*, **88**, 10, 625–633.

ROBERTS, R.G., ZHANG, P. & PEDERSEN, L.B. 1984. Remote reference magnetotelluric measurements (0.1–10 s) across the mylonite shear zone in South West Sweden. *Proc. Baltic Shield Symposium*, Department of Geophysics, University of Oulu, Rep. No. **8**, 328–339.

ROKITYANSKY, I.I. 1982. *Geoelectromagnetic Investigation of the Earth's Crust and Mantle*. Springer Verlag, Berlin, Heidelberg, New York, 378.

SCHOPPER, J.R. 1982. Electrical conductivity in moisture containing rocks. *In:* ANGENHEISTER, G. (ed.) Landolt-Bornstein Numerical data and functional relationships in Science and Technology, New Series, Vol. 1b, 276.

SCHWARZ, G., HAAK, V., MARTINEZ, E. & BANNISTER, J. 1984. The electrical conductivity of the Andean crust in northern Chile and southern Bolivia as inferred from magnetotelluric measurements. *J. Geophys.*, **55**, 169–178.

SHANKLAND, T.J. & ANDER, M.C. 1983. Electrical conductivity, temperatures and fluids in the lower crust. *J. Geophys. Res.*, **88**, 9475–9484.

SPEAR, F.S. & SELVERSTONE, J. 1983. Water exsolution from quartz: implications for the generation of retrograde metamorphic fluids. *Geology*, **11**, 82–85.

STERNBERG, B.K. 1979. Electrical resistivity of the crust in the southern extension of the Canadian Shield—layered earth models. *J. Geophys. res.*, **84**, 212–228.

SULE, P. & HUTTON, V.R.S. 1986. A broadband magnetotelluric study in S.E. Scotland—Data acquisition analysis and one-dimensional modelling, *Annales Geophysicae* **4**, B2, 145–156.

THOMPSON, B.G., NEKÜT, A. & KUCKES, A.F. 1983. Deep crustal electromagnetic sounding in the Georgian Piedmont. *J. Geophys. Res.* **88**, 9461–9473.

VANYAN, L.L., OKULESSKI, B.A. & SHILOVSKY, A.P. 1984. Two types of crustal conductive zones in ancient platforms. *Proc. of Baltic Shield Symposium*,

Dept. of Geophysics, University of Oulu, Finland, Report No. **8**, 107–109.

VASSEUR, G., BABOUR, K., MENVIELLE, M. & ROSSIGNO, J.C. 1977. The geomagnetic variation anomaly in the northern Pyrenees; study of the temporal variation. *Geophys. J.R. Astr. Soc.*, **49**, 593–607.

VOLL, G. 1983. Crustal xenoliths and their evidence for crustal structure underneath the Eifel volcanic district. *In:* FUCHS, K. *et al.* (eds) *Plateau Uplift*, Springer Verlag, Berlin, Heidelberg, 336–342.

WALTHER, J.V. & ORVILLE, P.M. 1982. Volatile productions and transport in regional metamorphism, *Contrib. Mineral. Petrol.*, **79**, 252–257.

VAN ZIJL, J.S.V. 1977. Electrical studies of the deep crust in various tectonic provinces of southern Africa. *In:* HEACOCK, J.G. (ed.) AGU Monogr., **20**, 470–500.

ZHANG, P., RASMUSSEN, T., KORJA, T., HJELT, S.E. & PEDERSEN, L.B. 1984. Preliminary results of MT measurements across the Oulu anomaly in 1983. *Proc. Baltic Shield Symposium*, Department of Geophysics, University of Oulu, Finland, Rep. No. **8**, 296–306.

ZHDANOV, M.S. *et al.* 1984. 2D model fitting of geomagnetic anomaly in the Soviet Carpathians, *Proc. of Baltic Shield Symposium*, Department of Geophysics, University of Oulu, Finland, Rep. No. **8**, 296–306.

VOLKER HAAK, Institut für Geophysikalische Wissenschaften, Freie Universität, Rheinbabenallee 49, D-1 Berlin 33, W. Germany.

ROSEMARY HUTTON, University of Edinburgh, Department of Geophysics, James Clerk Maxwell Building, Mayfield Road, Edinburgh EH9 3JZ.

The physical properties of layered rocks in deep continental crust

J. Hall

SUMMARY: Normal incidence seismic reflection profiles often show strong reflectivity—indicating sub-horizontal layering— in the deep continental crust, below poorly reflective upper crust. An example from the WINCH profile across the Caledonides of NW Britain shows necking of the lower crustal reflective layer near Islay. Combining the seismic data with gravity modelling constrains the lower crustal density to be about 3.1 Mg m^{-3} and the P-wave velocity to be about 7.3 km s^{-1}. This layer can be correlated with the deepest crustal layer of the parallel LISPB refraction profile where the layer has a similar P-wave velocity, Poisson's ratio of 0.25 and electrical conductivity of more than 0.003 S m^{-1}. These data suggest that a) the lower crust is heterogeneous, b) its mean composition is like that of very basic igneous rocks, and c) hydrated minerals and possibly free water are significant components. Aqueous fluids are shown to permeate deep crustal rocks even more effectively than they do sedimentary rocks and a mechanism of water capture in the lower crust is presented. Layering may result from interleaving of mantle and crust, fabric re-orientation due to shearing or, most likely, by intrusive underplating. Reflectivity may be enhanced by layer-variable water content.

Introduction

The aims of this paper are to provide estimates of, and then explain, the physical properties of deep, layered continental crust. The deep crust often shows strong reflectivity on seismic sections (Mueller 1977; Smithson 1978), well shown on much of the data collected by the British Institutions Reflection Profiling Syndicate (BIRPS) (Brewer *et al.* 1983; Brewer & Smythe 1984). A section of layered crust from the BIRPS Western Isles–North Channel (WINCH) profile is used as an example. This paper complements one (Hall 1986) which concentrates on the derivation of physical properties of the lower crust from the WINCH example. The findings are reported here, with the discussion amplified to include new work on permeability in the deep crust and the means by which water is preserved there.

The Western Isles–North Channel profile: 'WINCH'

Initial British deep reflection work was directed at the NW margin of the Caledonides. On the Moine and Outer Isles Thrusts profile (MOIST) (Smythe *et al.* 1982; Brewer & Smythe 1983; see Fig. 1), the two thrusts were clearly seen together with a deeper third—the Flannan Thrust—which appears to cut the crust–mantle boundary. The thrusts may have long histories and were certainly active in post-Caledonian sedimentary basin development, when they were reactivated as soles to listric normal faults in the extensional phase. The WINCH line (Fig. 1) was intended to follow up

MOIST by adding the beginnings of a three-dimensional view of the thrusts and then proceeding south-westwards across the foreland before turning to the southeast to cross all the major units of the British Caledonides except the Welsh shelf edge. Thus the profile would examine the marginal thrust belt to the northwest, the metamorphic Caledonides of the American (Laurentian) plate margin and the suture(s) of American and European continents, with whatever arc terranes may be trapped in between (Longman *et al.* 1979). Boundaries between the major tectonic units are often steep fault zones, which may have moved different ways at different times and include strike-slip movement (e.g. the Great Glen Fault).

A preliminary account of WINCH (Brewer *et al.* 1983) explains that several thrusts were seen in addition to confirmation of those seen in MOIST; very little evidence of steep tectonic boundaries is observed; sedimentary basins are often found in the hanging walls of thrusts; and the lower crust is variably reflective right across the section, while the upper crustal basement is invariably unreflective. The seismic reflection sections are available from the offices of the British Geological Survey in Edinburgh. A detailed intepretation of the middle section of the WINCH line (Hall *et al.* 1984) includes the major features shown in Fig. 2. This shows the possible nature of the obduction of the European continental margin against the Archaean foreland of the American continent with an intervening island arc (below the Midland Valley) shunted below the trench and ophiolite remnants exposed at the surface of the Southern Uplands (Hall *et al.* 1983). The wedging mecha-

From: DAWSON, J.B., CARSWELL, D.A., HALL, J. & WEDEPOHL, K.H. (eds) 1986, *The Nature of the Lower Continental Crust*, Geological Society Special Publication No. 24, pp. 51–62.

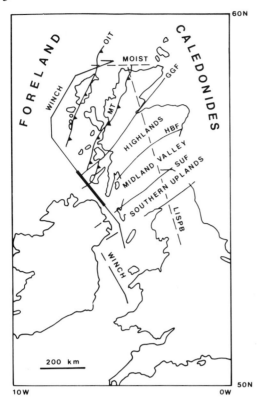

FIG. 1. Map showing location of MOIST and
WINCH deep seismic reflection lines, which run
from the Foreland of NW Britain across various
tectonic units of the Caledonides. OIT and MT are
the Outer Isles and Moine Thrusts, respectively.
HBF and SUF are the Highland Boundary and
Southern Uplands Faults. GGF = Great Glen Fault.
Thickened part of WINCH line is that examined in
this paper. Location of the LISPB refraction profile
is also shown.

nism inferred to have juxtaposed the Midland
Valley basement against the European continent
includes a deep-crustal northward-dipping thrust
which lies below the Southern Uplands and is, in
effect, the Iapetus Suture. A similar wedging
mechanism is believed to have operated to bring
the shelf-edge sediments of the Caledonide Dalradian basin against the foreland. Record quality
varies along the line but reflective lower crust is
discernible along most of the line and is terminated suddenly downwards at what is assumed to
be the crust–mantle boundary. This correlation
has been established on MOIST at its crossing
with the LISPB refraction profile (Bamford et al.
1978), and in the North Sea (Barton et al. 1984);
moreover the Moho delay time concurs roughly
with those from crustal refraction surveys conducted nearby, the Hebridean Margin Seismic
Experiment (Bott et al. 1979), for example.

It is surprising that after such a complex history
the variation in depth to the Moho should appear
so modest. It has been suggested (Hall et al. 1984)
that the density contrast across the Moho is so
large that significant relief on the Moho surface
implies gravitational stress which is largely
relaxed during later basin formation.

The Moho 'high'

There is one place along the WINCH line where
quite substantial relief of the Moho surface
appears to be present. This is between shot-points
12000 and 14000. In two-way time there is a
Moho 'high' here of over 1 s amplitude. This is
about an order of magnitude greater than any
pull-up effects between minor sedimentary
basins. Assuming a constant crustal velocity of 6
km s^{-1}, a first order estimate of the depth relief

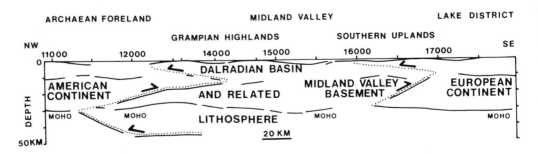

FIG. 2. Depth-converted, migrated section of part of the WINCH line showing principal deep reflectors.
Dotted lines indicate boundaries along which wedge obduction is believed to have occurred during the
closure of the Iapetus Ocean. The shot-point numbers at the top of the section show a jump of 1000 between
12000 and 14000, signifying a ship turn round between 12500 and 13600 (both marked). 100 shot points = 5
km. Note the Moho high at shot point 13600.

would be 3 km. If such relief is associated with a density contrast of say 500 kg m^{-3} between crust and mantle, then a gravity anomaly of 600 gu (60 mgal) would result. No such anomaly exists (Fig. 4): the Bouguer gravity rises steadily to the north-west with lows of the order of 100 gu (10 mgal) over known sedimentary basins.

Just above the Moho at its high, the lowermost crust is quite reflective, with the reflectivity diminishing rapidly both downwards across the Moho and upwards across a boundary picked as the top of the lower crustal reflective layer. Fig. 3 illustrates the record section at shot point 13600, on the high. The three labelled boundaries are clearly seen and correlated with the solid lines on Fig. 2. Moving both ways from shot point 13600 the Moho remains distinct but the top of the lower crustal layer less so. The latter rises to the south and may be picked with moderate certainty. To the north the reflective lower crust tapers (see Fig. 2) but becomes overlain by another (but less) reflective lower crust within the wedge of Archaean foreland. In Fig. 4 the best estimates of reflection times from the top of the lower crustal layer (or layers, where composite) and the Moho show an obvious inverse correlation, and because of that, a possible explanation of the lack of an observed gravity anomaly from the Moho high. If there is an appropriate density contrast across the top of the lower crustal reflective layer the depression in it may result in a gravity low which compensates the high from the Moho.

The argument is quantified in the complementary paper (Hall 1986). By assuming (i) a three layer upper lithosphere made of upper crust, reflective lower crust and upper mantle, (ii) constant density within each, and (iii) that perfect local isostatic compensation is achieved, it is possible to estimate the density of the reflective lower crustal layer, given the WINCH data and a velocity-density relationship.

The upper crustal density is taken to be 2750 ± 50 kg m^{-3} and mantle density 3350 ± 50 kg m^{-3}. Three different linear relationships of velocity with density were used in the calculations. There is a ± 25% uncertainty in the regression coefficient defining the inverse correlation of reflection times from the top and base of the lower crustal reflective layer. Combining all these factors and their uncertainties yields a lower crustal density of 3110 ± 70 kg m^{-3}, doubling the uncertainties in the assumed densities of upper crust and mantle, and varying further the velocity–density relationship, the lower crustal density remains within 3000–3250 kg m^{-3}. The corresponding P-wave velocity is in the range 7.0–7.5 km s^{-1}. Calculation of the gravity effect of such

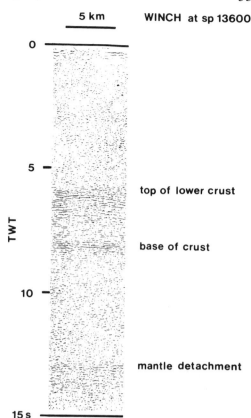

FIG. 3. A strip of the WINCH section at shot-point 13600 on the Moho high. Note the reflective lower crust with well-defined upper and lower boundaries. Mantle event forms part of the lower obduction wedge shown at the left of Fig. 2.

an earth model produces a good fit to the observed data (Fig. 4)—the mismatch being entirely attributable to known sedimentary basins, as indicated. The regional rise in gravity is then seen to be due to an overall rise in the mantle towards the NW.

Physical properties of the lower crust below the Caledonides

This example suggests that the layered lower crust has a P-wave velocity of about 7.3 km s^{-1} and a density of 3100 kg m^{-3}. If the structure persists 200 km along the Caledonoid strike to the NE, two other parameters become determinable. On the LISPB refraction profile (Bamford *et al.* 1978) across the Scottish Highlands (Figs 1, 5) a three-layered crust has a deepest layer with V_P about 7 km s^{-1} (similar to the estimate here) and a Poisson's Ratio of 0·249 ± 0·017 (Assumpcao &

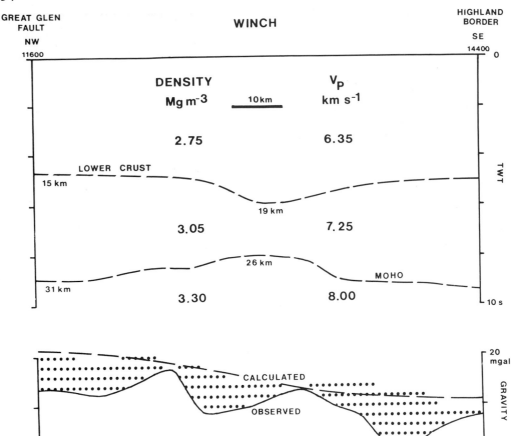

FIG. 4. Two way travel time (TWT) section of part of the WINCH line around shot-point 13600, showing inverse relationship of reflecting boundaries at top and base of the lower crust. Resulting velocity-density model shown gives depths to reflectors as indicated. Two-dimensional gravity model gives good fit of calculated to observed gravity (from Hipkin & Hussain 1983) corrected by dotted area for known sedimentary basins (Hall *et al.* 1984).

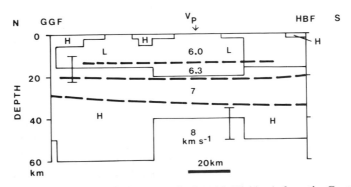

FIG. 5. Section of the LISPB refraction profile across the Scottish Highlands from the Great Glen (GGF) to the Highland Boundary Fault (HBF) based on Bamford *et al.* (1978). H and L denote blocks of high (>0.003 S m⁻¹) and low (<0.001 S m⁻¹) electrical conductivity as determined by Hutton *et al.* (1980). Error bars from Mbipom and Hutton (1983).

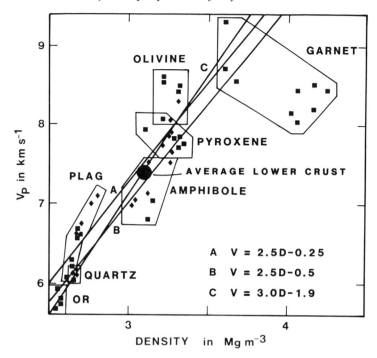

FIG. 6. Plot of P-wave velocity (V_P) against density (D) for common mineral groups at lower crustal conditions. Squares from single crystal data; diamonds from near mono-mineralic aggregates. Sources: Birch (1960, 1961), Anderson and Liebermann (1968), Simmons and Wang (1971) and Christensen (1982). OR = orthoclase; PLAG = plagioclase. Estimate of V_P, D for lower crust below Caledonides also shown.

Bamford 1978). A collinear magnetotelluric and magnetovariational survey (Hutton *et al.* 1980) shows electrical conductivity in the lower crust to be rather high, 0.003 S m^{-1} or more (Fig. 5). The upper and lower boundaries of the high conductivity layer are not well constrained (Mbipom & Hutton 1983) but could correlate with the top and base of the deepest crustal layer.

The correlation along strike must be viewed with caution. It has already been suggested by the author (Hall 1985) that a gravity lineament which may mark a major crustal boundary separates WINCH from LISPB. This merely reinforces the need for combining various geophysical techniques along 'geotraverses' so that correlations may be made more directly.

Interpretation

Consider first the P-wave velocity and density (Fig. 6). The estimate for the lower crust lies among those of isotropic aggregates of a variety of common rock-forming minerals under lower crustal conditions (Hall 1986). The velocity and density of an isotropic mixture of minerals is given closely by arithmetic means weighted in proportion to volume fractions. Possible average compositions of the lower crust include:

(a) 100% amphibole;
(b) 65% pyroxene + 35% quartz;
(c) 50% pyroxene + 50% feldspar; and
(d) combination of (a), (b) and (c).

All these indicate that the lower crust here is likely to have a basic composition, as suggested for lower crust elsewhere (Mueller 1977). Other compositions are not excluded by P-wave velocity and density, but ultramafics would need to be serpentinized to achieve low enough densities and would then have too high values of Poisson's Ratio; more acid compositions (Smithson 1978) would require garnet plus so much quartz that Poisson's Ratio would be too low.

The condition of the basic rocks may be evaluated from the additional properties, especially Poisson's Ratio and electrical conductivity (Table 1). Poisson's Ratios of 'dry' pyroxene granulites match satisfactorily provided there is only modest quartz content. Such rocks cannot explain the high electrical conductivity. Hydrated minerals (e.g. amphiboles) have higher electrical conductivity (Parkhomenko 1982) and are likely

TABLE 1. *Characteristic physical properties of isotropic aggregates of common minerals at 300°C and 0.6 GPa, compared with those of the lower crust*

Mineral group or crustal layer	P-wave velocity in km s^{-1}	Poisson's ratio	Density in Mg m^{-3}	Electrical conductivity in S m^{-1}
Quartz	6.2	0.08	2.65	10^{-5}
Feldspar	6.8	0.30	2.76	10^{-5}
Amphibole	7.3	0.29	3.2	$10^{-1}-10^{-8}$
Pyroxene	7.8	0.22	3.3	$10^{-2}-10^{-8}$
Olivine	8.4	0.26	3.3	10^{-5}
Lower Crust	7.3	0.25	3.1	10^{-2}

Values for minerals are for members of each group considered, from exposed rocks, likeliest components of the deep crust, thus around An_{35} for feldspar, hornblende for amphibole, a mix of clino- and ortho-pyroxene, and forsterite for olivine.
Sources as for Fig. 6 and also Parkhomenko (1982).

to be an essential component of a model satisfying the electrical conductivity of the deep crust. However such minerals tend to have high Poisson's Ratios so they can only constitute large proportions of the lower crust by violation of the assumption of isotropy. Regionally oriented amphibolite could form the deepest crustal layer (Hall & Simmons 1978).

It is unlikely that the minerals quoted can account for the observed electrical conductivity (high enough conductivity only occurs in amphibole types rare among the lower crust). More conductive minerals may explain the data provided they form an interconnected, conducting, network (e.g. graphite, Garland 1975) but examples in 'lower crust' now exposed do not seem as widespread as the lower crustal high conductivity layer appears to be (Hutton *et al.* 1980).

The remaining preferred solution is that the lower crust contains free or surface-bound water. Toussaint-Jackson (1984) has examined this aspect and notes that both free and surface-bound water appear to have a marked effect in increasing the conductivity of basic rocks. At low confining pressures, crystalline rocks often have porosities of 0.1–0.5%, for example in high grade Lewisian basement exposed in NW Scotland (Hall & Ali 1985). If such pore space were available to deep crustal brines then it is possible to estimate their effect on electrical conductivity. Molar NaCl has an electrical conductivity at 300°C and 6 kb of 60 S m^{-1} (Quist & Marshall 1968). Using Archie's Law (Brace 1977) an empirical estimate of the conductivity of a rock

with 1% porosity occupied by molar Na Cl would be 0.005 S m^{-1}: this matches the observed lower crustal conductivity (Mbipom & Hutton 1983).

There are a number of problems relating to water capture and retention in the lower crust which are addressed in the following paragraphs.

Permeability

Permeability in exposed, fresh samples of rock generated in the deep crust is due predominantly to transport along flat cracks. Transgranular cracks give high permeability compared to the more numerous, but often narrower grain-boundary and cleavage cracks. Preliminary measurements of the variation of permeability with confining pressure in a suite of Lewisian high grade metamorphic rocks demonstrate that confining pressure of the order of 200 bars reduces permeabilities of both transgranular and grain-dimension cracks by at least an order of magnitude (Fig. 7). This is in agreement with expectation from differential strain analysis which indicates that most crack porosity in these rocks resides in flat cracks closed by confining pressures of a few hundred bars. These experiments were conducted with low pore-fluid pressures. Permeability is really a function of effective stress, approximately equal to the excess of confining over pore pressure. Thus in the deep crust, the expectation from these data would be that once pore pressure drops to below a couple of hundred bars or so less than confining pressure permeability becomes negligibly small, less than 10^{-21} m^2

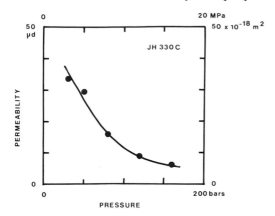

FIG. 7. Plot of permeability against confining pressure for a Lewisian granite. Permeability decreases with pressure due to crack closure. Measurements made with inert gas (nitrogen) and converted to equivalent water permeabilities.

(Walder & Nur 1984), and whatever water remains in the rock is then trapped.

This argument assumes that cracks in the deep crust have characteristics like those which are open in surface samples. Scanning electron microscopy (Fig. 8) shows that the same kinds of cracks, now sealed, were present in the rocks during uplift and were very much more abundant

than open cracks are now. The aspect ratios of the sealed cracks are difficult to measure but there is no obvious argument to suggest that these cracks when open had high aspect ratios requiring very much larger effective stresses to close them than required to close the open cracks. It is possible to estimate the permeabilities which the sealed cracks would have given the rock before sealing, on the assumptions that the cracks were open to the width of the present seal, and continued directly through the rock. The viscosity of water is roughly an order of magnitude lower at lower crustal temperatures: estimated permeabilities show that water may move around deep crustal rocks as readily as in excellent sedimentary rock aquifers, with permeabilities approaching a magnitude of 1 darcy (10^{-12} m^2).

Water capture and retention

Water may be able to move freely in deep crustal rocks provided pore pressure is within a hundred bars or so of confining pressure. The water may come from dehydration reactions of the prograde metamorphism of indigenous rocks, or it might be in transit from the mantle during thermal 'events' which may also bring magma into the crust. It is less likely to come directly at depths greater than 20 km from the dewatering of overthrust sediment, though it might come from

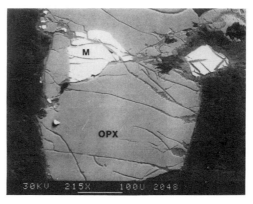

FIG. 8. Examples of sealed cracks in Lewisian rocks. These are much more abundant than open cracks and imply very high permeabilities before sealing. 100 micron scale bar at lower edge.
    (a) SEM view of JH 332, a Lewisian metabasite in amphibolite facies. Rock matrix is composed of amphibole (AMPH), plagioclase (F) and quartz (Q). A carbonate vein is rimmed with ilmenite, and the carbonate is partly replacing K-feldspar (K), which also occupies most of the small grain boundary cracks, together with ilmenite. This suggests at least 3 phases of permeation during uplift.
    (b) SEM view of JH 334, a two-pyroxene granulite. Here orthopyroxene (OPX), magnetite (M) and plagioclase (F) have been subjected to expansion cracks on uplift (clinopyroxene not shown is unaffected), and the permeating fluid has reacted with the grains with solution etching clearly along cleavages. Cracks are now sealed with magnetite (light phase) and ? chlorite.

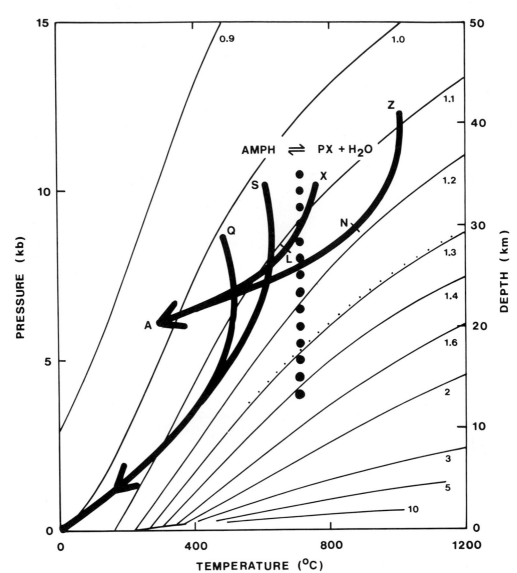

FIG. 9. Plot of the specific volume of water as a function of pressure and temperature, taken from Burnham *et al.* (1969). Thick lines show possible (P–T) paths of rocks now at surface and at lower crustal depths. L, N are points along paths from X and Z where water becomes trapped. Vertical line of dots is the metamorphic facies boundary between amphibolites and granulites and is characterized by the indicated reaction of dehydration of amphibole (AMPH) to pyroxene (PX). Isovolume line at 1.3 m³ Mg⁻¹ is that at which one particular amphibole-pyroxene reaction produces reactant volume equal to product volume. At higher pressure hydration produces amphibole of greater volume thus helping to prevent access of remaining water to more distant pyroxene (data from Fyfe *et al.* 1978, p. 138).

deep subducted ocean crusts. It is not unusual to find water in high grade, supposedly dry, metamorphic rocks. Saline fluid inclusions are preserved in such rocks (Touret & Dietvorst 1983), and only small amounts are required to explain the electrical conductivity. In the following paragraphs a mechanism for the capture of water is propounded and this leads to discussion of how water may then be retained in the lower crust.

Supposing there to be water in the lower crust at, say, the peak of the Caledonide events. How is it captured? Figure 9 shows a P–T diagram of the specific volume of water. The behaviour of water depends on how its specific volume varies relative to that of the minerals around it. Their specific volumes vary much less with both pressure and temperatures than does water and so their variation may be disregarded in this 'first order' argument.

Rocks now at the surface (pressure $= 0$, temperature $= 0$) in the Scottish Highlands have travelled on (P–T) paths from points such as Q and S (Wells 1979), possibly along the curved paths shown (such clockwise curves are typical of 'orogenic' paths where crustal thickening and subsequent erosion are accompanied by delayed thermal re-equilibrium (Royden & Hodges 1984)). Rocks now in the lower crust lie at points like A, and may have arrived there from points like X and Z. These lie in the granulite-facies field but pass through to amphibolite-facies at about 700°C (line of large dots) (Turner 1968, p. 366). As the rocks move away from points X and Z, water contained in them expands, building up pore pressure and hydrofracturing its way out of the system. This will continue until, at points L, N, the (P, T) paths parallel the isovolume lines for water. Beyond these points any water left will contract rapidly, pore pressure will drop, and consequently so will permeability. Water is then trapped. Diffusion through grains of such water away from low pressure traps will be on a scale of less than 1 m in 400 million years given typical solid diffusion coefficients (Fyfe *et al.* 1978, p. 117). It is possible that water may migrate more rapidly along grain boundaries and that low-pressure pockets at grain triple junctions will not survive for so long because of stress relaxation in the adjacent grains. But while it is not clear that low-pressure water can survive 400 million years since the Caledonide uplift, it might survive in the lower crust the 60 million years since Tertiary volcanism (where the trapping mechanism would be the same) and the formation of the adjacent continental margin, or it might survive by surface bonding along grain boundaries.

The reason for distinguishing paths from points X and Z is that L and N lie on opposite sides of the metamorphic facies boundary. Because the path from X passes this boundary before water is trapped, retrogression will be as widespread as the water flush will allow. The path from X traps water in rocks still in granulite-facies. Once trapped, the water, on subsequently crossing the boundary, can only react with pyroxene immediately around it and then amphibole formed may inhibit further migration. Thus this theory accounts not only for the preservation of water in the lower crust, but also for the common observation of patchy, localized, retrogression of granulite terrains, for example, in the Lewisian (Sutton & Watson 1951). Also because at higher levels the trapping point lies within retrogressive facies, all water is readily soaked up in rehydration reactions before being trapped and little is left to enhance electrical conductivity: the upper crust is resistive.

The retention of trapped water for substantial periods of geological time is difficult, as explained above. It is unlikely to remain at low pressure because of creep in surrounding grains. Indeed if it were to remain at low pressure it could not easily acquire the inter-connectivity required to provide electrical conductivity. Shankland & Ander (1983) suggest that deep water may be preserved at high pressure. This could occur if the fluid, having gradually increased in pressure due to creep in surrounding grains, creates enough permeability to flow upwards and then precipitate solid matter when temperature falls. This sealing mechanism is a characteristic of geothermal systems. The advantage of high-pressure retention is that inter-connection in the horizontal direction is likely and would then account for the observed electrical conductivity.

Reflectivity

Water preserved in the lower crust might vary in concentration from place to place because of repeated self sealings and water source distributions. Crack porosity would thus vary and this would cause seismic velocity variations (O'Connell & Budiansky 1974). Quite small crack porosities (0·5%) may yield variations in V_P of the order of 10% (Hall & Ali 1985). Reflectivity in the lower crust could be due to water accumulaton in particular horizons.

All this provides a rational explanation of deep crustal layering. Because of porosity and permeability variations, water is trapped in the lower crust on cooling after a thermal event in patchy concentrations. If the patches form sub-horizontal layers, then deep seismic reflections could be caused just by the variation in water content from layer to layer. The original reason for the local

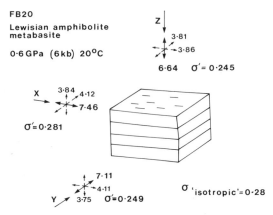

FB20

Lewisian amphibolite
metabasite

0·6 GPa (6 kb) 20°C

z

3·81
3·86
6·64   σ'= 0·245

X
3·84   4·12
7·46

σ'= 0·281

7·11
4·11
Y   3·75   σ'= 0·249

σ 'isotropic'= 0·28

FIG. 10. Summary of orthogonal set of seismic velocity measurements on a core of sample FB20, from a Lewisian dyke in amphibolite facies (Hall & Simmons, 1978). Block shows gneissose foliation (in X-Y plane) and amphibole lineation (parallel with X). Velocity values are given in km s^{-1} in sets of three (one P- and two S-wave velocities) divided according to propagation direction (X, Y or Z). Double headed arrows show vibration directions: dashes indicates S-wave, full indicates P-wave. σ is Poisson's Ratio: the isotropic value is calculated from means of V_P and V_S values; σ' is the Poisson's Ratio that might be estimated from the ratio of V_P to the faster of the V_S values in a given propagation direction. Measurements made at 0.6 GPa and 20°C. A boundary between crusts of different orientation of this one rock could give a strong reflection.

concentration may lie in lithological contrast so that this also could contribute to reflectivity.

In the particular example used, it may be that water was introduced into the crust as a volatile component in magmatic intrusion concomitant with that known to have invaded the overlying Dalradian basin late in its extensional development (Graham 1976).

Alternative explanations of the seismic layering are less attractive. Original complex lithological layering flattened by Mesozoic stretching seems unlikely to produce the degree of apparent flattening obtained given that the stretching appears to be small ($\beta < 2$) in this area. The reflectivity could be produced by anisotropy—amphibolite fabrics are known to be strongly oriented in shear zones—and the seismic velocities shown in Figure 10 indicate that sufficient velocity contrast could exist in one direction between a shear zone and

surrounding rock with a macroscopically random fabric but the same composition. The shear zones would need to be sub-horizontal to provide the observed reflections. However, neither of these alternative models, on its own, accounts for the high electrical conductivity. The problem of preserving water in granulite facies 'country rocks' has been mentioned above. While it may be possible to retain very saline waters in anatectic rocks (Touret & Dietvorst 1983), the preponderance of CO_2 inclusions in high grade rocks does suggest that water is often lost from granulites, not to be retained. It has been suggested here that water may be preserved in locally amphibolitized rock. Many of these difficulties are removed if the water is localized by post metamorphic intrusion of volatile magma. The sub-horizontal layering, the reflection coefficients and electrical conductivity are all readily satisfied by this model as quantified above.

Conclusions

The deep continental crust below part of the Caledonides may give some clues to the nature of the reflective lower crust encountered widely in deep reflection studies.

In this case, the reflective crust may have a P-wave velocity of over 7 km s^{-s} and a density of 3.1 Mg m^{-3}. It is likely to be associated with a Poisson's Ratio of about 0.25 and electrical conductivity greater than 0.003 S m^{-1}. From this is inferred that the deep crust has an average composition like that of very basic igneous rocks, may have been in granulite-facies but is likely to have been at least partly retrogressed to amphibolite-facies with free water possibly trapped at low pressure.

Reflectivity is probably produced by sill intrusion during underplating enhanced by variations in water content across horizons.

ACKNOWLEDGEMENTS: Shell Expro UK supported my work on the BIRPS data. The natural Environment Research Council (Grant GR3/2776) supported my work on permeability. Scanning electron microscopy and differential strain analyses were undertaken at MIT in collaboration with Gene Simmons: this work was also supported by NERC and by the Carnegie Trust for the Universities of Scotland. G. Gordon, R. McDonald and K. Roberts provided technical assistance in Glasgow.

References

ANDERSON, O.L. & LIBERMANN, R.C. 1968. Sound velocities in rocks and minerals: experimental methods, extrapolations to very high pressures, and results, *In*: MASON, W.P. (ed.) *Physical Acoustics IV B*, pp. 329–472.

ASSUMPCAO, M. & BAMFORD, D. 1978. LISPB-V. Studies of crustal shear waves, *Geophys. J.R. Astron. Soc.* **54**, 61–74.

BAMFORD, D., NUNN, K., PRODEHL, C. & JACOB, B. 1978. LISPB-IV Crustal structures of Northern Britain, *Geophys. J.R. Astron. Soc.* **54**, 43–60.

BARTON, P., MATTHEWS, D., HALL, J. & WARNER, M. 1984. The Mohorovicic Discontinuity seen on normal incidence and wide-angle seismic records, *Nature*, **308**, 55–56.

BIRCH, F. 1960. The velocity of compressional waves in rocks to 10 kbar, 1, *J. Geophys. Res.* **65**, 1083–1102.

——1961. The velocity of compressional waves in rocks to 10 kbar, 2, *J. Geophys. Res.* **66**, 2199–2224.

BOTT, M.H.P., ARMOUR, A.R., HIMSWORTH, E.M. & MURPHY, T. 1979. An explosion seismology investigation of the continental margin west of the Hebrides, Scotland, *Tectonophysics*, **59**, 217–31.

BRACE, W.F. 1977. Permeability from resistivity and pore shape. *J. Geophys. Res.* **82**, 3343–9.

BREWER, J.A., MATTHEWS, D.H., WARNER, M.R., HALL, J., SMYTHE, D.K. & WHITTINGTON, R.J. 1983. BIRPS deep seismic reflection studies of the British Caledonides—the WINCH profile. *Nature*, **305**, 206–10.

—— & SMYTHE, D.K. 1984. Moist and the continuity of crustal reflector geometry along the Caledonian–Appalachian orogen, *J. geol. Soc. London*, **141**, 105–20.

BURNHAM, C.W., HOLLOWAY, J.R. & DAVIS, N.F. 1969. Thermodynamic properties of water to 1000°C and 10,000 bars. *Spec. Paper Geol. Soc. Am. No.* 132, 96 pp.

CHRISTENSEN, N.I. 1982. Seismic velocities, *In*: CARMICHAEL, R.S. (ed.) *Handbook of Physical Properties of Rocks Vol. II*, pp. 1–228, CRC Press, Boca Raton, Florida.

FYFE, W.S., PRICE, N.J. & THOMPSON, A.B. 1978. *Fluids in the Earth's Crust*, Elsevier, Amsterdam, 393 pp.

GARLAND, G.D. 1976. Correlation between electrical conductivity and other geophysical parameters. *Phys. Earth Planet. Inter.* **10**, 220–30.

GRAHAM, C.M. 1976. Petrochemistry and tectonic significance of Dalradian metabasaltic rocks of the SW Scottish Highlands. *J. geol. Soc. London* **132**, 61–84.

HALL, J. 1984. Geophysical constraints on crustal structure in the Dalradian region of Scotland. *J. geol. Soc. London*, **142**, 149–55.

——1986. Nature of the lower continental crust: evidence from BIRPS work on the Caledonides. *In*: BARAZANGI, M. & BROWN, L. (eds) *Reflection seismology: the continental crust. Geodynamics Ser.* **14**, 223–31, Am. geophys. Un.

—— & ALI, M. 1985. Shear waves in a seismic survey of Lewisian basement: an extra control on lithological variation and porosity. *J. geol. Soc. London*, **142**, 149–155.

——POWELL, D.W., WARNER, M.R., EL-ISA, Z.H.M., ADESANYA, O. & BLUCK, B.J. 1983. Seismological evidence for shallow crystalline basement in the Southern Uplands of Scotland. *Nature*, **305**, 418–20.

—— & SIMMONS, G. 1979. Seismic velocities of Lewisian metamorphic rocks at pressures to 8 kbar: relationship to crustal layering in north Britain, *Geophys. J.R. Astron. Soc.* **58**, 337–47.

——BREWER, J.A., MATTHEWS, D.H. & WARNER, M.R. 1984. Crustal structure across the Caledonides from the WINCH seismic reflection profile: influences on the evolution of the Midland Valley of Scotland. *Trans. R. Soc. Edinburgh: Earth Sci.* **75**, 97–109.

HIPKIN, R.G. & HUSSAIN, A. 1983. Regional gravity analysis: North Britain. *Rep. Inst. Geol. Sci.* No. 82/10.

HUTTON, V.R.S., INGHAM, M.R. & MBIPOM, E.W. 1980. An electrical model of the crust and upper mantle in Scotland. *Nature*, **287**, 30–3.

LONGMAN, C.D., BLUCK, B.J. & VAN BREEMEN, O. 1979. Ordovician conglomerates and evolution of the Midland Valley. *Nature*, **280**, 578–81.

MBIPOM, E.W. & HUTTON, V.R.S. 1983. Geoelectromagnetic measurements across the Moine Thrust and the Great Glen in northern Scotland. *Geophys. J.R. Astron. Soc.* **74**, 507–24.

MUELLER, S. 1977. A new model of the continental crust. *In*: HEACOCK, J.G. (ed.) *The Earth's Crust, Geophys. Monog. Am. Geophys. Un.* No. 20, 754 pp.

O'CONNELL, R.J. & BUDIANSKY, B. 1974. Seismic velocities in dry and saturated cracked solids. *J. geophys. Res.* **79**, 5412–26.

PARKHOMENKO, E.I. 1982. Electrical resistivity of minerals and rocks at high temperature and pressure. *Rev. Geophys. Space Phys.* **20**, 193–218.

QUIST, A.S. & MARSHALL, W.L. 1968. Electrical conductances of aqueous sodium chloride solutions from 0 to 800° and at pressures to 4000 bars. *J. Phys. Chem.* **72**, 684 703.

ROYDEN, L. & HODGES, K.V. 1984. A technique for analyzing the thermal and uplift histories of eroding orogenic belts: A Scandinavian example. *J. geophys. Res.* **89**, 7091–7106.

SHANKLAND, T.J. & ANDER, M.E. 1983. Electrical conductivity, temperatures and fluids in the Earth's crust. *J. geophys. Res.* **88**, 9475–84.

SIMMONS, G. & WANG, H., *Single Crystal Elastic Constants and Calculated Aggregate Properties: a Handbook*. MIT Press, Cambridge, Mass., 1971.

SMITH, P.J. & BOTT, M.H.P. 1975. Structure of the crust beneath the Caledonian foreland and Caledonian belt of the north Scottish shelf region. *Geophys. J.R. Astron. Soc.* **40**, 187–205.

SMITHSON, S.B. 1978. Modelling continental crust: structural and chemical constraints. *Geophys. Res. Lett.* **5**, 749–52.

SMYTHE, D.K. *et al.* 1982. Deep structure of the Scottish Caledonides revealed by the MOIST reflection profile. *Nature*, **299**, 338–40.

SUTTON, J. & WATSON, J.V. 1951. The pre-Torridonian metamorphic history of the Loch Torridon and Scourie areas in the northwest Highlands, and its bearing on the chronological classification of the Lewisian. *J. geol. Soc. London.* **106,** 241–308.

TOURET, J. & DIETVORST, P. 1983. Fluid inclusions in high grade anatectic metamorphites. *J. geol. Soc. London.* **140,** 635–49.

TOUISSAINT-JACKSON, J.E. 1984. Unpubl. Ph.D. thesis, Univ. East Anglia.

WALDER, J. & NUR, A. 1984. Porosity reduction and crustal pore pressure development. *J. geophys. Res.* **89,** 11539–48.

WELLS, P.R.A. 1979. P-T conditions in the Moines of the Central Highlands, Scotland. *J. geol. Soc. London,* **136,** 663–71.

JEREMY HALL, Department of Geology, University of Glasgow, Glasgow, G12 8QQ.

Thermal gradients in the continental crust

D.S. Chapman

SUMMARY: A set of preferred geotherms for the continental crust is calculated using assumptions of steady state conductive heat transfer. The calculations use observed heat flow as the principal constraint, temperature and pressure dependent thermal conductivity functions, and a radiogenic heat generation profile that is consistent with a generalized petrological model of the lithosphere. Properties of these geotherms include: (1) a nearly constant gradient through the upper crust because the decrease in thermal conductivity at higher temperatures for felsic rocks counteracts the decreasing heat flow caused by crustal radioactivity; (2) divergence of the geotherms amounting to temperature differences of more than 500 K between rifts and shields at Moho depths, and (3) convergence of geotherms below 250 km in the asthenosphere as convective heat transfer dominates.

Emphasis is also placed on understanding the sensitivity of calculated temperatures to measureable properties. Computed geotherms are overall most sensitive to surface heat flow or reduced heat flow, and uncertainties in this quantity will ultimately limit the accuracy of temperature estimations for the deeper crust and lithosphere. Otherwise, the sensitivity to parameters such as heat production, thermal conductivity and its temperature dependence throughout the lithosphere and crustal thickness depends on depth and whether the region is characterized by high or low heat flow. Upper mantle heat production, pressure coefficients of thermal conductivity, and the position of the upper/lower crust boundary are relatively insensitive parameters for lower crustal temperature calculations.

Dynamic events accompanying crust forming processes lead to significant perturbations of the steady state static temperature field described above. Pressure–temperature pairs from granulite terrains plot near high temperature limits for those steady state geotherms and imply either a transient P–T–t path or a magmatic heat input from the mantle accompanying crustal thickening.

Temperature distributions within the lithosphere affect a wide variety of physical properties and processes. Rock density, electrical and magnetic properties and many seismic properties are temperature dependent. So are mineral phase boundaries, the rates of chemical reactions and many processes of rock deformation. For these reasons it is important to obtain as reliable an estimate of temperature within the lithosphere as possible.

For the uppermost few kilometres of the lithosphere, temperatures can be either measured directly in drillholes, or extrapolated confidently from measurement of surface heat flow and a knowledge of the behaviour of thermal properties with depth. For depths greater than about 10 km the actual temperature estimate is less certain and depends both on the model of heat transfer adopted and on accuracy of the thermal conductivity and heat production values estimated for the lower crust and uppermost mantle. The uncertainty of temperature estimates becomes progressively greater with greater depth.

To this day the most commonly cited temperature profiles for the lithosphere, especially in the petrologic literature, are those published 20 years ago by Clark & Ringwood (1964). The long standing popularity of the Clark & Ringwood geotherms is all the more remarkable because

their paper pre-dated the general acceptance of plate tectonics and sea floor spreading in the late 1960s, the discovery of the linear heat flow–heat production relationship in 1968 (which is a cornerstone of crustal geotherm calculations) and several important high temperature experiments concerning the thermal conductivity of mantle material. Furthermore, at that time there were only 73 continental heat flow measurements to serve as model constraints and to provide information on the thermal state of the continental lithosphere; the corresponding number at the end of 1984 exceeded 6,000.

Although the single steady state oceanic geotherm of Clark & Ringwood has had to be abandoned in favour of a family of transient geotherms (cf. Sclater & Francheteau 1971) related to the age of ocean floor and consistent with the model of sea floor spreading, the three continental geotherms of Clark & Ringwood (1964, their Fig. 2) are still in use. However some parameter values used by them, in particular high temperature thermal conductivity, have been shown by subsequent experiments (Schatz & Simmons 1972) to be more than 100% in error. Fortunately, geotherm calculations are not totally dependent on a single parameter, and thus through an acknowledged use of 'constraints imposed on the

From: DAWSON, J.B., CARSWELL, D.A., HALL, J. & WEDEPOHL, K.H. (eds) 1986, *The Nature of the Lower Continental Crust*, Geological Society Special Publication No. 24, pp. 63–70.

temperatures by considerations not of a purely thermal nature', Clark & Ringwood (1964, p. 53) arrived at a basic thermal state for shields and platforms which is not in great dispute.

The purpose of this paper is not so much to revise radically the Clark & Ringwood continental geotherms as to provide an updated rationale for the computation of geotherms in general. In doing so refinements in modelling lithospheric geotherms made by others over the last decade are used (Blackwell 1971; Roy *et al.* 1972; Lachenbruch & Sass 1977; Pollack & Chapman 1977). Of particular importance are recent compilations of thermophysical properties (Roy *et al.* 1981; Rybach & Cermak 1982; Cermak & Rybach 1982). A preferred family of geotherms is derived and presented together with parameter sensitivity analysis and a discussion of applicability. Support for the geotherm family is drawn from pyroxene geotherms derived from mantle xenolith studies. Finally the limitations of a steady state, static model based on conductive heat transfer are stressed and some alternative geotherm formulations are noted.

Geotherm models

Although both conductive and convective heat transfer may be important in determining temperature distributions in the lithosphere, especially in active orogenic regions, conductive processes mainly dominate in stable areas. Once an appropriate model of heat transfer is established, and boundary conditions are imposed, the calculation of temperature distributions is straightforward. The governing differential equation of time-dependent conductive heat transfer is:

$$-\text{div}\,(-k\,\text{grad}\,T)+A=\rho\,c\,\delta T/\delta t \qquad (1)$$

where T is temperature, t is time, A is volumetric heat production, k is thermal conductivity, ρ is density and c is specific heat.

For one dimensional heat transfer in a homogeneous and isotropic medium, equation (1) can be written as:

$$\frac{\delta^2 T}{\delta z^2} + \frac{A}{k} = \frac{1}{\alpha}\frac{\delta T}{\delta t} \qquad (2)$$

where $\alpha = k/\rho\,c$ is thermal diffusivity.

The most common simplification of (2), and one which is employed here involves steady state heat transfer in a medium with arbitrary heat production. This leads to Poisson's equation:

$$\frac{d^2 T}{\delta z^2} = -\frac{A}{k} \qquad (3)$$

Throughout the lithosphere heat production, A,

varies with depth following the distribution of radiogenic isotopes. Thermal conductivity k varies with composition and with both pressure and temperature. Analytic solutions for $T(z)$ can be found when the heat production $A(z)$ follows some simple analytical forms and when thermal conductivity is constant or temperature dependent (cf. Rybach 1981), however these requirements are restrictive. The algorithm incorporated here uses the solution of (3) in a layer of constant heat generation and constant thermal conductivity:

$$T(z) = T_T + \frac{q_T}{k}z - \frac{A\,z^2}{2\,k} \qquad (4)$$

where T_T and q_T are the temperature and heat flow respectively at the top of the layer where $z=0$.

If the layer has thickness Δz, then the temperature at, and heat flow through, the bottom of the layer (T_B, q_B) can be expressed in terms of the temperature and heat flow at the top of the layer (T_T, q_T) and properties (A,k) of the layer

$$T_B = T_T + \frac{q_T}{k}\Delta z - \frac{A\,\Delta z^2}{2k} \qquad (5)$$

$$q_B = q_T - A\,\Delta z \qquad (6)$$

Equations (5) and (6) are applied to successive layers, resetting T_T and q_T at the top of each new layer with the values T_B and q_B solved for the bottom of the previous layer. The layer thickness can be made small enough so that complicated and discontinuous distributions $A(z)$ are adequately approximated. Thermal conductivity effects $k(z,T)$ are incorporated and updated at each step in an iterative loop. In practice, heat production and thermal conductivity are described by piecewise continuous functions and the computations are carried out with a 0.1 km depth increment.

Thermophysical parameters

The gross pattern of thermal conductivity in the lithosphere is controlled by composition, to a lesser extent by temperature, and finally by pressure effects. Although the structural and compositional complexity of both the upper and lower crust (cf. Kay & Kay 1981) indicates a concomitant complexity in physical and thermal properties, it is necessary to generalize a structure for the purpose of computing geotherms. The generalized lithosphere model used here consists of an upper crust of granite to andesitic composition, a lower crust consisting of gabbro or granulite-facies metamorphic rocks and an ultra-mafic upper mantle.

Recent compilations of thermal conductivity determinations (Roy *et al.* 1981; Cermak & Rybach 1982) on several thousand rock samples provide a convenient summary from which an appropriate crustal thermal conductivity profile may be assembled. A value of 3.0 $Wm^{-1} K^{-1}$ is an appropriate value to use for a granite upper crust at conditions replicating the laboratory conditions (room temperature and atmospheric pressure) under which the conductivities were determined. The thermal conductivity of possible lower crustal rocks is constrained to a narrow range by the average values given by Cermak & Rybach (1982) for diorite (2.91 $W m^{-1} K^{-1}$), granodiorite (2.65 $W m^{-1} K^{-1}$), gabbro (2.63 $W m^{-1} K^{-1}$), amphibolite (2.46 $W m^{-1} K^{-1}$) and gneisses (2.44 $W m^{-1} K^{-1}$). Thermal properties of 'granulites' rarely appear in tabulations, but the thermal conductivity of rocks compositionally equivalent to granulites lies between 2.4 and 2.9 $W m^{-1} K^{-1}$. A value of 2.6 $W m^{-1} K^{-1}$ is therefore appropriate for generalized lower crustal rocks, but again only at laboratory measurement conditions.

Although in the past it has been common to consider thermal conductivity as being constant throughout the crust, it seems more appropriate now to use the growing experimental data base and to calculate thermal conductivity as a function of pressure and temperature directly in the geotherm computation. Thermal conductivity of most rocks at crustal conditions varies inversely with temperature and directly with pressure (or depth) according to the relation

$$k(T,z) = k_0 (1 + c z)/(1 + bT) \qquad (7)$$

where T is temperature in degrees Celsius, and b and c are constants, and k_0 is a conductivity measured at $0°C$ and one atmosphere pressure.

Details of the thermal conductivity profile for this average lithosphere model are as follows. For the upper crust, a temperature coefficient b of $1.5 \times 10^{-3} K^{-1}$ is selected, intermediate between granite and granodiorite. For the lower crust a temperature coefficient b of 1.0×10^{-4} is used. A pressure coefficient c of $1.5 \times 10^{-3} km^{-1}$ is used throughout the crust. Zero depth and zero temperature values k_0 are 3.0 and 2.6 $W m^{-1} K^{-1}$ for the upper and lower crust respectively. The relative effects of temperature and pressure depend on the particular geotherm and on depth in the crust. For the coolest thermal regime corresponding to a surface heat flow of 40 mW m^{-2}, the base of the upper crust attains a temperature of $199°C$ which lowers the surface thermal conductivity of 3.0 $W m^{-1} K^{-1}$ to 2.37 W

$m^{-1} K^{-1}$. The corresponding decrease in the upper crust associated with a 90 mW m^{-2} geotherm is from 3.0 to 1.74 $W m^{-1} K^{-1}$. In the lower crust the temperature coefficient is sufficiently small that the temperature and pressure effects nearly cancel. For the 40 mW m^{-2} geotherm, pressure effects are greater and conductivity increases slowly from 2.61 to 2.64 $W m^{-1} K^{-1}$ in the lower crust. For the 90 mW m^{-2} geotherm, temperature effects dominate, and conductivity decreases from 2.53 to 2.50 $W m^{-1} K^{-1}$ between 16 and 35 km. For the mantle portion of the lithosphere, the hypothetical mantle model of Schatz & Simmons (1972) is followed.

Although the structural and geochemical complexity inherent in the formation and consolidation of the continental lithosphere may well produce a complex vertical distribution of radioactive heat producers; it is again necessary to derive a generalized model for the calculation of geotherms. However any general model should adhere to the following constraints: (1) average levels of heat generation measured on surface felsic rocks cannot be sustained throughout the crust because the heat flow produced by such a vertical column would exceed the surface heat flow of 40 mW m^{-2} characteristic of shields; (2) a vertical distribution of heat generation in the crust should satisfy the linear heat flow–heat production relationship.

For the upper crust an exponentially decreasing heat production is used, $A(z) = A_0 \exp(-z/D)$ which is but one of various heat production distributions which satisfy the linear heat flow–heat production relationship

$$q_0 = q_r + D A_0 \qquad (8)$$

where q_0 is surface heat flow, q_r is reduced heat flow and A_0 is surface heat generation. D is a parameter with dimensions of length and characterizes the depth distribution $A(z)$. In order to parameterize the geotherm calculation in terms of surface heat flow, an empirical relationship that partitions the observed surface heat flow, with 40% being attributed to upper crustal radiogenic sources and 60% to deeper sources, is used (Pollack & Chapman 1977; Vitorello & Pollack 1981). Thus surface radioactivity A_0 can be related simply to surface heat flow using $A_0 = 0.4$ q_0/D. The exponential distribution is continued downwards until the heat production falls to the lower crust heat production A_L at $z = D + \ln(A_0/A_L)$, or until the Conrad discontinuity (assumed at 16 km in the general model) is reached.

Heat production in the lower crust is assumed to be constant in the general model. The actual value is best determined by comparison with probable lower crustal rock types and with measurements on lower crustal xenoliths. The latter illustrate the true variability which may exist in the lower crust: from 0.02 μW m^{-3} for pyroxene granulites found in the Delegate Pipe, Eastern Australia (Irving 1980), to 2.0 μW m^{-3} for metapelites found as xenoliths in the Massif Central, France (Dupuy *et al.* 1979). However these extreme values are not likely to represent the entire lower crust. When rock types are weighted as to their probable abundance in the lower crust, then the range of average heat production values is narrowed to about 0.3–0.7 μW m^{-3}. This range is also consistent with the estimate of Nicolaysen *et al.* (1981) who have calculated a lower crust heat production of 0.56 μW m^{-3} for the Vredefort area of South Africa based on an estimated lower crust composition of 50% leucogranofels and 50% mafic granulite. In this study the general model lower crust heat production is assumed to be 0.45 μW m^{-3}. Finally for the upper mantle section of the lithosphere a uniform heat production of 0.02 μW m^{-3} is assumed for the depleted upper mantle down to 200 km and 0.045 μW m^{-3} from 200 to 250 km, the depth limit for the calculation.

A family of steady state, static geotherms

A preferred family of geotherms for stable continental crust, parametric in surface heat flow, is shown in Fig. 1. The most important characteristic of this family of geotherms is the divergence from a common temperature at the surface. The geotherms converge again when they are truncated by the mantle 1300°C adiabat, the deepest convergence being at about 250 km for the 40 mW m^{-2} geotherm. Between these depths the limiting geotherms of 90 and 40 mW m^{-2} bracket a broad temperature–depth (equivalently P–T) space. Temperature differences at a common depth between these bracketing geotherms can be large: 500 K at 30 km depth and a maximum of 800 K at 60 km depth. Thus any notion of a single continental geotherm should be discarded in the same manner as the notion of a single oceanic geotherm has been. Instead, one must recognize the possible diversity of thermal states available for the continental crust and try to find the appropriate geotherm for a particular region of interest.

The geotherms in Fig. 1 are furthermore characterized by roughly constant gradients throughout the upper crust, lower but changing gradients throughout the lower crust and slowly decreasing gradients throughout the mantle portion of the lithosphere. The upper crustal gradients remain

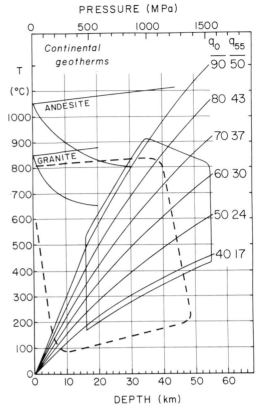

FIG. 1. A set of steady state-static geotherms for continental lithosphere. Geotherm parameter is characteristic surface heat flow q_0 in mW m^{-1}; heat flow at a depth of 55 km (q_{55}) is also shown. Solid lines marked by labels 'granite' and 'andesite' show possible solidii brackets for wet (minimum) and dry (maximum) melting in the crust. Heavy dashed line encloses PT field appropriate for metamorphic rocks; solid line polygon encloses steady state PT field of lower crust in stable regions. Lack of complete overlap in temperature space can be explained by dynamic crustal forming processes.

almost constant because the effect of decreasing heat flow due to crustal radioactivity is almost exactly counteracted by a change in thermal conductivity caused by temperature increase. For high heat flow regions this effect is pronounced and can lead to a maximum in the thermal gradient at the base of the upper crust.

A third characteristic feature of the geotherms (but not shown in Fig. 1) is the high temperature truncation of all geotherms for heat flow greater than about 40 mW m^{-2}. The truncation line represents an upper mantle adiabat which intercepts 1300°C at zero depth and has a gradient of

0.4 K km^{-1}. At depths greater than the truncation depth, heat transfer is presumed to be predominantly convective and the thermal gradient adiabatic. The 1300°C adiabat was chosen in preference to others because (1) it falls within the permissible temperature interval required by many oceanic and continental lithosphere thermal evolution models and (2) it bounds upper mantle temperature fields determined from xenoliths.

Also shown in Fig. 1 are representative solidii curves for the crust (cf. Myson 1981). The hottest thermal states characterized by 80 and 90 mW m^{-2} geotherms could provide for deep crustal melting under hydrous conditions. Dry melting in the crust seems to be precluded in regions where these steady state–static geotherms apply. Below the crust, the geotherms consistent with surface heat flow between 90 and 50 mW m^{-2} all intersect a mixed volatile mantle solidus but the 45 and 40 mW m^{-2} geotherms do not. This is consistent with the observation that low heat flow regions such as shields do not have pronounced seismic low velocity zones which are thought to be caused by partial melting.

Finally, one should recall the assumption of partitioning the source of continental heat flow with 40% being ascribed to upper crustal radioactive heat production and 60% to deeper sources, be they radioactive, advective or heat liberated from cooling the interior of the earth. Thus the reduced heat flow values implicit in the Fig. 1 geotherms are simply 0.6 q_0 and vary from 24 to 54 mW m^{-2} for the range of geotherms shown. Mantle heat flow is generally less than reduced heat flow; values of heat flow at a depth of 55 km for the Fig. 1 geotherms vary from 17 to 50 mW m^{-2}.

Average crustal radioactivity values implicitly assumed for these geotherms are very nearly the difference between q_0 and q_{55} divided by the 35 km crustal thickness used for computations. For most continental regions characterized by a heat flow of 50 to 70 mW m^{-2} the average crustal heat production assumed is 0.74 to 0.94 μW m^{-3}. These values fall within the range suggested independently for geochemical models of the continental crust (Smithson & Decker 1974; Haack 1983; Weaver & Tarney 1984).

Alternative assumptions about reduced heat flow and radioactivity distributions can be made and these assumptions may lead to somewhat different geotherms. To cover all possible cases one would need many suites of geotherms for different combinations of reduced and surface heat flow. However as our limited and imprecise knowledge of reduced heat flow precludes detailed knowledge I feel the averaging assumptions which result in a single set of geotherms are justified.

Parameter sensitivity

The geotherms are model dependent and thus it is instructive to examine the sensitivity of calculated temperatures to assumed values of model parameters. A summary of parameter sensitivity is given in Table 1. The single most important factor governing the computation of subsurface temperatures is surface heat flow. A change of 10% in

TABLE 1. *Parameter sensitivities for geotherm calculations*

Parameter	Value	Perturbation (%)	T (%) at 50 km shield	rift
heat flow q_0	40 mW m^{-2} (shield)	+10%	+13%	—
	80 mW m^{-2} (rift)	+10%	—	+11%
heat production:				
depth parameter D	8 km	+25%	+6%	+4%
lower crust A_1	0.45 W m^{-3}	+20	−5%	−1%
thermal conductivity:				
upper crust k_0	3.0 W m^{-1} K^{-1}	+20%	−8%	−8%
lower crust k_0	2.6 W m^{-1} K^{-1}	+20%	−5%	−1%
T dependence b	see text	100%	−4%	−9%
P dependence c	see text	100%	+2%	+2%
Moho depth	35 km	+20%	−2%	+1%
Conrad depth	16 km	+25%	+1%	+3%

surface heat flow corresponds to temperature changes of between 10% and 15% at deeper crustal and upper mantle depths. This justifies the parameterization of geotherms in terms of their corresponding heat flow. But it also indicates a confidence limit for actual temperatures in the crust, because at present it is difficult to determine heat flow for specific regions to much better than about 10% uncertainty.

Heat production within the crust is also a sensitive parameter, but less than surface heat flow. A variation of 25% in D produces about a 5% change in temperatures at all depths. A 20% variation in lower crustal heat production has a stronger effect on lower lithosphere temperatures but this sensitivity diminishes for higher heat flow values. The difference between a uniform heat production in the lower crust and a two step model (with the same average value) as proposed by Nicolayson *et al.* (1981) for the Vredefort area of South Africa is 16 K at 35 km for the 40 mW m^{-2} geotherm and about twice that for the 80 mW m^{-2} geotherm. Upper mantle radioactivity is unimportant for crustal temperatures.

Thermal conductivity at all levels in the lithosphere has an important effect on computed temperatures, but the effects are complicated by the apparently strong temperature dependence of conductivity in some depth regions, and the observation that temperature effects are negative for most crustal rocks but positive in the mantle. Thus an underestimation of upper crustal thermal conductivity produces an overestimation of temperature at the Moho, but this in turn produces an overestimation of mantle conductivity (through temperature dependence) with a compensating lower thermal gradient. Thus errors in conductivity have a built in negative feedback. Table 1 shows that it is important to consider temperature dependence of thermal conductivity, especially for high heat flow geotherms. Pressure effects are less important, amounting to about one third of temperature effects and having an opposite effect on calculated geotherms.

Crustal thickness is another parameter which has a significant influence on deep lithosphere temperature calculations, especially for low heat flow regions. This arises through the lower crust radioactivity. For a heat production of 0.45 μW m^{-3} each kilometre of additional lower crust reduces the mantle heat flow by 0.45 mW m^{-2} or by about 2% of the heat flow at that depth, and the effect is manifested in lower temperatures predicted at greater depth. The position of the upper/lower crust boundary has less importance because of the model of the upper crustal radioactivity assumed, and the consequent lack of heat production contrast at that depth.

Thermal conditions for some dynamic crustal processes

The geotherms discussed previously are appropriate for static, steady state conditions dominated by conductive heat transfer. However in active geodynamic processes, the thermal state of the crust is more often dominated by advective heat transfer and transient thermal phenomena. This section describes some examples of such thermal states and concludes with a speculation about the deep crustal P–T conditions inferred from petrological studies of granulite terrains.

In general the downward transport of material or fluid decreases temperature or depresses isotherms and upward transport increases temperatures at a given level. A continental rifting example where both processes act simultaneously, but at different crustal levels, has been described by Lachenbruch *et al.* (1985). During rifting the crust is physically stretched, thinned, and the lower crust either underplated with a mantle derived melt or injected with dikes (Fig. 2 (a)). The stretching underplating and magma injection all constitute transport of material and heat toward the surface in the lower crust and results in elevated lower crust temperatures. At the same time isostatic considerations require surface subsidence. If the sediment supply is sufficient the rift trough is continually being filled as shown schematically in Fig. 2 (a); the downward transport of sediment decreases temperatures in the sedimentary section. The result of both of these processes is a steady state but dynamically supported sigmoid shaped geotherm shown in Fig. 3, which may be quite different

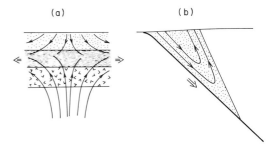

FIG. 2. Schematic diagrams showing possible geodynamic processes involving significant heat advection (a) active rifting involving both rapid sedimentation at the surface and magmatic addition to the lower crust. Solid lines with arrows show stream lines for material transport. (b) circulation of material in accretionary prism driven by subduction. P–T trajectory for particle travelling along stream line is shown by curve 2 in Fig. 3.

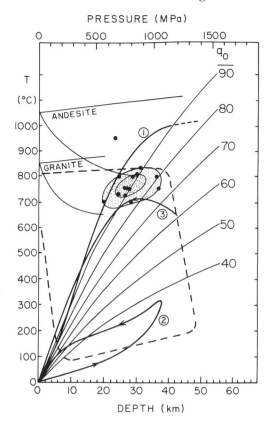

FIG. 3. Geotherms and P–T trajectories resulting from dynamic crustal processes superimposed on the static geotherms of Fig. 1. Curve 1 is a geotherm which may exist in a rift environment as shown in Fig. 2(a) (after Lachenbruch *et al.* 1985). Curve 2 is a possible p–T trajectory in an accretionary prism as shown in Fig. 2 (b) (after Wang & Shi 1984). Curve 3 is a generic P–T trajectory resulting from overthrusting and subsequent erosion (after England & Richardson 1977). Solid dots show P–T conditions inferred for granulite terrains (plotted from Table 5 of Bohlen *et al.* 1983).

from a steady state static rift geotherm (90 mW m^{-2}).

A second example of large-scale material transport and its thermal consequences is drawn from subduction processes. Wang & Shi (1984) have recently calculated some hypothetical P–T–t trajectories for material within an accretionary prism of a subduction zone. They assume that the prism acts as a viscous fluid within which an eddy current is shear driven by the subducting lithosphere (Cloos 1982). Although the Wang & Shi (1984) model may not be accurate in detail, the general mechanism of rapid burial and exhuma-

tion provides a rational explanation for high pressure–low temperature conditions (see Fig. 3) recorded in metamorphic rocks but seemingly outside of the P–T fields covered by static steady state geotherms.

A third example is drawn from the work of England & Richardson (1977) in tracing the P–T–t trajectories of material involved in burial by overthrusting and subsequent uplift and erosion. Both overthrusting and erosion processes produce a transient temperature field. The actual thermal history of any rock suite depends on the speed and depth of burial by overthrusting, the burial time prior to uplift and the erosional history of the terrain. A generic P–T–t trajectory in this process is shown by curve 3 in Fig. 3. Material buried to depths of 30–40 km may experience an extended period (10–20 Ma) of warming and then subsequent cooling along the transient geotherms elevated by the transport of material rebounding toward the surface.

It is instructive to superimpose upon this selection of static as well as dynamic thermal states, P–T conditions for a number of granulite terrains (Bohlen *et al.* 1983). Conditions for these terrains span both a restricted pressure range of 700–1000 MPa and a restricted temperature range 700–850°C. If the granulite terrain P–T field records a stable, conduction dominated thermal state then those terrains were all characterized by 90 mW m^{-2} geotherms typical of rift provinces. However the extension of these elevated conduction dominated geotherms through another 30–40 km of underlying crust suggests wholescale melting conditions within the lower crust except in the most stringent anhydrous conditions. One alternate hypothesis which avoids the lower crust melt problem is to consider that the granulite P–T field arises in an advection dominated thermal field similar to curve 1 of Fig. 3. The magmatic thickening both provides an advective heat transport which diminishes the lowermost crustal thermal gradient and produces the required thickened crust. A second alternative hypothesis is suggested by the P–T trajectory 3 in Fig. 3. Thermal modelling of thrust events and subsequent erosion episodes (cf. England & Richardson 1977) indicate that although the crust may initially exist at a variety of thermal states, deep burial and subsequent rapid uplift produce P–T trajectories which have T maxima in or near the P–T fields recorded in granulites. The very existence of granulite facies metamorphic rocks at the surface today requires a transient dynamic process of exhumation. It is not unreasonable to suppose that an appropriate metamorphic mineral assemblage recorded an equally transient thermal field.

ACKNOWLEDGEMENTS: This work was conducted in part while I was supported on sabbatical leave by the Alexander von Humboldt Foundation. I also acknowledge travel support from the Alfred Wegener Foundation which enabled me to attend the London meeting on 'The Nature of the Lower Crust'. Bill Powell and Sean Willett read the manuscript and made suggestions for its improvement.

References

BLACKWELL, D.D. 1971. The thermal structure of the continental crust. *In: The Structure and Physical Properties of the Earth's Crust*, Geophys. Mon. Ser. Amer. Geophys. Union, 14 (AGU, Washington, DC) 169–184.

BOHLEN, S.R., WALL, V.J. & BOETTCHER, A.L. 1983. Experimental investigation and application of garnet granulite equilibria. *Contrib. Mineral. Petrol.* **83**, 52–61.

CERMAK, A. & RYBACH, L. 1982. Thermal conductivity and specific heat of minerals and rocks. *In:* ANGENHEISTER, G. (ed.) *Landolt-Bornstein Numerical Data and Functional Relationships in Science and Technology*, New Series, Group V, vol. 16, Springer-Verlag, Berlin.

CLARK, S.P., Jr. & RINGWOOD, A.E. 1964. Density distribution and constitution of the mantle. *Rev. Geophys.* **2**, 35–88.

CLOOS, M. 1982. Flow melanges: numerical modeling and geological constraints on their origin in the Franciscan subduction complex, California. *Geol. Soc. Am. Bull.* **93**, 330–345.

DUPUY, C., LEYRELOUP, A. & VERNIER, J. 1979. The lower continental crust of the Massif Central (Bournac, France)—with special references to REE, U and Th composition, evolution, heat flow production. *Physics and Chemistry of the Earth*, **11**, 401–415.

ENGLAND, P.C. & RICHARDSON, S.W. 1977. The influence of erosion upon the mineral facies of rocks from different metamorphic environments. *J. Geol. Soc. Lond.* **134**, 201–213.

HAACK, U. 1983. On the content and vertical distribution of K, Th, and U in the continental crust. *Earth Planet. Sci. Lett.* **62**, 360–366.

IRVING, A.J. 1980. Geochemical and high pressure experimental studies of garnet pyroxenite and pyroxene granulite xenoliths from the Delegate basaltic pipes, Australia, Petrology & geochemistry of composite ultramafic xenoliths in alkali basalts & implications for magmatic processes within the mantle. *Am. J. Sci.* **280**, 389–426.

KAY, R.W. & KAY, S.M. 1981. The nature of the lower continental crust: inferences from geophysics, surface geology, and crustal xenoliths. *Rev. Geophys. Space Phys.* **19**, 271–297.

LACHENBRUCH, A.H. & SASS, J.H. 1977. Heat flow in the United States and the thermal regime of the crust. *In:* HEACOCK, J.G. (ed.) *The Earth's Crust—Its Nature and Physical Properties.* Geophys. Monogr. Ser. Amer. Geophys. Union 20, (AGU, Washington, DC), 503–548.

—— SASS, J.H. & GALANIS, S.P. 1985. Heat flow in southernmost California and the origin of the Salton Trough. *J. Geophys. Res.* **90**, 6709–6736

MYSEN, B. 1981. Melting curves of rocks and viscosity of rock-forming melts. *In:* TOULOUKIAN, Y.S., JUDD, W.R. & ROY, R.F. (eds) *Physical Properties of Rocks and Minerals*, Vol. II of McGraw-Hill/CINDAS Data Series on Material Properties McGraw-Hill, New York, 361–407.

NICOLAYSEN, L.O., HART, R.J. & GALE, N.H. 1981. The Vredefort radioelement profile extended to supracrustal strata at Carletonville, with implications for continental heat flow. *J. Geophys. Res.* **86**, 10653–10661.

POLLACK, H.N. & CHAPMAN, D.S. 1977. On the regional variation of heat flow, geotherms and lithosphere thickness. *Tectonophysics*, **38**, 279–296.

ROY, R.F., BECK, A.W. & TOULOUKIAN, Y.S. 1981. Thermo-physical properties of rocks, *In:* TOULOUKIAN, Y.S., JUDD, W.R. & ROY, R.F. (eds), *Physical Properties of Rocks and Minerals*, Vol. II-2 of McGraw-Hill/CINDAS Data series on Material Properties. McGraw-Hill, New York, 409–502.

—— BLACKWELL, D.D. & DECKER, E.R. 1972. Continental heat flow. *In:* ROBERTSON, E.C. (ed.) *The Nature of the Solid Earth.* McGraw-Hill, New York, 506–543.

RYBACH, L. 1981. Geothermal systems, conductive heat flow, geothermal anomalies. *In:* RYBACH, L. & MUFFLER, L.J.P. (eds) *Geothermal Systems: Principles and Case Histories.* Wiley, Chichester, 3–36.

—— & CERMAK, V. 1982. Radioactive heat generation in rocks. *In:* ANGENHEISTER, G. (ed.) *Landolt-Bornstein Numerical Data and Functional Relationships in Science and Technology*, Group V, Vol. 1a, Springer-Verlang, Berlin, 353–371.

SCHATZ, J.P. & SIMMONS, G. 1972. Thermal conductivity of earth materials at high temperatures. *J. Geophys. Res.* **77**, 6966–6983.

SCLATER, J.C. & FRANCHETEAU, J. 1970. The implication of terrestrial heat flow observations on current tectonic and geochemical models of the crust and upper mantle of the earth. *Geophys. J. Roy. Astron. Soc.* **20**, 509–542.

SMITHSON, S.B. & DECKER, E.R. 1974. A continental crustal model and its geothermal implications. *Earth Planet. Sci. Lett.* **22**, 215–225.

VITORELLO, I. & POLLACK, H.N. 1980. On the variation of continental heat flow with age and the thermal evolution of continents. *J. Geophys. Res.* **85**, 983–995.

WANG, C.Y. & SHI, Y.L. 1984. On the thermal structure of subduction complexes: a preliminary study. *J. Geophys. Res.* **89**, 7709–7718.

WEAVER, B.L. & TARNEY, J. 1984. Empirical approach to estimating the composition of the continental crust. *Nature*, **310**, 575–577.

D.S. Chapman, Dept. of Geology and Geophysics, University of Utah, Salt Lake City, Utah 84112, USA.

Diversity in the lower continental crust

J.F. Dewey

SUMMARY: Arguments based upon tectonic style and process suggest that the lower continental crust is laterally exceedingly inhomogeneous on the scale of tens and hundreds of kilometres. At least twenty basic variations exist apart from unmodified Archaean nuclei. Some 80% of the continental crust was generated by 2.5×10^9 years and most of this has been modified and/or redistributed by tectonic and sedimentary processes. Two general secular trends have occurred to increase homogeneity in the continental crust; addition of mafic igneous rocks to the lower crust, and differentiation, in both extensional and shortening environments, into a more mafic, refractory lower crust and an upper, wetter, more silicic, crust enriched in lithophile and radiogenic elements.

Introduction

The lower oceanic crust is well-exposed on land in thin ophiolite sheets of young oceanic lithosphere obducted onto continental margins. By contrast, direct samples of the lower continental crust are available more rarely as xenoliths in kimberlites and other volcanic suites, in the core of the Vredefort Dome (Nicolaysen *et al.* 1981), a probable meteorite impact structure, and where thrust restacking of a previously stretched and thinned crust has occurred (Ivrea, Finero, Ronda, Balmuccia, Beni Bousera; Dewey 1982; Kornprobst 1976; Reuber *et al.* 1982). This limited sample indicates great diversity and, unlike the oceanic sample, probably bears the difficult-to-decipher imprint of processes that occurred during the events that led to its exposure.

Geophysical investigations have allowed investigation of a much wider global sample. Heatflow modelling indicates a lower crust depleted in radiogenic elements, electrical conductivity measurements a conductive middle and lower crust (Shankland & Ander 1983), and narrow angle reflection profiling (COCORP and BIRPS) that the lower crust is commonly strongly layered. Inversion of seismic velocities to composition from refraction studies and laboratory measurements (Jackson & Arculus 1984) suggests also that the lower crust is commonly more mafic. This is especially so if small amounts of water are present but, conversely, moderate quantities of garnet would allow a more silicic bulk composition. Lastly, if the continental crust has grown by the addition of volcanic arcs of intermediate average composition, a more silicic differentiated observed upper crust implies a more mafic lower crust (Taylor & McClennan 1981; Weaver & Tarney 1984).

The broad conventional wisdom that emerges from these studies (Smithson 1978; Dewey & Windley 1981; Kay & Kay 1981) is that the continental crust is commonly, perhaps pervasively, differentiated into a more refractory granulitic lower crust and a more silicic radiogenic upper crust (Heier & Adams 1965), that a high P-wave velocity lower crust, below a Conrad discontinuity, is seismically reflective (and therefore layered or laminated; Hale & Thompson 1982; Kaila *et al.* 1981), electrically conductive and more mafic.

The high electrical conductivity may be caused by hydrated minerals, saline or carbonic waters in shear zones and zones of vein and grain boundary fracture porosity, oxides, sulphur and other volatiles, and graphite in veins and in graphitic schists. Over most of the continents, temperatures in the lower crust are too low to enhance conductivity significantly. Consequently, partial melt zones cannot be a cause except in regions of very high heatflow and/or thick crust or where the lower crust is melted by the injection of large amounts of mafic magma.

Some possible causes of lower crustal layering and lamination are as follows; underthrusted sediments, extensional fabrics and low angle normal shear zones, low angle thrust fabrics (Jones & Nur 1984) and mylonite zones (Fountain *et al.* 1984), compositional gneissic banding, layered intrusions, intrusive sheets, magmatic underplating (McKenzie 1984), Moho thrust-imbrication (Matthews & Hirn 1984), partial melt/restite zones, sheets of kinzigite and graphite schist, and fluids in horizontal crack-arrays. It is argued below that none of these is a sole cause of lower crustal reflectivity.

Rheologically, the general downward increase in temperature and confining pressure results in an increase in grain size, ductility, strain penetration and homogeneity, and metamorphic grade, and a decrease in water and radiogenic nuclides. However, rheological change is not continuous nor uniform. The thickness of an upper crustal seismogenic layer (Sibson 1983) in which brittle

From: DAWSON, J.B., CARSWELL, D.A., HALL, J. & WEDEPOHL, K.H. (eds) 1986, *The Nature of the Lower Continental Crust*, Geological Society Special Publication No. 24, pp. 71–78.

discontinuous strain occurs, is controlled by the intersection of fracture and creep envelopes for quartz which, in turn, are controlled by strain rate, geothermal gradient and the presence or absence of water (Meissner & Strehlau 1982; Dewey *et al.* 1986). Below the seismogenic layer, the middle and lower crust deforms by creep except at very high strain rates in dry cold lithologies. For a range of likely geological strain rates and geothermal gradients, ductile deformation will begin below depths from about 5 to 20 km, assuming that quartz is the dominant mineral phase. If mafic mineral phases dominate the lower crust, brittle deformation may persist to those depths or constitute a second brittle seismogenic layer. The thrusting seen to Moho depths on COCORP (Wind River) and BIRPS (Flannan and Outer Isles) profiles suggests a dry refractory if not a mafic, lower crust in those areas.

The rather limited evidence now available suggests that the lower continental crust is vertically and laterally inhomogeneous on the scale of at least tens and probably hundreds of kilometres. Direct observation shows the upper crust to be exceedingly inhomogeneous at all scales. Upper crustal structure and composition is, to a large extent, typical of and dependent upon the tectonic environment(s) in, and processes by, which the crust was made and modified (Dewey & Windley 1981). The consequent lateral variation is likely to be reflected also in the lower crust. In this paper, the problem of the nature of the lower crust is approached from a tectonic standpoint. Tectonic arguments suggest that there are at least twenty-three types of continental crust, with an associated great variation in the structure and composition of the lower crust.

Crustal types

Archaean

By the end of the Archaean, about 2.5×10^9 years, about 80% of the present volume of continental crust had been generated (Dewey & Windley 1981). Portions of this Archaean continental protolith, unmodified except for uplift and denudation, form a few percent of continental area, whereas most has been modified in a wide range of Proterozoic and, to a lesser extent, Phanerozoic tectonic settings. The origin of the Archaean crust has been extensively debated and is unclear; hypotheses range from an origin by the generation and amalgamation of volcanic arcs to severe modification by intense meteorite bombardment (Grieve 1980). Hence, it is difficult to draw inferences about the Archaean lower crust

from arguments based upon tectonic modelling and, therefore, the question of the nature of the unmodified Archaean lower crust is avoided here, beyond the observation that it appears to be non-reflective, non-layered, and non-mafic, perhaps with flat isoclinal structure (Drummond 1983; Gibbs *et al.* 1984; Smithson 1978).

Crustal generation

Crustal generation occurs in volcanic arcs (Fig. 1, 1–4), subduction–accretion prisms (Fig. 1, 6–9) and by sedimentary accumulation on oceanic lithosphere (Fig. 1, 10–11). Primary ensimatic arc generation (Yermakov 1981), probably usually nucleated on oceanic fracture zones (1) and plateaux (3) (Casey & Dewey 1984), yields mafic lower crust consisting of oceanic foundation and early arc boninites (Reagan & Meijer 1984) overlain by blob-geometry intermediate crust consisting of calc-alkaline volcanics intruded by calc-alkaline plutons. The deep intrusion of mafic igneous rocks as giant differentiated sill complexes (Coward *et al.* 1982) will produce melting or partial melting of more silicic protoliths and the consequent upper crustal injection of minimum melting granites. In intra-arc pull-aparts (2, 4) lower crustal horizontal lamination may be induced by stretching. Subduction-accretion (6–9) involves the structural accumulation of a 30 km silicic sedimentary upper and middle crust or greenschist and sub-greenschist facies above a lower crustal mafic foundation of subcreted remnant oceanic crust. Water will be present to lower crustal depths and the lower part of the silicic crust is likely to be characterized by sub-horizontal structure and fabrics. More complicated blob geometries with a large mafic component will typify the lower parts of accretionary prisms where substantial volumes of seamounts and oceanic plateaux are accreted (7). Possibly, the Ivrea Zone crust was generated in this way (Sills & Tarney 1984). Ridge/trench interaction beneath accretionary prisms (8) results in mafic injection into and partial melting of the accretionary prism and a consequent trend towards lower crustal dehydration and the upward movement of volatiles and minimum melting silicic magmas. A similar trend will occur where an arc environment is superposed on an accretionary prism (9) (Hudson & Plafker 1982) or other older silicic crust (5). 5 and 9 are therefore possible sites for lower crustal layered complexes with anorthositic and ultramafic differentiates. Sediment accumulation on oceanic lithosphere either adjacent to an arc (10) or as a delta prism (11) can yield a 24 km crust consisting of 16 km of volcaniclastics of intermediate composition (10) or silicic sediments overly-

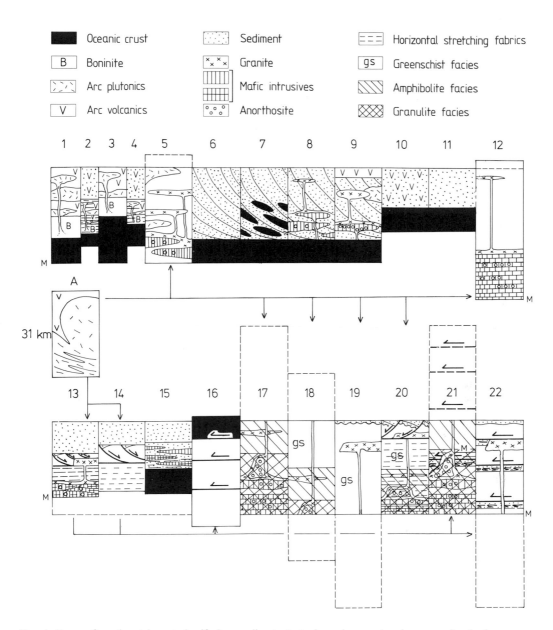

Fig. 1. Types of continental crust classified according to tectonic environment and process. A—Archaean protolith. 1—Volcanic arc nucleated on oceanic crust; 2—intra-arc pull-apart in 1; 3—volcanic arc nucleated on oceanic plateau; 4—intra-arc pull-apart in 3; 5—continental margin arc; 6—subduction-accretion prism; 7—subduction accretion prism with clipped-off seamounts and oceanic crustal slivers; 8—subduction-accretion prism with trench-ridge interaction; 9—subduction accretion prism overprinted by volcanic arc; 10—back-arc basin; 11—oceanic delta; 12—crust underplated by mafic igneous rocks; 13—crustal extension to $\beta > 2$; 14—crustal extension to $\beta < 2$; 15—ridge blanketed by sediment; 16—thrust stacked crust with high-level ophiolite nappe; 17–19—crusts thickened by vertical stretching; 21 and 22—thickened by thrust stacking; 17 and 21—denuded back to normal thickness; 18—denuded and delaminated back to normal thickness; 19 and 22—delaminated back to normal thickness; 20—stretched back to normal thickness.

ing an 8 km mafic lower crust (11). The latter may be an important thin crust environment as infilled remnant oceanic 'holes' in collisional orogens. During the earliest phases of oceanic growth, immediately following continental splitting, a spreading ridge may be blanketed by a thick sedimentary prism, as in the Colorado Delta at the northern end of the Gulf of California. The continental margin transitional crust thus generated (15) will be some 24 km thick with an oceanic mafic lower crust passing upwards into an extended complex of mafic sills and metasediments overlain by underformed sediments (Saunders *et al.* 1982).

Crustal modification

Modification of continental crust occurs in two principal tectonic environments—extension (13, 14) and shortening (16–22). Lithospheric extension causes crustal cooling and hardening and retrograde metamorphism. Extended crusts are characterized by sediments overlying a faulted layer beneath which ductile extension occurs (13, 14). The amount of stretching (β) determines the relative thicknesses of the three layers and the role of syn-extensional igneous rocks. Where β is sufficiently large to allow fertile mantle at the peridotite solidus to rise to partial melting depths, basaltic volcanism may be accompanied by basaltic underplating with partial melting of the extended lower crust and the rise of silicic minimum melting liquids to freeze in upper crustal flat-based batholiths (Lynn *et al.* 1981; Verma *et al.* 1984).

Lithospheric shortening causes crustal warming and softening and prograde metamorphism. Crustal shortening may be accomplished by thrust stacking of a previously thinned crust (16, 21, 22) or, possibly, by vertical stretching (Dewey & Burke 1973; Dewey *et al.* 1986). Thrust stacking enables sediments and water to reach lower crustal depths and is likely to generate sub-horizontal shortening structures and fabrics in the lower crust. Crustal thickening, by either mechanism, accompanies lithospheric thickening and either progressive heating (Fig. 2, path a) as the lithospheric root is converted into peridotite at its solidus or rapid heating if the lithospheric root is delaminated (that is detached as a slab from the overlying crust). Delamination to the base of a thickened crust (Bird 1978) is a mechanism by which fertile mantle at 1330°C would cause rapid heating, the injection of mafic igneous rocks into the lower crust and very rapid uplift and denudation (Fig. 2, path c). Most orogens show a regional late kinematic phase of rapid

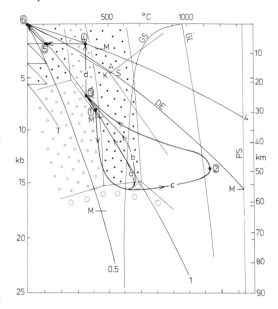

FIG. 2. Pressure/depth–temperature plot showing some possible P/T paths for lower crustal rocks en route to a final exposure at the surface of the earth. A—andalusite facies; GL—granite liquidus; GS—granite solidus; K—kyanite facies; M—Moho; PS—peridotite solidus; S—sillimantite facies; solid squares–pumpelleyite facies; triangles—greenschist facies; small open circles—blueschist facies; solid circles—amphibolite facies; large open circles—eclogite facies; white—granulite facies; 0.5, 1, 4—geothermal gradients corresponding with those surface heatflow values in HFU; DE—geothermal gradient resulting from lithospheric delamination to the base of a crust thickened to 62 km with 5 km of denudation accompanying delamination and uplift; T—geothermal gradient resulting from multiple thrust stacking A material point starts at (1) near the base of a normal thickness crust, then follows a crustal thickening path a, b, or c (discussed in text), finally returning to a lower crustal site (3) in a crust returned to normal thickness by denudation. From (3) the point rises (d) to (4), during crustal stretching by a β value of 4, followed by cooling to (5) then reaches the surface (6) by uplift and denudation during thrusting.

heating, uplift and denudation and mafic injection, which can only be accounted for by some form of delamination. Lithospheric thinning induced by thermal equilibration is much too slow to allow temporal overlap between shortening and heating. Lower crustal heating, particularly by delamination, leads to the upward flux of volatile, lithophile and radiogenic elements with minimum melting granites and thus crustal differentiation into a dry, refractory, granulitic, anorthosite-rich lower crust and a wet upper amphibolitic/greenschist upper crust (Dewey & Burke 1973). The eventual nature of the lower crust in zones of crustal shortening depends not only upon the shortening mechanism but also upon the mechanism(s) by which the thickened crust is returned to normal thickness. This may be by denudation (17) in which case, the lower crust consists of high-grade rocks with a large mafic component. If large volumes of lower crust are delaminated with the mantle portion of the lithosphere, the lower crust will be of greenschist facies with granites and preserved upper crustal sediments (19). Between the end members of denudation and delamination, transitional combinations are possible (18). A third possibility is lithospheric extension (20) as in, for example, the Basin and Range, the Aegean, the Pannonian Basin and the Tyrrhenian. Stretching, like denudation, preserves higher-grade rocks but they remain in the lower crust with superposed stretching fabrics.

Lastly, basaltic underplating may affect not only stretched and delaminated, shortened, crusts but may be a pervasive continental phenomenon beneath plateau basalt provinces such as the Karroo and Drakensburg (Cox 1980) or beneath plateaux without surface volcanism (McKenzie 1984). The injection of large volumes of mafic rock into the lower crust of such provinces provides a heat source for lower crustal melting and the rise of minimum melting liquids. Random mafic injections into the lower crust is an effective way of generating plateaux with upper crustal granites, such as the Jos Plateau in Nigeria. The mechanism encounters some difficulties as a general one for plateau uplift in that some plateaux, such as the Colorado, do not have the expected higher heatflow and do not appear to be characterized by upper crustal silicic magmatism.

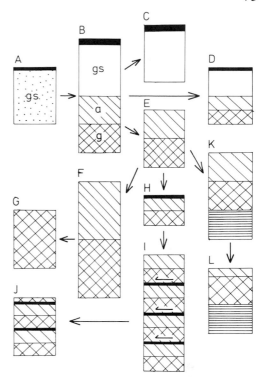

FIG. 3. Methods by which high-grade metamorphic rocks reach the surface of the earth. A— undifferentiated greenschist–facies protolith; B—crust thickened by vertical plane strain stretching; C, D, E—crust returned to normal thickness by, respectively, delamination, stretching and denudation; F—denuded crust thickened again by vertical plane strain; G—crust returned to normal thickness by denudation; H—crust thinned by vertical plane stretching; I—thinned crust stacked by thrusting; J—thrust-thickened crust returned to normal thickness by denudation; K—crust underplated and uplifted by mafic underplating; underplated crust returned to normal thickness by denudation. Black—sediment; gs—greenschist facies; a—amphibolite facies; g—granulite facies; horizontal lines—underplated mafic igneous rocks.

Discussion

The crustal generation and modification modes outlined above may be superposed to give complicated tectonic pathways and sequence linkages, a simple example of which is shown in Fig. 2. This, in turn, means that the continental crust must be laterally and vertically heterogeneous. A further degree of complex lateral inhomogeneity is induced by terrain transfer and assembly (Coney *et al.* 1980). Terrain accretion at the margins of large long-lived oceans will generate an exceedingly complicated and rapid lateral variation in crustal types bounded by steep transform and more gently-dipping thrust boundaries. However, a general temporal trend will exist from simpler undifferentiated crusts with wet low-grade lower crusts to highly differentiated crusts with drier higher-grade lower crusts. Differentiation by upward flow of volatiles, radiogenic nuclides and silicic liquids is accomplished by crustal warming, which occurs where mafic magmas intrude the lower crust and in regions of lithospheric shortening, especially where enhanced by delamination. The high electrical conductivity, perhaps suggesting fluid-filled cracks, of some lower crusts with a complicated upper crustal history is somewhat enigmatic in that a system of cracks interconnected to the surface for a long time would lead to rapid fluid loss by upward flow. A possible explanation could be that mid-crustal ductile strain forms an effective seal zone between upper and lower crustal brittle zones (Dewey *et al.* 1986) so that fluids could be trapped at pressures approaching lithostatic for long periods of time. Fluid recharging of the lower crust can occur only in convergent tectonic regimes by subduction or the thrust-stacking of thinned continental crust. The latter has affected much of the continental crust and is, therefore, an important mechanism for recycling large amounts of water through the crust. Furthermore, thrust stacking is an effective way of enabling sediment to reach lower crust depths.

A second general trend is towards a more mafic lower crust which in turn increases its strength, and hence enhances brittle deformation and the consequent fluid-filled cracks. Such crusts, generated by the accumulation of sediments and/or volcanics on an oceanic foundation, have a ready-made mafic lower crust. Underplating, that is the injection of mafic magmas into the lower crust (Fig. 1; 8, 9, 12, 13, 15, 17, 21), gives a lower crust younger than the upper crust and generates silicic minimum melts (Patchett 1980), which rise into the upper crust with volatiles. Two types of anorthosite may be associated with the resulting mafic lower crust; those in layered complexes (Kaila & Bhatia 1981) and those occurring as restites following the extraction of minimum melts (Dewey & Burke 1973; Taylor *et al.* 1984). Most crustal generation and modification mechanisms lead to a layered lower crust and an upper crust with a blob geometry. Lower crustal layering results especially from mafic underplating and ductile fabric development whereas upper crustal amorphous and blob geometries with scattered reflecting surfaces below layered stratigraphical

sequences result from discontinuous brittle deformation and pluton intrusion (Rickard 1984; Rickard & Ward 1981). All large-scale deformation styles, except transcurrent, yield horizontal to sub-horizontal lower crustal structures and fabrics. A mid-crustal Conrad discontinuity between a lower layered, and an upper unlayered, crust is the logical consequence of multiple tectonic modification in a mature crust.

The exposure, at the surface of the earth, of high-grade rocks that were generated in the lower crust may be accomplished in several ways (Fig. 3). Starting with an undifferentiated crust (A), shortening and thickening leads to differentiation into an upper greenschist, a middle amphibolitic, and a lower granulitic crust (B). Delamination of the lower half of the crust (C) and the consequent intense heating would lead to a second post-kinematic phase of granulite and amphibolite facies metamorphism superposed on the lower part of the delaminated crust. Stretching of the thickened crust (D) leads to a similar vertical metamorphic facies arrangement to (C) but with superposed stretching fabrics and structures. Denudation alone would eventually expose amphibolite facies rocks at the surface (E). The exposure of granulites may be achieved in one of three ways. First, a second phase of shortening and thickening (E) followed by denudation (G) brings granulites to the surface with a crust consisting wholly of granulites. Second, the most efficient and likely method is by stretching (H), thrust stacking (I) and denudation (J) such that high grade and low grade rocks are interleaved throughout the crust. This allows several chances of granulite exposure depending upon the level of denudation. Third, a progressive random and episodic underplating of mafic igneous rocks (Cox 1980; McKenzie 1984) in several stages, provides an elevator mechanism (K, L) for thickening the continental crust (Rodgers *et al.* 1984) and for eventually exposing granulite facies rocks.

References

BERBERIAN, M. 1983. The southern Caspian: a compressional depression floored by a trapped, modified oceanic crust. *Can. J. Earth Sci.*, **20**, 163–183.

BIRD, P. 1978. Initiation of intracontinental subduction in the Himalayas. *J. Geophys. Res.*, **83**, 4975–4987.

CASEY, J.F. & DEWEY, J.F. 1984. Initiation of subduction zones along transform and accreting plate boundaries, triple junction evolution and fore-arc spreading centres—implications for ophiolitic geology and obduction. *In*: GASS, I.G., LIPPARD, S.J. &

SHELTON, A.W. (eds) *Ophiolites and Oceanic lithosphere*, Geol. Soc. Lond. Spec. Publ. p. 269–290.

CONEY, P.J., JONES, D.L. & MONGER, J.W.N. 1980. Cordilleran suspect terranes. *Nature*, **288**, 329–33.

COWARD, M.P., JAN, M.Q., REX, D., TARNEY, J., THIRLWALL, M. & WINDLEY, B.F. 1982. Geotectonic framework of the Himalaya of N. Pakistan. *J. geol. Soc. London*, **139**, 299–308.

COX, K.G. 1980. A model for flood basalt vulcanism. *J. Petrol.*, **21**, 629–650.

DEWEY, J.F. 1982. Plate tectonics and the evolution of the British Isles.*J. geol. Soc. London*, **139**, 371–412.

——& BURKE, C.A. 1973. Tibetan, Variscan and Precambrian basement reactivation: products of continental collision. *J. Geol.*, **81**, 683–692.

—— HEMPTON, M.R., KIDD, W.S.F., SAROGLU, F. & SENGOR, A.M.C. 1986. Shortening of continental lithosphere: the neotectonics of Eastern Anatolia— a young collision zone. *In*: COWARD, M.P. & RIES, A.C. (eds) *Collision Tectonics*. Geol. Soc. London Spec. Publ. No 19, 3–36.

—— & WINDLEY, B.F. 1981. Growth and differentiation of the continental crust. *Phil. Trans. Roy. Soc. London*, **301**, 189–206.

DRUMMOND, B.J. 1983. Detailed seismic velocity/depth models of the upper lithosphere of the Pilbara Craton, N.W. Australia. *Bur. Miner. Resource, J. Aust. Geol. Geophys.*, **8**, 35–51.

FOUNTAIN, D.M., HURICH, C.A. & SMITHSON, S.B. 1984. Seismic reflectivity of mylonite zones in the crust. *Geology*, **12**, 195–198.

GIBBS, A.K., PAYNE, B., SETZER, T., BROWN, L.D., OLIVER, J.E. & KAUFMAN, S. 1984. Seismic-reflection study of the Precambrain crust of central Minnesota. *Bull. Geol. Soc. Am.*, **95**, 280–294.

GRIEVE, R.A.F. 1980. Impact bombardment and its role in proto-continental growth on the early Earth. *Precambrian Research*, **10**, 217–248.

HALE, L.D. & THOMPSON, G.A. 1982. The seismic reflection character of the continental Mohorovicie Discontinuity. *J. Geophys. Res.*, **87**, 4625–4635.

HAMILTON, W. 1981. Crustal evolution by arc magmatism. *Phil. Trans. Roy. Soc. London*, **301**, 279–291.

HEIER, K.S. & ADAMS, J.A.S. 1965. Concentration of radioactive elements in deep crustal material. *Geoch. Cosmoch. Acta*, **29**, 53–61.

HUDSON, T. & PLAFKER, G. 1982. Palaeogene metamorphism of an accretionary flysch terrane, eastern Gulf of Alaska. *Bull. Geol. Soc. Am.*, **93**, 1280–1290.

JACKSON, I. & ARCULUS, R.J. 1984. Laboratory wave velocity measurements on lower crustal xenoliths from Calcutteroo, South Australia. *Tectonophysics*, **101**, 185–197.

JONES, T.D. & NUR, A. 1984. The nature of seismic reflections from deep crustal fault zones. *J. Geophys. Res.*, **89**, 3153–3171.

KAILA, K.L. & BHATIA. 1981. Gravity study along the Kawali-Udipi deep seismic sounding profile in the Indian Peninsular Shield: some inferences about the origin of anorthosites and the eastern Ghats orogeny. *Tectonophysics*, **79**, 129–143.

—— MURTY, P.R.K., RAO, V.K. & KHARCTCHKO, G.E. 1981. Crustal structure from deep seismic soundings along the Konya. II (Kelsi-Loni) profile in the Deccan Trap area, India. *Tectonophysics*, **73**, 365–384.

KAY, R.W. & KAY, S.M. 1981. The nature of the lower continental crust: inferences from geophysics, surface geology and crustal xenoliths. *Rev. Geophys. Space Phys.*, **19**, 271–297.

KORNPROBST, J. 1976. Signification structurale des peridotites dans l'orogene betico-rifain: arguments tires de l'etude des detritus observes dans les sediments palaeozorques. *Bull. Soc. geol. France*, **18**, 607–618.

LYNN, H.B., HALE, L.D. & THOMPSON, G.A. 1981. Seismic reflections from the basal contacts of batholiths. *J. Geophys. Res.*, **86**, 10633–10638.

MATTHEWS, D.H. & HIRN, A. 1984. Crustal thickening in Himalayas and Caledonides. *Nature*, **308**, 497–498.

McKENZIE, D.P. 1984. A possible mechanism for epeirogenic uplift. *Nature*, **307**, 616–618.

MEISSNER, R. & STREHLAU, J. 1982. Limits of stresses in continental crusts and their relation to the depth frequency distribution of shallow earthquakes. *Tectonics*, **1**, 73–89.

NICOLAYSEN, L.O., HART, R.J. & GALE, N.H. 1981. The Vredefoert radioelement profile extended to supracrustal strata and Varletonville, with implications for continental heatflow. *J. Geophys. Res.*, **86**, 10653–10661.

PATCHETT, P.J. 1980. Thermal effect of basalt on continental crust and crustal contamination of magmas. *Nature*, **283**, 559–561.

REAGAN, M.K. & MEIJER, A. 1984. Geology and geochemistry of early arc-volcanic rocks from Guam. *Bull. Geol. Soc. Am.*, **95**, 701–713.

REUBER, I., MICHARD, A., CHALOUAN, A., JUTEAU, T. & JERMOUMI, B. 1982. Structure and emplacement of the Alpine-type peridotites from Beni Bousera, Rif, Morocco: a polyphase tectonic interpretation. *Tectonophysics*, **82**, 231–251.

RICKARD, M.J. 1984. Pluton spacing and the thickness of crustal layers in Baja California. *Tectonophysics*, **101**, 167–172.

—— & WARD, P. 1981. Palaeozoic crustal thickness in the southern part of the Lachlan orogen deduced from volcano and pluton-spacing geometry. *J. Geol. Soc. Aust.*, **28**, 19–32.

RODGERS, J.J.W., DABBAGH, M.E., OLSZEWSKI, W.J., GAUDETTE, H.E., GREENBERG, J.K. & BROWN, B.A. 1984. Early poststabilization sedimentation and later growth of shields. *Geology*, **12**, 607–609.

SAUNDERS, A.D., FORNARI, D.J. & MORRISON, M.A. 1982. The composition and emplacement of basaltic magmas produced during the development of continental-margin basins: the Gulf of California, Mexico. *J. geol. Soc. London*, **139**, 335–346.

SHANKLAND, T.J. & ANDER, M.E. 1983. Electrical conductivity, temperatures and fluids in the lower crust. *J. Geophys. Res.*, **88**, 9475–9484.

SIBSON, R.H. 1983. Continental fault structure and the shallow earthquake source. *J. geol. Soc. London*, **140**, 141–767.

SILLS, J.D. & TARNEY, J. 1984. Petrogenesis and tectonic significance of amphibolites interlayered with metasedimentary gneisses in the Ivrea Zone, Southern Alps, Northwest Italy. *Tectonophysics*, **107**, 187–206.

SMITHSON, S.B. 1978. Modelling continental crust: structural and chemical constraints. *Geophys. Res. Lett*, **5**, 749–753.

TAYLOR, S.R., CAMPBELL, I.H., McCULLOCH, M.T. & McCLENNAN, S.M. 1984. A lower crustal origin for massif anorthosites. *Nature*, **311**, 372–374.

—— & McCLENNAN, S.M. 1981. The composition and evolution of the continental crust; rare element evidence from sedimentary rocks. *Phil. Trans. Roy. Soc. London*, **301**, 381–399.

VERMA, R.K., SARMA, A.U.S. & MUKHOPADHYAY, M. 1984. Gravity field over Singhbhum, its relationship to geology and tectonic history. *Tectonophysics*, **106**, 87–107.

WEAVER, B.L. & TARNEY, J. 1984. Empirical approach to estimating the composition of the continental crust. *Nature*, **310**, 575–578.

YARMAKOV, V.A. 1981. The role of volcanism in the transformation of the earth's crust at continental margins with reference to Kamchatka. *Tectonophysics*, **77**, 95–132.

J. F. DEWEY, Dept. of Earth Sciences, Parks Road, Oxford OX1 3PR, UK.

Continental lithosphere strength: the critical role of lower crustal deformation

N.J. Kusznir & R.G. Park

SUMMARY: A thermo-rheological model of lithosphere deformation, incorporating the elastic, ductile and brittle behaviour of lithosphere material, has been used to examine intraplate continental lithosphere strength, brittle–ductile transition depth and flexural rigidity. These parameters are critically dependent on crust and mantle rheology and consequently on geothermal gradient, crustal thickness and lower crustal composition.

For lithosphere subjected to a lateral tectonic force, creep in the lower crust and mantle leads to stress release, and the subsequent stress redistribution generates stresses in the upper lithosphere sufficient to cause brittle fracture. The extent of creep in the lower crust and mantle and the degree of upper lithosphere stress amplification (which together determine bulk lithosphere strength) increase with geothermal gradient. For significant lithosphere extension, under maximum likely levels of available tectonic stress, a lithosphere surface heat flow of 60 $mW\ m^{-2}$ or greater is required, while for compressive lithosphere deformation, heat flow must exceed 75 $mW\ m^{-2}$. Similarly flexural rigidity increases with increase in the thermal age of the lithosphere at the time of loading.

The depth of the brittle–ductile transition decreases with increase in geothermal gradient. For a limited range of gradients (expressed by heat flow $q = 50$–$55\ mW\ m^{-2}$) multiple brittle–ductile transitions may exist in the middle and lower crust and upper mantle, with important tectonic implications (e.g. for intra-crustal detachments and crust–mantle decoupling).

Lithosphere strength in extension and compression, and flexural rigidity, are both controlled by the quartzo–feldspathic rheology of the crust for thermally young lithosphere and by the olivine rheology of the mantle for older lithosphere. Lithosphere strength is therefore critically influenced by the thickness of the crust (decreasing with increase in crustal thickness) and by the composition of the lower crust, particularly for lithosphere with intermediate heat flows.

Introduction

The strength of a piece of continental lithosphere plays a decisive role in controlling its tectonic, and consequently its geological evolution. Lithosphere strength controls both the initiation and character of extensional or compressional deformation (Kusznir & Park 1984). It also controls the subsequent mechanical response of the lithosphere under flexure, when loaded by sediments or orogenic thrust sheets. Both the absolute strength of a section of lithosphere and its strength relative to adjacent lithosphere are important.

The bulk strength of the lithosphere results from the combined effects of the strength of each individual component. The variation in strength with depth controls the vertical distribution of stress in both the ductile and brittle deformation fields and, as a consequence, the depth of the brittle/ductile transition. This vertical distribution of strength defines the critical force which must be applied to the lithosphere in order to produce geologically significant strain rates in both tension and compression. The vertical distribution of ductile and brittle strength is also important in controlling the flexural strength of the lithosphere.

While brittle strength is controlled primarily by lithostatic pressure and increases with depth, ductile strength is strongly controlled by temperature and consequently decreases with depth because of the geothermal gradient. For tectonically stable lithosphere, the upper region of brittle deformation and the middle and lower regions of ductile deformation will be separated by a region of competent elastic lithosphere (Fig. 1). However for lithosphere undergoing tectonic deformation, the elastic core is destroyed and the regions of brittle/failure and ductile deformation intersect. The depth of the brittle–ductile transition so defined, and the lateral and flexural strength of the lithosphere, are critically controlled by the rheology of the lower crust and upper mantle, which is in turn controlled by *temperature, lower crustal composition* and *crustal thickness*. These parameters, through their influence on *lower crustal rheology* play a critical role in controlling the extensional, compressional and flexural strength of the lithosphere.

In Fig. 2A, strain rates of some minerals and rocks relevant to ductile lithosphere deformation

From: DAWSON, J.B., CARSWELL, D.A., HALL, J. & WEDEPOHL, K.H. (eds) 1986, *The Nature of the Lower Continental Crust*, Geological Society Special Publication No. 24, pp. 79–93.

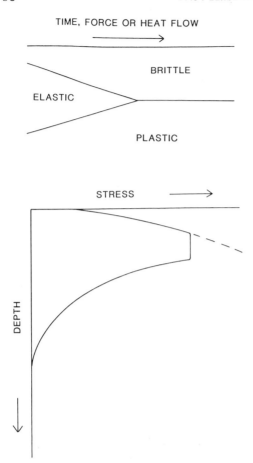

FIG. 1. A) Diagrammatic representation of the regions of brittle, elastic and plastic behaviour in viscoelastic lithosphere subjected to an applied lateral force. With increase in time, force or heat flow the regions of brittle and plastic deformation expand and eventually coalesce at the brittle-ductile transition when *whole lithosphere failure* occurs.

B) Idealized stress-depth curve at a given time after the application of a lateral force. The top part of the curve is determined by the brittle failure curve and the bottom part by the downward decrease in strength due to decreasing viscosity. The straight part of the curve is the elastic core of the lithosphere.

are shown as a function of temperature. The curves, based on experimental data, assume a dislocation creep mechanism and correspond to a stress difference $\sigma_1-\sigma_3$ of 50 MPa (0.5 kb). Temperature is seen to have a marked effect on strain rates—an increase from 500°C to 700°C for wet olivine, for example, increases the strain rate from $c.$ 10^{-20} s^{-1} to 10^{-14} s^{-1}. At the same temperature, quartz is significantly weaker than olivine—at 400°C, the strain rates of wet quartz and olivine are $c.$ 10^{-13} s^{-1} and $c.$ 10^{-25} s^{-1} respectively. The strain-rate/temperature curves for pyroxene (with the exception of diopside) fall in a similar range to those of olivine and dunite.

It should be noted that the strain-rates shown in Fig. 2A have been extrapolated to many orders of magnitude lower than those at which the laboratory experiments were conducted. The relative strength of minerals and rocks undergoing ductile deformation, as shown in Fig. 2A, is however consistent with the relative strengths indicated by the examination of deformed rocks of exhumed middle and lower crust.

In the upper crust, quartz (probably wet) controls the ductile deformation while in the mantle the controlling mineral is olivine. The rheology of the lower crust is most likely to be controlled by the ductile deformation of plagioclase since, in deformed samples of lower crustal material, plagioclase appears to have deformed more readily than pyroxene (or olivine). Quartz, while weaker than plagioclase, is unlikely to be present in sufficient quantity to control the deformation. In the absence of published data for plagioclase, a tentative temperature–strain rate relationship is shown on Fig. 2A on the basis of its apparent relative strength. A lower crustal composition of 10% dry quartz is used for the calculations of this paper which probably slightly underestimates the strength.

Typical strain rates for significant tectonic deformation should be in the range 10^{-16} s^{-1} to 10^{-15} s^{-1} (10^{-15} s^{-1} produces 30% strain in 10 Ma). For quartz subjected to a stress of 50 MPa, a temperature of $c.$ 300°C would be needed to achieve this strain rate, while for wet olivine/ dunite a temperature of $c.$ 600°C is required. In Fig. 2B temperature is plotted as a function of depth for various geothermal gradients (characterized by surface heat flow) including typical curves for continental shield and basin-and-range regions, and for average continental lithosphere.

It is clear that the critical temperatures required to generate geologically significant strain rates are reached at varying depths, depending on the geotherm. For an average continental geotherm (corresponding to a surface heat flow of 60 mW m^{-2}) these critical temperatures for quartz and olivine are reached at depths of $c.$ 17 km and 40 km respectively—i.e., at mid-crustal and upper mantle levels. These weak zones will control the initiation of the deformation and the subsequent kinematic behaviour of the lithosphere resulting from an applied force.

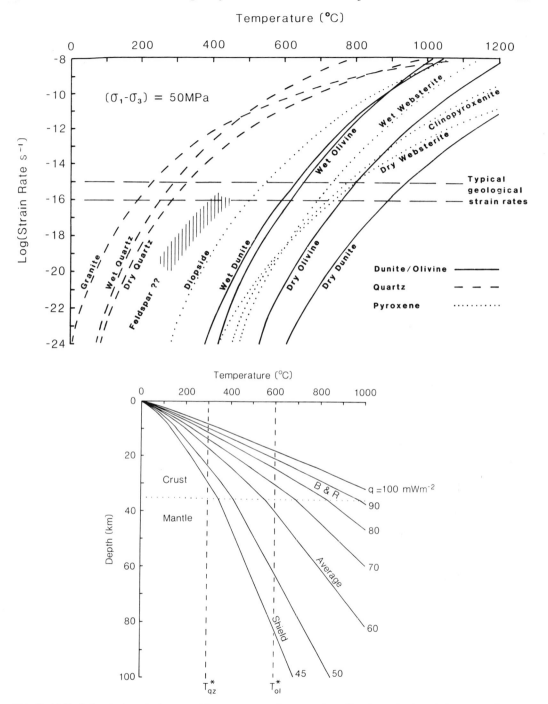

FIG. 2. A) Variation in log strain rate with temperature for a number of rocks and minerals important in lower crustal/upper mantle deformation. The curves are derived from experimental data as follows: granite (Carter *et al.* 1981), quartz (Koch *et al.* 1980), diopside (Avé Lallement 1978), dunite (Post 1977), dry olivine (Goetze 1978), wet olivine (Bodine *et al.* 1981), clinopyroxenite (Kirby & Kronenberg 1984), websterite (Avé Lallement 1978).

B) Temperature–depth plot for a range of surface heat flows. Typical Basin-and-Range and continental shield values are shown and also the average continental heat flow. T^*_{qz} and T^*_{ol} are the critical temperatures required to produce significant strain rates in quartz and olivine respectively. These values are reached in normal lithosphere with $q=60$–70 mW m^{-2} at depths of 13–17 km and 31–40 km respectively.

The mathematical model of intraplate lithosphere deformation

The mathematical model examines the elastic, ductile and brittle response of intraplate lithosphere to lateral applied stress, and takes into account the redistribution of stress within the lithosphere caused by brittle failure and by ductile deformation. The primary sources of applied tectonic stress, termed renewable forces by Bott & Kusznir (1984) arise from plate boundary forces and isostatically compensated loads and persist even when intraplate deformation occurs. A fundamental feature of the model, consequently, is the conservation of the horizontal tectonic force applied to the lithosphere, and this gives the equation:

$$\int_0^L \sigma_x \, dz = \text{constant}$$

where L is lithosphere thickness and σ_x is horizontal stress in the direction of the applied force. The perpendicular horizontal axis is labelled y and the vertical axis z. The assumption that all infinitesimal layers of the lithosphere are welded together gives:

$$d\dot{\varepsilon}_x/dz = 0$$

This assumption will only apply during the initiation of intraplate deformation. Once significant horizontal deformation occurs, involving low angle tectonic fabrics, this condition will be violated.

Each infinitesimal component of the lithosphere is assumed to behave as a Maxwell viscoelastic material in which strain may be relieved by brittle deformation. Strain and stress are consequently linked by the equation:

$$\varepsilon_x = \frac{1}{E}(\sigma_x - \sigma_x^0) - \frac{v}{E}(\sigma_y - \sigma_y^0)$$

$$- \frac{v}{E}(\sigma_z - \sigma_z^0) + \varepsilon_x^v$$

and similarly for ε_y and ε_z, where ε is total strain, σ is stress, ε^v is ductile strain and σ^0 the initial stress used to incorporate brittle failure. E and v are Young's modulus and Poisson's ratio respectively.

The assumption of plane strain in the y axis and the condition that the vertical stress, σ_z, arising from the applied force is zero, gives the additional equations $\varepsilon_y = 0$ and $\sigma_z = 0$. The manipulation and integration of the above equations (described in detail in Kusznir (1982)) gives the following equations for the behaviour of σ_x and σ_y with depth and time.

$$\sigma_x = \int_0^t \left(\frac{1}{L} \int_0^L k\dot{\varepsilon}_v dz - k\dot{\varepsilon}_v \right) dt' - \frac{1}{L} \int_0^L \sigma_x^0 \cdot dz + \sigma_x^0$$

$$\sigma_y = \int_0^t \left\{ v\sigma_x - E\frac{(2\sigma_y - \sigma_x)}{6\eta} \right\} dt' + \sigma_y^0 - v\sigma_x^0$$

where $k = E/(1-v^2)$,

$$\varepsilon_v = (\sigma_x(2-v) - \sigma_y(1-2v))/6\eta$$

and η is apparent viscosity.

Brittle failure which is predicted by Griffith theory (Griffith 1924) as modified by McClintock & Walsh (1962) is incorporated into the model using the initial stress terms. The values of the coefficient of friction, μ, the tensile strength, T_0 and σ_{gc} the critical stress to close cracks used to predict failure are 0.5, 20 MPa and 100 MPa respectively. The choice of these parameters and the use of initial stress to model brittle failure is discussed in Kusznir & Park (1984).

Ductile deformation in the upper crust is assumed to be controlled by dislocation creep in wet quartz while that in the mantle is assumed to be controlled by a combination of dislocation creep and Dorn law creep in olivine. A 50% concentration of quartz is assumed for the upper crust. As explained earlier the rheology of the lower crust is assumed to correspond to that of dry quartz with a 10% concentration, which probably slightly underestimates the strength. The flow laws used, which are strongly stress and temperature dependent, are as follows:

Wet Quartz
$$\dot{\varepsilon} = 4.36 \, [\exp(-19332/T)](\sigma_1 - \sigma_3)^{2.44} \, s^{-1}$$

Dry Quartz
$$\dot{\varepsilon} = 0.126 \, [\exp(-18245/T)](\sigma_1 - \sigma_3)^{2.86} \, s^{-1}$$

Olivine
$$\dot{\varepsilon} = 7 \times 10^{10} \, [\exp(-53030.3/T)](\sigma_1 - \sigma_3)^3 \, s^{-1}$$
$$\text{for } (\sigma_1 - \sigma_3) < 2 \, \text{kbar}$$

$$\dot{\varepsilon} = 5.7 \times 10^{11}$$

$$\exp\{-55556/T[1 - (\sigma_1 - \sigma_3)^2/85]\}s^{-1}$$

$$\text{for } (\sigma_1 - \sigma_3) > 2 \, \text{kbar}$$

where $(\sigma_1 - \sigma_3)$ is measured in kbar. Wet and dry quartz flow laws are taken from Koch et al. (1980) while those of olivine are taken from Bodine et al. (1981) which are based on Goetze (1978) and Post (1977). For the crustal rheologies, the creep rates are scaled to reflect the proportion of quartz, assumed to be the dominant ductile mineral present. Decreasing the proportion of quartz has the effect of increasing the ductile strength.

The strong temperature dependence of lithosphere rheology means that the geothermal gra-

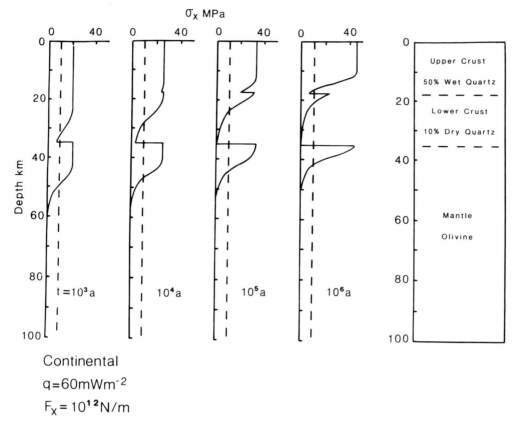

FIG. 3. Stress–depth plots at various times after application of a force of -10^{12} N m^{-1} to continental lithosphere with a surface heat flow of 60 mW m^{-2}. Note discontinuities corresponding to rheology changes at upper/lower crust and lower crust/mantle boundaries.

dient is very important in determining the variation of ductile strength with depth. Geotherms have been calculated using a steady-state geotherm model in which the geothermal gradient is characterized by the surface heat-flow. The values of thermal conductivity and radiogenic heat productivity, which vary with depth through the crust and mantle, are as given by Pollack & Chapman (1977).

Variation of strength with depth and the resulting stress distribution

Fig. 3 shows the calculated stress–depth relationship at various times after application of a compressional force of 10^{12} N m^{-1} to continental lithosphere with a surface heat flow of 60 mW m^{-2}. As time progresses, ductile creep in the lower lithosphere results in the dissipation of stress in the lower lithosphere and its transfer to the upper

lithosphere where stress levels consequently increase. This process of stress amplification allows large levels of stress to be reached in the cooler non-ductile upper lithosphere sufficient to generate brittle failure. By 1 Ma after the application of the initial force, stress levels in the upper crust and mantle have increased by a factor of four. The two stress–depth discontinuities correspond to the changes in rheology associated with composition changes at the Moho and between the upper and lower crust.

In Fig. 4, stress is shown as a function of depth for lithosphere with heat flow 70 mW m^{-2} at a time of 1 Ma after the application of a tectonic force and shows the effect of varying the sign and magnitude of the force (the effect of variation in heat flow is discussed in the next section). With a tensional force of 10^{12} N m^{-1} extensive brittle deformation has occurred in the upper crust while with a compressive force of the same magnitude none has yet taken place (lithosphere material is

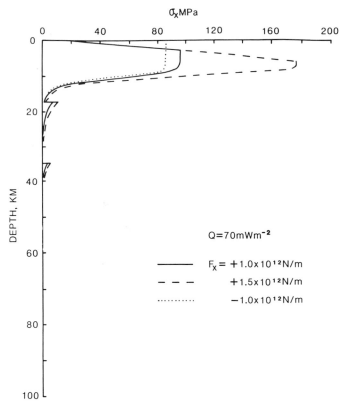

FIG. 4. Stress–depth plot at 1 Ma after application of forces of $\pm 1.0 \times 10^{12}$ N m^{-1} and $+1.5 \times 10^{12}$ N m^{-1} to continental lithosphere with a surface heat flow of 70 mW m^{-2}. Note that the stress release by brittle failure in the upper crust is associated with the tensional forces but not with the compressional. With the larger tensional force of $+1.5 \times 10^{12}$ N m^{-1} the elastic core has almost disappeared. With a slightly larger force (or longer time) whole lithosphere failure would occur.

stronger in compression than in tension). The brittle failure in the upper crust has also generated greater stress amplification in the remaining elastic core of the lithosphere compared with the compressive force. Increasing the magnitude of the tensile force to 1.5×10^{12} N m^{-1} has the effect of increasing the level of stress in the middle crust, above the elastic–ductile transition, and consequently of deepening the zone of brittle fracture. A small, dominantly elastic, core still remains however and the horizontal strains within the model ($< 1\%$) are not yet geologically significant.

If the applied force were increased further or if the stress amplification were allowed to increase with time (or if the heat flow were increased—see below), the expansion of the regions of brittle failure and ductile deformation would lead to the destruction of the elastic core. This development, called *Whole Lithosphere Failure* (WLF) by Kusznir (1982) and Kusznir & Park (1984) after which all parts of the lithosphere will deform either in a

ductile or brittle fashion, is an essential precondition for the development of large horizontal lithosphere strains leading to significant tectonic deformation.

After WLF, in both extension and compression, large tectonic strains will occur which will involve significant displacements along inclined dip-slip structures or flat detachment horizons. Such deformation violates the assumption of the mathematical model that $d\dot{\varepsilon}_x/dz = 0$. The mathematical model is consequently only applicable to the initiation of intraplate deformation up to WLF.

The effect of the geothermal gradient on lithosphere strength

While brittle deformation is obviously important, it is ductile deformation which exerts the greatest control on lithosphere strength and this ductile

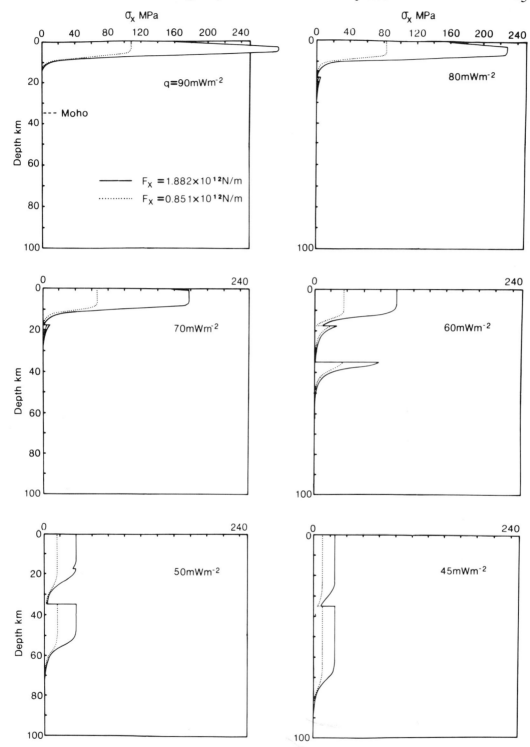

FIG. 5. The effect of the thermal gradient on lithosphere strength. Stress-depth plots at 1 Ma after application of compressional forces of -0.851×10^{12} N m^{-1} and -1.882×10^{12} N m^{-1} to lithosphere with various geothermal gradients corresponding to surface heat flows ranging from 45 to 90 mW m^{-2}. Note that for high heat flow values, the stress is entirely concentrated in the upper crust but that the strength discontinuities at the middle and base of the crust are most pronounced in the intermediate geotherm range (50–60 mW m^{-2}).

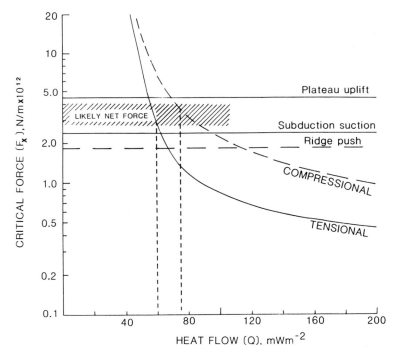

FIG. 6. Critical force (compressional and tensional) required to produce whole lithosphere failure in 1 Ma plotted against heat flow. Also shown are estimated values of tectonic forces arising from ridge push (compressional) and subduction suction and plateau uplift (tensional). The likely range in available net force $(2.8–4.0 \times 10^{12}$ N m$^{-1})$ requires a minimum heat flow of *c.* 60 mW m^{-2} for tensional failure and *c.* 75 mW m^{-2} for compressional failure.

deformation is critically controlled by temperature (see Fig. 2A) through the temperature-dependent rheology. As a consequence, lithosphere temperature structure would be expected to have a most important influence on the strength of the lithosphere and on the distribution of stress with depth. In Fig. 5, stress–depth relationships are shown at a time 10^6 years after the application of a compressive force to lithosphere with a range of geothermal gradients corresponding to surface heat flows between 45 and 90 mW m^{-2}. The value of compressive force used here is the calculated resultant tectonic force in intraplate continental lithosphere arising from the combination of the compressive ocean ridge-push force and the tensional force associated with an isostatically compensated passive continental margin. Two estimates of this force are given, 1.882 and 0.85×10^{12} N m^{-1} (see Bott & Kusznir 1986).

For low heat flows (e.g. $q=45$ mW m^{-2} which corresponds to continental shield lithosphere) the stress is carried to greater depths in the lithosphere and little stress amplification occurs. With increase in heat flow, the degree of stress release

by ductile deformation increases, the stress is concentrated increasingly in the upper lithosphere and large levels of stress amplification occur. For high heat flows (e.g. $q=90$ mW m^{-2} corresponding to Basin and Range lithosphere) stress amplification by a factor of fifteen has occurred in the upper lithosphere resulting in the brittle deformation penetrating to greater depth. A competent elastic core still exists in the upper crust however—WLF has not yet occurred.

For lithosphere models with heat flows in the range 50–60 mW m^{-2} a pronounced low-stress and low-viscosity zone exists in the lower crust. This is due to the weaker quartz rheology of the lower crust directly overlying the stronger olivine rheology of the mantle. With increase in heat flow however, this low-stress, low-viscosity sandwich in the lower crust disappears as the olivine mantle becomes weaker. Such low-stress, low-strength regions can also occur within the crust wherever vertical variations in composition and therefore rheology occur. This is illustrated in Fig. 5 for the $q=60–80$ mW m^{-2} models. These low strength regions represent probable sites for the development of detachment horizons. For a vertically

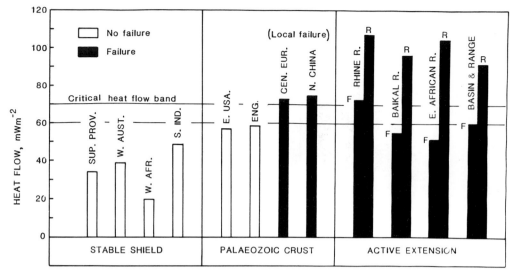

FIG. 7. Comparison of heat flow values and tectonic status for a number of representative regions grouped into: stable shield (Precambrian cratons), Palaeozoic orogenic crust, and areas of active extensional rifting. Heat flow data, in mW m^{-2}, are: Superior Province[1] 34±8, West Australia[1] 39±8, West Africa (Niger)[1] 20±8, South India[1] 49±8, E. USA[1] 57±17, England and Wales[1] 59±23, Central Europe (Bohemian massif)[1] 73±18, Northern China[2] 75±15, Rhine graben[3] 107±35 (rift) and 73±20 (flanks—Rhenish massif), Baikal rift[3] 97±22 (SE flank 55±10), East African rift[3] 105±51 (flanks 52±17), Basin and Range Province[1] 92±33 (E. flank, Colorado plateau[4] 60). Data from Vitorello and Pollack (1980)[1], Pollack and Chapman (1977)[2] and Morgan (1982[3], 1984[4]). For regions undergoing active extension both Rift (R) and Flank (F) heat flows are given.

heterogeneous layered crust, several potential low strength horizons may occur. Their existence and relative strength and importance will depend on the geothermal gradient; high gradients increasing the importance of the potential detachment horizons in the upper crust.

As the geothermal gradient of the lithosphere is increased, the critical force which must be applied to the lithosphere to generate WLF decreases. This is shown in Fig. 6 where the critical force required to generate WLF within 1 Ma is plotted against heat flow for both compressive and tensional deformation. This critical force corresponds to the bulk strength of the lithosphere. As would be expected, the lithosphere is substantially stronger in compression than tension.

Also shown in Fig. 6 are estimated values of tectonic forces arising from plate boundaries and isostatically compensated loads (see Bott & Kusznir 1984). The tensile forces are those arising from subduction suction and plateau uplift, while the compressive force is that generated by ridge push. The likely range of the maximum net resultant tectonic force is also shown. Significant lithosphere extension requires a heat flow in excess of

c. 60 mW m^{-2} while significant compressional deformation requires a heat flow in excess of *c.* 75 mW m^{-2}.

These model predictions may be compared with heat flow data and current tectonic status of a number of representative regions (Fig. 7) divided into (a) stable continental shields (Precambrian cratons) (b) regions of Palaeozoic orogenic crust and (c) thermally active regions undergoing current or recent rifting and vulcanicity. None of group (a) with heat flow values in the range 34–49 mW m^{-2} shows signs of significant tectonic deformation. In group (b) with heat flow values in the range 57–75 mW m^{-2} local extensional failure occurs in Central Europe (Rhine–Ruhr rift system) and in North China (Shansi graben system) both with heat flows of over 70 mMw m^{-2}. In group (c) high heat flows (92–107 mW m^{-2}) are associated with the active rifts but lower heat flows occur in the flanking areas of these regions of active extension.

The model can be strictly applied only to the initial stage of rifting, starting from 'normal' lithosphere. Once significant crustal thinning has taken place, both the geometry and the rheology

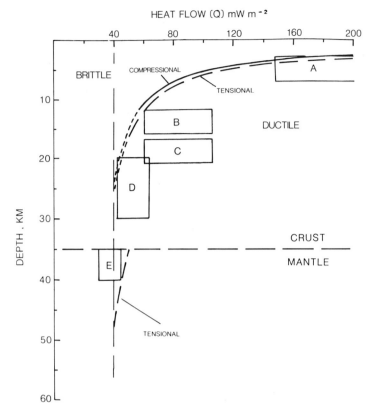

FIG. 8. The effect of the geotherm on the depth of the brittle–ductile transition. Depth of the brittle–ductile transition plotted against surface heat flow. The depth is calculated at whole lithosphere failure after 1 Ma and is slightly greater for tension than for compression. For heat flow values less than *c.* 55 mW m^{-2} a second brittle–ductile transition curve is present in the olivine rheology of the mantle. The force levels required to generate WLF in lithosphere with heat flows < 60 mW m^{-2} (compressional) and < 50 mW m^{-2} (tensional) are unrealistically high and the corresponding sections of the curves are dotted.

The boxes represent observational data from Sibson (1982) showing the maximum depth of earthquake hypocentres in regions of known heat flow which may also be used to estimate the depth of the brittle–ductile transition. A—Geysers and Clearlake Highlands, B—Central California and Coso Range, C—Wasatch Front, D—Eastern USA and Canada, E—Sierra Nevada.

of the lithosphere are changed by the emplacement of warmer asthenospheric material in the lower part of the lithosphere (cf. McKenzie 1978), and by the development of vulcanicity. The flank values may be taken as minimum estimates and the rift values as maximum estimates of the heat flow at the time of initiation of rifting.

Taking the data as a whole, there appears to be a critical heat flow band between 60 and 70 mW m^{-2} above which significant tectonic deformation may take place, which is consistent with the predictions of the model.

The model can also be used to predict the depth of the brittle–ductile transition which depends critically on the geothermal gradient. In Fig. 8 the

predicted depth of the brittle–ductile transition is shown to decrease with increase in surface heat flow. The brittle–ductile transition depths corresponding to WLF by 1 Ma are slightly greater for extensional than for compressional deformation. For lithosphere with heat flow less than *c.* 55 mW m^{-2} two brittle–ductile transitions are apparent; one in the quartz rheology crust and the other in the olivine mantle.

Also shown in Fig. 8 is observational data of the depth of the brittle–ductile transition based on the maximum depth of earthquake hypocentres in regions of known heat flow. The data are taken from Sibson (1982). The observational data show the same general decrease in brittle–

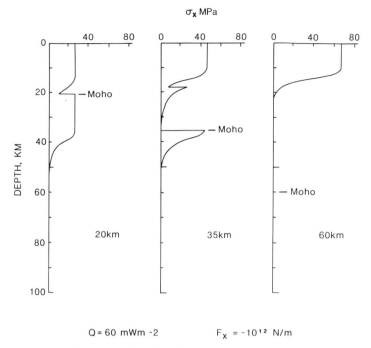

FIG. 9. Horizontal stress as a function of depth for lithosphere models with crustal thickness of 20, 35 and 60 km. The stresses correspond to those at 1 Ma following the application of a compressive force of -10^{12} N m^{-1} to lithosphere with heat flow 60 mW m^{-2}. The effect of increasing crustal thickness is to enhance the contribution of the weaker quartz rheology and thereby to weaken the lithosphere—concentrating stress in its upper non-ductile part.

ductile transition depth with increase in heat flow as the model, but predict a slightly deeper transition depth, suggesting that the model requires a stronger rheology within the lower crust or a weaker brittle failure criterion. The use of a stronger plagioclase rheology in the lower crust, as suggested previously, would increase the brittle–ductile transition depth.

The effect of crustal thickness and lower crustal composition on lithosphere strength

Crustal thickness affects the vertical distribution of stress in the lithosphere (and therefore its strength) because of its influence on the relative contributions of the weaker quartz and stronger olivine rheologies. This is shown in Fig. 9 for lithosphere with heat flow of 60 mW m^{-2} subjected to a compressive force of 10^{12} N m^{-1}. With increase in crustal thickness to 60 km, the weaker quartz rheology extends to greater depth, resulting in an increased transfer of stress to higher levels of the lithosphere, with a $\times 6.5$ stress amplification. For a decreased crustal thickness of 20 km, the effect of the stronger olivine rheology below 20 km results in lower levels of stress amplification ($\times 2.5$) and the stress is carried to greater depths.

In Fig. 10 the critical force required to give WLF in 1 Ma for extensional tectonic deformation is shown as a function of heat flow for various crustal thicknesses. Increasing the crustal thickness causes a decrease in the effective strength of the lithosphere. For continental lithosphere with an average heat flow of $q = 60$ mW m^{-2}, the effect of increasing crustal thickness from 20 to 60 km is to decrease the strength of the lithosphere by a factor of four.

Also shown in Fig. 10 are the 'end-member' curves for lithosphere extensional strength corresponding to wholly olivine and wholly wet quartz rheologies. These are the upper and lower limiting curves of lithosphere strength predicted by the weakest and strongest rheological models. For lithosphere with low heat flow, lithosphere strength is controlled by the olivine rheology of the mantle, while for high heat flow, the strength

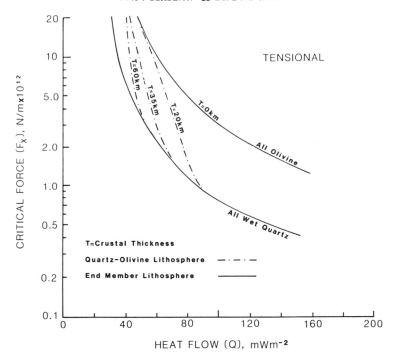

Fig. 10. Critical force required to produce tensional Whole Lithosphere Failure in 1 Ma as a function of heat flow. Increasing the thickness of the crust enhances the contribution of the quartz rheology and thereby weakens the lithosphere.

is controlled by the wet quartz rheology of the upper crust. The position of the 'cross-over' between the two end-member strength curves which takes place with increase in heat flow, is strongly controlled by the thickness of the crust.

The effect of replacing the rather weak quartz rheology of the lower crust by a stronger plagioclase (or even pyroxene) rheology would have a similar effect to decreasing the crustal thickness and would shift the strength curve upwards towards the olivine end-member curve. The effects of variation in lower crustal composition on strength will be pronounced for heat flows up to about 80 mW m^{-2}.

The control of flexural rigidity by lithosphere temperature structure and crustal thickness

When loaded by sediments, volcanics or thrust sheets, the lithosphere responds partly by flexure and its flexural strength may be defined by the flexural rigidity parameter, $D = ET_e^3/12(1-v^2)$ where T_e is the effective elastic lithosphere thickness, E is Young's modulus and v Poisson's ratio.

Flexural rigidity D controls both the amplitude and lateral wavelength of the flexural displacements resulting from an imposed load. Low values of D (low flexural strength) give relatively high-amplitude, short-wavelength flexures while larger values of D give lower-amplitude, longer-wavelength flexures.

Observations of lithosphere flexural rigidity in both oceans and continents show that the effective elastic flexural thickness of the lithosphere is appreciably less than seismic estimates. The flexural strength of the lithosphere has also been shown to be strongly controlled by the thermal state of the lithosphere (Bodine *et al.* 1981; Karner & Watts 1983). Our model of lithosphere strength can be used to demonstrate how the flexural rigidity is controlled by lithosphere rheology, which in turn depends on the geothermal gradient (or thermal age of the lithosphere), on crustal thickness and on lower crustal composition.

In Fig. 11 flexural rigidity for continental lithosphere is shown as a function of time elapsed since loading for lithosphere with a range of temperature structures characterized by the thermal age of the lithosphere (Vitorello & Pollack 1980; Morgan 1984). Model geotherms have been

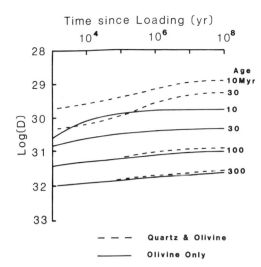

FIG. 11. Flexural rigidity, D (dyne cm) plotted against time since loading for lithosphere of various thermal ages. Flexural rigidity decreases with time since loading and increases with thermal age at loading. The effect of increasing the olivine component in the lithosphere is to give a greater and more rapidly equilibrating flexural rigidity.

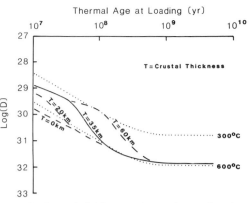

FIG. 12. Flexural rigidity, D (dyne cm) as a function of the thermal age of the lithosphere at the time of loading, for lithosphere models with various crustal thicknesses. Curves correspond to an elapsed time of 100 Ma after loading. Flexural rigidity increases with increase in thermal age at time of loading. The effect of increasing crustal thickness is to weaken the flexural strength of the lithosphere. Also shown are the predictions of the simple thermo-elastic model for the 300 and 600°C isotherms.

calculated using a cooling plate model. With increase in lithosphere thermal age (the time since the last major thermo-tectonic event) the lithosphere progressively cools. The procedure for the calculation of flexural rigidity and the transient thermal model are described in greater detail by Kusznir & Karner (1985).

Flexural rigidity can be seen to decrease with the time elapsed since loading, rapidly at first but then equilibrating to a nearly steady-state value by 1–10 Ma. As would be expected, lithosphere flexural rigidity increases with increase in thermal age; hot, thermally young lithosphere having much lower flexural rigidities than thermally older, cooler lithosphere. The rate of equilibration of flexural rigidity decreases with increase in thermal age.

Curves are compared in Fig. 11 for lithosphere models with a quartz crust and olivine mantle rheology and an all-olivine rheology. The effect of increasing the olivine content of the lithosphere is to increase greatly the flexural strength of the lithosphere. Lithosphere rheology can also be seen to affect the rate of equilibration of flexural rigidity after loading. The all-olivine model shows a more rapid equilibration of flexural rigidity than the model with a quartz rheology crust. In the case of flexure by sediment loading during basin formation, the rate of equilibration of flexural rigidity has an important influence on

basin stratigraphy. Rapidly equilibrating flexural rigidities give an on-lap stratigraphy while very slow equilibration gives off-lap. Through their control on lithosphere rheology therefore, lithosphere temperature, crustal thickness and crustal composition may each exert an important control on equilibration rates and thus on basin stratigraphy.

In Fig. 12 flexural rigidity is plotted as a function of thermal age for lithosphere models with various crustal thicknesses. The curves correspond to a time after loading of 100 Ma. Lithosphere flexural strength is seen to increase with increase in thermal age and coolness of the lithosphere. Crustal thickness also exerts considerable control on the flexural rigidity. For lithosphere with a thermal age of 100 Ma an increase in crustal thickness from 35 to 60 km lowers the flexural rigidity by a factor of ten.

Also shown in Fig. 12 are the flexural rigidities predicted by the simple thermo-elastic model (e.g. Watts *et al.* 1982) in which the elastic thickness of the lithosphere is defined by a single isotherm controlling the elastic–ductile transition. For thermally young, hot lithosphere the flexural rigidities predicted by the thermo-rheological model described in this paper are controlled by the quartz rheology of the crust and by the depth of the 300–400°C isotherm. For thermally older, cooler lithosphere flexural rigidity is controlled by the olivine rheology of the mantle and the 600°C isotherm. Crustal thickness essentially controls the cross-over between the quartz (300°C) and olivine (600°C) curves which takes place as the flexural rigidity increases with increase in thermal age.

Conclusion

Lithosphere strength in compression, tension and flexure has been investigated using a mathematical model which incorporates the elastic, ductile and brittle response of lithosphere material. Lithosphere strength is shown to be critically controlled by the ductile deformation of the lithosphere and consequently by its rheology. Lithosphere rheology itself is primarily controlled by temperature gradient and composition, both of which in turn control the distribution of stress with depth through the lithosphere.

The geothermal gradient determines the depth of the brittle–ductile transition (or transitions). Increasing the geothermal gradient (or heat flow) decreases the depth of the brittle–ductile transition and the thickness of the upper non-ductile lithosphere. For lithosphere subjected to an applied tectonic force, the level of stress within the upper brittle–elastic lithosphere and the extent of the brittle failure in this region are critically controlled by the depth of the brittle–ductile transition and therefore by the geothermal gradient. The strength of the lithosphere consequently increases with increase in heat flow. Lithosphere with a crustal thickness of 35 km, subjected to the maximum likely levels of tectonic force, will only experience significant extensional deformation for heat flows of 60 mW m^{-2} or greater and compressional deformation with heat flows of 75 mW m^{-2} or greater.

The flexural strength of the lithosphere is similarly controlled by the geothermal structure. Flexural rigidity increases with decrease in heat flow (or increase in thermal age) at the time of loading. Through the flexural rigidity, the geothermal gradient therefore controls the amplitude and wave length of flexures resulting from sedimentary or thrust sheet loading.

The quartzo–feldspathic rheology of the crust is significantly weaker than the olivine rheology of the mantle. For high geothermal gradients, the brittle–ductile transition will be in the quartz crust, while for lower gradients it will be in the olivine mantle. As a consequence the strength of the lithosphere in extension, compression and flexure is controlled by the weaker quartz rheology for high heat flows (thermally young lithosphere) and the stronger olivine rheology for cooler (thermally older) lithosphere. For a limited range of heat flow ($q = 50$–60 mW m^{-2} for a crustal thickness of 35 km) two or more brittle–ductile or elastic–ductile transitions may exist giving pronounced low-viscosity, low-stress zones in the lower crust. These low-stress, low-strength regions are the obvious sites for the development of mid and lower crustal detachment horizons. The pronounced low-strength region above the Moho may play an important tectonic role in crust–mantle decoupling.

The strength of the lithosphere is strongly influenced by the relative contribution of the weaker quartz and stronger olivine rheology to the lithosphere as a whole and will therefore decrease with increase in crustal thickness. Thickening the crust from 35–60 km can decrease flexural rigidity by a factor of ten.

The composition of the lower crust in particular has a very important influence on lithosphere strength. The effect of substituting a stronger rheology (e.g. plagioclase or pyroxene instead of dry quartz) has a similar effect to reducing crustal thickness, i.e. lithosphere strength increases. Lateral variations in lower crustal composition may thus give rise to significant lateral variations in the strength of the lithosphere. These effects will only be important however, for intermediate geothermal gradients (heat flows in the range $q = 50$–80 mW m^{-2}) where the brittle–ductile transition lies in the lower crust.

The mathematical model described and used in this paper uses estimates of the ductile and brittle strength of lithosphere material obtained by extrapolating laboratory data considerably beyond their laboratory range. This is particularly true for the strain-rate data. In addition the compositions and rheologies adopted to represent the ductile behaviour of the crust are rather over simplistic. As a consequence caution should be exercised against regarding the results of the model as giving precise stress-depth and lithosphere strength determinations. Nevertheless the results of the model provide a useful estimate of the strength of the lithosphere and indicate how it is controlled by the geothermal gradient, crustal thickness and crustal composition.

References

Avé Lallement, H.G. 1978. Experimental deformation of diopside and websterite. *Tectonophys.*, **48**, 1–27.
Bodine, J.H., Steckler, M.S. & Watts, A.B. 1981. Observations of flexure and the rheology of the oceanic lithosphere. *J. geophys. Res.* **86**, 3695–3707.
Bott, M.H.P. & Kusznir, N.J. 1984. Origins of

tectonic stress in the lithosphere. *Tectonophys.* **105**, 1–14.

—— & —— 1986. Tectonic Stress: its origin and distribution in the lithosphere, in preparation.

CARTER, N.L., ANDERSON, D.A., HANSEN, F.D. & KRANZ, R.L. 1981. Creep and creep rupture of granitic rocks. *In*: CARTER, N.L., FRIEDMAN, M., LOGAN, S.M. & STEARNS, D.W. (eds), *Mechanical Behaviour of Crustal Rocks*. Am. Geophys. Union Monograph, **24**, 61–82.

GOETZE, C. 1978. The mechanisms of creep in olivine. *Phil. Trans. R. Soc. A.* **288**, 99–119.

GRIFFITH, A.A. 1924. Theory of rupture. *Proc. first int. Congr. Applied Mechanics, Delft*, **A221**, 163–198.

KARNER, G.D. & WATTS, A.B. 1983. Gravity anomalies and flexure of the lithosphere at mountain ranges. *J. Geophys. Res.* **88**, 10449–10477.

KIRKBY, S.H. & KRONENBERG, A.K. 1984. Deformation of clinopyroxenite: evidence for a transition in flow mechanisms and semibrittle behaviour. *J. geophys. Res.* **89**, 3177–3192.

KOCH, P.S., CHRISTIE, J.M. & GEORGE, R.P. 1980. Flow law of 'wet' quartzite in the α-quartz field. *Trans. Am. geophys. Un.* **61**, 376.

KUSZNIR, N.J. 1982. Lithosphere response to externally and internally derived stresses: a viscoelastic stress guide with amplification. *Geophys. J. R. astr. Soc.* **70**, 399–414.

—— & PARK, R.G. 1984. Intraplate lithosphere deformation and the strength of the lithosphere. *Geophys. J. R. astr. Soc.* **79**, 513–538.

—— & KARNER, G.D. 1985. Flexural Rigidity of the Continental Lithosphere; its rheological and temperature dependence. In press.

MCCLINTOCK, F.A. & WALSH, J.B. 1962. Friction on Griffith cracks under pressure. *Proc. fourth U.S. Nat. Congr. Applied Mechanics*, 1015–1021.

MCKENZIE, D. 1978. Some remarks on the development of sedimentary basins. *Earth planet. Sci. Lett.* **40**, 25–32.

MORGAN, P. 1982. Heat flow in rift zones. *In*: PALMASON, G. (ed.), *Continental and Oceanic rifts*. Am. Geophys. Union, Geodynamics Series, **8**.

—— 1984. The thermal structure and thermal evolution of the Continental Lithosphere. *Phys. Chem. Earth*, in press.

POLLACK, H.N. & CHAPMAN, D.S. 1977. On the regional variation of heat flow, geotherms and lithosphere thickness. *Tectonophys.* **38**, 279–296.

POST, R.L. 1977. High temperature creep of Mt Burnet Dunite. *Tectonophys.* **42**, 75–110.

SIBSON, R.H. 1982. Fault zone models, heat flow, and the depth distribution of earthquakes in the continental crust of the United States. *Bull. seism. Soc. Am.* **72**, 151–163.

VITORELLO, I. & POLLACK, H.N. 1980. On the variation of continental heat flow with age and the thermal evolution of continents. *J. geophys. Res.* **85**, 983–995.

WATTS, A.B., KARNER, G.D. & STECKLER, M.S. 1982. Lithosphere flexure and the evolution of sedimentary basins. *Phil. Trans. R. Soc. Lond.* **A305**, 249–281.

N.J. KUSZNIR & R.G. PARK, Department of Geology, Keele University, Keele, Staffs ST5 5BG.

Metamorphism and crustal rheology—implications for the structural development of the continental crust during prograde metamorphism

K. Weber

SUMMARY: Based on its metamorphic development it can be argued that the orogenic continental crust during prograde metamorphism consists of three layers of contrasting bulk rheology: a lower layer of granulite-facies rocks able to carry relatively high flow stresses; a middle layer of migmatites and amphibolite-facies rocks of very low mechanical strength and an upper layer of low-grade and very low-grade metamorphic rocks, the so-called 'brittle' crust in the geophysical sense with higher yield strength possibly increasing upward. There are presumably main décollement horizons between the lithospheric mantle and lower crust, in the roof of a granulite-facies lower crust and along the boundary between the middle and upper crust, giving rise to large-scale displacements and the formation of nappe structures of different metamorphic rocks. Granulite-facies rocks of the lower crust will tend to form large-scale folds which pierce the overlying crust during increasing amplification. Further crustal shortening may produce granulite-facies nappes from strongly amplified large-scale fold structures leading to inverse metamorphism which is generally accompanied by granulite-facies nappe complexes.

Introduction

The continental crust model generally accepted today consists of an upper 'brittle' and a lower 'ductile' crust. This geophysical division of the crust is for example based on the differences in seismicity observed. Seismicity is high in the upper crust and low or absent in the deeper crust (Meissner & Strehlau 1982; Meissner 1981).

The depth of the boundary between the brittle and ductile crust is contingent on local geothermal gradients. The boundary is usually deeper in old shields than in younger crusts, and its position is particularly high in continental rift zones as for example in the basin and range province. This geophysical division of the present crust, which is essentially based on differences in the elastic properties of rocks under various PT-conditions, is of only limited importance in understanding the development of orogenic structures.

During the formation of tectonic structures the elastic properties of rocks play no or only a very subordinate role. Concerning geological strain rates, the rheologic behaviour of rocks is dominantly determined by the viscous properties, and the possible magnitude of stress is therefore strongly strain-rate dependent. This is also true for upper crustal levels in which the rocks are geophysically brittle, but the geological structures in folded mountains are ductile.

On the other hand, brittle structures like agmatites formed at high temperatures in the roofs of granite plutons can also be seen in very deep crustal levels, so that the geophysical crustal model is also limited here based on geological or petrological data (Fig. 2a).

Considering the petrological aspects of crustal rheology there are numerous possible weakening and hardening mechanisms which have been treated in the relevant literature and which concern mainly solid state deformation mechanisms. This paper deals with the relationship between crustal rheology and crustal structure with special regard to fluids during prograde metamorphism.

Fluids in the earth's crust

Fluids will be treated in a broad sense as all low-viscosity phases regardless of whether they are aqueous solutions, fluids above the critical temperature or melts. Such phases possess no shear strength, particularly at geological strain rates, and are thus not able to transfer differential stresses.

The understanding of the mechanical effects of fluids is based on the concept of effective normal stress, which was introduced by Terzaghi (1923) into soil mechanics and first applied to tectonic problems by Hubbert & Rubey (1959) and later by numerous other scientists (Murrell & Ismail 1976; Paterson 1978; Fyfe et al. 1978; Weber 1980; Etheridge et al. 1983, 1984).

The concept of effective normal stress is founded on the experimentally proven fact that the magnitude of normal stress transmitted from grain to grain in a rock body is reduced by the fluid pressure acting in the intergranular space.

From: DAWSON, J.B., CARSWELL, D.A., HALL, J. & WEDEPOHL, K.H. (eds) 1986, *The Nature of the Lower Continental Crust*, Geological Society Special Publication No. 24, pp. 95–106.

This can be very simply demonstrated using the example of vacuum-packed peanuts. When closed, the package behaves as a solid body; when opened, this property is lost.

The magnitude of the mean effective normal stress $\bar{\sigma}_e$ is

$$\bar{\sigma}_e = \bar{\sigma} - P \text{ with } \bar{\sigma} = \frac{\sigma_1 + \sigma_2 + \sigma_3}{3}$$

and P the fluid pressure, which is essentially a hydrostatic pressure. In the case of non-cohesive materials, mean effective stress approaches zero, when the mean effective stress $\bar{\sigma}_e$ equals P. Under these conditions the material possesses no strength and this also occurs at very high fluid pressures with $P_{H_2O} = P_{tot}$ in deep crustal levels. Thus, the Mohr-Coulomb failure criterion can be applied to non-cohesive materials. The shear strength τ decreases with increasing fluid pressure and approaches zero, when the mean stress equals zero.

In the case of cohesive materials a parabolic envelope is generally assumed based on the Griffith theory. Negative effective normal stresses can occur here and lead to the formation of tension fractures, when the smallest effective normal stress σ_{3e} equals the tensile strength of the material; or shear fractures can be opened, when a shear strength lower than the cohesion τ_0 has been reached (Weber 1980).

It is generally assumed that prograde metamorphic reactions under very low-grade to medium-grade metamorphic conditions take place with $P_{H_2O} = P_{tot}$ and that in most tectonites these metamorphic reactions proceed contemporaneously with deformation. This assumption has important consequences for the rheologic behaviour of metamorphic rocks, the deformation mechanisms effective during metamorphic processes and the structures developed at different crustal levels.

Diagenetic to low-grade metamorphic rocks

Under diagenetic conditions high pore pressures are very common and have been detected through oil drillings in sediments of the Early Tertiary and locally up to the Upper Cretaceous. In addition to numerous causes, which are summarized for example by Gretener (1969) and Weber (1980) compaction disequilibria play a decisive role in the formation of abnormally high pore-fluid pressures.

Compaction disequilibria result when the dewatering of a sediment does not keep up with the increasing overburden pressure caused by younger sediments. Compaction disequilibria can also be increased by superimposition of tectonic stresses. In Fig. 1 the permeability coefficient of various argillaceous sediments is plotted against the porosity. The permeability coefficient has the unit s^{-1} and is a direct measure of dewatering rate of a porous sediment. The permeability coefficient in most argillaceous sediments with porosities smaller than 0.4 is less than 10^{-8} cm s^{-1}. Compaction disequilibria arise when the permeability coefficient is smaller than the sedimentation rate or a corresponding rate of subsidence. An increase in overburden corresponding to a subsidence rate of $3 \cdot 10^{-9}$ cm s^{-1}, i.e. 100 m per million years, is sufficient for producing a compaction imbalance in sediments with porosities smaller than 0.4–0.3 and thus abnormally high pore-fluid pressures.

The low permeability of fine-grained argillaceous sediments is presumably based on the large internal surface and thus on the high amount of water adsorbed to clay mineral surfaces, particularly montmorillonite and illite. According to Grim (1968), water molecules are very strongly attached to clay mineral surfaces up to distances of c. 10 Å so that they cannot move freely. This leads to a 20–30% reduction in the effective porosity with internal surface area to volume ratios of $(250–400)10^6$ m^{-1} (Weber 1980). This could be a fundamental cause for the very low permeability of argillaceous sediments with porosities under 0.3–0.4. The mobilization of this non-liquid adsorbed water supposedly results largely through grain coarsening during higher temperature diagenesis and very weak metamorphism. However, in this stage substantial quantities of water must also migrate through the rock, documented by the widespread pressure solution phenomena particularly in strained rocks.

Permeability measured in drill holes in crystalline rocks are, according to Brace (1984), in the range of 10^{-18} to 10^{-13} m^2 (10^{-3} to 10^2 mD). Most frequent values range between 10^{-15} to 10^{-13} m^2 ($1–10^2$ mD). Permeabilities inferred from earthquake migration and other large-scale crustal phenomena are in the range of $10^{-16}–10^{-14}$ m^2 (10^{-1} to 10 mD) according to Brace (1984). These values are thus in the same range as the permeabilities measured in drill holes. The permeabilities of low porosity shales range from 10^{-20} to 10^{-18} m^2 ($10^{-5}–10^{-3}$ mD) and are thus about four to five orders of magnitude lower than the permeabilities of most crystalline rocks. Therefore, argillaceous sediments overlying crystalline rocks may act as a relatively impermeable barrier and may promote the formation of abnormally high fluid pressures in deeper crustal levels.

Medium-grade metamorphic rocks

Medium-grade metamorphic rocks are character-

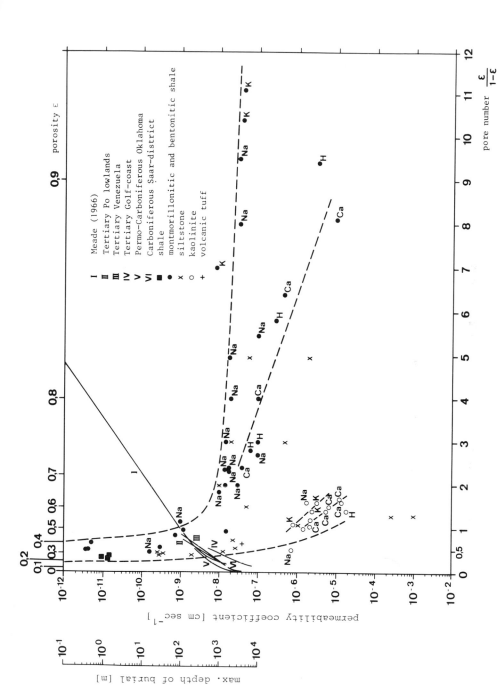

FIG. 1. Relationship between porosity and permeability coefficients of clays, shales and siltstones with various cation charges and porosity/depth (max. depth of burial) relationships from boreholes (Data from Meade, 1966; Hedberg, 1936; Weber, 1975; Kézdi, 1969).

ized by the prevalence of hydrous minerals like phyllosilicates and amphiboles. Low effective stresses during prograde metamorphism are indicated by weak or absent lattice-preferred orientation of quartz and feldspar, relatively coarse grain size and well equilibrated grain boundaries, unless such rocks have suffered younger deformation under retrograde metamorphic condition.

The general assumptions that prograde metamorphic phase equilibria have taken place at $P_{H_2O} = P_{tot}$ and that regional metamorphism proceeded contemporaneously with deformation have important consequences for the rheological properties and the effective deformation mechanisms.

In the course of an amphibolite-facies part of a PT-path σ_{eff} will change from low values during subsidence and prograde metamorphic conditions due to dewatering processes and reaction enhanced ductility (White & Knipe 1978; Rubey 1983) to higher values, when peak metamorphic conditions have been exceeded during synorogenic uplift. In most cases peak metamorphic grain fabrics are not (or not completely) preserved because strain proceeds during orogenic uplift. However, with decreasing temperature this retrograde metamorphic strain is concentrated more and more in separate shear zones, having unaffected domains between them.

The assumption of $P_{H_2O} = P_{tot}$ implies brittle behaviour during deformation which cannot be observed macroscopically in the present rocks. Microscopically, strongly annealed secondary fluid inclusions could be interpreted as evidence of an earlier stage of brittle deformation on the grain scale. However, microfractures do not necessarily give sufficient evidence of abnormally high pore-fluid pressure in the range of $P_{H_2O} = P_{tot}$ since microfractures can also be produced during ductile deformation, particularly by subcritical crack growth under the influence of stress corrosion (Atkinson 1982, 1984). Furthermore the residence time of individual grains or grain boundaries will possibly be short and metamorphic reactions during prograde metamorphism will be fast relative to geological strain rates (Etheridge et al. 1983). Therefore, it can be assumed that brittle deformation which has taken place on the grain scale (intragranular fractures, grain boundary fractures) became later completely obliterated by recrystallization and grain growth.

Walther & Orville (1982) have calculated the rate of metamorphism and volatile production in the temperature interval between 400°C and 600°C. The rate of volatile production of 10 km of average pelitic rock is 30–300 Mg m^{-2} in 1 Ma. This amount of volatiles must pass through each square metre of crust overlying the 400°C isotherm. They conclude that fracturing must take place because permeability of the unfractured rock is not sufficient to accomodate the flux of volatiles produced by prograde metamorphism. Micro- and macrofractures in combination with the crack-seal mechanism (Durney & Ramsay 1974; Ramsay 1980) are regarded as possible channel-ways for the metamorphic fluids.

A further observation might be interpreted as an indication of low effective stresses. In amphibolite facies rocks and migmatites early fold generations, i.e. folds which are formed under prograde metamorphic conditions, are generally small-scale folds. The dominant parameters, which control the wavelength of fold structures, are the mechanically effective layer thickness and the competency differences in the folded layers. At geological strain rates the contrast in competency can be understood as the viscosity differences between mechanically effective layers of different lithology. According to Ramberg (1963) the dominant wavelength Wd is

$$\mathrm{Wd} = 2\pi h (2\mu_1/3\mu_0)^{\frac{1}{3}}$$

with h the layer thickness and μ the viscosity ($\mu_1 > \mu_0$). Wavelength refers to the initial small-amplitude buckles or to the arc length of more mature folds. In Ramberg's model single layers are welded together and give somewhat smaller wavelength/thickness ratios than the free-slip model of Biot (1957):

$$\mathrm{Wd} = 2\pi h (\mu_1/6\mu_0)^{\frac{1}{3}}.$$

Other models (e.g. Fletcher 1977) give slightly different results particularly with higher viscosity ratios (Ramberg 1979).

A possible explanation for the dominance of small-scale folding under early prograde amphibolite facies conditions is the reduction of effective stresses. This can easily be understood in the case of migmatites where the presence of intergranular melt produces low effective stresses which promote layer-parallel slip. The same effect will have intergranular fluids produced by dewatering reactions.

It seems that layer-parallel slip has been active during small-scale folding under amphibolite-facies conditions. The dominant type of folds formed at this metamorphic stage are internal folds. Internal folding is basically a process of disharmonic folding which presumes the existence of décollement and layer-parallel slip or layer-parallel shear at least along the décollement planes (Fig. 2c,d). In contrast, the formation of similar folds by superimposition of homogeneous strain on an initially buckled layer (Ramsay 1967) presumes the absence of layer-parallel slip or

FIG. 2. (a) Agmatite with fragments of Khomas schists at the rim of a granite. Damara Orogen of Namibia, Skeleton Coast, north of Möve Bay. (b) Migmatic gneis showing brittle fabrics inside the leucosomes. Damara orogen of Namibia, Skeleton Coast. (c) and (d) Internal fold in migmatic gneiss. Sunn Møre Unit, Norwegian Caledonides.

shear. Small viscosity differences and/or high effective stresses are therefore prerequisites for similar folding.

The different style of deformation under prograde metamorphic conditions—large-scale folding and thrusting in the sedimentary sequences of very low-grade to low-grade metamorphic belts and dominantly early small-scale folding in the underlying medium grade metamorphic rocks—must lead to the development of a prominent décollement between the two contrasting crustal levels. Therefore, this boundary or boundary zone will be a possible source of low-grade metamorphic nappe complexes.

Furthermore, stress relaxation in the deeper crust has the important consequence of a stress amplification in the upper 'brittle' crust (Kusznir & Park, this vol.; Kusznir & Bott 1977; Mithen 1982), which may contribute to decoupling effects between brittle and ductile crustal levels.

High-grade metamorphic rocks

Migmatites

At regional metamorphic temperatures above approximately 650°C felsic or pelitic rocks may partially melt. As a rule, formation of migmatitic melts occurs syntectonically and they intrude parallel to metamorphic layering, i.e. along the planes of weakness. Migmatitic fabrics can be best explained as resulting from hydraulic fracturing by partial melts under negative effective stresses (Fig. 2b). Therefore, only very low deviatoric stresses could have been effective. This is supported by the fact that at such high temperatures the solid phases are not likely to be strong enough to support significant deviatoric stresses. An effect of low effective stresses is the dominance of brittle deformation also under high-grade metamorphic conditions. This becomes especially clear along intrusion contacts of granitic or granodioritic melts in the form of agmatic structures (Fig. 2a). The angular fragments of wall rocks are the result of brittle fracturing. Yet, angular fragments of host rocks are also frequently found in migmatites.

Experimental deformation of partially melted granite carried out by van der Molen & Paterson (1979), Arzi (1978) and Paquet *et al.* (1981) has shown a great decrease in strength with increasing melt fraction of about 25–35 vol. %. The flow stress is less than 1 MPa. Although the fraction of grain boundaries observed to bear melt was determined by van der Molen & Paterson to be more than 90% for 10 vol. % melt and decreases to 84% and 63% for 7 vol. % and 5 vol. %, melt, respectively, a remarkable strength of up to about 100 MPa at a strain rate of $10^{-5}\,s^{-1}$ and confining pressure of 300 MPa has been observed at less than 10 vol. % of melt fraction. They conclude that these rocks are substantially disaggregated as a result of this widespread presence of melt at grain boundaries. This means that the initial polycrystalline solid has been largely disaggregated into an assemblage of grains with negligible cohesion between them. Under these circumstances the effective stresses should be zero.

The ability of the partially melted rock to support compressive differential loading under experimental conditions has been explained by van der Molen & Paterson with the effects of initial compaction and dilatancy hardening. In the initial stage of straining, the grains will generally be moved towards each other in the direction of shortening, squeezing out intervening melt by dilatancy pumping until a continuous framework of contacts is established that can support positive deviatoric stresses. Since the compacted granular assemblage will have a fairly high packing density, further straining will involve dilatancy and thus lowering of pore pressure.

Whether these effects are active under natural conditions, particularly at much lower strain rates, is questionable. It must be assumed that the experimental deformation of partially melted granite was carried out under the presence of vapour. Vapour-present melting at least at moderate pressures and water activity are accompanied by a volume decrease and may lead to an increase in effective stresses. In contrast, vapour-absent melting of crystalline hydrate-bearing assemblages commonly involves a volume increase (Burnham 1978; Clemens & Wall 1981), which has the effect of decreasing the effective stresses.

Mass transfer by mobile silicate melts produced during high-grade metamorphism is readily demonstrable on both mesoscopic and regional scales. Migmatites, which are commonly developed in metasedimentary sequences, are characterized by separation of biotite-rich melasomes and quartz-feldspar-rich leucosomes. Dilatancy hardening is highly unlikely to be present in biotite-rich palaeosomes under low effective stress. Therefore, the total fluid pressure should rapidly readjust to lithostatic pressure.

Grain fabrics of migmatites are typified by general features similar to medium-grade metamorphic rocks, i.e. increasing grain size due to grain boundary migration and secondary recrystallization, particularly in the case of quartz. Further signs of low effective stresses are weak or absent lattice-preferred orientation and the absence of other indications of crystal plasticity.

Granulite facies rocks

In contrast, high-grade rocks composed dominantly of water-free minerals such as granulite-facies rocks formed at $P_{H2O} < P_{tot}$ commonly reveal strong lattice-preferred orientation (Behr 1961; Starkey 1979; Lister & Dornsiepen 1982). Such granulites, of both felsic and mafic compositions, have well-developed foliations and are relatively fine-grained and show un- or weakly-equilibrated grain boundaries. They were probably formed by the deformation of formerly coarser-grained assemblages which are sometimes preserved as inclusions in garnet. Garnet fabrics particularly of felsic granulites, for example the classic Saxony 'core' granulites (Behr 1961), are best described as high-grade blastomylonites formed under conditions of high effective and relatively high deviatoric stresses. The transformation of white mica into K-feldspar additionally reduces the fabric anisotropy. The aforementioned fabric characteristics as well as results of experimental deformation of pyroxene, amphibole, dry quartz and other granulite-facies minerals indicate significantly higher flow stresses in comparison to those affecting wet amphibolite-facies rocks. This will have important implications for the structural development of orogenic belts.

Today, there is wide agreement that the lower crust consists largely of granulite-facies rocks with possibly more mafic granulites in its lower part and more felsic granulites in its upper part (Kay & Kay 1980; Den Tex 1982; Upton *et al.* 1983; Christensen & Fountain 1975). Granulites can be formed in different tectonic environments. As shown by Den Tex (1982), van Calsteren & Den Tex (1978) Weber & Behr (1983) and Weber (1984, 1984a), medium-pressure granulites can be formed in the lower crust under crustal stretching conditions in continental rift environments under the influence of high mantle-derived heat flow. During the continental collision stage of crustal shortening particularly high-pressure granulites may be formed in the lower part of a tectonically thickened pile of continental crust and by a corresponding increase in radiogenic heat production.

Independent of whether granulite facies rocks have been formed during a preorogenic stage of continental rifting or during the continental

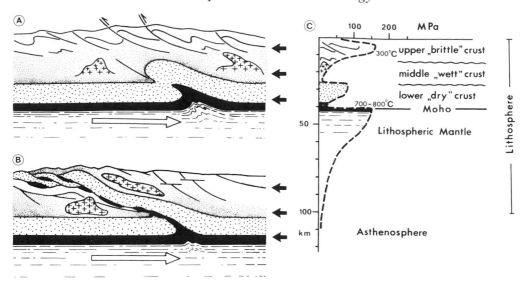

FIG. 3. Diagrammatic sketches of the development of granulite-facies nappe complexes (A and B) and of steady-state flow stresses (C) in the continental lithosphere during prograde metamorphism. (Further explanation in the text.)

collision stage, high viscosity differences must be assumed between (dewatered) granulite-facies and (dewatering) amphibolite-facies rocks and particularly migmatites.

Three layer crustal model

Founded on the preceding arguments it can be assumed that a *prograde* metamorphic orogenic crust will be composed of at least three layers of different bulk rheology (Fig. 3).

1 A lower layer of granulite-facies rocks representing the lower crust *sensu stricto* which is able to carry bulk flow stresses in the range of 50 to 100 MPa or more depending on composition and temperature.

2 A middle layer composed of migmatites and amphibolite facies rocks, with very low flow stresses particularly in its lower part where migmatites are dominant.

3 And finally an upper layer of low-grade and very low-grade rocks, the so-called brittle crust in the geophysical sense, with yield strength possibly increasing upward.

Of course, such a rheological three-layer model of orogenic continental crust is too simple in view of its great structural diversity. Some principal features, however, can be explained quite well.

Due to its relatively high strength and high anisotropy the uppermost layer, which is composed mainly of sedimentary and weakly meta-

morphosed metasedimentary sequences, reacts during orogenic shortening by folding and thrusting on various scales. This style of deformation is usually preserved in the external zones of orogens, for example in the Rhenohercynian zone of the European Variscides. Vergence, fabric symmetry and traceable transport point to relative movements of higher levels in the direction of the orogenic foreland.

During the prograde metamorphism the most pronounced rheological boundary is to be expected between the migmatites and underlying granulite-facies rocks. This boundary is not steady-state: it will move upward with the rising metamorphic front.

The transformation of white mica + quartz into K-feldspar has the effect of reducing the fabric anisotropy and increasing the mechanically effective layer thickness. Using Biot's or Ramberg's equations (see above) a thick and relatively competent granulite layer overlain by a low viscosity material will form large scale fold structures during crustal shortening. Assuming a viscosity contrast of about one hundred, a 2 km thick granulitic layer will have an initial wavelength of about 32 km. During its amplification, such large-scale fold structures will pierce the overlying crust in the form of a large 'injective' fold producing diapir-like or brachyanticlinal structures like the Saxony Granulite Mountains, or more complex high-grade gneiss domes like the granulites of the Limpopo Belt and the Adirondack Mountains. Further crustal shortening may produce granulite-facies nappes from such amplified large scale asymmetrical fold structure. This mechanism might be an explanation for the fact that granulite-facies nappe complexes are very frequently, if not generally, accompanied by inverted metamorphism. This especially applies to dominantly felsic granulites like the Saxony Granulite Mountains, the granulites of the Waldviertel in Lower Austria and the widespread leptinitic gneisses of the Bohemian Massif and the Massif Central. The felsic granulites of the Saxony Granulite Mountains—containing approximately 5% mafic granulites (Behr 1980)—yield a density of 2.64–2.68 Mg m^{-3} (Kopf 1968, 1976). Buoyancy forces will promote the uprise of such felsic granulites during fold amplification and intracrustal nappe displacement processes.

In detail, the mechanisms of deformation and uprise of granulite-facies fold and nappe structures are complex, and a simplified rheological three-layer crustal model can only be satisfactory for a first approximation. In talking about crustal rheology and deformation one at least has to take into account the rheological properties of lithospheric mantle. Similar to mafic granulites of the lower crust, minerals like olivine, pyroxene and garnet, yielding high steady-state creep stress also at elevated temperatures, represent the stress-supporting minerals. In comparison to a granulite facies granodioritic to granitic lower crust with quartz and feldspar as the stress supporting minerals, mafic granulites and peridotites will carry much higher flow stresses than felsic granulites (Fig. 3). It is then to be anticipated that there is a well-developed, geodynamically effective rheological boundary between a quartz-feldspar rich lower continental crust and a peridotitic lithospheric mantle (Weber & Behr 1983; Weber 1984). Lamellae of mafic granulites inside the lower crust will produce a more or less complex transition zone between crust and mantle. Many felsic granulites contain more or less isolated, strongly sheared nodules and lenses of mafic granulites which must be related to formerly more continuous layers of mafic intrusive rocks.

During large scale folding and thrusting of lower crustal granulite-facies rocks, a buoyancy-controlled process will be active which separates dense mafic rocks from less dense felsic rocks. As a rule, the peridotitic upper mantle itself will not be incorporated in the large-scale folding process, although some of the large Alpine and older peridotite massifs could have a similar early history.

Except for a few large-scale mafic granulite complexes (Ivrea, Calabria) mafic granulites and eclogites usually form lenses—frequently

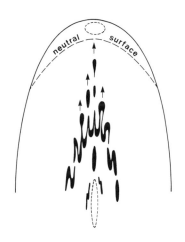

FIG. 4. Diagrammatic sketch of the formation of nodules or lenses of competent mafic granulites inside a buckled layer of felsic granulites. (Further explanation in the text.)

strongly sheared—in granulite-facies and also amphibolite-facies rocks. Tectonic emplacement is probable and the following mechanisms might have been active during large-scale folding and thrusting.

1 Mafic granulites form slices and lenses along the main thrust plane, as shown schematically in Fig. 3.

2 Buckling of a thick layer will be accompanied by tangential–longitudinal strain (Ramsay 1967). Compression in the inner arc and stretching in the outer arc increases with increasing distance from the neutral surface and thus with the increasing thickness of the folded layer. Stress and strain in the hinge zone of a thick folded layer are extremely inhomogeneous under the conditions of tangential–longitudinal strain. Therefore, a pressure gradient will be produced from the inner to the outer arc, which acts as the driving force for the transport of nodules and lenses of more competent mafic granulites into higher crustal levels (Fig. 4).

It is highly improbable that longitudinal–tangential strain represents the only or dominant deformation mechanism during the outlined large-scale folding process. Under high-grade conditions a substantial amount of homogeneous strain will be superimposed on tangential–longitudinal strain and diminish the effective pressure gradient between the inner and outer arc. Otherwise, the rheological properties of rocks are highly strain-rate sensitive and the strain in the inner arc of a buckled layer increases exponentially during tangential–longitudinal strain (Ramsay 1967, page 399). As long as the thick folded layer of granulite facies rock is more competent than the overlying amphibolite facies rock or migmatite, tangential–longitudinal strain will be active and a pressure gradient will exist from the inner to the outer arc of the fold hinge.

Conclusions

Based on a 3-layer rheological model for synorogenic continental crust during prograde metamorphism, three prominent décollement horizons can be expected:
—the crust–mantle boundary as a décollement zone of high grade metamorphic lower crustal rocks,
—the transition zone from dry granulite-facies to wet amphibolite-facies crust,
—the transition zone from dominantly small-scale folded amphibolite-facies crust to dominantly large-scale folded and imbricated 'brittle' crust.
These décollement horizons are prominent sources of nappe complexes of the different metamorphic rocks.

The granulite-facies rocks of the lower crust can build intracrustal nappes which reach the surface through late-orogenic uplift and erosion. However, they can also be brought to the surface through very extensive transport as shown in Fig. 3, or can overthrust weak metamorphic rocks as supercrustal nappe complexes as in the case of the central Iberian catazonal complexes of Tras os Montes (Hirn *et al.* 1982) and Cabo Ortegal (Bard *et al.* 1980) or the Münchberger Gneismasse (Behr *et al.* 1982; Franke 1984). In such nappe complexes a distinct retrograde-metamorphic, tectonic overprinting is observed.

The transport of granulite-facies nappe complexes into the upper crust is more likely for felsic granulites than for mafic granulites for isostatic reasons. Therefore, the frequency of felsic granulite-facies nappes is not representative of the composition of the lower crust. Mafic granulites may perhaps be more abundant in the lower crust than in the granulite-facies nappe complexes. Décollement horizons in the transition zone from the middle to lower crust may correspond to the Conrad discontinuity. According to this interpretation the Conrad discontinuity is regarded as a tectonically strongly overprinted, metamorphic boundary zone.

The décollement horizon between the upper and middle crust is viewed as a source of weakly metamorphic, supracrustal nappes which, however, themselves can be overthrust by higher metamorphic, crystalline nappes during orogenic crustal shortening. Furthermore, gravitational nappes can be derived from these supracrustal nappes, e.g. the Upper Devonian graywacke nappes of the southern Rheinische Schiefergebirge (Engel *et al.* 1983; Weber & Behr 1983) or the supracrustal nappes of the Montagne Noire (Engel *et al.* 1978, 1983). However, crystalline nappes which already overthrust weak metamorphic rocks in high crustal levels can also be incorporated in these gravitational transports as proven by Behr *et al.* (1982) and Franke (1984) for the Münchberger Gneismasse for example.

From the present P velocity-depth function in Central Europe Mueller (1977) has derived a lithological crustal model which is similar to the synorogenic rheomorphic crustal model presented here. Below a near surface low-velocity layer a zone of positive velocity gradient is found up to a depth of about 10 km which is assumed to consist of metamorphic rocks such as gneisses and schists. It is underlain by a low-velocity zone which is composed of granitic intrusions. The middle part of the crust is characterized by a relatively low average velocity and has been related to magmatites by Mueller (1977). This middle part of the crust is followed by a high-velocity zone topped by the C-discontinuity. It seems probable that the higher velocity of this zone is caused by a larger amount of mafic rocks, whereas the lower velocity of the lower crust is related to dominantly felsic granulites.

One has to bear in mind that the present seismic crustal structure represents a postorogenic feature, i.e. a structure which developed after extensive stacking processes had taken place (Giese *et al.* 1983; Weber 1984; Behr *et al.* 1984).

Petrological data from the present outcrop levels in the central European Variscides indicate that the late tectonic history of the Hercynian orogen is characterized by high temperature metamorphism (Zwart & Dornsiepen 1978; Weber & Behr 1983; Blümel 1982). the geothermal gradient increases from Ordovician to Permian time from about $20°C$ km^{-1} to about $80°C$ km^{-1} (Zwart & Dornsiepen 1978; Weber & Behr 1983; Buntebarth 1982; Buntebarth *et al.* 1982).

For the time of the subsequent Permo-Carboniferous magmatism the generation of rhyolitic magmas requires a high temperature at the crust–mantle boundary. Therefore, it seems reasonable that the present seismic crustal structure might be largely related to this late orogenic metamorphic history. Without input of water the granulitic lower crust will be largely preserved. Inside the amphibolite facies crust renewed migmatization could have taken place. The huge amounts of late to post-tectonic S-type granites in the Hercynian crust must be related to this high temperature metamorphic event. It seems possible to me that the high velocity zone representing the C-discontinuity level might result from a larger amount of mafic melt-residuals.

The P-T path of metamorphic rocks formed during an orogenic cycle is always better documented in its younger stage than in the older one. The rheomorphic petrology of rocks exposed at the present earth's surface dominantly represents relatively late stages of an orogenic P-T- and deformation-path. Therefore, it seems possible that the present crustal structure dominantly reflects the late orogenic history. Fabrics which are indicative of deformation mechanisms active during early stages of deformation and metamorphism will be more or less extinguished and must be inferred from scarce relics, experimental data and theoretical considerations. A geophysical division of the continental crust into an upper 'brittle' and a lower 'ductile' crust, which is well in accord with seismicity data (Meissner & Vetter 1979; Meissner 1981; Meissner & Strehlau 1982), is too simple for understanding orogenic crustal structures.

ACKNOWLEDGEMENTS: I owe my thanks to H. Ahrendt, H.-J. Behr, W. Franke for many helpful discussions and two unknown reviewers for their constructive criticism.

References

ARZI, A.A. 1978. Critical phenomena in the rheology of partially melted rocks. *Tectonophys.* **44**, 173–184.

ATKINSON, B.K. 1982. Subcritical crack propagation in rocks: theory, experimental results and applications. *J. Struct. Geol.* **4** (1), 41–56.

—— 1984. Subcritical crack growth in geological materials. *J. Geophys. Res.* **89**, 4077–4114.

BARD, J.P., BURG, J.P., MATTE, P. & RIBEIRO, A. 1980. La chaine hercynienne d'Europe occidentale en termes de tectonique de plaques. *Publications du 26ᵉ Congrès Géologique International*, Colloq. **6**, 233–246.

BEHR, H.-J. 1961. Beiträge zur petrographischen und tektonischen Analyse des Sächsischen Granulitgebirges. *Freiberge Forschhft.* C **119**, 9–146.

—— 1980. Polyphase shear zones in the granulite belts along the margins of the Bohemian Massif *J. struct. Geol.* **2**, 249–254.

—— ENGEL, W. & FRANKE, W. 1982. Variscan Wildflysch and nappe tectonics in the Saxothuringian Zone (Northeast Bavaria, West Germany). *Am. J. Sci.* **282**, 1438–1470.

—— ENGEL, W., FRANKE, W., GIESE, P. & WEBER, K. 1984. The Variscan belt in Central Europe: Main structures, geodynamic implications, open questions. *Tectonophysics*, **109**, 15–40.

BIOT, M.A. 1957. Folding instability of a layered viscoelastic medium under compression. *Proc. R. Soc. London, A*. **242**, 444–454.

BLÜMEL, P. 1982. Aufbau, metamorphose und geodynamische Deutung des Variszischen Grundgebirges in Mereich der Bundesrepublik. *Jb. Ruhr-Universität Bochum*, 169–201.

BRACE, W.F. 1984. Permeability of crystalline rocks: new in situ measurements. *J. geophys. Res.* **89**, 4327–4330.

BUNTEBARTH, G. 1982. Zur Paläogeothermik im Permokarbon der Saar-Nahe-Senke. *Nachr. dt. geol. Ges.*, **27**, 44.

—— KOPPE, J. & TEICHMÜLLER, M. 1982. Palaeogeothermics in the Ruhr Basin. *In*: OERMAK, V. &

HAENEL, E. (eds), *Geothemics and Geothermal Energy*, Schweiozertbart, Stuttgart, 45–53.

BURNHAM, C.W. 1978. Magmas and hydrothermal fluids. In: BARNES, H.L. (ed.) *Geochemistry of Hydrothermal Ore Deposits*. Wiley Interscience, New York, 71–136.

CALSTEREN, VAN, P.W. & DEN TEX, E. 1978. An early Palaeozoic continental rift in Galicia (W-Spain). *In*: RAMBERG, J.B. & NEUMANN, E.R. (eds) *Tectonics and Geophysics of Continental Rifts*, 125–132. Reidel, Dordrecht.

CHRISTENSEN, N.I. & FOUNTAIN, D.M. 1975. Constitution of the lower continental crust based on experimental studies of seismic velocities in granulite. *Bull. Geol. Soc. Am.* **86**, 227–236.

CLEMENS, J.D. & WALL, V.J. 1981. Origin and crystallization of some peraluminous (s-type) granitic magmas, *Can. Mineralogist*, **19**, 111–131.

DEN TEX, E. 1982. Dynamothermal metamorphism across the continental crust, *Fortschr. Min.* **60**, 1, 57–80.

DURNEY, D.W. & RAMSAY, J.G. 1973. Incremental strains measured by syntectonic crystal growths. *In*: DE JONG, K. & SCHOLTEN, R. (eds) *Gravity and Tectonics*, Wiley, New York, 67–96.

ENGEL, W. FEIST, R. & FRANKE, W. 1978. Syonogenic gravitational transport of the Montagne Noire (S. France). *Z. dt. geol. Ges.* **129**, 461–472.

—— FRANKE, W., GROTE, C., WEBER, K., AHRENDT, H. & EDER, W. 1983. Nappe tectonics in the southestern part of the Rheinische Schiefergebirge. In: MARTIN, H. & EDER, W. (eds), *Intracontinental Fold Belts—Case Studies in the Variscan Belt of Europe and the Damara Orogen of Namibia*, Springer Verlag Berlin, Heidelberg, 267–288.

ETHERIDGE, M.A., WALL, V.J. & COX, S.F. 1984. High fluid pressures during regional metamorphism and deformation: implications for mass transport and deformation mechanisms. *J. geophys. Res.* **89**, 4344–4358.

——, —— & VERNON, R.H. 1983. The role of the fluid

phase during regional metamorphism and deformation. *Metamorphic Geol.* **1**, 205–206.

FLETCHER, R.C. 1977. Fluid dynamics of viscous buckling applicable to folding of layered rocks. *Tectonophys.* **39**, 593–606.

FRANKE, W. 1984. Variszischer Deckenbau im Raume der Münchberger Gneismasse—abgeleitet aus der Fazies, Deformation und Metamorphose im umgebenden Paläzoikum. *Geotekton. Forsch.* **68**, I–II, 1–253.

FYFE, W.S. 1976. Chemical aspects of rock deformation. *Phil. Trans. R. Soc. London, A* **238**, 221–226.

—— PRICE, N.J. & THOMPSON, A.B. 1978. *Fluids in the Earth's Crust.* Developments in Geochemistry 1, 383 pp., Elsevier.

GIESE, P., JÖDICKE, H., PRODEHL, C. & WEBER, K. 1983. The crustal structure of the Hercynian Mountain System—a model for crustal thickening by stacking. *In*: MARTIN, H. & EDER, W. (eds) *Intracontinental Fold Belts*. Springer Verlag, Berlin, Heidelberg, 405–426.

GRETENER, P.E. 1969. Fluid pressure in porous media—its importance in geology; a review. *Bull. Canadian Petrol. Geol.* **17**, 255–295.

GRIM, R.E. 1968. *Clay Mineralogy*. McGraw-Hill, New York, London, Toronto, 596 pp.

HEDBERG, H.D. 1936. Gravitational compaction of clays and shales. *Amer. J. Sci.* **31**, 241–287.

HIRN, A., SENOS, L., SAPIN, M. & MENDES VICTOR, L. 1982. High to low velocity succession in the upper crust related to tectonic emplacement: Tras os Montes—Galicia (Iberia), Brittany and Limousin (France). *Geophys. J. R. astr. Soc.* **70**, 1–10.

HUBBERT, M.K. & RUBEY, W.W. 1959. Role of fluid pressure in mechanics of overthrust faulting, Part I. *Bull. Geol. Soc. Amer.* **70**, 115–166.

KAY, R.W. & KAY, S.M. 1980. Chemistry of the lower crust: inferences from magmas and xenoliths. *In*: *Continental Tectonics*. Studies in Geophysics, National Academy of Sciences, Washington, 135–150.

KOPF, M. 1968. Zur petrophysikalischen Untersuchung und Abgrenzung von metamorphen Para- und Orthogneisen. *XXIII Int. Geol. Congr.* Vol. **4**, 237–251.

—— 1976. The use of petrophysics in petrography. *Z. geol. Wissenschaften* H. 7, 1049–1068.

KUSZNIR, N.J. & BOTT, M.H.P. 1977. Stress concentration in the upper lithosphere caused by underlying visco-elastic creep. *Tectonophys.* **43**, 247–256.

—— & PARK, R.G. this vol. Continental lithosphere strength: the critical role of lower crustal deformation.

LISTER, G.S. & DORNSIEPEN, U.F. 1982. Fabric transitions in the Saxony granulite terrain. *J. Struct. Geol.* **4**, No. 1, 81–92.

MEISSNER, R. 1981. Bruch-Kriechprozesse in kontinentaler Kruste. *Z. dt. geol. Ges.* **131**, 591–603.

—— & STREHLAU, J. 1982. Limits of stresses in continental crusts and their relation to the depth-frequency distribution of shallow earthquakes. *Tectonics* **1**, 73–89.

—— & VETTER, U. 1979. Rheologic properties of the lithosphere and application to passive continental margins. *Tectonophys.* **59**, 369–380.

MITHEN, D.P. 1982. Stress amplification in the upper crust and the development of normal faulting. *Tectonophys.* **83**, 259–273.

MUELLER, S. 1977. *A new model of the continental crust.* Geophysical Monograph 20. The Earth's Crust. Am. Geophys. Union, Washington, D.C. 289–317.

MURRELL, S.A.F. & ISMAIL, I.A.H. 1976. The effect of decomposition of hydrous minerals on the mechanical properties of rocks at high pressures and temperatures. *Tectonophys.* **31**, 207–258.

PAQUET, J., FRANCOIS, P. & NEDELEC, A. 1981. Effect of partial melting on rock deformation; experimental and natural evidences on rocks of granitic compositions. *Tectonophys.* **78**, 545–565.

PATERSON, M.S. 1978. *Experimental Rock Deformation—The Brittle Field.* Springer Verlag, Berlin, 254 pp.

RAMBERG, H. 1979. Folding of a single viscous layer: exact infinitesimal-amplitude solution—discussion. *Tectonophys.* **56**, 321–326.

—— 1963. Fluid dynamics of viscous buckling applicable to folding of layered rocks. *Bull. Am. Ass. Petrol. Geol.* **47**, 484–505.

RAMSAY, J.G. 1967. *Folding and Fracturing of Rocks*. McGraw-Hill. 568 pp.

—— 1980. The crack-seal mechanism of rock deformation. *Nature*, **284**, 135–139.

RUBEY, D.C. 1983. Reaction-enhanced ductility: The role of solid-solid univariant reactions in deformation of the crust and mantle. *Tectonophys.* **96**, 331–352.

STARKEY, J. 1979. Petrofabric analysis of Saxony granulites by optical and X-ray diffraction methods. *Tectonophys.* **58**, 201–219.

TERZAGHI, K. 1923. Die Berechnung der Durchlässigkeitsziffer des Tones aus dem Verlauf der hydrodynamischen Spannungserscheinungen. *Sitz. Akad. Wissensch., Wien, Math.- Naturwisse. kl. Abt. IIa*, **123**, 125–138.

UPTON, B.J.G., ASPEN, P. & CHAPMAN, N.A. 1983. The upper mantle and deep crust beneath the British Isles: evidence from inclusions in volcanic rocks. *J. Geol. Soc.* **140**, 105–122.

VAN DER MOLEN, J. & PATERSON, M.S. 1979. Experimental deformation of partially-melted granite. *Contrib. Min. Petrol.* **70**, 299–318.

WALTHER, J.V. & ORVILLE, P.M. 1982. Volatile production and transport in regional metamorphism. *Contrib. Min. Petrol.* **79**, 252–257.

WEBER, K. 1980. Anzeichen abnormal hoher Porenlösungsdrucke am Beginn der Faltung im Rheinischen Schiefergebirge. *Z. dt. geol. Ges.* **131**, 605–625.

—— 1984. Variscan events: early Palaeozoic continental rift metamorphism and late Palaeozoic crustal shortening. *In*: HUTTON, D.H.W. & SANDERSON, D.J. (eds) *Variscan Tectonics of the North Atlantic Region*. Geol. Soc. Spec. Publ. No. 14, 3–22.

—— 1984a. Variation in tectonic style with time (Variscan and Proterozoic systems) *In*: HOLLAND, H.D. & TRENDALL, A.F. (eds) *Patterns of Change in Earth Evolution*. Springer Verlag, Berlin, 371–386.

—— & BEHR, H.J. 1983. Geodynamic interpretation of the mid-European Variscides. *In*: MARTIN, H. & EDER, F.W. (eds) *Intracontinental Fold Belts.* Springer Verlag, Berlin, 427–469.

WHITE, S.H. & KNIPE, R.J. 1978. Transformation- and reaction-enhanced ductility in rocks. *J. Geol. Soc. London*, **135**, 513–516.

ZWART, H.J. & DORNSIEPEN, U.F. 1978. The tectonic framework of Central and Western Europe. *Geologie and Mijnbouw* **57,** 627–654.

K. WEBER, Institut für Geologie und Dynamik der Lithosphäre, Universität Göttingen, Goldschmidtstr. 3, D-3400 Göttingen.

The role of deformation, heat, and thermal processes in the formation of the lower continental crust

S.A.F. Murrell

SUMMARY: Because as much as ninety per cent of the continental crust originated in Archaean times its structure has been modified by prolonged, though punctuated, thermal and tectonic processes. In spite of considerable heterogeneity arising from a complex tectono-thermal history, continental lower crust is now revealing significant geophysical structures (low-velocity layers, high-conductivity layers, high seismic reflectivity). It is suggested that these structures are a consequence of tectono-thermal metamorphism which produces trapped layers of high pressure aqueous pore fluids. These pore fluids also result in deep seismic activity. A brief review of crustal tectonics and crustal growth based on concepts mainly due to Tarling (1978) leads to the postulate that the nature of the lower continental crust depends on the age of the last significant tectono-thermal event (Archaean, Proterozoic, and Phanerozoic structures differing significantly) due to the continuing reduction in the Earth's internal energy and in the vigour of tectonic activity of the oceanic lithosphere ('sea-floor spreading') associated with mantle convection. The Phanerozoic has been the era of plate tectonics, marked by deformation at the margins of lithosphere plates or at sutures or by stretching and rifting within plates, where the deformation has caused crustal metamorphism. In the Proterozoic the continental crust was relatively much more stable than it was either in Archaean or Phanerozoic times, and it was transformed more by thermal effects of widespread igneous activity than by interactive plate margin tectonics. The Archaean was the period in which the major part of the continental crust was formed by igneous differentiation associated with mantle convection, and by sedimentary processes. The primal character of the processes at this time, together with the greater geothermal energy, distinguishes them from the similar processes of the Phanerozoic era. During a prograde metamorphic event dehydration of a particular crustal layer will tend to cause a build up of volume and pressure of entrapped pore water, and under such 'undrained' conditions tectonic stress and dilatancy-hardening will cause cataclastic flow distributed throughout the layer. Thick fluid-filled layers of this type could produce the geophysical characteristics observed in the lower crust. Where 'drained' conditions occur strain-softening may result under tectonic stress, leading to the formation of localized faults and fractures, which may be too thin to be geophysically detectable. The drainage of pore fluids from a metamorphically dehydrated layer due to tectonic fracturing (e.g. at an ocean–continent margin) could also cause the layer to show less contrast of geophysical properties, and make it undetectable. This might explain some of the apparent discontinuity of deep crustal structures.

Heat and tectonics in the formation of the lower continental crust

Whereas the oceanic crust is formed largely by a thermal process during sea-floor spreading, together with some subsequent retrograde metamorphism due to cooling and to reaction with sea-water, the continental crust has, in contrast, been formed by prolonged thermal, tectonic, and sedimentary processes over much of the Earth's history, involving both retrograde and prograde metamorphism of both igneous and sedimentary components. As a consequence the lower continental crust has had a long history of hot working. The purpose of this paper is to outline briefly the main geophysical features, origin and history of the continental crust, then to consider the nature of the hot working processes to which the lower crust has been subjected, and finally to consider the implications of this for the nature of the lower crust.

Geophysical features of the lower continental crust

The heterogeneity of the continental crust is well-known (Kay & Kay 1981; Green 1977). In some areas, however, the lower continental crust does appear to have some definite structure, revealed by such features as zones of high electrical conductivity (Mitchell & Landisman 1971), relatively high seismic reflectivity (Smithson et al. 1977; Brewer & Oliver 1980; Brewer et al. 1983) or lowered seismic velocity (Landisman et al. 1971; Mueller 1977). These are dealt with in other papers presented in this volume.

From: DAWSON, J.B., CARSWELL, D.A., HALL, J. & WEDEPOHL, K.H. (eds) 1986, *The Nature of the Lower Continental Crust*, Geological Society Special Publication No. 24, pp. 107–117.

To those working in rock mechanics, the question immediately prompted is: could these geophysical features be associated with pressurized pore fluids due to metamorphic dehydration reactions? This was earlier suggested by Mitchell & Landisman (1971), Landisman *et al.* (1971) and Mueller (1977). Burkhardt *et al.* (1982) have demonstrated experimentally a sharp reduction in P- and S-wave velocities as a consequence of dehydration reactions in rocks under undrained conditions at high pressures and temperatures (e.g. ~ 350°C for a zeolite reaction and ~ 500–600°C for chlorite reactions).

Another important geophysical property of the crust is its seismicity. In thermally old continental areas seismicity may extend to depths of perhaps 60 km, extending into the uppermost mantle, and in subducted lithosphere it may extend to depths of 40–50 km below the uppermost surface (Murrell 1984). The mechanical implications of these observations will be further discussed below, but they do strongly suggest that pore fluids may exist in the lithosphere down to uppermost mantle levels, and therefore by implication that such pore fluids can and do exist in the lower crust. Murrell & Ismail (1976a) (see also Murrell 1985) have demonstrated experimentally a sharp reduction in strength and the occurrence of brittle faulting as a consequence of dehydration reactions in rocks under undrained conditions at high pressures and temperatures (e.g. ~ 300°C for a brucite reaction and ~ 400–600°C for chlorite and serpentine reactions, see Fig. 1 (iii) below).

Evolution of the lithosphere and crust

The tectonic and thermal consequences of the evolution of the Earth's lithosphere and crust, in response to thermal convection in the mantle, have been discussed recently by Tarling (1978, 1980, 1981), and Piper (1982) has presented and reviewed the palaeomagnetic data relevant to the tectonic state of continental crust in Precambrian times. In Tarling's view lithosphere evolution reflects the marked reduction in the Earth's internal heat and in the rate of radiogenic heat production since the Earth was formed, and he sees several stages in this evolution. The evolution of the crust is discussed by a number of geologists in the book edited by Tarling (1978), and also by Brown & Mussett (1981) and Windley (1984). The conclusions of Brown & Mussett (pp. 203–4) and of Windley (pp. 354–9) are broadly similar to those of Tarling. Tarling's concepts are used as a basis for discussion.

Pre-Archaean and Archaean times

In the earliest, pre-Archaean (> 3.9 Ga), stage of Earth's history additional internal energy was available, so exceptionally vigorous mantle convection was required to remove the heat. There is no remaining evidence of segregated crust from this period.

However, by the time the earliest surviving crustal rocks were formed, in the Archaean, there is evidence from 3.8 Ga old sediments that running water was present, indicating surface temperatures similar to those of today.

During the Archaean (from 3.9 Ga to *c.* 2.5–2.7 Ga) continental crust was formed which still survives, and possibly 90% of the present continental area had formed by the time the Archaean–Proterozoic boundary was reached (Tarling 1978; Dewey & Windley 1981; Brown & Mussett 1981). The Archaean gneiss terrains required for their formation that the hydrous oceanic crust and mantle in the oceanic lithosphere formed by partial melting of the upwelling mantle should be cool enough and thick enough to be able to survive during subduction to a deep level (Tarling (1980) suggests 100–120 km increasing to 200 km) from which the release of water during dehydration would enable the secondary differentiation process to take place (Ringwood 1974, 1975) which would give rise to crust of granodioritic composition. Once formed these low density rocks could not be subducted back into the mantle.

Both the granite-gneiss terrains and the greenstone belts of the Archaean (dominantly basic volcanic and volcanic-derived sediments) experienced strong thrusting and nappe formation (Tarling 1978), probably related to the production of calc-alkaline magma in a collision (? subduction) environment (Bridgwater *et al.* 1978; Brown & Mussett 1981). Goodwin (1978) suggests that the granite-gneiss terrain was produced by granitization, metamorphism and intrusion of supra-crustal rocks.

Proterozoic times

The palaeomagnetic record begins only at ~ 2.8 Ga ago, close to the Archaean–Proterozoic boundary, however, Piper (1982) shows that there are some suggestions that by then the continental crust had already consolidated into the Supercontinent which appeared to have existed in the earlier Proterozoic (see also Brown & Mussett 1981).

In earlier Proterozoic times the tectonic style of the continental crust showed a marked change from Archaean times (Tarling 1978, 1980; Brown

& Mussett 1981; Windley 1984). Wide areas of continental crust became stabilized, and conditions existed for the formation of thick (> 10 km) layers of platform sediments, deep intra-cratonic troughs, and large dyke swarms.

An important tectonic feature of the late Archaean–early Proterozoic continental crust is the occurrence of shear belts (Sutton 1978) up to 2000–3000 km in length, which Piper (1982) shows to run parallel to the long axis of the Supercontinent. These belts are narrow, dip steeply, and probably penetrate the full crustal thickness. They have a strong planar, mylonitic fabric, and have been a locus for diffusion of fluids, as well as being a continuing source of crustal weakness (Sutton 1978; Piper 1982).

From the mid-Proterozoic (∼ 2.1 Ga ago) down to the beginning of the Phanerozoic, tectonism in the continental crust was concentrated in mobile belts much wider than the shear belts, but still with a trend parallel to the long axis of the Supercontinent (Piper 1982).

Phanerozoic times

The palaeomagnetic data suggest that the process of disruption of the Proterozoic Supercontinent began at ∼ 1.1 Ga ago (Piper 1982), with the rotation of the Fennoscandian Shield relative to the Laurentian Shield, while retaining at the same time their contiguity. This was preceded at a slightly earlier date by the widespread intrusion of dolerite dyke swarms. There is some evidence of other rifting episodes about this time, however Piper (1982) concludes that the main resultant effect was only to shorten the long axis of the Supercontinent from ∼ 220° to ∼ 180° of arc. At this time the Gondwanaland structure was established and then survived down to lower Cambrian times.

By the end of lower Cambrian times Piper (1982) shows that at least four major continental blocks were moving relative to one another, and the Phanerozoic era of plate tectonics, with continental rifting and sea-floor spreading, had begun. This was accompanied by a marked change in the character of igneous activity, with a new access of upper mantle material in the new ocean basins and at the subduction zones. At the same time the present tectonic style marked by continental rifting, subduction zones, and collision (or suture) zones was established.

Nature and strength of the lithosphere and crust

Both the vigour (Tarling 1978), and the style (Piper 1982) of mantle convection changes as the Earth's heat supply diminishes, and Tarling (1978, 1980, 1981) suggests that the tectonism and

tectonic style of the crust are determined by these factors and by the relative thickness and strength of the continental and oceanic lithospheres. Tarling suggests that rifting and dispersal of the continental lithosphere occurred in the Phanerozoic when the older and thicker parts of the oceanic lithosphere became stronger than the continental lithosphere. In mid-Proterozoic times continental lithosphere strength was so much greater than oceanic lithosphere strength that continental rifting was inhibited (Brown & Mussett 1981, p. 204, and Windley 1984, p. 156).

Tarling (1978, 1980) suggests that in the Archaean both continental and oceanic lithosphere were thin, so that continental geotherms were steep and similar to oceanic geotherms. He envisages an initial rapid growth in continental lithosphere thickness in the earlier Proterozoic, which he associates with the stabilization of garnet granulite facies rocks in the lower crust, followed by a continued thickening of both the continental and the oceanic lithosphere. The continental and oceanic geothermal gradients diverged during this period—the oceanic lithosphere remaining thin with a steep geothermal gradient.

There is an alternative view (Davies 1979) that Archaean continental lithosphere thickness was already similar to Phanerozoic because of major differences in continental chemistry and rheology compared with oceanic lithosphere. Support has been claimed for this from recent work on diamond inclusion thermo-barometry (Boyd *et al.* 1985). The differing views on Archaean lithosphere are discussed by Brown & Mussett (1981), who also point out that granulite-facies rocks were already forming in the Archaean (in contradistinction to Tarling's suggestion, see above).

In summary, however, there is a broad concensus that the Phanerozoic is characterized by continental rifting, the Proterozoic was a time of continental stabilization, and the Archaean was the time in which the major part of the geochemically distinct continental crust and lithosphere were constructed by igneous and sedimentary processes driven by mantle convection.

Tarling (1980) made the interesting suggestion that continental lithosphere formation caused the major part of the Earth's surface heat loss by convective processes to become concentrated in oceanic areas (involving advection and subduction of oceanic lithosphere), and that this was the factor that caused continental crust to become relatively more stable in the Proterozoic.

Contrast between oceanic and continental lithosphere and crust

The present oceanic crust and lithosphere is

geologically young and was formed rapidly by the process of sea-floor spreading, so that it has a well-characterized relatively homogeneous structure. The continental crust, by contrast, has been formed and transformed over a large part of the Earth's history and has been subjected to a succession of tectonic and igneous events (see Tarling 1978; Piper 1982; Brown & Mussett 1981; Windley 1984; and the review above). Consequently, continental crust is heterogeneous in structure. Tarling's (1978, 1980) model of lithospheric and crustal evolution provides a framework within which to discuss the nature of the continental lower crust in terms of the physical processes to which it has been subjected. The purpose of the rest of this paper is to consider, within this framework, the inter-relationships between tectonic stress, thermal events, metamorphism, and the structure of the lower crust, using experimental studies of rock deformation under metamorphic conditions as a guide.

Stages in continental crust formation and tectonism

According to Piper's (1982) models and the discussion above we can envisage the following stages in continental crust formation: (1) the formation of the earliest surviving crust during the Archaean, culminating in the massive formation of crust (perhaps 90% of present crustal area) by latest Archaean times; (2) the stabilization of continental crust and lithosphere at the beginning of the Proterozoic, so that it possibly retained its integrity as a single Supercontinent until the beginning of the Phanerozoic, interrupted only by incipient fracturing in the late Proterozoic; (3) and finally the fracturing and fragmentation of the Supercontinent, associated with the formation of new ocean basins followed by their closure and the collision and suturing of continental fragments in accordance with the 'Wilson cycle' (Roberts & Gale 1978).

Relationship between continental crust structure and stages of tectonism

Lithosphere and crust is much weaker in extension than in compression, and orogenic belts form by compression only on continental margins or in areas where the lithosphere and crust have been previously thinned and weakened by stretching (Murrell 1986). The orogenic process is greatly assisted by magmatism associated with dehyd-

ration of subducted plates, which produces new continental crust, and by the related increase in arc and back-arc heat flux.

In the Archaean the rate of formation of continental crust and lithosphere was much more rapid than subsequently (e.g. see Brown & Mussett 1981, Fig. 10.2), and the processes of plating, welding, or suturing together with igneous differentiation and intrusion must have been far more frequent and widespread than in the Proterozoic or Phanerozoic.

In the Proterozoic the stabilization of the continental crust and lithosphere caused rifting to be confined to the thin oceanic lithosphere until a late stage. Orogenic deformation on the continental margins would have been limited also because the oceanic lithosphere was thin and weak (Tarling 1980). So the transformation of the continental crust at this time was largely associated with anorogenic igneous intrusion (Windley 1984).

Finally, in the Phanerozoic rifting ceased to be confined to oceanic areas as the strength of the oceanic lithosphere increased. Fragmentation of the continents by rifting was accompanied by subduction and collision processes at some continental margins, leading to compressive orogenic deformation with crust and lithosphere thickening and large-scale crustal metamorphism (Murrell 1986).

Relationship between metamorphism, strength, and crustal structure

In the absence of metamorphism the strength of rocks (whether or not they have pre-existing faults) increases rapidly with pressure, and therefore with depth of burial, unless a high pore pressure exists (as may be the case in upper crustal sediments). On the other hand, the increase of temperature at greater depths leads (again, in the absence of metamorphism) to increased ductility and eventually to a lower plastic flow stress. (See Murrell 1977, 1984, 1985a; Murrell & Tsang 1984; Goetze & Evans 1979; Kirby 1983.)

Metamorphism which causes evolution of water (i.e. dehydration), creates high pore pressures and lowers the effective stresses, thereby weakening and embrittling rocks, but the deformation may be accompanied either by strain-softening (Fig. 1 (i), (iii)) and the formation of faults on which displacements are concentrated, or by strain-hardening (Fig. 1 (i–iii)) with a distributed style of mass deformation due to cataclastic flow (Ismail & Murrell 1976; Murrell & Ismail 1976a; Murrell 1985—for photographs

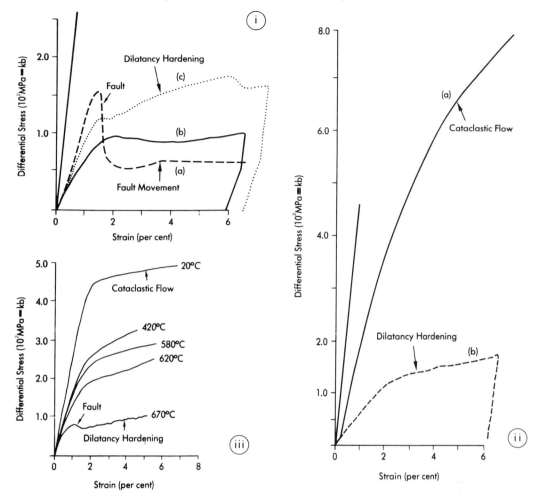

FIG. 1. Stress–strain curves illustrating strain-softening associated with brittle faulting, and strain-hardening associated with distributed cataclastic flow under dry and undrained, water-saturated conditions.

(i) Tests on sandstone. (a) dry, 0.17 kb confining pressure—strain-softening; (b) saturated, 0.35 kb; (c) saturated, 1.10 kb—note strain-hardening due to dilatancy.

(ii) Tests on sandstone at 4.14 kb confining pressure. (a) dry—strain-hardening due to cataclastic flow; (b) saturated—strain-hardening due to dilatancy, note low strength.

(iii) Tests on chloritite at 5.52 kb confining pressure, undrained, and at various temperatures. Note strain-hardening at 20°C due to cataclastic flow and at 670°C due to dilatancy (following an initial strain-softening instability). (Based on figures from Ismail & Murrell 1976, and Murrell & Ismail 1976a, reprinted with permission of Editors, *Geophys. J. R. astr. Soc.* and *Tectonophys.*)

illustrating cataclastic flow in gypsum, serpentinite, and chloritite see Murrell & Ismail 1976a, Fig. 28, and in sandstone see Murrell 1965, Fig. 14c).

Thus metamorphism will strongly influence the strength of the crust and its style of deformation. The latter may in turn influence the rate of loss of water from the crust.

Rock which has been subjected to granulite-grade metamorphism and has entirely lost all water (and other volatiles) will retain high effective stresses at depth and will therefore remain strong unless high temperature reduces the plastic flow stress. However, little is known from direct experiments about the mechanical behaviour of granulite under lower crustal conditions.

Partial melting also greatly weakens rocks (Murrell & Ismail 1976b; Van der Molen & Paterson 1979; Murrell 1985), and this will be important where crustal anatexis occurs.

Metamorphism, zones of pressurized pore fluids, and metamorphic textures

From the point of view of this symposium the questions that rock mechanics might raise are: firstly, the possible existence of zones of high pore-fluid pressure (especially high pore-water pressure) created in the lower crust by infiltration from water saturated crust at higher levels or by metamorphic or igneous processes; secondly, the mechanisms by which pore fluids move in the crust; and thirdly, the texture of the rocks produced by prograde metamorphism.

The likely role of high pore pressures in the formation and movement of thrusts and nappes was recognized by Hubbert & Rubey (1959), and has been emphasized again recently (Murrell 1981, 1986; Westbrook *et al.* 1982). However, the creation of these high pore pressures, which must approach the confining pressure if they are to achieve their greatest effect (Murrell 1981), requires the rocks containing them to be sealed (encapsulated) by less permeable rocks. Obviously metamorphism which creates pressurized pore water thereby increases the permeability of the rock. Dilatancy due to brittle deformation also increases the permeability, and when a pore fluid is present this dilatancy will either cause diffusive flows of the fluid into or out of the rock (when the rock experiences 'drained' conditions, i.e. it is not sealed) or else will cause the pore pressure to drop, which results in dilatancy-hardening (when the rock is in an 'undrained' condition) (Ismail & Murrell 1976; Brace & Martin 1968, see Fig. 1 above). During a prograde phase of metamorphism rising temperature will tend to further increase permeability due to thermal expansion of water (or other volatiles) which is released, whereas during a retrograde phase in which cooling occurs permeability will be reduced as dissolved minerals are precipitated from pore fluids. At higher crustal levels low permeability may be produced by consolidation and diagenesis as well as by precipitation from cooling pore fluids infiltrated from overpressured pore fluid zones at deeper levels.

The relationships between permeability, dilatancy, and prograde or retrograde metamorphism, and more generally the effect of confining pressure, temperature, and strain conditions on these relationships, are matters which require a great deal more experimental (and theoretical) investigation (see Batzle & Simmons 1977; Murrell 1985).

The subject of the textures produced by prograde metamorphism in which water is evolved has been addressed recently (Murrell 1985). It is suggested that during burial metamorphism of a hydrated rock distributed deformation (associated with dilatancy-hardening and cataclastic flow) will tend to take place at the peak of the metamorphism, and this will be accompanied by grain-growth and recrystallization of the new mineral phases. A fabric will be produced which reflects the deviatoric stress state, though little is known about this from experiment (see Rutter & Peach 1984). However, this stage may be preceded and followed by stages of fracture and vein formation as water evolves by dehydration from deeper or shallower rocks.

Two modes of metamorphism are envisaged. One arises from lithosphere deformation (thinning by stretching or thickening by shortening, see Murrell 1985, 1986), and this is particularly applicable to Phanerozoic events. The other arises from igneous intrusion, which in some cases is associated with subduction. Fabrics will be produced during metamorphism, which will depend on the regional or local tectonic stresses. These fabrics, which will affect the seismic reflectivity of the lower crust, are discussed in more detail below.

Seismic and flexural behaviour of the lower crust

In the absence of deep drilling into the lower crust (which is now projected, Anderson 1984) we have to rely on geophysical observations to determine conditions at these depths. We discuss here seismicity, which is indicative of some form of strain-softening, and crust and lithosphere flexure, which is indicative of elastic or plastic deformation.

Crustal seismicity occurs at mid- to upper-crust depths in thermally younger crust but may extend throughout the crust and into the uppermost mantle in cratonic areas (Murrell 1984; Murrell & Tsang 1984; Murrell 1977). At these levels the strain-softening process involves brittle faulting (either by extending or forming a fault, or by causing movement, opposed by friction, on a pre-existing fault). There are good reasons for believing that in most cases high pore pressures, which reduce the effective stresses, are involved in the faulting process. However, under undrained conditions, although the strength may be low when a pore fluid is present, deformation may be accompanied by strain-hardening due to dilatancy, and seismic faulting will not occur (Ismail & Murrell 1976). Deformation under these conditions is by cataclastic flow.

Cataclastic flow can also occur at high confining pressures in the absence of pore-fluids, when the deviatoric stress required to cause brittle faulting is exceeded by the deviatoric stress due to friction on a fault (Murrell 1965; Edmond & Murrell 1972, see Fig. 1 above). This is also aseismic, but the deviatoric stresses required are very large and it seems unlikely that cataclastic flow will occur naturally in the absence of pore fluids (Murrell 1984, 1985).

The seismicity cut-off in the lower crust or uppermost mantle is often thought to be due to enhanced ductility of the rock, which is generally ascribed to a reduction of the plastic flow stress caused by elevated temperature. However, an alternative possibility is that there is a transition from a strain-softening to a strain-hardening regime associated with dilatancy-hardening during cataclastic flow. Crystal plasticity would be the likely deformation mechanism in dehydrated rock, such as crustal granulite or upper mantle dunite, at sufficiently high temperatures and low strain rates. If water is present in the rock then cataclastic flow is a likely deformation mechanism.

The elastic thickness of the lithosphere can be calculated from flexure observations (Walcott 1970; Watts *et al.* 1975; Murrell 1976; Caldwell & Turcotte 1979; Karner *et al.* 1983). In this case Karner *et al.* have shown that the flexural rigidity (proportional to the cube of the elastic thickness) increases with the thermal age of the lithosphere at the time of loading, and this increase is related to the cooling and thickening of the lithosphere. They show that the depth limit of elastic behaviour coincides with the 450°C geotherm in both continental and oceanic lithosphere (continental lithosphere behaviour is more complex, Willett *et al.* 1985). The data reviewed by Murrell (1984) shows that the seismic cut-off boundary occurs at a similar temperature. Although there is no experimental evidence of steady state creep in silicate rocks at temperatures as low as 450°C, there is evidence of transient creep following Andrade's law, indicating a significant contribution of thermal recovery (annealing) processes to the creep (Murrell & Chakravarty 1973; Murrell 1976).

It therefore seems possible that both plastic and cataclastic deformation processes could occur in the lower crust. However, it is very difficult to think of an example of a lower crustal rock tectonized by plastic deformation alone. In the lower crust it seems likely that deformation is by a combination of a cataclastic process initiated during the earlier stages of prograde metamorphism, with subsequent plastic processes associated with fine-grained new mineral phases or cataclasized inert phases. The plastic processes would tend to be diffusive in character, and would tend to be accelerated by the presence of hot, pressurized water released during metamorphism.

Structures in the lower continental crust produced by deformation and metamorphism ('hot working')

Stretched crust

Lithosphere stretching leading to crustal thinning will result in an initial retrograde phase of metamorphism as the crust cools. As a basin forms on the thinned crust and becomes filled with sediments the original crust is buried and transformed into new lower crust as increasing pressures and temperatures produce a later stage of prograde (burial) metamorphism (see Fig. 2(a)). Water from the basin will become incorporated into the upper part of the original crust, which forms the basement to the new sediment infill. The balance between the retrograde and prograde phases of metamorphism will depend on the depth in the original crust.

Insofar as the geotherms are relatively flat-lying, zones of pressurized pore water formed by metamorphism might tend to be sub-horizontal. These might localize the soles of faults. At the same time, however, normal faults formed during stretching will also tend to be filled with water (or other fluids) from zones of pressurized pore fluids from all levels of the crust, since faulting at the relatively low levels of deviatoric stress thought to exist can only take place when effective stresses are low (Murrell 1977, 1981).

During compaction and diagenesis and during prograde metamorphism dehydration under undrained conditions will tend to result in distributed cataclastic deformation leading to the formation of a texture (Murrell 1985). A similar but more marked texture would be formed in fault zones.

Shortened crust

In the case of lithosphere shortening the consequent crustal thickening will result in prograde metamorphism as the crust heats up (in this case burial of crust is due to tectonic deformation rather than sedimentation). Retrograde metamorphism will follow as a result of uplift and erosion (Murrell 1986, and see Fig. 2b below).

Once again water released during the prograde phase will tend to be trapped during the retrograde phase, and will tend to occur in flat-lying zones, including sub-horizontal or low-inclina-

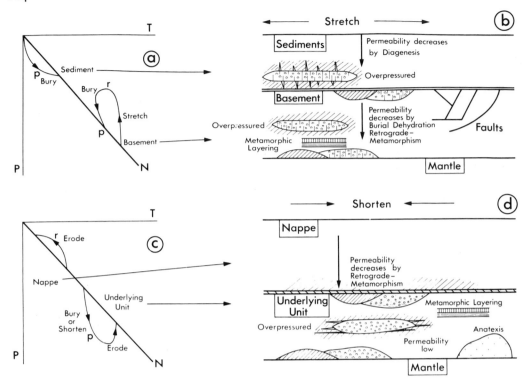

FIG. 2. Metamorphism due to lithosphere deformation illustrated by paths in pressure (P)—temperature (T) space, with cartoons suggesting crustal structures produced by this tectono-thermal process.

(a) Metamorphism due to crustal stretching leading to thinning and basin formation.

(b) Cartoon of crustal structures produced by stretching.

(c) Metamorphism due to crustal shortening leading to thickening and mountain-building followed by erosion.

(d) Cartoon of crustal structures produced by shortening.

Legend: p, prograde metamorphism; r, retrograde metamorphism; N, normal geotherm

Note: over-pressured layers or lenses may have sub-vertical or sub-horizontal hydraulic fractures or veins in stretched or shortened crust respectively.

tion fault zones and nappe soles, and zones characterized by a texture produced by cataclastic deformation and the subsequent growth of new mineral phases (Murrell 1985). In collision zones another source of deep, pressurized pore-water is in geosynclinal sediments trapped and tectonically buried (Murrell 1986).

Structures and textures due to 'hot working'

The formation of pressurized pore water during metamorphism and during the burial and consolidation of saturated sediments will tend to produce microscopic or macroscopic hydraulic fractures (e.g. veins) oriented perpendicular or sub-perpendicular to the least compressive principal stress (the orientation depending on the magnitude of the deviatoric stress, see Jaeger & Cook

1976). Thus in stretched crust the hydraulic fractures will tend to be vertical or sub-vertical, and in shortened crust they will tend to be horizontal or sub-horizontal.

The fractures may be rotated during subsequent deformation, and this must be a major factor during cataclastic flow. In fault zones which have been strongly sheared a strongly planar texture will be produced.

Migration of hot pore-water, followed by cooling, will result in mineral precipitation, which may produce a laminated structure, and could also produce sufficient acoustic contrast to cause good seismic reflectivity (Mueller 1977; Smithson *et al.* 1977).

Cartoons illustrating the structures suggested are given in Fig. 2b (stretched crust) and Fig. 2d (shortened crust).

Geophysical observation of zones of pressurized pore water in the crust and upper mantle

Zones in which minerals have lost bound water by metamorphism or which have been invaded by water evolved during this process can obviously be thick, especially under undrained conditions when loss of water is small due to the low permeability of surrounding rocks. Under these latter conditions a more uniform, cataclastic texture will be produced.

With sufficient water present in such thick zones there would be good seismic and electrical conductivity contrast with surrounding rock formations containing less water. In stretched crust, from the arguments above, vertical or sub-vertical hydraulic fractures in these hydrous zones, which would have regional parallelism (determined by the regional stress pattern), would give the zones some vertical planar anisotropy which could perhaps be detected geophysically. In shortened crust horizontal or sub-horizontal hydraulic fractures would give the hydrous zones a horizontal planar anisotropy.

Although fault zones, many of which will have a horizontal or sub-horizontal orientation, would produce similar (perhaps stronger) geophysical contrasts over short distances, these zones will tend to be initially thin because of their origin from a strain-softening process. They might not, therefore, be observable geophysically. However, on the assumption that the deformation processes involve low effective stresses because of the presence of highly-pressurized pore water evolved during metamorphism, it is possible that there could be a switch from undrained conditions in which widespread, homogeneous cataclastic deformation would take place to drained conditions in which strain-softening and fault formation would occur and then again to undrained conditions with a return to widespread, homogeneous cataclastic deformation. Such switches from undrained to drained conditions and vice versa would depend on the rate of strain, the temperature, the permeability (and permeability contrasts), and the effects of deformation and temperature (e.g. through solubility changes) on permeability. Therefore some fault zones might be thin, but others might be thick because of the occurrence of substantial cataclastic deformation due to episodes of undrained deformation.

Lateral termination of zones of pressurized pore water

These zones will be terminated at the margins of zones of metamorphism and, hence, at the margins of basins and of orogenic belts. In addition they will be terminated where there are highly permeable conduits, such as major inclined fault zones, through which pore water could drain away.

Where lithosphere and crustal stretching has led to continental rifting and sea-floor spreading it might be expected that in the continent–ocean transition zone extensive fracturing and faulting of the crust, due both to stretching and to isostatic adjustment, would tend to lead to drainage of water out of the hydrous zones, though this will also be affected by the complex metamorphic conditions (associated with the changing thermal conditions), which will affect both the evolution of water and the crustal permeability.

Lower-crustal structure and crustal age

From the arguments above, although as much as 90% of continental crustal area was formed during Archaean time, large areas were affected by igneous events during Proterozoic times, and in the Phanerozoic rifting and orogeny have affected many parts of the crust but in relatively narrow belts.

The structures postulated would be expected to characterize the era of plate tectonics, comprising the Phanerozoic and the latest Proterozoic, and also perhaps any crust which has remained stabilized since Archaean times.

On the other hand, crust which has remained stabilized since the last Proterozoic igneous event to which it was subjected might be expected to show a different type of structure. Although hydrous zones would have been formed by thermal metamorphism they would tend to be wrapped around batholith-like bodies, and would not have the flat horizontal, sub-horizontal, or inclined lens or layer shape postulated.

Conclusions

1 It is suggested that the structure of the lower continental crust has been established by prolonged thermal and tectonic processes which caused metamorphism leading to the creation of zones of highly-pressurized pore water.

2 Following Tarling (1978) and Piper (1982) it is suggested that there have been three principal periods of crustal structure formation:
(a) The Archaean (3.9 to 2.5 Ga), during which perhaps 90% of the crust was formed, involving thermal and tectonic processes associated with vigorous mantle convection;

(b) the Proterozoic, during which large areas of the crust became stabilized but were affected by igneous activity; and

(c) the Phanerozoic, during which substantial areas of crust in relatively narrow belts have been re-worked by thermal and tectonic processes, involving rifting, orogeny, and suturing and associated magmatism.

3 Generalization of McKenzie's (1978) hypothesis of lithosphere deformation by stretching to form basins to encompass shortening to form orogenic belts at continental margins (Murrell, 1985a) leads also to the concept that the thermal perturbations due to lithosphere deformation (by stretching or shortening) give rise to specific patterns of metamorphism. Such metamorphism, associated with lithosphere deformation, is thought particularly to characterize the Phanerozoic and perhaps also the Archaean periods of crustal structure formation.

4 The evolution of water during metamorphism generates microscopic and macroscopic hydraulic fractures oriented perpendicular or sub-perpendicular to the least compressive principal stress, and these will therefore be initially in vertical planes in stretched crust and in horizontal planes in shortened crust. Lowering of the effective stresses by the pore-water pressure enables large scale deformation to take place by either: (a) strain-softening and faulting in thin zones under drained conditions; or (b) strain-hardening and cataclastic deformation in thick zones under undrained conditions. Faulting may be seismic, and it is postulated that seismic faulting in the lower crust and uppermost mantle is associated with pressurized pore fluids (especially water) released during metamorphism.

5 It is suggested that seismically reflective, low velocity, and high conductivity sub-horizontal and inclined layers or lenses observed in the crust and uppermost mantle are likely to be thick zones of cataclastically deformed rock containing large quantities of highly-pressurized pore water released by metamorphism. A laminated structure arising from migration and cooling of pore water leading to mineral precipitation could also be responsible for high seismic reflectivity.

References

ANDERSON, R.N. 1984. Continental interior in view. *Nature*, **311**, 512.

BATZLE, M.L. & SIMMONS, G. 1977. Geothermal systems: rocks, fluids, fractures. *In*: HEACOCK, J.G. (ed.) *The Earth's Crust: its Nature and Physical Properties*. American Geophysical Union: Washington, D.C., 233–242.

BRACE, W.F. & MARTIN, J.R. 1968. A test of the law of effective stress for crystalline rocks of low porosity. *Int. J. Rock Mech. Min. Sci.* **5**, 415–426.

BOYD, F.R., GURNEY, J.J. & RICHARDSON, S.H. 1985. Evidence for a 150–200 km thick Archaean lithosphere from diamond inclusion thermobarometry. *Nature*, **315**, 387–9.

BREWER, J.A. & OLIVER, J.E. 1980. Seismic reflection studies of deep crustal structure. *Ann. Revs. Earth Planet. Sci.* **8**, 205–230.

—— *et al.* 1983. BIRPS deep seismic reflection studies of the British Caledonides. *Nature*, **305**, 206–210.

BRIDGWATER, D., COLLERSON, K.D. & MYERS, J.S. 1978. The development of the Archaean gneiss complex of the North Atlantic region. *In*: TARLING, D.H. (ed.) *Evolution of the Earth's Crust*. Academic Press: London, 19–69.

BROWN, G.C. & MUSSETT, A.E. 1981. *The Inaccessible Earth*. George Allen & Unwin: London. pp. 235.

BURKHARDT, H., KELLER, F. & SOMMER, J. 1982. Compressional and shear wave velocities in metamorphic rocks under high pressures and temperatures. *In*: SCHREYER, W. (ed.) *High-Pressure Researches in Geoscience*. E. Schweizerbart'sche: Stuttgart, 47–65.

CALDWELL, J.G. & TURCOTTE, D.L. 1979. Dependence of the thickness of the elastic oceanic lithosphere on age. *J. geophys. Res.* **84**, 7572–6.

DAVIES, G.F. 1979. Thickness and thermal history of continental crust and root zones. *Earth. Planet. Sci. Lett.* **44**, 231–8.

DEWEY, J.F. & WINDLEY, B.F. 1981. Growth and differentiation of the continental crust. *Phil. Trans. R. Soc. Lond.* **A301**, 189–206.

EDMOND, O. & MURRELL, S.A.F. 1972. Experimental observations on rock fracture at pressures up to 7 kbar and the implication for earthquake faulting. *Tectonophysics*, **13**, 71–87.

GOETZE, C. & EVANS, B. 1979. Stress and temperature in the bending lithosphere as constrained by experimental rock mechanics. *Geophys. J. R. astron. Soc.* **59**, 463–478.

GOODWIN, A.M. 1978. The nature of Archaean crust in the Canadian Shield. *In*: TARLING, D.H. (ed.) *Evolution of the Earth's Crust*. Academic Press: London, 175–218.

GREEN, A.R. 1977. The evolution of the Earth's crust and sedimentary basin development. *In*: HEACOCK, J.G. (ed.) *The Earth's Crust: its Nature and Physical Properties*. American Geophysical Union: Washington, D.C., 1–17.

HUBBERT, M.K. & RUBEY, W.W. 1959. Role of fluid pressure in mechanics of overthrust faulting, I. *Bull. geol. Soc. Am.* **70**, 115–166.

ISMAIL, I.A.H. & MURRELL, S.A.F. 1976. Dilatancy and the strength of rocks under undrained conditions. *Geophys. J. R. astr. Soc.* **44**, 107–134.

JAEGER, J.C. & COOK, N.G.W. 1976. *Fundamentals of Rock Mechanics*. Chapman & Hall: London.

KARNER, G.D., STECKLER, M.S. & THORNE, J.A. 1983. Long-term thermo-mechanical properties of the continental lithosphere. *Nature*, **304**, 250–253.

KAY, R.W. & KAY, S.M. 1981. The nature of the lower continental crust: inferences from geophysics, surface geology and crustal xenoliths. *Rev. Geophys. Space. Phys.* **19**, 271–297.

KIRBY, S.H. 1983. Rheology of the lithosphere. *Rev. Geophys. Space Phys.* **21**, 1458–1487.

LANDISMAN, M., MUELLER, S. & MITCHELL, B.J. 1971. Review of evidence for velocity inversions in the continental crust. *In*: HEACOCK, J.G. (ed.) *The structure and physical properties of the Earth's crust*. American Geophysical Union: Washington, D.C., 11–34.

McKENZIE, D.P. 1978. Some remarks on the development of sedimentary basins. *Earth. Planet. Sci. Lett.* **48**, 25–32.

MITCHELL, B.J. & LANDISMAN, M. 1971. Electrical and seismic properties of the Earth's crust in the southwestern Great Plains of the U.S.A. *Geophysics*, **36**, 363–381.

MUELLER, S. 1977. A new model of the continental crust. *In*: HEACOCK, J.G. (ed.) *The Earth's crust: its nature and physical properties*. American Geophysical Union: Washington, D.C., 289–317.

MURRELL, S.A.F. 1965. The effect of triaxial stress systems on the strength of rocks at atmospheric temperature. *Geophys. J. R. astr. Soc.* **10**, 231–281.

—— 1976. Rheology of the lithosphere-experimental indications. *Tectonophysics*, **36**, 5–24.

—— 1977. Natural faulting and the mechanics of brittle shear failure. *J. geol. Soc. Lond.* **133**, 175–189.

—— 1981. The rock mechanics of thrusts and nappes. *In*: McCLAY, K.R. & PRICE, N.J. (eds) *Thrust and Nappe Tectonics*. Geol. Soc. Lond. Spec. Publ. No. 9. Blackwell Scientific Publications, Oxford, 99–109.

—— 1984. Earthquake source physics and mechanical processes in the lithosphere. *In*: *Continental Seismicity and Earthquake Prediction*. Seismological Press: Beijing, pp. 676–687.

—— 1985. Aspects of relationships between deformation and prograde metamorphisms which causes evolution of water. *In*: THOMPSON, A.B. & RUBIE, D.C. (eds) *Metamorphic reactions: Kinetics, textures and deformation*. (Adv. in Physical Geochemistry, Vol. 4). Springer-Verlag: New York, 211–241.

—— 1986. The mechanics of tectogenesis in plate collision zones. *In:* COWARD, M.P. & RIES, A.C. (eds) *Collision Tectonics*. Geol. Soc. Lond. Spec. Publn. No. 19, 95–111.

—— & CHAKRAVARTY, S. 1973. Some new rheological experiments on igneous rocks at temperatures up to 1120°C. *Geophys. J. R. astr. Soc.* **34**, 211–250.

—— & ISMAIL, I.A.H. 1976a. The effect of decomposition of hydrous minerals on the mechanical properties of rocks at high pressures and temperatures. *Tectonophysics*, **31**, 207–258.

—— & ISMAIL, I.A.H. 1976b. The effect of temperature on the strength at high confining pressure of gradodiorite containing free and chemically-bound water. *Contrib. Mineral. Petrol.* **55**, 317–330.

—— & TSANG, S.H. 1984. Rock mechanics, earthquake mechanisms, and earthquake prediction. *In*: RIKITAKE, T. (ed.), *Proceedings 1979 Earthquake Prediction Symposium*. UNESCO: Paris, 70–90.

PIPER, J.D.A. 1982. Movements of the continental crust and lithosphere–asthenosphere systems in pre-Cambrian times. *In*: BROSCHE, P. & SUNDERMAN, J. (ed.), *Tidal Friction and the Earth's Rotation*. Springer-Verlag: Berlin, 253–321.

RINGWOOD, A.E. 1974. Petrological evolution of island arc systems. *J. geol. Soc. Lond.* **130**, 183–204.

—— 1975. *Composition and Petrology of the Earth's Mantle*. McGraw-Hill: New York, pp. 249–322.

ROBERTS, D. & GALE, G.H. 1978. The Caledonian–Appalachian Iapetus ocean. *In*: TARLING, D.H. (ed.), *Evolution of the Earth's Crust*. Academic Press: London, 255–342.

RUTTER, E. & PEACH, C.J. 1984. Experimental 'syntectonic' hydration of basalt. *In*: HENDERSON, C.M.B. (ed.), *Progress in Experimental Petrology. 1981–1984*. N.E.R.C.: Swindon, 249–260.

SMITHSON, S.B., SHIVE, P.N. & BROWN, S.K. 1977. Seismic velocity, reflections, and structure of the crystalline crust. *In*: HEACOCK, J.G. (ed.), *The Earth's crust: its nature and physical properties*. American Geophysical Union: Washington, D.C., 254–270.

SUTTON, J. 1978. Proterozoic of the North Atlantic. *In* TARLING, D. H. (ed.), *Evolution of the Earth's Crust*. Academic Press: London, 239–254.

TARLING, D.H. 1978. Plate tectonics: present and past. *In*: TARLING, D.H. (ed.), *Evolution of the Earth's Crust*. Academic Press: London, pp. 361–408.

—— 1980. Lithosphere evolution and changing tectonic regimes. *J. geol. Soc. Lond.*, **137**, 459–467.

—— 1981. The tectonic evolution of the Earth's surface and changing lithospheric properties. *In*: DAVIES, P.A. & RUNCORN, S.K. (eds), *Mechanisms of Continental Drift and Plate Tectonics*. Academic Press: London, 61–73.

VAN DER MOLEN, I. & PATERSON, M.S. 1979. Experimental deformation of partially melted granite. *Contrib. Mineral. Petrol.* **70**, 299–318.

WALCOTT, R.I. 1970. Flexural rigidity, thickness, and viscosity of the lithosphere. *J. geophys. Res.* **75**, 3941–3954.

WATTS, A.B., COCHRAN, J.R. & SELZER, G. 1975. Gravity anomalies and flexure of the lithosphere: a three-dimensional study of the Great Meteor Seamount, northeast Atlantic. *J. geophys. Res.* **80**, 1391–8.

WESTBROOK, G.K., SMITH, M.J., PEACOCK, J.H. & POULTER, M.J. 1982. Extensive underthrusting of undeformed sediment beneath the accretionary complex of the Lesser Antilles subduction zone. *Nature*, **300**, 625–8.

WILLETT, S.D., CHAPMAN, D.S. & NEUGEBAUER, H.J. 1985. A thermo-mechanical model of continental lithosphere. *Nature*, **314**, 520–3.

WINDLEY, B.F. 1984. *The Evolving Continents*. 2nd Edition. Wiley: Chichester, 399 pp.

STANLEY ARTHUR FRANK MURRELL, Department of Geological Sciences, University College London, Gower Street, London WC1E 6BT.

Intraplate seismicity induced by stress concentration at crustal heterogeneities—the Hohenzollern Graben, a case history

K. Fuchs

SUMMARY: The intraplate seismicity of the Hohenzollern Graben in southwest Germany is concentrated within a small region on the northern end of long shear lineaments extending from the Alps into their northern foreland. Strong shocks with magnitudes not more than 5.6 started in 1911 and repeat about every 30 years with a tendency to migrate northward and at the same time to become shallower. Recent seismic reflection and refraction work have suggested that the Hohenzollern Graben is located in the southeastern border region of an anomalous low velocity, probably high temperature body, near the Urach geothermal anomaly. It is also suggested that the top of the lower crust dips sharply from this body to the southwest. The hypocentres of the deepest earthquakes become shallower as the lineaments approach and traverse the Hohenzollern Graben in the direction towards the Urach body. From all these observations taken together, it appears that the release of seismic energy is concentrated into the area of the Hohenzollern Graben by an extensive lower crustal heterogeneity. The stress concentration in the brittle upper crust is induced by the thickening of the lower crust with its ductile behaviour. The viscosity of the lower crust in the Urach body has suffered further reduction by increased temperatures. Future research has to clarify the dominant cause of the stress concentration: is it the heterogeneity in material or the heterogeneity in thermal properties?

Earthquakes in the middle of continents are one of the phenomena to which the International Lithosphere Program devotes special attention. The location of intraplate seismicity cannot be understood in terms of plate tectonics. The lower crust appears to be free from earthquakes since creep release of stress is the dominating feature here (Chen & Molnar 1983; Meissner & Strehlau 1982). Yet it transpires that the structure of the lower crust governs the places where earthquakes occur in the middle of continents. Lateral variations of the thickness of the lower crust can modulate an otherwise homogeneous regional stress field, thinning the shallow brittle part of the crust and causing stress concentrations which ultimately will overcome the resistance to fracture. The intraplate seismicity of the Hohenzollern Graben region in southwestern Germany where lateral crustal heterogeneities have been discovered recently will serve as an example for this process.

The geological setting

The Hohenzollern Graben (abbreviated to HZGR in the following) is located in the southwestern part of the Federal Republic of Germany,

about 100 km SSW of Stuttgart (Fig. 1). It is characterized by Hercynian basement rocks covered by gently dipping Mesozoic strata. The graben is 28 km in length and 1.5 km in width and has subsided only 100–110 m. The thickness of the sedimentary cover is estimated to be about 1200 m. The northern rim of the graben is formed by a distinct fault, while its southern margin can be observed as a series of faults. There is no evidence of tectonic movement since the upper Miocene (Illies 1982). Therefore, the intraplate seismicity of the HZGR is not connected with recent tectonics of the graben.

Instead the seismicity is believed to be related to the position of the HZGR in the regional stress field, governed by the Alpine orogeny in the south and the Rhine Graben taphrogenesis in the west. Illies (1982) suggested, as the origin of the seismic activity in the HZGR area, the blocking of a sinistral shear motion on a bundle of N-S trending lineaments by the subsurface structure of the HZGR. He suggested that these shear lineaments were activated a few million years ago when the central European stress field rotated anticlockwise and strike slip motions were initiated along the master faults of the upper Rhine Graben. However, the *en enchelon* structure of the Rhine Graben impeded this motion. Instead the conti-

From: DAWSON, J.B., CARSWELL, D.A., HALL, J. & WEDEPOHL, K.H. (eds) 1986, *The Nature of the Lower Continental Crust*, Geological Society Special Publication No. 24, pp. 119–132.

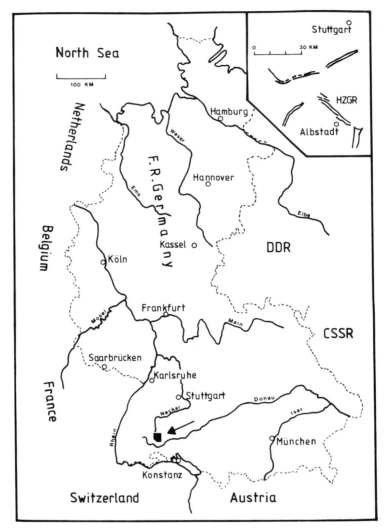

FIG. 1. The Hohenzollern Graben (HZGR) is located in southwestern Germany (F.R.), indicated by the black arrow. A blow up is shown in the right upper corner. The HZGR is the NW–SE striking feature north of the town of Albstadt.

nuing orogeny of the Alps formed a new pattern of shear motions in its foreland in Southern Germany to the east of the Rhine Graben. These lineaments which originate in the Alps extend for several hundred kilometres, well into southern Germany. They also intersect the HZGR.

There is further evidence that the seismic activity is not related to rifting of the HZGR. The recent activity starts below a depth of 2 km (Turnovsky 1981), i.e. below the small wedge of the HZGR.

The seismicity of the Hohenzollern Graben

The occurrence of seismic activity in the region of the HZGR can be traced back to about 1850 (Fiedler 1954). The scientific investigation of the seismicity in the western Swabian Alps started after the 1911–1913 earthquake series. Its largest event, on 16 November 1911, has a surface wave magnitude $M_L = 5.6$. It is the largest seismic event in central Europe north of the south-Alpine

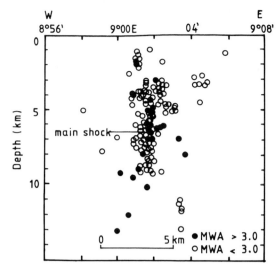

FIG. 2. Hypocentre depth distribution of the 1978 series projected into an E–W section. The E–W extent of the seismic zone is only 3–4 km (Turnovsky 1981). MWA = local magnitude Wood-Anderson.

seismic belt. In 1933 the first seismological station (Meßstetten, MSS), was installed close to the epicentral area (Hiller 1934). The seismic activity reached a new climax between 1934 and 1944; the two strongest events in May 1943 caused considerable damage. After 1945 the network of seismic stations in this area was considerably expanded and improved (Berckhemer & Hiller 1960). Kinematic studies revealed the predominant left lateral strike-slip character of the motions on the fault planes (Schneider *et al.* 1966; Schick 1968, 1970). There is a tendency to northward migration of the seismic activity (Brenner 1961).

After an almost quiet period of about 10 years, seismic activity started again in 1969 as a series climaxed on 3 September 1978 with the large earthquake MWA = 5.7 followed by a series of aftershocks (Haessler *et al.* 1980; Scherbaum 1980; Turnovsky 1981).

An important characteristic of seismic activity in the HZGR region is the distribution of hypocentres in a relatively narrow vertical zone with a thickness of only a few kilometres. This is revealed by a study of aftershocks of the 1978 series (Fig. 2; Turnovsky 1981).

Crustal structure of the Hohenzollern Graben region

Information on the crustal structure of the HZGR region was obtained during seismic studies in a site survey of the geothermal research well Urach 3 (Haenel 1982), about 50 km from the HZGR. It revealed strong lateral variations of crustal structure. The survey was carried out with both refraction and reflection experiments. Several refraction lines traversed the region of the HZGR (Fig. 3). The combined interpretation of these data revealed the presence of strong lateral inhomogeneities meeting in the vicinity of the HZGR.

(I) An anomalous upper crustal body of low velocity was detected at the site of the Urach well both from refraction (Prodehl *et al.* 1982) and reflection surveys (Bartelsen *et al.* 1982; Trappe 1983) with its south-western margin extending into the region of the HZGR. Low P-velocity combined with high density are special characteristics of the body required to match both travel times and amplitude observations.

(II) The lower crust to the W of the HZGR is obviously slower and more homogeneous than the crust to the NE.

(III) Alternative models for the uprise of the lower crust in the vicinity of the HZGR give an estimate of the range of possible structures compatible with the seismic observations (Fig. 4; Gajewski & Prodehl 1984). The models show that the currently available data do not constrain the structure sufficiently. However, the existing surveys do show that in the region of the HZGR strong lateral heterogeneities exist. In the following their significance for the release of seismic energy will be discussed.

Figures 4a–d show the most important record sections from the refraction experiment which provide constraints on the structure of the 'Urach body' of low P-velocity to the NE of the HZGR (Gajewski & Prodehl 1984). Phase 'b' is the wide-angle reflection from the bottom of the body, observed to the NE from both shot points U1 (Fig. 4b) and U2 (Fig. 4a); however, towards the SW it is only recorded from U1 (Fig. 4c) but not from U2 (Fig. 4d). This latter section is characteristically void of all phases beween 'a_3' (i.e. the arrival from the crystalline upper crust) and 'c' (reflection from the Moho). The absence of phase 'b' on the profile U2-SW is the most important though indirect information on the southwestern margin. The model is also corroborated by the amplitudes of phase 'a_3' propagating though the upper crust. On top of the 'Urach body' this phase decays strongly, reaching distances of hardly 50–60 km. It is screened off by the top of the 'Urach body'. In contrast, to the SW from shotpoint U2 this phase is propagated to distances larger than 100 km.

The model presented in Fig. 5 is concordant

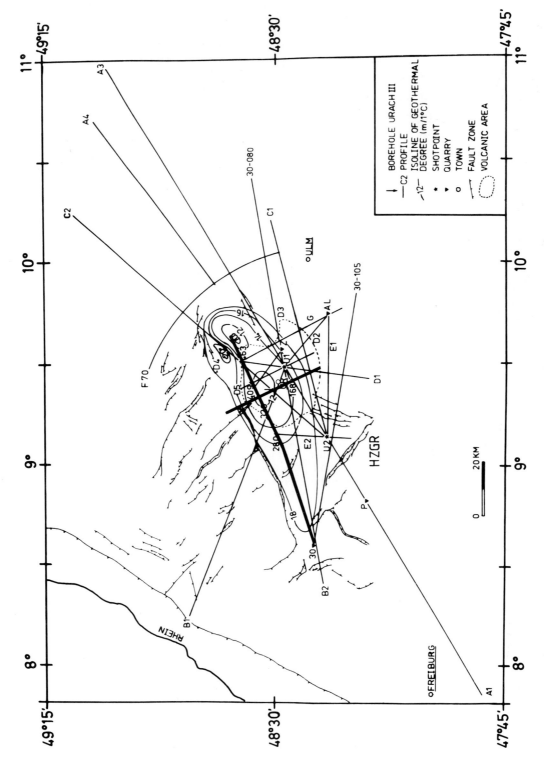

FIG. 3. Deep seismic sounding profiles in the neighbourhood of the geothermal research well Urach 3. Thin straight lines: refraction profiles; thick crossing lines: reflection profiles.

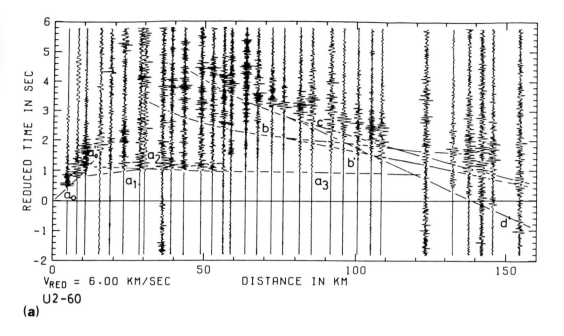

(a)

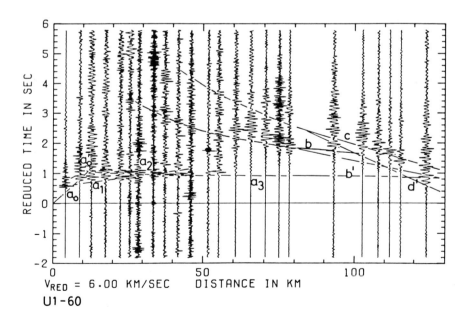

(b)

FIG. 4. (a–b) Record sections from the refraction seismic survey of the Urach area of lines crossing the HZGR: (a) U2-60: shotpoint U2 to NE; (b) U1-60: shot point U1 to NE. The positions are given in Fig. 3. The letters signify various seismic phases (see text). Reduction velocity 6.0 km/s (Gajewski & Prodehl 1984). 60 is the shot-station azimuth.

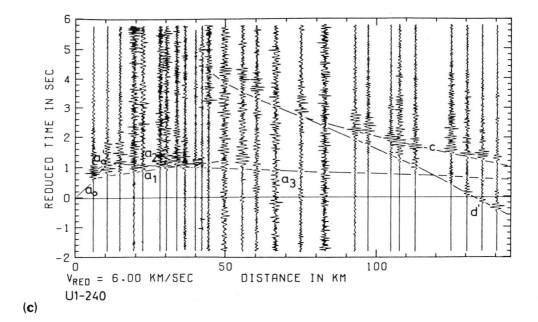

(c)

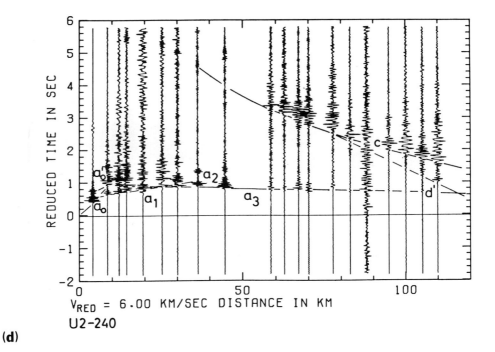

(d)

FIG. 4. (c–d) Record section from refraction seismic survey of the Urach area of lines crossing the HZGR: (c) U1-240: shotpoint U1 to SW; (d) U2-240: shotpoint U2 to SW. Compare Fig. 4a–b. 240 is the shot-station azimuth.

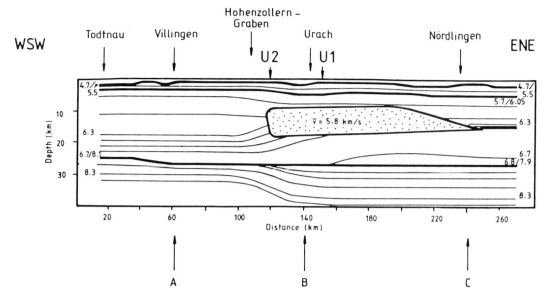

FIG. 5. Models of lateral crustal heterogeneities in the neighbourhood of the Hohenzollern Graben (HZGR) (Gajewski & Prodehl 1984).

with the observed record sections shown in Fig. 4a–d. The 'Urach body' is bordered to the SW by a local rise of lower crustal material. The synthetic seismogram section together with the corresponding ray diagram through the crust SW of shot point 240, i.e. to the SW of the 'Urach body', is shown in Fig. 6. Clearly phase 'b' is absent in this section. This zone to the SW of the 'Urach body' appears to be characterized by the absence of noticeable crustal heterogeneities. Also strong vertical gradients cannot exist, otherwise phases of type 'b' would have been generated. It is expected that the seismic experiments carried out in the HZGR region in August 1984 will lead to a better resolution of the crustal structure.

Details of the structure of the anomalous 'Urach body' have been established by a reflection experiment with combined steep- and wide-angle observations (Bartelsen *et al.* 1982; Meissner *et al.* 1983; Trappe 1983). The reflection profile (see Fig. 3) did not reach the HZGR, but it provided independent evidence for the presence of the low velocity body and delineated its lateral extent to the west of Urach. The bottom of the low velocity body is more clearly recognized than its top in the near vertical reflections. From the sparsity of reflections from the top of the 'Urach body' and by matching acoustic impedances, Gajewski & Prodehl (1983) arrived at a high density for the low velocity body. Although this

conclusion is rather tentative, the petrological implications of such an anomalous body are a challenge to future field experiments.

The stress-field in the region of the Hohenzollern Graben

The phenomenon of intraplate seismicity in the region of the HZGR can only be understood if the stress field is well determined in this area. Three independent methods have been applied in the region of the HZGR: fault plane solutions of local earthquakes, wall breakouts in the Urach well, and *in situ* determinations by the methods of over-coring as well as by hydrofracturing.

Fault plane solutions of earthquakes in the Swabian Alps (Ahorner *et al.* 1972; Ahorner & Schneider 1974) reveal an extremely uniform stress field with a horizonal axis of compression in the direction N155°E. For the earthquake of 3 September 1978 (MWA = 5.7) near Albstadt, Turnovsky (1981) derives a solution with almost vertical fault planes (Fig. 7); the active sinistral fault plane striking N22.5°E. From spectral analysis, stress drops of about 1 MPa are found to be typical for nearly all events in the region of the HZGR.

In situ stress determinations in the HZGR have been performed by a group from the Geological

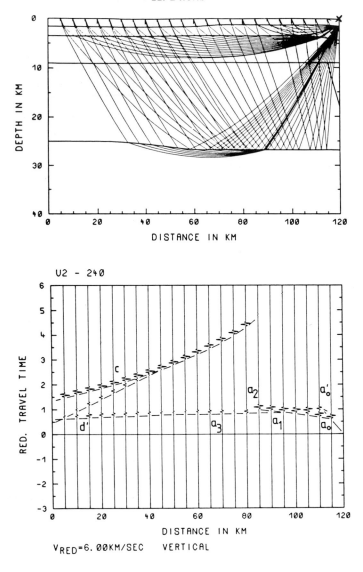

FIG. 6. Ray diagram (top) and synthetic seismogram section (bottom) corresponding to profile U2-240 (compare Fig. 4d; Gajewski & Prodehl 1984).

Institute of Karlsruhe University (Greiner 1974, 1978; Illies 1974, 1982; Baumann & Becker 1983), predominantly with the door-stopper method which measures strain-release of stressed media during over-coring. These measurements reveal the largest horizontal principal stress direction to be between 130° and 152°N with 1.7 MPa compressional stress. The smallest horizontal principal stress amounts to −0.2 to −0.5 MPa dilatational stress (Greiner 1974). The most noteworthy result of the absolute determination

of the stress is the large value of the horizontal shear stress (1.0–1.2 MPa). This value clearly exceeds that of the adjacent regions. Measurements with triaxial cells for a complete determination of the stress tensor are currently in progress (Baumann & Becker 1983; Baumann 1984). Furthermore two stations with continuous digital recording of the stress components (sampling rate 1 h⁻¹) have been installed in the region of the HZGR using the Gloetzl pressure cell and the Maihak sensor (vibrating wire) in 30 m deep bore

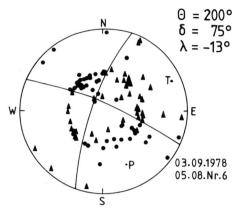

$$\Theta = 200°$$
$$\delta = 75°$$
$$\lambda = -13°$$

03.09.1978
05.08.Nr.6

FIG. 7. Fault plane solution of the earthquake on 3 September 1978 (MWA = 5.7) in the HZGR from short period observations (Turnovsky 1981). MWA = local magnitude Wood-Anderson. Triangles indicate compressional first motion, circles dilatational.

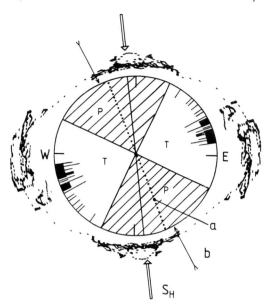

FIG. 8. Comparison of stress directions determined by three different methods in or near the geothermal research well Urach 3: S_H is the direction of the horizontal main compressional direction as determined from the continuous mapping of wall breakouts (depicted outside the circle). The short radial lines within the circle display the directions of 65 discrete breakouts. From this the mean horizontal compression is estimated at N162°E (dotted line, marked a). The figure contains also a representative fault-plane solution from the Swabian Alb (Turnovsky & Schneider 1981) (P = pressure, T = tension). The b arrow bisects the P-quadrant and is the direction of maximum compression if the fault planes are without friction. The discrepancy between the directions of S_H and of b is 29° and can be explained by a coefficient of internal friction of about 0.6.

holes. Although some correlations with local seismic events are indicated it is still difficult to establish such a relation with certainty since the instruments are still settling and instrumental noise is still high (Baumann & Becker 1983; Baumann 1984).

The direction of the axes of the main stresses as obtained by the hydrofracturing method near the HZGR structure (Rummel & Jung 1975) are concordant with the directions derived by the doorstopper method in this region.

Wall breakouts in deep wells have recently been introduced as indicators of the orientation of crustal stress tensors (Babcock 1978; Bell & Gough 1979). The breakouts are oriented normal to the direction of the maximum horizontal compressional stress. This method has been applied to the crystalline part of the geothermal well Urach 3 (Bluemling *et al.* 1983) using a new technique in which the breakouts are not sampled discretely but continuously. In Fig. 8 a wall-contour plot is shown together with a typical fault plane solution for the Swabian Alps. The direction of the compressional axis derived from the breakouts is N173° ± 10°E, deviating about 15° from the direction bisecting the angle between the two fault planes. This is consistent with a coefficient of internal friction $\mu = 0.6$. Breakouts have also been analysed from televiewer images of the deformed wall (Fig. 9). The results corroborate the direction derived from the 4-arm caliper data, however, with higher resolution.

Intraplate seismicity and crustal heterogeneities

What is the mechanism that triggers the release of seismic energy at a particular place in a rather uniform crustal stress field and on a small element of a several hundred of kilometres long bundle of shear lineaments? This is the key question for our understanding of the relatively strong seismicity occurring in the rather small, confined area at the HZGR.

The strong lateral heterogeneities detected by deep seismic sounding demand a closer examination of the spatial distribution of hypocentres in the region of the HZGR. Fig. 10 displays this

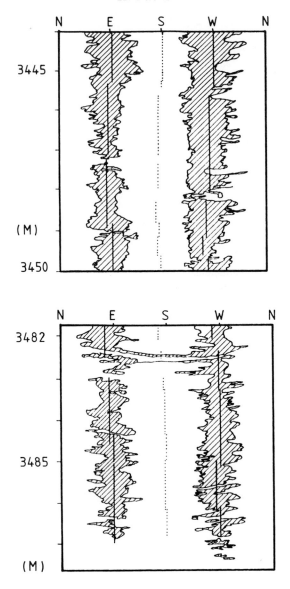

FIG. 9. Televiewer data from two depth ranges of the deep geothermal Urach 3 well. The hatched areas signify the azimuthal distribution of the breakouts. The dotted line corresponds to the resultant direction of maximum compressional stress (Bluemling, contribution to Alfred Wegener Stiftung 1984).

distribution during the period 1911–1978. The position of the corresponding epicentres and the orientation of the N–S sections into which the hypocentres have been projected are shown in Fig. 11. The hypocentres of the strong earth-quakes of 1911, 1943 and 1978 are specially marked. In the northern part, the earthquakes are all less than 10 km deep. It can be seen that the

envelope of the deepest events climbs steeply and with a slight curvature from south to north from focal depths of about 20 km to about 10 km. Although an apparent effect from the improve-ment of the network of seismic stations cannot be excluded we take this observation as support of our working hypothesis.

It is unlikely that this steep ascent is a purely

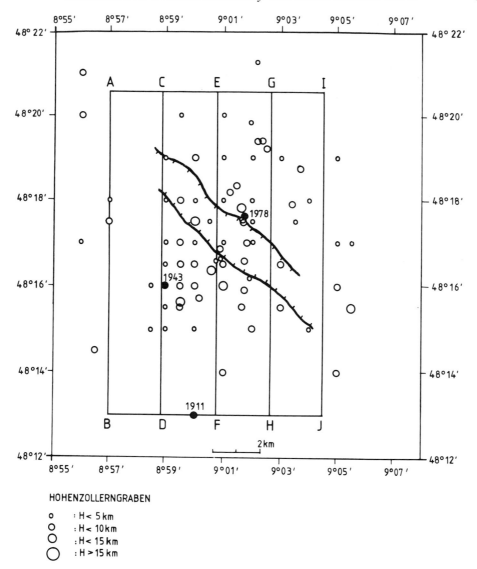

HOHENZOLLERNGRABEN

○	: H < 5 km
○	: H < 10 km
○	: H < 15 km
○	: H > 15 km

FIG. 10. Distribution of epicentres in the HZGR region for the period 1911–1978. The hypocentres are projected into vertical sections through the N–S lines AB, CD, etc. (see Fig. 11).

geothermal effect. Although the increase of temperature towards the geothermal anomaly of Urach tends to move the brittle–ductile transition to shallower depth the isothermal surface at the border of the Urach anomaly will not rise as steeply as the envelope of the deepest hypocentres.

Therefore, it is suggested that in addition to the influence of temperature the position of the brittle–ductile transition is strongly influenced by

the lateral changes in physical properties of the crust in the HZGR region. This can either be achieved by the ascent of the lower crust from SW to NE as in model A of Gajewski & Prodehl (1983) (see Fig. 5); and/or by the anomalously heated body of the Urach geothermal anomaly with a possibly strongly reduced viscosity due to thermal heating and the presence of fluid phases. As a consequence, the effective cross-section of the brittle part of the crust is thinned by a factor

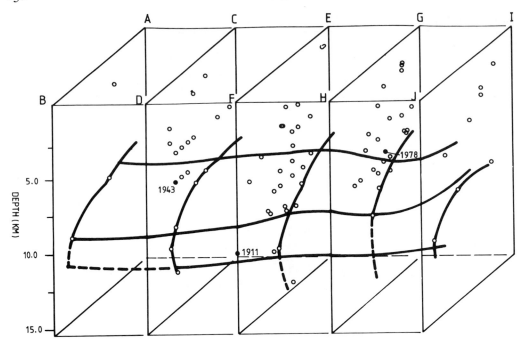

FIG. 11. Spatial projection of hypocentres in the HZGR region for the period 1911–1978. The hypocentres have been projected onto the vertical N–S sections AB, CD, etc. (see Fig. 10). The net of thick lines marks the envelope of the deepest hypocentres. Note its rise to the North.

of about two approaching the HZGR from the south. This leads to a stress amplification (Kusznir & Bott 1977) which may elevate the stress above the fracture level, particularly on the region of the HZGR.

Intraplate seismicity linked to lateral crustal heterogeneities has been reported for other regions as well as the HZGR. A similar phenomenon can also be observed in the New Madrid rift zone (Zoback *et al.* 1980; Illies 1982; Mooney *et al.* 1983; Peters & Mooney 1983). In this region there appears to be a correlation between the trend of the epicentres and the shape of the 7.3 km/s layer of the lower crust.

Conclusions

The currently available information on crustal structure, physical properties and distribution of hypocentres has led to the formation of a working hypothesis on the cause of intraplate seismicity in a small confined zone at the HZGR. The presence of lateral crustal heterogeneities amplifies a regionally uniform stress field locally by a thinning of the brittle zone in the upper crust. The effect may

be amplified by a neighbouring thermal anomaly. This hypothesis should be tested in the future by a denser network of seismic profiles, both of reflection and refraction surveys. Also the number of permanent and temporary seismic stations should be increased to improve the reliability of hypocentre determination of local earthquakes. The ultimate goal, of course, would be a deep drill hole into the crust of the HZGR. It will also allow us to distinguish whether the stress concentration occurs mainly because of the nearby geothermal anomaly or because of compositional heterogeneities.

ACKNOWLEDGEMENTS: The presite survey of the Hohenzollern Graben region is a joint effort of geoscientists from all disciplines. A project study (Alfred Wegener Stiftung 1984, see vol. 2 for the location Hohenzollern Graben) was prepared by E. Althaus, F. Bauer, H. Baumann, H.-J. Bayer, P. Bluemling, T. Farkas-Jandl, K. Fuchs (ed.), D. Gajewski, H. Langer, K. Lindner, H. Maelzer, J. Neuberg, C. Prodehl, C.-D. Reuther, G. Sattel, F. Scherbaum, R. Schick, G. Schneider, R. Stellrecht, F. Quiel, J. Walther, K.W. Zippelt. This study received financial support from the German Research Society (Deutsche Forschungsgemeinschaft) within the KTB

(Continental Deep Drilling Program) with funds made available by the BMFT (Federal Ministry for Research and Technology) and within the SFB 108 'Stress and Stress Release in the Lithosphere', a special research program at the University of Karlsruhe sponsored by the Deutsche Forschungsgemeinschaft as a German contribution to the International Lithosphere program. M. Berry offered a number of critical comments in

several aspects of this paper and kindly helped to improve the English. All this support is gratefully acknowledged. This paper is Contribution No. 77 of the SFB 108 of Karlsruhe University.

The late Henning Illies was the first to propose the Hohenzollern Graben as a possible site for a deep continental drill hole in a seismically active area (Illies 1982).

References

AHORNER, L., MURAWSKI, H. & SCHNEIDER, G. 1972. Seismotektonische Traverse von der Nordsee bis zum Apennin. *Geol. Rundschau*, **61**, 915–942.

—— & SCHNEIDER, G. 1974. Herdmechanismen von Erdbeben im Oberrheingraben und in seinen Randgebirgen, *In*: ILLIES, H. & FUCHS, K. (eds), *Approaches to Taphrogenesis*, Schweizerbart, Stuttgart, 104–117.

ALFRED-WEGENER S. (ed.). 1984. *Statusbericht des Kontinentalen Tiefbohrprogramms (KTB) der Bundesrepublik Deutschland.* Bonn, 5 volumes, ca. 1400 p.

BABCOCK, E.A. 1978. Measurement of subsurface fracturing from dipmeter logs. *Bull. Am. Assoc. Petrol. Geol.* **62**, 1111–1126.

BARTELSEN, H., LUESCHEN, E., KREY, T., MEISSNER, R., SCHMOLL, H. & WALTER, C. 1982. The combined seismic reflection-refraction investigation of the Urach Geothermal Anomaly. *In*: HAENEL, R., (ed.), *The Urach Geothermal Project*, Schweizerbart, Stuttgart, 231–245.

BAUMANN, H. 1984. Aufbau und Meßtechnik zweier Stationen zur Registrierung von Spannungsänderungen im Bereich des Hohenzollerngrabens—1. Resultate. *Oberrhein. Geol. Abhandlungen*, **33**, 1–14.

—— & BECKER, A. 1983. Einrichtung einer Dauermeßstelle zur Registrierung von Spannungsänderungen im Hohenzollerngraben und Spannungsmessungen im südlichen Rheingraben. SFB-108 Berichtsband 1981–1983, Universitaet Karlsruhe, 23–40.

BELL, J.S. & GOUGH, D.I. 1979. Northwest-southwest compressive stress in Alberta: evidence from oil wells. *Earth Planet. Sci. Lett.*, **45**, 475–482.

BERCKHEMER, H. & HILLER, W. 1960. Kurzperiodische Stationsseismographen mit traegerfrequenzverstaerkter und mechanischer Registrierung. *Z. Geophys.* **26**, 1–8.

BLUEMLING, P., FUCHS, K. & SCHNEIDER, T. 1983. Orientation of the stress field from breakouts in a crystalline well in a seismic active area. *Phys. Earth Planet. Inter.*, **33**, 250–254.

BRENNER, K. 1961. Die Verlagerung der Bebenherde und des Herdmechanismus im Gebiet der Südwestalb seit em Jahre 1911. Diplomarbeit. TH Stuttgart, 109 pp.

FIELDER, G. 1954. Die Erdbebentaetigkeit in Südwestdeutschland in den Jahren 1800 bis 1960. Dissertation, TH Stuttgart, 152 p.

CHENG, WANG-PING & MOLNAR, P. 1983. Focal depths of intracontinental and intraplate earthquakes and their implications for the thermal and mechanical properties of the lithosphere. *J. Geophys. Res.*, **88**, 4183–4212.

GAJEWSKI, D. & PRODEHL, C. 1983. Die Struktur von Kruste und oberem Erdmantel im Bereich des Schwäbischen Jura, abgeleitet aus seismischen Messungen. SFB-108 Berichtsband 1981–1983, Universitaet Karlsruhe, 177–208.

—— & —— 1984. Crustal structure beneath the Swabian Jura, SW-Germany from seismic refraction investigations. Submitted to *J. Geophys.*

GREINER, G. 1974. In-situ stress measurements—first results. *In*: ILLIES, H. & FUCHS, K. (eds), *Approaches to Taphrogenesis*, Schweizerbert, Stuttgart, 118–121.

—— 1978. Spannungen in der Erdkruste — Bestimmung und Interpretation am Beispiel von in-situ Messungen im süddeutschen Raum. Dissertation, Univesitaet Karlsruhe, 192 p.

HAENEL, R. 1982. *The Urach Geothermal Project.* Schweizerbart, Zurich.

HAESSLER, H., HOANG-TRONG, R., SCHICK, R., SCHNEIDER, G. & STROBACH, K. 1980. The September 3, 1978, Swabian Jura earthquake. *Tectonophysics*, **68**, 1–14.

HILLER, W. 1934. Eine Erdbebenkarte im Gebiet der Schwaebischen Alb. *Z. Geophys.* **9**, 229–234.

ILLIES, H. 1974. Intra-Plattentektonik in Mitteleuropa und der Rheingraben. *Oberrhein. Geol. Abh.*, **23**, 1–24.

—— 1982. Der Hohenzollern Graben und Intraplattenseismizitaet infolge Vergitterung lamellaerer Scherung mit einer Riftstruktur. *Oberrhein. Geol. Abh.*, **31**, 47–78.

KUSZNIR, N.J. & BOTT, M.H.P. 1977. Stress concentration in the upper lithosphere caused by underlying visco-elastic creep. *Tectonophysics*, **43**, 247–256.

MEISSNER, R. & STREHLAU, J. 1982. Limits of stresses in the continental crust and their relation to the depth-frequency distribution of shallow earthquakes. *Tectonophysics*, **1**, 73–89.

——, LUESCHEN, E. & FLUEH, E.R. 1983. Studies of the continental crust by near-vertical reflection methods; a review. *Phys. Earth Planet. Inter.*, **31**, 363–376.

MOONEY, W.D., ANDREWS, M.C., GINZBURG, A., PETERS, D.A. & HAMILTON, R.M. 1983. Crustal structure of the Northern Mississippi Embayment and a comparison with other continental rift zones. *Tectonophysics*, **94**, 327–348.

PETERS, D. & MOONEY, W.D. 1983. Tectonic evolution of the Mississippi Embayment. (in preparation).

PRODEHL, C., EMTER, D. & JENTSCH, M. 1982. Seismic refraction studies of the geothermal area of Urach, Southwest Germany. *In*: CERMÁK, V. & HAENEL, R.

(eds). *Geothermics and Geothermal Energy*. Schweizerbart, Stuttgart. 227–283.

RUMMEL, F. & JUNG, R. 1975. Hydraulic fracturing stress measurements near the Hohenzollern Graben structure SW-Germany. *PAGEOPH*, **113**, 321–330.

SCHERBAUM, F. 1980. Untersuchungen zur Struktur der P- und S-Phase in Epizentralgebiet, Dissertation, Universitaet Stuttgart, 76 p.

SCHICK, R. 1968. Untersuchungen über die Bruchausdehnung und Bruchgeschwindigkeit bei Erdbeden mit kleinen Magnituden (M 4). *Z. Geophys.*, **34**, 267–286.

—— 1970. A method determining source parameters of small magnitude earthquakes. *Zeitschr. f. Geophys.*, **36**, 205–224.

SCHNEIDER, G., SCHICK, R. & BERCKHEMER, H. 1966. Fault-plane solutions of earthquakes in Baden-Wuerttemberg. *Z. Geophys.*, **32**, 383–393.

TRAPPE, H. 1983. Eine Auswertung von Weitwinkelreflexionen auf dem Profil Urach und eine Korrelation zu den Ergebnissen der Steilwinkel-Reflexionsseismik, Diplomarbeit, Universitaet Kiel.

TURNOVSKY, J. 1981. Herdmechanismus und Herdparameter der Erdbebenserie 1978 auf der Schwaebischen Alb, Dissertation, Universitaet Stuttgart, 109 p.

—— & SCHNEIDER, G. 1981. The seismotectonic character of the September 3, 1978, Swabian Jura earthquake series. *Tectonophysics*, **83**, 151–162.

ZOBACK, M.D., HAMILTON, R.M., GRONE, A.J., RUSS, D.P., MCKEOWN, F.A. & BROCKMAN, S.R. 1980. Recurrent intraplate tectonism in the New Madrid seismic zone. *Science*, **209**, 4460, 971–976.

K. FUCHS, Geophysikalisches Institut, Universitaet Karlsruhe, Hertzstrasse 16, D-7500 Karlsruhe 21, FRG.

Crustal structure and evolution of the central Australian basins

K. Lambeck

SUMMARY: The basins of central Australia formed over a period of 600 Ma or more from late Proterozoic to Carboniferous time. These basins appear to overlie crust of normal thickness whereas the crust of the intervening arches has been uplifted and reduced in thickness. The Moho, in consequence, varies in depth by up to ± 10 km about its mean depth, over horizontal distances of 50 km or less. The stresses associated with this lower crustal structure are of the order of a few kbars and extend throughout the crust. These stresses have existed since at least late Proterozoic time, and because the Moho undulations appear to follow the near surface deformations, this rules out the existence of a mechanically weak lower crust. The region is out of isostatic equilibrium and this state must be maintained by either a mechanically strong lithosphere of more than 150 km thickness or by horizontal compressive stresses, of the order of a few kbars. The former possibility is ruled out by the available mantle flow laws and geotherms proposed for the central Australian shield. In the latter case the horizontal compressive forces have also played the key role in shaping the basins and crustal structure. The horizontal compression must act down to at least the Moho and must characterize the dominant state of stress since late Proterozoic time.

Introduction

The tectonics of central Australia are dominated by east–west trending structural elements comprising sedimentary basins whose histories span time from the late Proterozoic to the early Carboniferous, and exposed cratons of early and middle Proterozoic ages (Fig. 1). The basins extend down to a depth of 10 km or more and were formed in an intracratonic environment while the crust between the basins has been subjected to significant uplift. Major orogenies occurred, also in an intracratonic environment, in early Cambrian and early Carboniferous time. It is known from gravity observations that this basin and arch structure is out of local isostatic equilibrium and this raises at least three questions. What is the structure below these arches and basins, and does lateral structure extend into the lower crust and upper mantle? How did this structure evolve over the long time intervals suggested by the sedimentary records? What is the stress field implied by these structures and how is it maintained through time? Answers to these questions are of relevance to understanding the physical properties of the crust, the state of stress in the crust and the consequences of this stress state. Only tentative answers to some of these questions can be given because the available geophysical and geological information is still incomplete. Nevertheless some conclusions can be drawn with only minimal reservations, these are: (i) Variations in Moho depth of 10–15 km occur over horizontal distances of less than 50 km and more abrupt offsets in depth may also occur. (ii) The seismic P-wave velocity–density relation for the lower crust and upper mantle suggests a proportionality constant of about 5–7 (km s^{-1}) (Mg m^{-3})$^{-1}$ rather than about three as is often used for the Birch law. (iii) The differential stresses within the crust are a few kbars and these have persisted for at least the last 300 Ma. These stress differences occur throughout the crust and there is no evidence for a mechanically weak lower crust. (iv) The non-isostatic state is maintained by horizontal compressive forces that have also shaped the present basins and arches.

The geological setting

The central Australian basins under discussion here are the Amadeus, Officer and Ngalia Basins (Fig. 1), basins that are separated by the metamorphics and granites of the Musgrave and Arunta Blocks.

The Proterozoic history of the central Australian region has been discussed by Plumb (1979a,b), Shaw et al. (1984) and Black et al. (1983). It appears that the cratonization of the region was largely complete by about 1400 Ma ago although some igneous and tectonic activity continued up to about 1100–1000 Ma ago. This was followed by a period of tectonic tranquility during which a widespread deposition of sediments occurred over much of central Australia: the Heavitree Quartzite of the Amadeus Basin, the Vaughan Springs Quartzite of the Ngalia Basin and the Townsend Quartzite of the Officer Basin. The mechanism leading to this deposition has not yet been examined in detail although a preliminary analysis suggests that thermal subsi-

From: DAWSON, J.B., CARSWELL, D.A., HALL, J. & WEDEPOHL, K.H. (eds) 1986, *The Nature of the Lower Continental Crust*, Geological Society Special Publication No. 24, pp. 133–145.

133

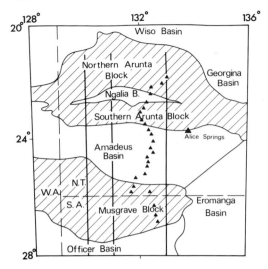

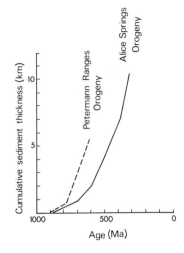

FIG. 1. Location map of the central Australian basins and exposed blocks. NT = Northern Territory; SA = South Australia; WA = Western Australia. The three north–south lines indicate the location of the gravity profiles illustrated in Figure 3 and the small triangles locate the sites at which travel time anomalies have been measured (cf. Figs 3 and 4).

FIG. 2. Approximate rates of sediment accumulation in the Amadeus Basin. The dashed line is for the southern part of the basin and accumulation essentially ceased after the Petermann Ranges Orogeny 600 Ma ago. The solid line is for the northern part of the basin and accumulation here ceased after the Alice Springs Orogeny in early Carboniferous time.

dence mechanisms may be relevant here. We ignore this aspect for the present and consider this sequence as a convenient starting point for modelling the subsequent history of the basin evolution in an intracratonic environment.

The geology of the Amadeus Basin is described in Wells *et al.* (1970), that of the Ngalia Basin in Wells & Moss (1983) and that of the Officer Basin in Krieg *et al.* (1976), Pitt *et al.* (1980) and Jackson & van den Graaff (1981). The southern part of the Amadeus Basin and the Officer Basin contain mainly late Proterozoic sediments deposited in shallow water or continental environments in a fairly continuous sequence but with the rates of deposition apparently increasing through time (Fig. 2). Both basins have very similar depositional histories. During this interval the Musgrave Block was subjected to uplift as indicated by the felspar and biotite ages of Gray & Compston (1978). The evolution of these two basins culminated with the Petermann Ranges Orogeny of Early Cambrian age, an event that deformed both the southern and northern margins of the Musgrave Block (Jackson & van de Graaff 1981; Forman 1966).

Late Proterozoic, post-Heavitree, sequences are rather thin in both the northern part of the Amadeus Basin and the Ngalia Basin and much of the evolution occurred in Palaeozoic time with

a similar sequence of events as occurred during the Proterozoic phase: continuous deposition of sediments in shallow water or continental environments, an increasing rate of deposition with time, significant uplift of that part of the Arunta between the two basins and the culmination by a major orogeny: the Alice Springs Orogeny of late Devonian to early Carboniferous age along the northern margin of the Amadeus Basin (Armstrong & Stewart 1975) and, at about the same epoch, the Mt Eclipse Orogeny along the northern margin of the Ngalia Basin (Wells & Moss 1983).

The Petermann Ranges and Alice Springs Orogenies resulted in extensive folding, faulting and metamorphism of both the sediments and basement rocks. The deformation at the time of the two orogenies is believed to have involved the crust as a whole through the development or reactivation of major thrust faults (Forman & Shaw 1973; Milton & Parker 1973; Wells & Moss 1983; Shaw *et al.* 1984). The early Carboniferous orogenies mark the end of substantial tectonic processes in central Australia. Some subsequent evolution of the basins and surrounding blocks has taken place but at a rate and on a scale that appears to be insignificant in comparison with the earlier history (Burek & Wells 1978; Wells *et al.* 1970; Wells & Moss 1983).

Geophysical observations

The presently available geophysical evidence for the crustal and upper mantle structure is limited to gravity and seismic travel time anomalies. Early seismic reflection surveys by the Bureau of Mineral Resources estimated the depth of the Moho to be about 35–40 km beneath both the Amadeus and Ngalia Basins (Mathur 1976). Systematic deep crustal seismic reflection surveys have been shot recently by the Bureau of Mineral Resources but are not yet available. The gravity anomalies and terrain elevations are illustrated in Fig. 3 in the form of three north–south profiles. The Bouguer anomalies vary by up to 180 mgal along these profiles, with the maximum values occurring over the blocks and the minimum values occurring at the margins of the basins. These values cannot be interpreted within the conventional framework of isostasy since the variations in elevation across the region as well as lateral density variations in the upper crust are relatively small (Lambeck 1983; Stephenson & Lambeck 1985a).

The second geophysical observation comprises seismic travel time anomalies (Lambeck & Penney 1984). Fig 4 illustrates some results for an array of instruments located across the basins and blocks as illustrated in Fig. 1. The residuals in Fig. 4 are for waves originating from several localities. The Fiji–Tonga anomalies are for waves arriving from a direction that is approximately parallel to the tectonic strike of the region (except for the northernmost station where the tectonic structure is oblique to the line). In a general way, these travel times are early for stations on the Arunta and Musgrave Blocks compared with stations in the basins, with the difference reaching nearly 0.5 s between the northern Amadeus Basin and the Arunta Block and 0.7 s between the southern Arunta and Ngalia Basin.

The travel time residuals for the earthquakes from Japan and the Marianas (Fig. 4) result in significantly different results for some of the stations when compared with the Fiji–Tonga results. Likewise, the Macquarie Ridge earthquakes are distinctly different although these residuals are the least well-determined, being based on a small number of shallow crustal earthquakes. Observed anomalies in azimuth cannot be attributed to station mislocations or other observational errors (Lambeck & Penney, 1984). Instead, these results can be interpreted in terms of steeply dipping surfaces of large velocity contrast, particularly the Moho, beneath some of the stations. The result for station 20, for example, is consistent with a shallow Moho dipping southwards by more than 20° although this

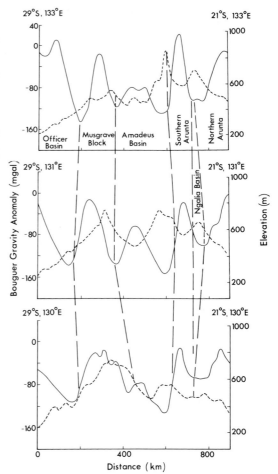

FIG. 3. Gravity (solid lines) and elevations (dashed lines) along three north–south profiles across the basins and blocks (see Fig. 1 for the locations of these profiles but note that the latter extend to 29° S).

interpretation is not unique. A preliminary examination of new travel times collected from a more closely spaced array between sites 17 and 24 and from a second parallel array confirms the substantial variations illustrated in Fig. 4 in the travel times across the northern Amadeus Basin and southern Arunta Block.

Heat flow observations in central Australia are sparse and no measurements have been made in the basement rock of the area in question. Hyndman (1967) measured a surface heat flow of 63 mW m^{-2} (1.5 heat flow units; h.f.u.) for a site south of Alice Springs in the Amadeus Basin sediments. A measurement in the Ngalia Basin, gave 56 mW m^{-2} (Cull & Denham 1979). Crustal heat production observations are not available for these sites and the nearest surface measure-

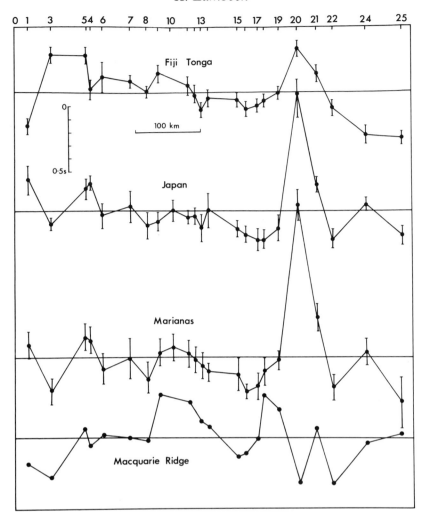

FIG. 4. Observed seismic travel time anomalies across central Australia (solid circles with error bars) for earthquake waves generated in four different regions and recorded at the sites located in Fig. 1 (the most southerly site is site 1, the most northerly is site 25).

ment is for a site 500 km to the north of Alice Springs, for which Sass & Lachenbruch (1979) report 3.9 μW m^{-2} (9.3 heat generation units; h.g.u.). This compares with an average surface heat production of 5 μW m^{-3} for the central Australian shield (Sass & Lachenbruch 1979).

Geotherms for the area have been estimated by Sass & Lachenbruch (1979) and by Cull & Conley (1983) using different observations and assumptions and the results are consistent with a model of a stable continental crust of high average radioactivity and little contemporary tectonic activity. The Sass & Lachenbruch geotherms are given for different values of surface heat flow q_0 and surface heat generation A_0 (Fig. 5). The values of $A_0 = 0$ and $q_0 = 0.65$ h.f.u. are believed to be appropriate for the Archaean shield while the other two curves are more appropriate for the central shield province and for central Australia. Cull & Conley's geotherm is for a surface heat flow of 75 mW m^{-2} and it may therefore overestimate the temperatures although this is consistent with the generally higher geothermal gradients reported for the Amadeus Basin oil wells by Gorter (1984) than the values given by Cull & Conley.

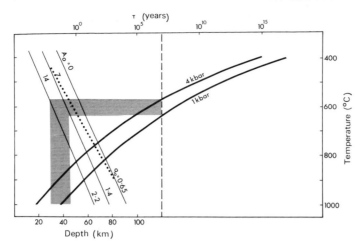

FIG. 5. Effective mantle relaxation time τ as a function of temperature and two stress-differences (1 and 4 kbar) for wet olivine according to Chopra and Paterson (1981) and, left, three geothermal profiles according to Sass and Lachenbruch (1979). q_0 is the surface heat flow (in heat flow units) and A_0 is the surface radioactivity (in heat generation units). The lowest gradient is for the Archaean shield of Western Australia. The other two gradients span a range of values that are consistent with surface geology, heat flow and other geophysical observations for the central Australian shield. The shaded region spans the range of temperatures and depths greater than which stress-differences of 1–4 kbars relax with a time constant of 10^7 years or less according to the Sass-Lachenbruch geotherms. The dotted line is the geotherm proposed by Cull and Conley (1983) for the central Australian region with a surface heat flow of 75 m Wm^{-2}.

Crustal structure

The average density of the basin sediments is about 2.65 Mg m^{-3} (Froelich & Krieg 1969; Schroder & Gorter 1984) while densities in the Arunta and Musgrave Blocks range from about 2.6 Mg m^{-3} for some granites to about 3.1 Mg m^{-3} for mafic granulites (Mutton & Shaw 1979; Smith 1979). The latter are restricted to narrow zones, probably less than 30 km wide in the Arunta Block (Mutton & Shaw 1979) and less than 15 km wide in the Musgrave Block (Smith 1979; see also Mathur 1976) and contribute to only the shorter wavelengths in the anomalous gravity field. On average, the observed lateral density differences in the upper crust are about 0.2 Mg m^{-3} or less and these would have to extend to depths well below 10 km if they are to explain the observed gravity values.

Proterozoic sediments in the basins have an average P-wave velocity of 5.5 km s^{-1} while the Palaeozoic sediment velocities average about 4.8 km s^{-1} (Froelich & Krieg 1969). Velocities for Arunta upper crustal rocks vary from 5.1–6.0 km s^{-1}. If an average upper crustal velocity of 5.5 km s^{-1} is adopted, then departures from this mean value are about ± 0.5 km s^{-1} and for a 10 km path the travel time anomaly is about ± 0.16 s. Here also, the observed quantity cannot be wholly explained by lateral contrasts in upper crustal seismic velocities.

In so far as the required variations in density and seismic velocity are several times greater than those occurring in the upper crust, an alternative interpretation is that they arise from variations in the depth of the Moho. For a density contrast of 0.4 Mg m^{-3} across the Moho and an average Moho depth of 35 km, the gravity anomaly fixes an upper limit of about ± 14 km for the Moho depth undulation. The seismic travel time anomaly for seismic waves arriving from a direction that is parallel to the tectonic strike (i.e. from Fiji–Tonga, Fig. 4) is

$$\delta t = \delta h \sec \alpha \ (v_m - v_c)/v_m v_c,$$

where δh represents the thickness of the region of anomalous velocity and α ($\simeq 30°$) is the angle of incidence of the emerging ray. Also v_c (P-wave velocity of the lower crust) $\simeq 6.5$ km s^{-1} (Hales & Rynn, 1978), and v_m (P-wave velocity of the mantle) $\simeq 8.3$ km s^{-1}. Therefore

$$dh = (\delta t/2) \cos \alpha \ v_m v_c/(v_m - v_c) \simeq 12.5 \ \delta t \text{ km,}$$

for δt in seconds. With $\delta t = 0.7$ s (Fig. 4), the fluctuation in Moho depth is $dh \simeq \pm 9$ km compared with $dh = \pm 14$ km deduced from the gravity observations.

By adopting a lower crustal density or velocity

in these calculations it is assumed that any intermediate crustal layer across which a velocity contrast may occur is undeformed. If, instead, such surfaces follow the Moho undulation, it is necessary to take the velocity or density contrasts between the upper crust and mantle. For example, with v_c equal to the upper crustal velocity of 6.0 km s^{-1} (Hales & Rynn, 1978), $dh = 9.4\,\delta t$ km. The difference with the earlier estimate is too small to distinguish between the two models of crustal structure using only the present data set and we cannot establish whether intermediate crustal layers are horizontal or parallel to the Moho.

Gravity represents a weighted volumetric integral of anomalous density while the travel time anomalies represent a line integral of anomalous velocity along a specific ray path and a simple relation between the two will not exist in regions of rapid lateral variation in crustal structure. For a smoothly varying interface, the predicted relation between gravity variation δg and travel time anomaly δt, both measured at site i, is

$$\delta g_i = -2\pi G \frac{\rho_m - \rho_c}{v_m - v_c} v_m v_c \cos \alpha \, \delta t_i$$

where ρ_m, ρ_c are the average densities of the upper mantle and lower crust respectively. The observed relation is $\delta g_i = -263\,\delta t_i$ mgal (Lambeck & Penney 1984) and

$$\frac{1}{v_m v_c} \frac{v_m - v_c}{\rho_m - \rho_c} = -2\pi G \cos \alpha \frac{\delta t_i}{\delta g_i} =$$
$$0.137\ (\mathrm{Mg\ m^{-3}})^{-1}\ (\mathrm{km\ s^{-1}})^{-1}$$

Of the quantities on the left hand side of this equation the best known is probably v_m. If intermediate crustal layers are deformed parallel to the Moho then $v_c = 6.0$ km s^{-1} and $dv/d\rho \simeq 6.8$ (km s^{-1}) (Mg m^{-3})$^{-1}$. If intermediate crustal layers are undeformed, $v_c \simeq 6.5$ km s^{-1} and $dv/d\rho \simeq 7.4$ (km s^{-1}) (Mg m^{-3})$^{-1}$. These ratios are about twice the values used in the typical Birch law for upper mantle materials and are also somewhat higher than given by the Nafe–Drake relation for sedimentary, igneous and metamorphic rocks of densities greater than about 2.4 Mg m^{-3} (Nafe & Drake 1963).

A Moho undulation model consistent with the gravity and Fiji–Tonga travel times is illustrated in Fig. 6 (see Lambeck & Penney 1984). This model is also generally consistent with the azimuth variations observed in the travel times (Fig. 4). At station 1 for example, the Japan events are early while the Macquarie ridge and Fiji–Tonga events are late, pointing to a Moho which is thicker than average and which dips southwards. The residuals for station 3 indicate a shallow Moho dipping northwards. The anomalies observed at station 20, on the southern Arunta, point to a Moho below the site that is shallow and southward dipping. This model has many affinities with models proposed by Forman & Shaw (1973), based mainly on the close correlation between metamorphic grade of surface rocks and gravity anomaly values and on structural geological observations, and by Mathur (1976), based mainly on gravity data. The major faults depicted in this model are based on seismic reflection surveys to 5 seconds, on surface structural geology, on magnetic and gravity anomaly modelling and on the rapid azimuthal variations in travel time anomalies observed at some stations. The continuation of these faults down to the Moho is, however, speculative.

The state of stress

The geophysical observations indicate that the crust of central Australia is very much out of isostatic equilibrium (Fig. 3) and that the anomalous structure giving rise to this state occurs mainly at the crust–mantle interface. Significant departures from a hydrostatic stress-state have therefore existed throughout the crust and upper-mantle since Proterozoic time in the south and since at least late Palaeozoic time in the north. An approximate estimate of the maximum stress-difference is given by

$$\sigma_{max} = \beta(\rho_m - \rho_c)g\,dh$$

where dh is the amplitude of the undulation of the Moho and the parameter β is a function of the degree of regionality of the support of the stress. If the region were in local isostatic equilibrium $\beta = 1$. With $(\rho_m - \rho_c) = 0.4$ Mg m^{-3} and $dh = 10$ km, $\sigma_{max} \simeq 800$ bars (80 MPa). This represents a lower limit to the maximum stress difference in the crust and upper mantle (Jeffreys 1959). Because the region is out of local isostatic equilibrium β is larger, and model dependent, and σ_{max} may be 3–5 times greater or about 1.5–4 kbar (Lambeck 1983; Stephenson & Lambeck 1985a).

For internal or surface loads, whose wavelength greatly exceeds the thickness of the layer supporting the load, any readjustment to the isostatic state is controlled more by the buoyancy force than by the strength of the crust, with the latter controlling the degree of regionality of the response to the buoyancy force. Basins, associated with the large negative gravity anomalies and mass deficit at depth, should be rebounding and the arches, associated with the relative gravity highs and mass excess, should be subsiding at rates that are largely controlled by the viscosity of the mantle, at a rate that is rapid when compared

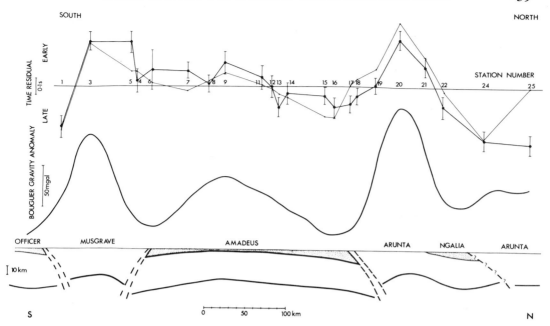

FIG. 6. Crustal structure deduced from the gravity anomalies and differential travel time residuals (lower part of figure). The top curves represent the observed (solid circles plus error bars) and model-predicted (thin line) travel times for earthquakes from the Fiji–Tonga region. The observed Bouguer gravity anomalies along the profile are also shown.

with the time elapsed since the last major tectonic events. There is, however, no evidence for such large scale differential vertical movements since Carboniferous time and the total amount that appears to have occurred is only a small fraction of what would occur if the adjustments were unimpeded by other forces.

One explanation for the lack of regional rebound is that the density anomalies are supported by a thick and strong lithosphere such that the rebound takes place with a wavelength that is very much greater than the typical width of a basin and arch. Any rebound will then involve the area as a whole, as may be implied by the generally high mean elevation of the central Australian region when compared with surrounding areas. The wavelength of the regional rebound response is given approximately by four times the flexural parameter l defined as

$$l^4 = D/(\rho_m - \rho_c)g$$

where D is the effective flexural rigidity, defined in terms of an effective thickness of the lithosphere H as

$$D = \mu H^3/6(1-\nu).$$

ν is Poisson's ratio and μ the rigidity or shear modulus. For differential vertical rebound between arches and basins to be insignificant, l must exceed the combined width of a basin or arch, or $l > 300$ km. This is turn requires that $D > 10^{25}$ N m and, with $\nu > = 0.5$ and $\mu = 2 \times 10^{10}$ N m^{-2} (assuming a relaxed shear modulus of about half the seismic value, Minster & Anderson 1980), $H > 150$ km.

The effective time constant τ of mantle deformation can be defined as

$$\tau = \eta/\mu = \sigma/\dot{\varepsilon}\mu$$

where η is the effective Newtonian viscosity, σ is the stress difference and $\dot{\varepsilon}$ is the strain rate, itself a function of σ and temperature T. Of interest here is mainly the state of stress in the upper mantle and the wet olivine flow law of Chopra & Paterson (1981, 1984) is adopted. The effective time constant is then greater than 10^7 years if temperatures are kept below about 620°C for a differential stress of 1 kbar (Fig. 5), or below about 570°C for a differential stress of 4 kbar. For the Cull & Conley (1983) geotherms these temperatures are reached at depths of about 50 km and the mantle below this depth will not support stress differences of 1 kbar on time scales of the order of 10^7 years and longer. For the Sass &

Lachenbruch central shield geotherms this temperature is reached between about 30 and 45 km or at about 60 km depth for the Archaean geotherm. These estimates do require some cautionary comment because the flow laws have been extrapolated beyond the temperature limits of 1200–1300°C of the experimental conditions, but Paterson's (1976) observation, that if a certain strain rate occurs in the laboratory it will occur under geological conditions at a lower temperature, is also valid. The flow law used here is for wet rather than dry olivine, the rationale being that Chopra & Paterson (1984) found that very small amounts of water, less than 0.01 weight per cent, have a very significant weakening effect. What these results indicate is that stresses of the order of 1 kbar cannot be supported on the required time scales by the upper mantle beneath the central Australian province. The model of a Moho of variable depth frozen into a thick lithosphere such that the isostatic state cannot be reached at wavelengths of a few hundred kilometres and less, is inconsistent with the flow law and geotherm models.

Flow laws relevant for lower crustal materials are largely unknown (e.g. Kirby 1984) but it is widely held that the strength of these materials is less than that of olivine (e.g. Brace & Kohlstedt 1980; Chen & Molnar 1983). Both the Sass–Lachenbruch and Cull–Conley geotherms for the crust indicate that the temperatures exceed half the melting point temperature (both expressed in K) at depths greater than about 30–40 km, and that the lower crust may also be expected to creep when subjected to kbar level stress-differences on time scales of 10^7–10^8 years.

An alternative model is one in which a mechanical balance has been reached between the buoyancy force, the near-surface load, the strength of the crust and an in-plane or horizontal compressive force. The magnitude of the requisite horizontal forces is of the order of a few kbars and they must have persisted in parts of the crust for much of the time elapsed since the Petermann Ranges Orogeny, and certainly since the Alice Springs Orogeny. These stresses would be restricted to that part of the crust and upper mantle where temperatures are sufficiently low, less than about 500°C, for the effective relaxation time constant to be in excess of about 10^8 years.

The origin of the horizontal force remains enigmatic although there is considerable evidence that predominantly north–south horizontal compressive forces have been more the norm than the exception in shaping the central and northern Australian continental crust since late Proterozoic time (e.g. Plumb 1979b; Plumb *et al.* 1981). Certainly the present state of the continent is one in which compressive stresses dominate (Lambeck *et al.* 1984).

Tectonic evolution

The central Australian region has a tectonic history which goes back to at least middle Proterozoic time although the basin histories started at about 900 Ma ago with the deposition of the Heavitree Quartzite and its correlates in the Ngalia and Officer Basins. The gravity observations and the sedimentation record (Fig. 2) rule out basin formation models that are based on passive sediment loading, thermal processes or crustal stretching (Lambeck 1984), but because the present structure appears to be maintained by horizontal compressive forces it is reasonable to consider whether these forces have also led to the present structure.

Buckling of a homogeneous elastic crust is excluded because of the large magnitude of the requisite force, of the order of 40 kbars (e.g. Heiskanen & Vening Meinesz 1958), but this is not the solution sought. Instead, the problem is one of finding the deformation of an inhomogeneous crust with a viscous element in its rheology and with the surface load evolving through time by erosion and sedimentation (Lambeck 1983).

Mathematically, the problem is reduced to the solution of an equation balancing a number of forces; the surface load and buoyancy (with both being time dependent due to erosion and sedimentation), the elastic and viscous forces characterizing the plate's resistance to deformation, and the in-plane horizontal force. To obtain tractable analytical solutions a number of simplifying assumptions must be introduced. The rheology must be reduced to a minimum set of parameters defining the elastic and viscous response; a simple erosion and sedimentation model must be adopted and a simplified horizontal compression model is required. Each of these aspects of the model must be defined by the minimum number of parameters because the number of observational constraints on the evolution is also small.

An appropriate rheological model for the lithosphere would be one of several coupled layers. The first layer, representing the uppermost crust, could be modelled as an elastic layer of low strength or as a viscoelastic layer in which stress relaxation occurs predominantly by microseismicity. The second layer, representing the middle crust, could be modelled as an essentially elastic layer and the next layer, the lower crust and upper mantle, would, in view of the discussion in the preceding section, be a viscoelastic layer. This would be particularly so 10^9 years ago if the basin initiation followed immediately upon thermal events of middle Proterozoic time. A gross simplification of this model is the homogeneous Maxwell solid characterized by an effective flexural rigidity D and an effective time constant τ, parameters which pertain to the depth-averaged properties only. The relaxation time τ is therefore a measure of macroscopic time-dependent deformation in the upper and middle crust by brittle failure, and of microscopic creep at greater depth by the more conventional creep mechanisms appropriate for the higher temperature conditions of the lower crust and upper mantle. The plate overlies the asthenosphere which, on the time scale of 10^4 years and longer can be treated as fluid (Fig. 5).

The simplest surface loading model is to assume that the amount of material deposited or eroded at any time t is proportional to the deformation, and the observation that the sediments have been deposited mostly in shallow water environments means that the proportionality factor is near unity. A somewhat more realistic model is one in which the rate of erosion at any time is proportional to elevation with a time constant that may be of the order of 100–300 Ma (Stephenson 1984; Stephenson & Lambeck 1985a). For the in-plane force model to have some validity, another time-constant (or constants) would have to be introduced to describe either a decaying force or a periodic force, but considerable trade-off occurs between the various parameters and at present the only model that has been considered in detail is one with a uniform and constant Maxwell rheology, erosion and sedimentation proportional to elevation or subsidence, and a constant in-plane force (Lambeck 1983). This model appears to be capable of producing very significant deformations, even for high lithospheric viscosities, when compressive stresses of a few kbars operate for periods of time of the order of 10^8 years or more. Here it suffices to emphasize: (i) when the plate is subjected to a horizontal compression, some of the original deformations are magnified elastically by amounts that are, *inter alia*, a function of the wavelength of the original load, (ii) if the compression is maintained the deformations of some of the wavelengths grow in time due to the viscous element in the layer, and (iii) erosion of uplifting areas and deposition of sediments into the downwarped areas further aids the deformation by effectively reducing the buoyancy force in the subsiding areas and increasing this force in the uplifting areas. The parameters required to produce subsidence consistent with the geological data are $D \simeq (5-10)10^{22}$ N m, $\tau = 25-50$ Ma, and

$\sigma = 150-400$ MPa (Lambeck 1983; Stephenson & Lambeck 1985a).

The parameters D, τ and σ are not wholly independent. Similar basin evolution sequences can be computed if D is increased provided that σ is also increased or that τ_v is decreased. D and τ_v are particularly interdependent and a better constrained quantity is their product $D\tau_v \simeq (15-30)10^{23}$ N m Ma. These values are in general agreement with estimates obtained for other regions (Stephenson & Lambeck 1985b; Sleep & Snell 1976; Beaumont 1981; Stephenson 1984). The difficulty lies in interpreting these effective parameters in terms of crustal properties, for not only do they represent depth integrated rheological parameters as discussed above, they also represent time integrated parameters from 1000 Ma ago, soon after the last major thermal events in central Australia, to the present. With the above definition for D, and relaxed moduli of $\mu \sim 2 \times 10^{10}$ N m^{-2}, and $v = 0.5$, the effective elastic thickness of the plate H is 20–25 km for $D = (5-10)10^{22}$ N m. This is the average effective thickness of the mechanical lithosphere and is consistent with the Sass–Lachenbruch geotherm and the notion that crustal materials will deform when temperatures exceed about one half the melting point temperature. These values, are, however, less than the depth to the Moho of about 35–40 km as estimated from seismic reflection surveys at two localities. Because the gravity and seismic data indicate that the Moho participates in the deformation, this difference in thickness can perhaps be attributed to the depth dependence of the rheology, with the upper crust being of relatively low strength. Alternatively, the difference is a consequence of the basins being shaped at a time of warmer and weaker lithosphere so that the present Moho has adjusted partly to the high stress-differences. In this case the Moho undulations will not wholly reflect the surface deformations of the crust.

As the deformation proceeds, the bending stresses increase with time and failure occurs locally when the stress differences exceed the strength of the crust. The stress pattern is such that, should failure occur, it will be in the form of thrust faulting near the surface, with the block and part of the sediments being thrust steeply over the basin. Where failure actually occurs will depend largely on the lateral structure of the crust, and any pre-existing zones of weakness, away from the maximum stress difference locations, may be reactivated first when stress differences there exceed the local strength of the crust.

The final predicted shallow crustal cross-section across the basins and blocks is illustrated in

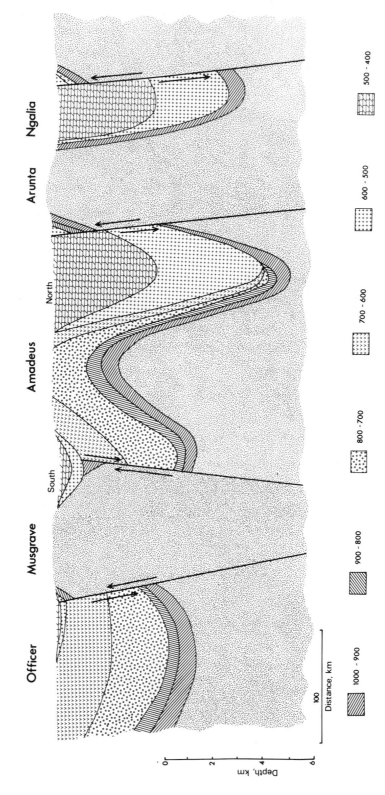

FIG. 7. Predicted north–south cross-section, from the Officer Basin to the northern Arunta, through the upper crust according to the horizontal compression model of Lambeck (1983). The Musgrave and Arunta blocks have been subject to uplift of 10–15 km since Late Proterozoic time. The Moho is predicted to be deeper than average below the basins and shallower than average below the exposed basement.

Fig. 7 (Lambeck 1983, 1984). Maximum basin subsidence is about 10 km for the northern Amadeus Basin and the basins are flanked by large thrust faults that appear to steepen with depth and which may extend down to, or perhaps beyond, the Moho. Significant uplift is predicted, of the order of 10–15 km and more, for both the Musgrave and Arunta Blocks from Cambrian time to the early Carboniferous, uplifts which would be superimposed on vertical movements that may have occurred during earlier Proterozoic tectonic events. The Moho is predicted to follow these expressions of surface deformation, being uplifted by 10–15 km beneath the blocks and depressed by 5–10 km beneath the basins, a structure which is consistent with the structure deduced from the seismic and gravity data.

Conclusion

The geophysical observations of the central Australian basins and arches point to (i) a crust that is grossly out of isostatic equilibrium, (ii) significant lateral variations in density and seismic velocity at the base of the crust and possibly in the upper mantle, and (iii) the existence of stress differences in the crust and possibly the upper mantle that have been of the order of a few kbars throughout Phanerozoic time. These observations raise a number of questions which are relevant to understanding tectonic evolution within an intracratonic environment and to understanding crustal properties. Some preliminary answers to these questions have been suggested above. In particular, the structure below the basins and arches extends down to at least the Moho, although a detailed mapping of the undulations and abrupt changes in depth of this surface has not yet been possible. The existence of these undulations requires either a cold and rigid lithosphere of mechanical thickness in excess of 150 km, or a balance of the vertical loading and buoyancy forces by horizontal compression and elastic stresses. The suggested present-day geotherms for central Australia rule out the first possibility and this model is even less relevant to earlier times when the lithospheric temperatures may have been higher than the present values.

The compressive force that formed the structures and supported them through time is estimated to have been of the order of a few kbars and to have acted down to at least the depth of the Moho, because this latter surface appears to have been deformed almost as much as the upper surface of the crust. The upper crust appears to be able to support these stresses without brittle failure because seismicity has not been observed in the areas of steepest gravity field gradients. The lower crust also appears capable of supporting these stresses and, in central Australia at least, there is no evidence for a mechanically weak lower crust.

Each tentative answer rests on a number of assumptions which are not strongly anchored in either a sound observational basis or an understanding of lower crustal processes. The central Australian example is not unique in this respect and many of the questions raised by the proposed model for this area are also relevant to other studies of continental tectonics.

To observe and understand the lower crust, improved information on the upper crustal structure is required in the form of better stratigraphic information in the basins and more structural and chronological control on the deformation and uplift history of the arches. The densities and seismic velocities of the basin sediments are essential in separating upper and lower crustal contributions to the gravity and seismic observations. The deformation of the sediment sequences in the basins, for example, contain information on the forces that may have operated in the crust as a whole.

The rheological implications of the geophysical observations are geotherm dependent and this is perhaps the least well constrained geophysical quantity of the lower crust and mantle. Not only does the computed geotherm depend on reliable and representative heat flow measurements, it also rests on a number of assumptions about the depth distribution of thermal conductivities and heat generation and about the mechanisms of heat transport.

Progress has been made in understanding the rheology of quartz under laboratory conditions, but the extrapolation to geological conditions and to the lower crust remains obscure. Conclusions based on such extrapolations are only indicative of possible responses of the crust to loading and cannot be used to distinguish categorically between alternative models. The rheology of olivine under laboratory conditions is more amenable to study but the recent work of Chopra & Paterson indicates that olivine is weaker than indicated by many previous experimenters.

An essential question raised by the observations of central Australia is whether significant deformation has occurred since the last orogeny in early Carboniferous time. Limited geological and geomorphological observations indicate that such deformations have not been significant but, here also, further work is required.

References

ARMSTRONG, R.L. & STEWART, A.J. 1975. Rubidium-strontium dates and extraneous argon in the Arltunge Nappe Complex, Northern Territory. *Geol. Soc. Aust. J.* **22**, 103–115.

BEAUMONT, C. 1981. Foreland basins. *Geophys. J. R. Astron. Soc.* **65**, 291–329.

BLACK, L.P., SHAW, R.D. & STEWART, A.J. 1983. Rb-Sr geochronology of Proterozoic events in the Arunta Inlier, central Australia. *BMR J. Aust. Geol. Geophys.* **3**, 35–42.

BRACE, W.F. & KOHLSTEDT, D.L. 1980. Limits on lithospheric stress imposed by laboratory experiment. *J. Geophys. Res.* **85**, 6248–6252.

BUREK, P.J. & WELLS, A.T. 1978. Tectonomagnetism and magneto-stratigraphy of the late Proterozoic tillite marker caps in the Ngalia Basin, central Australia. *Res. School Earth Sci. Ann. Rept.* 54–58.

CHEN, W.-P. & MOLNAR, P. 1983. Focal depths of intracontinental and intraplate earthquakes and their implications for the thermal and mechanical properties of the lithosphere. *J. Geophys. Res.* **88**, 4183–4214.

CHOPRA, P.N. & PATERSON, M.S. 1981. The experimental deformation of dunite. *Tectonophysics*, **78**, 453–473.

—— & —— 1984. The role of water in the deformation of dunite. *J. Geophys. Res.* **89**, 7861–7876.

CULL, J.P. & CONLEY, D. 1983. Geothermal gradients and heat flow in Australian sedimentary basins. *BMR J. Aust. Geol Geophys.* **8**, 329–337.

—— & DENHAM, D. 1979. Regional variations in Australian heat flow. *BMR J. Aust. Geol. Geophys.* **4**, 1–13.

FORMAN, D.J. 1966. Regional geology of the south-west margin, Amadeus Basin, central Australia. *BMR Aust. Bull.* **87**.

—— & SHAW, R.D. 1973. Deformation of the crust and mantle. *BMR Aust. Bull.* **144**.

FROELICH, A.J. & KRIEG, E.A. 1969. Geophysical-geologic study of northern Amadeus Trough, Australia. *Am. Assoc. Petrol. Geol. Bull.* **53**, 1978–2004.

GORTER, J.D. 1984. Source potential of the Horn Valley siltstone, Amadeus Basin. *APEA J.* **24**, 66–100.

GRAY, C.M. & COMPSTON, W. 1978. A rubidium-strontium chronology of the metamorphism and prehistory of central Australian granulites. *Geochim. Cosmochim. Acta*, **42**, 1735–1747.

HALES, A.L. & RYNN, J.M.W. 1978. A long range controlled source seismic profile in northern Australia. *Geophys. J. R. Astron. Soc.* **55**, 633–644.

HEISKANEN, W.A. & VENING MEINESZ, F.A. 1958. *The Earth and Its Gravity Field.* McGraw Hill, New York.

HYNDMAN, R.D. 1967. Heat flow in Queensland and Northern Territory, Australia. *J. Geophys. Res.* **72**, 527–539.

JACKSON, M.J. & VAN DEN GRAAFF, W.J.E. 1981. Geology of the Officer Basin. *BMR Aust. Bull.* **206**.

JEFFREYS, H. 1959. *The Earth* (4th ed.). Cambridge University Press, Cambridge.

KIRBY, S.H. 1984. Introduction and digest to the special issue on chemical effects of water on the deformation and strength of rocks. *J. Geophys. Res.* **89**, 3991–3995.

KRIEG, G.W., JACKSON, M.J. & VAN DEN GRAAFF, W.J.E. 1976. Officer Basin. *In*: LESLIE, R.B., EVANS, H.J. & KNIGHT, C.L. (eds), *Economic Geology of Australia and Papua New Guinea.* **3**, 247–253. Monogr. 7. Aust. Inst. Min. Metall.

LAMBECK, K. 1983. Structure and evolution of the intracratonic basins of central Australia. *Geophys. J. R. Astron. Soc.* **74**, 843–886.

—— 1984. Structure and evolution of the Amadeus, Officer and Ngalia Basins of central Australia. *Aust. J. Earth Sci.* **31**, 25–48.

—— & PENNEY, C. 1984. Teleseismic travel time anomalies and crustal structure in central Australia. *Phys. Earth Planet. Int.* **34**, 45–56.

——, MCQUEEN, H.W.S., STEPHENSON, R.A. & DENHAM, D. 1984. The state of stress within the Australian continent. *Annales Geophysicae*, **2**, 723–741.

MATHUR, S.P. 1976. Relation of Bouguer anomalies to crustal structure in southwestern and central Australia. *BMR J. Aust. Geol. Geophys.* **1**, 277–286.

MILTON, B.E. & PARKER, A.J. 1973. An interpretation of geophysical observations of the northern margin of the eastern Officer Basin. *Q. Geol. Notes, Geol. Surv. South Aust.* **46**, 10–14.

MINSTER, J.B. & ANDERSON, D.L. 1980. Dislocations and nonelastic processes in the mantle. *J. Geophys. Res.* **85**, 6347–6352.

MUTTON, A.J. & SHAW, R.D. 1979. Physical property measurement as an aid to magnetic interpretation in basement terrains. *Bull. Aust. Soc. Explor. Geophys.* **10**, 79–91.

NAFE, J.E. & DRAKE, C.L. 1963. Physical properties of marine sediments. *In*: HILL, M.N. (ed.), *The Sea.* **3**, 794–818, Wiley, N.Y.

PATERSON, M.S. 1976. Some current aspects of experimental rock deformation. *Phil. Trans. Roy. Soc., Lond.* **A283**, 163–172.

PITT, G.M., BENBOW, M.C. & YOUNGS, B.C. 1980. A review of recent geological work in the Officer Basin, South Australia. *Aust. Petrol. Explor. Assoc. J.* **20**, 209–220.

PLUMB, K.A. 1979a. The tectonic evolution of Australia. *Earth Sci. Rev.* **14**, 205–249.

—— 1979b. Structure and tectonic style of the Precambrian shields and platforms of northern Australia. *Tectonophysics*, **58**, 291–325.

——, DERRICK, G.M., NEEDHAM, R.S. & SHAW, R.D. 1981. The Proterozoic of northern Australia. *In*: HUNTER, D.R. (ed.), *Precambrian of the Southern Hemisphere.* 205–307, Elsevier, New York.

SASS, J.H. & LACHENBRUCH, A.H. 1979. Thermal regime of the Australian continental crust. *In*: MCELHINNY, M.W. (ed.), *The Earth: Its Origin, Structure and Evolution.* 301–351, Academic Press.

SCHRODER, R.J. & GORTER, J.D. 1984. A review of the recent exploration and hydrocarbon potential of the Amadeus Basin, Northern Territory. *APEA J.* **24**, 19–41.

SHAW, R.D., STEWART, A.J. & BLACK, L.P. 1984. The Arunta Inlier: A complex ensialic mobile belt in central Australia. Part 2: Tectonic History. *Aust. J. Earth Sci.* **31,** 457–484.

SLEEP, N.H. & SNELL, N.S. 1976. Thermal contraction and flexure of mid-continent and Atlantic marginal basins. *Geophys. J. R. Astron. Soc.* **45,** 125–154.

SMITH, R.J. 1979. *An interpretation of magnetic and gravity data in the Musgrave Block in South Australia.* Rept. No 79/103, Dept. Mines Energy, South Australia.

STEPHENSON, R. 1984. Flexural models of continental lithosphere based on the long-term erosional decay of topography. *Geophys. J. R. Astron. Soc.* **77,** 385–413.

—— & LAMBECK, K. 1985a. Isostatic response of the lithosphere within-plane stress: Application to central Australia. *J. Geophys. Res.* **90,** 8581–8588.

—— & LAMBECK, K. 1985b. Erosion-isostatic rebound models for uplift: An application to southeastern Australia. *Geophys. J. R. Astron. Soc.* **82,** 31–55.

WELLS, A.T., FORMAN, D.J., RANDFORD, L.C. & COOK, P.J. 1970. Geology of the Amadeus Basin, central Australia. *BMR Aust. Bull.* **100.**

—— & MOSS, F.J. 1983. The Ngalia Basin, Northern Territory. *BMR Aust. Bull.* **212.**

KURT LAMBECK, Research School of Earth Sciences, Australian National University, Canberra, Australia.

Petrology and geochemistry of the lower continental crust: an overview

R.W. Kay & S. Mahlburg Kay

SUMMARY: Continental lower crust comes in a limited number of types that are best distinguished tectonically. Lower crust is formed primarily in oceanic or continental zones of lithospheric plate convergence and secondarily in regions of intracontinental rifting. Formation processes involve magmatic intrusion of mantle-derived basalt and structural burial of pre-existing upper crust by thrust or reverse faults. In both cases, metamorphism, crustal melting and ductile deformation result from the accompanying heating and release of volatiles. A compositional conundrum exists for both arcs and continental rifts since the new continental crust has a basaltic rather than an andesitic (mean crustal) composition. Possible solutions to this compositional disparity include formation of more silicic lower crust in the Archaean, and lower crustal delamination which could preferentially remove tectonically immature basaltic crust.

Introduction

The total mass, age distribution and composition of continental crust are three fundamental properties of the earth. We focus on composition in this paper, and in particular on lower crustal composition, which is far less certain than upper crustal composition. First, we will review the geophysical, petrological and geochemical rationales behind lower crustal compositions proposed in the last three decades. The differences in the approaches used in these studies outline the nature of the problem and the end-member possibilities. In the following sections, we will discuss the role of arc magmas in producing net crustal additions, the role of intracrustal melting and crystallization and the heat sources required to produce intracrustal melting, and the role of volatiles in the lower crust.

Lower crustal composition: is there a consensus?

In 1954 Poldervaart convened a symposium on the crust of the earth, and reserved for himself the topic of composition. He calculated the mean compositions of crustal lithologies in four crustal subdivisions: oceanic, suboceanic, young folded belts and continental shield. Poldervaart's (1955) average continental crustal composition has 59% SiO_2 and is 'andesitic'. Incorporation of the suboceanic portion yields a total crust with about 56% SiO_2. Sediments comprise only 6% of Poldervaart's total crust. Other compositions used correspond to igneous rocks (plutonic and volcanic) that were not recovered from great depth or

even crystallized at great depth. The biggest single unknown in Poldervaart's calculation (it represents 40% of the total) is the lower crust. He assumes a basaltic composition and a gabbroic mineralogy, relying on a generalized match of density and seismic velocity.

Ronov & Yaroshevsky (1969) presented their crustal composition as a revision of Poldervaart's model. When faced with the necessity of estimating the composition of the lower crustal 'basaltic shell' of the continental crust they arbitrarily assumed equal amounts of basaltic and granitic components. They recognized that the rocks are often metamorphic, but used igneous compositions. A somewhat more siliceous bulk crustal composition than Poldervaart's results. In summarizing seven author's estimates (their Table 9), the continental crustal average ranges from 58 to 63% SiO_2, which is 'andesitic'. The range reflects, almost entirely, the choice of lower crustal composition. Progress in understanding the chemical composition of the crust was limited by various authorities' ability to do any more than make 'reasonable' estimates of proportions of common rock types in mid and lower crustal regions. This contrasts with the extensive data base and careful weighting procedure used to arrive at an upper crustal composition that represents only a third of the total.

More recent estimates of lower crustal composition are tied more directly with geophysical data, exposed lower crustal sections, and lower crustal rocks recovered in xenoliths. Smithson (1978), Smithson & Brown (1977) and Smithson et al. (1981) emphasize the constraints of mean crustal seismic velocity (e.g. Christensen & Fountain 1975; Christensen 1979) and of crustal reflection properties on lower crustal composi-

From: DAWSON, J.B., CARSWELL, D.A., HALL, J. & WEDEPOHL, K.H. (eds) 1986, *The Nature of the Lower Continental Crust*, Geological Society Special Publication No. 24, pp. 147–159.

tion. A key assumption in these arguments is that the mineralogy of the lower crustal rocks is appropriate to granulite-facies conditions. Thus, both density and seismic velocities of basaltic composition lower crustal rocks are higher than in Poldervaart's models. Basaltic composition— as garnet granulite—is ruled out on this basis, and an 'andesitic' composition (59% SiO_2) is deduced. Here, it is inappropriate to use the word 'andesite' however, because the granulites have much lower U and Th contents than extrusive andesite with the same SiO_2 content (e.g. Smithson & Heier 1971; Heier 1979). The major element composition of lower crust is quite variable, both locally, as indicated by crustal reflectors, and regionally, as indicated by differing crustal velocity profiles.

Exposed sections of lower crust have been described from several orogenic belts (Fountain & Salisbury 1981). Lower crustal lithologies in the Ivrea zone (Italy) and the Fraser Range (Australia) are predominantly mafic, while those from the Musgrave Range (Australia) and the Karila Group (Western Africa) are predominantly intermediate to silicic. These rocks are all in the granulite metamorphic facies. Fountain & Salisbury (1981) point out that the change to increasingly lower metamorphic grade (amphibolite, then greenschist) at shallower crustal levels occurs in all proposed lower crustal sections. This change in metamorphic grade, rather than compositional change, seems to be the unifying feature of all the sections. These exposed sections suggest a compositionally diverse lower crust which includes both meta-igneous and metasedimentary rocks ranging in composition from mafic to felsic. The tectonic event that has exposed middle and lower crust is thought to be overthrusting at a continental collision zone. Thus, a rather special class of continental crust may be exposed by such a process, so extrapolation of the exposed sections to the crust as a whole should be avoided. However, the compositional diversity is noteworthy in even such a tectonically restricted environment.

The xenolithic crustal baggage brought to the surface by magmas and fluidized solids as they traverse the crust provides data about the lower crust from a wide range of crustal sections and tectonic environments (e.g. Dawson 1977; Rogers & Hawkesworth 1982; Kay & Kay, 1981, 1983; Padovani *et al.* 1982; McCulloch *et al.* 1982). Sample biasing is not so much in areal coverage, but in vertical representation—the samples brought from different crustal horizons have been scrambled. Of interest in the study of xenolithic fragments is the methodology outlined by McGetchin & Silver (1972). They used an inverse size–depth relation to order fragments from a

serpentinite diatreme in Utah. The results of their study are also novel: they interpreted the abundant retrograde (low temperature) hydration of mafic granulites as a pervasive *in situ* hydration. Seismic velocities and densities of these mafic rocks are much lower then their granulite-facies anhydrous counterparts. Thus, McGetchin & Silver (1972) propose that the lower crust in the region is mafic. Kay & Kay (1981, 1983, 1985a) have pointed to the widespread abundance of basic granulites in xenolith populations from areas as diverse as island arcs, continental and oceanic rifts and cratons. At face value, this would imply a mafic lower crust. In contrast, xenolithic fragments from basalts of the Rio Grande rift contain abundant aluminous garnet granulite xenoliths that appear to have clay-rich sedimentary protoliths. Padovani & Carter (1977) believe that the anhydrous granulites are residues complementary to granites that have migrated to shallower crustal levels. A lower crust of granulites with intermediate silica content and a non-igneous composition is indicated by Padovani & Carter (1977).

Over the past decade, Taylor (e.g. Taylor & McLennan, 1981) has calculated lower crustal composition by using an entirely different procedure. He assumes an andesitic bulk crustal composition (58% SiO_2) and uses observed upper crustal composition to calculate lower crustal composition by mass balance. An andesite bulk composition is chosen because magmatic additions of new crust at island arcs, the only volumetrically important site of new crustal addition, are assumed to be andesite. The result is a magmatically differentiated crust whose upper third is further differentiated by sedimentary processes and whose lower crust is 'restite', or residue complementary to upper crust granodiorite, rather than igneous melt, as in Ronov's and Poldervaart's models.

Finally, study of lower crustal terrains that correspond to addition sites of crustal mass (generally, magmatic arcs) has led Weaver & Tarney (1980 a,b) and others to postulate a quartz diorite (andesitic composition) lower crust in these terrains. Analogy between Archaean (Lewisian terrain, Scotland) and recent (Chilean Andes) terrains allows them to generalize their conclusions to much of Earth history. Weaver & Tarney emphasize the importance of lower crustal igneous intrusions, in contrast to residues from melting and fractionated crystals.

This selective summary illustrates the diversity of opinion and the main trends of thought about the chemistry of lower continental crust. Is there any emerging consensus? Apparently not between different workers or even within the statements of individual workers. For instance, Ringwood (1979) summarizes his view on crustal composition by accepting Taylor's (1977) andesite model, yet implicitly rejects Taylor's lower crustal composition (54% SiO_2) in favour of acidic-intermediate rocks in the garnet granulite facies, which would have 58% SiO_2 or more.

Lower crustal composition: primary magmas, intracrustal processes, and the andesite model

The andesite model is based on the postulate that bulk crustal composition and bulk volcanic arc composition are andesitic. Lower crustal melting or crystallization (e.g. Hamilton 1981) leaves a less silicic lower crust complementary to a more silicic upper crust. In the mass balance of Taylor (1977) and Taylor & McLennan (1979, 1981) andesite equals 1/3 upper crust plus 2/3 lower crust. The following aspects of the andesite model are of particular interest: the bulk composition of present additions of new crust at volcanic arcs, the state of stress and heat budget for intracrustal differentiation, and Precambrian lower crustal granulite compositions. The discussions of Tarney & Windley (1977, 1979) and Weaver & Tarney (1980a,b) are complementary to this discussion.

The andesite model finds its original rationale in the observation that andesites and compositionally similar quartz diorites are common volcanic and plutonic rock types at magmatic arcs. But in many oceanic arcs, basalt is at least as common as andesite among extrusives. Among intrusives, only the very shallowest one are exposed. Although these are dominantly 'andesitic', there is good reason to postulate that deeper, and in particular lower, crustal intrusives are gabbroic and even ultramafic (e.g. Conrad & Kay 1984; Kay & Kay, 1985a). The situation is analogous to that which we face in estimating the bulk composition of the oceanic crust. Crustal addition or growth is only concerned with primary compositions, and the commonest mid-oceanic ridge basalts are not primary: they include many highly differentiated samples, and the plutonic section of the oceanic crust contains the crystalline residue implied by the differentiation. In the arc case, it is equally certain that the common basalts and andesites that reach the shallowest arc levels do not originate directly from the mantle (e.g. they are not primary compositions). In the Aleutian arc (see Fig. 1), we have identified olivine tholeiite as a primary arc lava and have proposed that the early fractionated phases (olivine and clinopyroxene) have accumulated at Moho depth and represent newly formed upper mantle. The magma composition that results from this frac-tionation is high-Al basalt, which represents the composition added to the crustal section of the arc. Thus, for arcs with even the smallest proportion of high-Al basalt, and especially for arcs with olivine tholeiitic basalts (representing the little-fractionated, near-primary basalt) we find little support for the andesite model.

If we change the bulk composition of new arc from andesite (e.g. Tatsumi 1982) to high-Al basalt, shallow level silicic volcanic and plutonic rocks can be derived by a combination of crystal fractionation from a high-Al basalt and assimilation of a low melting fraction from the crust (DePaolo 1981), but mass balance makes it difficult to avoid a mafic lower crust. Furthermore, if the bulk composition of the earth's continental crust is andesitic, the crust formation process is not duplicated in arcs, as noted by Kay (1980), Smithson *et al.* (1981), Karig & Kay (1981), Anderson (1982), and Kay & Kay (1985a). Weaver & Tarney (1980a,b) implicitly recognize this problem by postulating crustal underplating of quartz dioritic magma—which would be a primary melt as it occurs immediately over the Moho.

The trace element and minor element content of the primary arc tholeiites is variable, reflecting mantle processes, the subduction process being the most important. If we admit that some elements of the primary magmas in arcs have travelled down with the subducted oceanic plate, under the arc and into the zone of arc magma generation (level B recycling, Fig. 1), a redefinition of new crustal addition is required. From the primary composition we will have to subtract the recycled components added at various levels (Kay 1985). When this is done, Kay (1980) and Karig & Kay (1981) note that not only is the bulk composition of the primary magmas inconsistent with creation of an andesitic crust, but the unrecycled K content is much too low (see also White & Patchett 1984).

The actual crustal section of a particular arc depends on the pre-existing crust (oceanic and cratonic are end-members) at the arc magmatic axis, and on the structural response of the arc. By isolating the primary magma argument, we have neglected discussion of the other parts of the equation—intracrustal differentiation and assimilation—which have relevance to the composition of the lower crust and to the contrast in upper and lower crustal compositions. We discuss two crustal end-members: thin oceanic crust and thick continental crust; and two stress regimes: extensional and compressional.

In the case of oceanic crust, we ask whether any is trapped at mid to lower crustal levels of arcs, buried and downwarped by overlying arc volca-

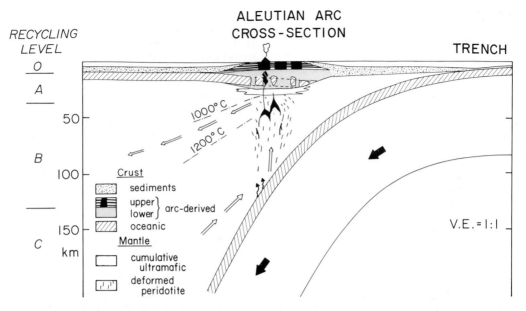

FIG. 1. Cross section showing crustal formation at a convergent plate margin (after Kay 1985 and Kay & Kay 1985a). Plate geometry and crustal thickness are those of the central Aleutian island arc. In the lower crust, note the trapped oceanic crust and the igneous cumulates, complementary to differentiated volcanic and plutonic rocks of the upper crust. Crustal recycling levels (O, A, B, C, after Kay 1985) refer to erosion; intracrustal melting, and sedimentary underplating in the forearc region; melting at about 100 km under arc volcanoes; and long time scale (billion years) mantle recycling. Net addition of crustal material or of any particular element must be calculated from the gross rate by correcting for recycled components.

nic and plutonic rocks (as shown in Figure 1). This situation would occur only for arcs that are under compression on the arc axis, for there the initial arc basement would be oceanic crust (note that in map view the site of the magmatic arc is perhaps 50–100 km from the plate boundary itself, which generally is not simple oceanic crust—Karig 1982). This trapped oceanic crust has escaped its normal fate of subduction, to become incorporated into new continental crust. A calculation in the Aleutian arc (Kay & Kay 1985a) indicates that this crust partially hydrated and metasomatized by circulating ocean water and capped by pelagic sediment, could be a significant component of the arc column. Evidence for trapped oceanic crust has been reported from both active arcs: (the Aleutians, based on xenolithic samples of mafic and ultramafic rock, Kay & Kay, 1985a), and exposed lower crustal sections (the Ivrea zone, Pin & Sills, 1986, this volume).

For the other end-member, arc basement of thick continental crust, the presence of the pre-existing crust is not doubted, nor is its assimilation by rising arc magmas (e.g. Moorbath et al. 1981; Vollmer & Norry 1983; Oxburgh & McRae

1982; Thompson et al. 1982; Harmon et al. 1984). Clearly, the bulk compositions of crustally contaminated lavas and shallow intrusives are not the compositions of new additions to the crust—they are products of intracrustal (A-level, Fig. 1) recycling. The western U.S. and the Andes serve as examples (DePaolo & Farmer 1984; Hawkesworth et al. 1982; James 1982; Thorpe et al. 1984). This crustal assimilation yields magmas with isotopic mixing characteristics that furnish much information on underlying crust, but more importantly, provides evidence for residual rocks (restites) in the lower crust. New additions of basaltic (?) melt to the crust are an obvious choice for the heat source required in any igneous differentiaiton by melting of the crust. If the pre-existing crust is intermediate in composition, voluminous silicic melts or silicic contamination end-members that mix with more basic melts would result. A silicic melt (more dacitic) formed by melting a basaltic composition lower crust is also possible—as suggested by Conrad (1983) for a rift environment (see below). In any case the proportion of new to old crust is a free variable and may be reflected in the diverse crustal sections described by Fountain & Salisbury (1981). Smithson *et al.* (1981) and Pitcher (1979) have noted that the seismic velocity, gravity and field geology of the Sierra Nevada (USA) and the Peruvian Andes indicate a more 'basaltic' root zone for these continental arcs. Thorpe *et al.* (1981) have proposed that basaltic magmas have ponded at the Andean Moho, and Herzberg *et al.* (1983) outline a density stratification model that limits the penetration of basaltic melts into silicic crust.

The state of stress in the arc crust—particularly the mid to lower crust—may have an important influence on the crustal response to the magmatism (e.g. Kay *et al.* 1982). The rapid uplift of magmatic zones along the Andean axis (Altiplano) with contemporaneous silicic volcanism may indicate that under compression, all but the least dense magmas are trapped in the crust. Heat supplied by magmatic intrusions may have raised the brittle–ductile transition to shallow levels and have facilitated the rapid Neogene structural thickening (Isacks 1985). Over this short time period, thickening by magmatic addition alone would require an excessively high magmatic addition rate (see Karig & Kay 1981; Reymer & Schubert 1984).

A lower limit on the magmatic addition rate is provided by the eruption and intrusion rate of secondary (crustally-derived) magmas. Except in regions with abnormally hot crust that has been structurally imbricated (e.g. Molnar *et al.* 1983), heat released by basaltic primary magma seems to be the most likely heat source for crustal fusion, and basalt of mass at least equal to the mass of secondary magma is implied. Often, the identification of the primary magma as basalt is based on indirect evidence. The widespread occurrence of basaltic parents to secondary magmas in arcs with thin crust together with the independence of magma generation processes in the mantle and the thickness of overlying crust, argues for a similar bulk composition (although not trace element and isotopic composition) or parental magmas in all arcs.

In summary, present arc systems, the site of addition of new crust, do not support an arc model for generation of bulk crust of andesitic composition, and for calculation of lower crustal composition by mass balance. Where the identification of primary magma has been made it appears to be olivine tholeiite, and the composition added to the crust high-Al basalt. Oceanic crust incorporated in the arc section (Kay & Kay 1985a, Fig. 1) must also be inventoried as new continental crust. Taken in a broader context, the crustal composition formed at present arcs is too basic, and the crustal formation rate is too low to form the whole crust. Furthermore, crustal growth models with significant sediment recy-

cling into deeper mantle regions (level C, Fig. 1) run the risk of basifying crustal composition since the Archaean. One solution to this situation is to postulate that crust formed from the mantle in the Archaean, which in most models is a substantial fraction of crustal mass and in some models the whole mass (Armstrong 1981a,b), was more silicic than crust formed at present. Modern arcs have shown us that in order to identify the composition of newly added crust (that coming from the mantle) a primary magma must be identified (A-level, or intracrustal recycling corrected, Fig. 1) and subduction-related recycled elements must be identified (B-level recycled components). Examination of Archaean crustal terrains must be done with these principles in mind. The analogy of some Archaean terrains to present day island arcs has been claimed (Bickle *et al.* 1983; Tarney *et al.* 1977, 1979; Weaver & Tarney 1980a,b). An effort must be made to establish any Archaean-post Archaean compositional contrast, perhaps due to different plate-tectonic styles (see Dewey & Windley, 1981).

Intracrustal melting and crystallization: vertical tectonics and chemical structure

a) Granitoids and the restite model

We have seen that as mantle-derived magmas cool, they precipitate minerals, and crystalline residue accumulates in the lower crust. In addition, heat released by this cooling and crystallization causes partial melting of pre-existing lower crust, yielding 'granitic' melts. Often the 'granites' mix with or contaminate the basalt or its differentiated daughter magmas. Sometimes the granites segregate and rise as a water-undersaturated granitic crustal melt. When this does occur, a passive 'stoping' mechanism may require an extensional stress regime.

The chemistry of granitic melts provides the most readily accessible information about the lower crust in the same way that basalt chemistry provides information about the mantle. Chappell (1984) emphasized a fundamental distinction between S-type granitoids which have a sedimentary crustal protolith and I-type granitoids, which have an igneous crustal protolith. In most cases, incomplete segregation of partial melt has occurred by a porous flow mechanism and the intrusives are mixtures of melt and entrained unmelted residue (or restite).

Chappell identifies the most mafic bulk composition in a series of cogenetic granitoids as the

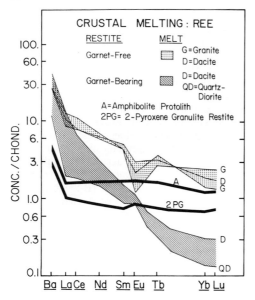

FIG. 2. The rare earth content, depicted as multiples of the concentrations in a chondritic meteorite (Leedy), of silicic magmas derived by melting of continental crust. The light rare earth to heavy rare earth ratio, of which the La/Yb ratio is good index, reflects source REE content, mineralogy of residual minerals and percentage of melting. A shift in restite mineralogy from two pyroxene granulite to garnet-bearing or (to a lesser extent) hornblende-bearing rock, will cause partial melts to have higher La/Yb ratios, everything else held constant. Higher water content will cause melting at lower temperatures, favouring garnet and hornblende-bearing restite mineralogies. Presence of feldspar on the restite causes the metal to have a negative Eu anomaly. Several igneous rocks (granites, dacites, quartz diorite) have been segregated from garnet-free and garnet-bearing crustal sources, as shown. Data are from Arth and Barker (1976): dacite porphyry from the Raton-Clayton volcanic field, New Mexico, Arth and Hanson (1972): Saganaga quartz diorite, northern Minnesota; O'Brien (1983) granite xenolith, Buell Park diatreme, Arizona; and Conrad (1983) dacite from McDermitt Caldera, Nevada. The amphibolite (A) and two pyroxene granulite (2PG) are xenoliths from Buell Park diatreme that may represent protolith amphibolite, and restite granulite complementary to the granulite from the same locality (O'Brien 1983).

source composition. The restite itself is considerably more mafic, and for the more segregated compositions (richer in melt, poorer in restite) restite remains in the lower crustal source. Melting of restite, following withdrawal of granitoid, yields A-type granitoids, which are hot and dry and often F and Cl-rich (Collins *et al.* 1982). The

mineralogy of the restite that is entrained with melt changes in response to pressure release during intrusion and therefore is not indicative of the source mineralogy. Trace element studies on rhyolites or on restite-poor granitoids (those near the partial melt composition) may provide the best information on source mineralogy. In particular sources that are hornblende-bearing and especially those that are garnet-bearing may be identified by the heavy REE depletion of granitoids and silicic volcanic rocks derived from them (Arth & Hanson 1972; Barker & Arth 1976; Arth & Barker 1976; Nicholls & Harris 1980). In contrast, (see Fig. 2) melts originating from sources with lower (or no) modal garnet and hornblende (corresponding to melting in a hot and dry crust) have higher heavy REE. Examples are from silicic magmas of bimodal centres, the Scourian (Pride & Muecke 1980, 1981) and the Colorado Plateau xenoliths (O'Brien 1983). The latter example matches granitoids and residual granulites in a vertical crustal column using xenolithic inclusions in 'kimberlite' diatremes in the Colorado Plateau. In using trace elements to assess source area signatures, care must be taken to account for accessory phases which can significantly affect trace element distributions (e.g. Miller & Mittlefehldt 1982).

The application of melt-restite models to volcanic rocks with crustal origin is limited to those few with mafic clots that have been at equilibrium with the more silicic host. Often, mixing is important in the intermediate compositional range, but the mixing end-members are melts, and therefore do not have a melt-restite relationship. Rather, the end-members are basaltic, andesitic, or rhyolitic liquids, with their characteristic phenocrysts—which are easily recognized compositionally (e.g. Kay & Kay 1985b).

Volcanic rocks in bimodal suites from the McDermitt caldera (Nevada) can be related to crustal melting. Conrad (1983, 1984) suggests that crystal-liquid fractionation has controlled the dacite-rhyolite trend, but no more basic magma type appears to be parental to the dacite. Conrad proposes that the dacite is a primary crustal melt, which he models as the melting of a two-pyroxene gabbro (itself a recent basaltic crustal addition along the axis of the rift zone on which the caldera lies).

b) Heat Sources

Often the source rocks of crustal melts are not exposed. The granitoids intrude cooler country rocks, therefore melting occurs at deeper, hotter crustal levels. The heat source is treated somewhat mysteriously by Chappell (1984). He does not favour a basaltic intrusive heat source because he sees very few exposed basalts. This is not as definitive as it appears, for it is difficult to see how a molten basalt could penetrate through a less dense silicic pluton. Pitcher (1979) shows large amounts of basaltic melt at depth under the Andes, yet the surface plutons are quite silicic. The alternative chosen by Chappell is a continuously elevated geotherm over the time of granitoid intrusion. This has no present-day analogy and we think that it is unlikely in the absence of magmatic intrusion from the mantle. The thermal time constant for the crust is only several tens of millions of years and, in any case, the heat flux would have to furnish the heat of fusion continuously. As a general rule we believe that heat necessary for fusion of crust is furnished dominantly by intrusion of mantle-derived (i.e. basaltic?) melts into the crust, or perhaps at the crust–mantle boundary.

S-type granitoids in the forearc of the Aleutian arc, Alaska, provide a particularly illuminating illustration of the source rocks and heat sources. The 60 my granites, as exposed on Kodiak and adjacent islands, cut slates of an accretionary prism at a time when a spreading oceanic ridge intersected the arc. The granitoids themselves have a small component of basaltic origin and a large component of sedimentary origin. Hill *et al.* (1981) interpret the granitoids as partial melts of accretionary prism greywackes with the abnormal heat provided by basaltic intrusion at the base of the accretionary prism. This is not normally a site of magmatic activity, therefore the special tectonic circumstance is required.

Chappell (1984) has noted that the source compositions of S-type granitoids are more variable than those of I-type granitoids. This parallels the observation that sediments are more diverse than basic igneous rocks. He holds that the basic igneous rock sources are deeper than the sediment sources and calls upon pre-existing underplating of basalt to furnish the gabbroic source compositions for his I-type granitoids. This is a similar mechanism to the one called for by Conrad (1983).

Probably the best case for a structural cause for crustal melting has been made for large (300 km^2) S-type leucogranites in the Himalayas by LeFort (1981). Heat sources considered by Molnar *et al.* (1983) are overthrusting (which heats rocks of the lower plate), radiogenic heating, and frictional heating along the thrust. A combination of unusually high temperatures (for a craton) in the underthrust Indian plate, and a frictional contribution is sufficient to raise the temperature to 650°C (which is above the granite minimum) at 20 kilometres. Important limitations to the calcula-

tion are the uncertainties in thermal conductivity values and the maximization of temperature by the exclusion of melting and metamorphic reaction heat sinks. We recognize that relatively cool water-rich granites like the Himalayan ones appear to have melted without any increase in mantle heat flow or any temperature increase due to asthenospheric upwelling (perhaps accompanying lithospheric delamination). However, granitoids and volcanic rocks caused by crustal melting are usually hotter (often 800°–900°, Hildreth 1981) than the Himalayan leucogranites. The thermal models of Molnar *et al.* illustrate the extreme difficulty in achieving this high a crustal temperature by purely structural processes.

c) Summary

Two end-member interpretations govern our view of the implications of crustally-derived magmatism for the lower crust.

(A) If shallow silicic magmas result from crystal fractionation of basaltic parent magmas then a readily calculated volume of mafic crystal sediment is present at depth in arcs. Kay & Kay (1985a) estimate a relative mass ratio of 2 grams of fractionate crystals for every gram of calc-alkaline pluton (58% SiO_2 average) in the Aleutian arc. For more silicic plutons, the fraction of crystal cumulate will be greater. The actual crystal fractionation mechanism probably involves side and roof plating of liquidus phases in a convecting magma chamber. The magma chamber will filter ascending magmas by density; denser basaltic magmas entering the base of the magma chamber will rise no further, and will not erupt. Evidence for the existence of a series of magma compositions representing a liquid line of descent is a prerequisite to this interpretation. This evidence is seen in both volcanic and plutonic rocks. The cumulates themselves have been recovered in several active magmatic arcs (e.g. Conrad & Kay 1984).

(B) If the shallow silicic magmas result from partial melting of pre-existing crust, we may postulate at least an equal volume of basaltic material intruding the crust to serve as a heat source. We note, as does Chappell (1984), that a heat source for melting I and S granitoids remains a fundamental unaddressed question in his crustal model. In his Australian example, he proposes that a high thermal gradient existed in the crust of the Lachlan fold belt throughout the time of crustal melting. However, modern analogies argue that in places where sustained thermal perturbations exist in the crust, transfer of adequate heat from the mantle via basaltic intrusion accompanies asthenospheric shallowing and is the main cause of crustal heating. For instance, the thinning of lithosphere caused by the rise of hot mantle to shallow levels, as in the Basin and Range (United States), is accompanied by basaltic volcanism.

Volatiles in the crust— horizontal tectonics

The importance of volatiles in lower crustal rocks far outweighs their total percentage. Among the major volatile species are H_2O (possibly with high salinity), CO_2 and perhaps CH_4. Water in particular exerts primary control over rheology (including melt migration), melting, reaction rate and metasomatic transfer in the lower crust. Probably the two most important observations relevant to the volatile content of the lower crust are the occurrence of deep crustal (> 20 km) zones with high electrical conductivity, and the widespread occurrence of liquid CO_2 inclusions in granulites from the lower crust. The first observation may best be explained by interconnected saline pore fluid (Lee *et al.* 1983; see review of Kay & Kay 1981). Because CO_2 is a non-conductive fluid, the second observation seems to contradict the first. However, the immiscibility of brine and CO_2 (Sisson *et al.* 1981), and the high solubility of H_2O but low solubility of CO_2 in granitoid melts under lower crustal conditions suggests several possible mechanisms for separating the two volatile phases. A third observation, the commonly low U and Th content of lower crustal granulites, has been directly linked to removal of these elements as dissolved species in water derived from dehydration of amphibolites during granulite facies metamorphism.

A deep regional electrically conductive layer occurs in the United States under the Adirondacks, Blue Ridge Mountains, and Central Wisconsin, in South Africa under the Kapvaal Craton, and presumably within many other regions of cool, tectonically inactive continental crust (see review and references in Kay & Kay 1981). Oxide and silicate mineral conductivities (even for minerals with bound water) appear to be too low, but an intergranular fluid or graphite seem to be good possibilities to explain this conductive layer. Archie's law conductivity would demand at least a 1% interconnected porosity of a saline (chloride or sulphate-rich) hydrothermal fluid. Lee *et al.* (1983) report experiments which indicate unexpectedly high conductivity in water–gabbro mixtures. This enhanced conductivity is due to conductive surface layers on the minerals. The halogens and sulphate could be bound in mineral

phases or could be in a brine, if the water content exceeded that necessary to form the conductive surface layers. Conductivity increases with temperature, therefore ambient lower crustal temperatures are of interest. In many cases (e.g. Adirondacks) the temperature is no higher than 300–400°C at the top of the conducting layer. If significant deep circulation of meteoric water occurs at shallower crustal depths, then the temperatures should be cooler (note however, unexpectedly high temperatures in the Kola drill hole, Kozlovsky 1984).

The CO_2 that fills many of the primary fluid inclusions in granulite facies crustal and, for that matter, mantle tectonites (e.g. Touret & Dietvorst 1983), implies a low water partial pressure in lower crustal granulites. With such a desiccating fluid phase, it is not surprising that the only volatile-bearing phases in these rocks are CO_2, SO_3 and halogen-rich (biotite, hornblende, scapolite), and H-poor. If these rocks are residual, complementary to upper crustal granitoids, then these granitoids melted in a water-undersaturated condition. Water present in the lower crustal protolith dissolved in the melt, and was removed by it. The CO_2 was not soluble and remained in the source. Conversely, we can recognize granitoids and extrusive equivalents that are water undersaturated as high temperature ones, those most likely to reach shallow crustal levels. For the thermal problem of crustal fusion, this implies a temperature considerably higher than the granite minimum ($\simeq 650°C$) at lower crustal depths, probably in excess of 800°C (see Bohlen *et al.* 1983; Hildreth 1981). As argued above, modern crust with temperatures greater than 800°C at mid to lower crustal levels occurs in regions with thin lithosphere associated with upwelling asthenospheric mantle that yields basaltic melts. Thus heat released by crystallization of basaltic intrusions is directly responsible for crustal melting.

Granulites commonly have distinctively low concentrations of the heat producing elements U and Th as well as Rb and to a lesser extent K (Heier 1979; Smithson & Heier 1971; Jaupart *et al.* 1981). This is true for granulites across the compositional spectrum from granitic to gabbroic, and includes both meta-igneous and metasedimentary rocks. If U, Th and Rb were in the protoliths of the granulites, they have escaped by one of two mechanisms: they were dissolved in either a water-fluxed granite or a solute-rich aqueous fluid that migrated to shallower levels. The water was either present in or passed through the protolith. In the first case, the water might have been originally bound in hydrous phases, and subsequently released by breakdown of these hydrous phases as temperature increased. If permeability exists, migration of the water (and its dissolved solutes) to higher crustal levels occurs (e.g. Glikson & Lambert 1976; Gray 1977; Holt & Wightman 1983; Heier 1979), leaving a fluid-depleted and U-, Th- and Rb-depleted granulite. For this mechanism to work, both the U, Th and K content of the fluid, and the fraction of fluid to rock must be high (Mueller 1967). For instance, to remove 5 ppm Th from an amphibolite with 2% bound water, the Th concentration would have to be 250 ppm Th in the aqueous phase. This is clearly too high, exceeding the Th solubility by perhaps a factor of 100. In order for fluid transfer to work, a circulating system would have to be established, as in the circulation of meteoric water through a cooling pluton. Element transfer is still solubility-controlled, but a much lower concentration is necessary (see Albarede 1975; Buntebarth 1976). If such circulation occurs in the crust, and indications from the Kola Peninsula deep drill hole indicate circulation to at least 11 km, then the thermal profiles calculated from heat flow and conductivity are wrong: temperatures would decrease with circulation.

We regard the mobility of Th in an aqueous phase with great scepticism, as no adequate solubility has been demonstrated. Th is much more likely removed in a melt—in particular in melts that are not zircon saturated, as is the case for many granitoids (Harrison & Watson 1983). The removal of U at the same time would be likely. Concentrations of U and Th are high (several tens of ppm) in many granitoids and their extrusive equivalents, and the percentage of melt phase necessary for segregation is high—due to the high viscosity of granitic melt (McKenzie 1984). So we seen no obstacle in postulating the removal of U, Th and Rb by a melt phase, especially in view of the common occurrence of shallow level granitoids and silicic volcanic rocks with high concentrations of these elements (as an extreme example, standard granodiorite GSP-1 has 110 ppm Th).

The volatiles in lower crustal rocks have several sources. Like the rocks themselves, the volatiles can originate in the mantle or from the earth's surface. The mantle-derived volatiles, perhaps most clearly identified by rare gas signature (He, Ar, Xe isotopes), arrive in the crust dissolved in basaltic magmas, which commonly contain 1–3% water (in the case of arc magmas, Anderson 1982), and dissolved carbon dioxide (in the case of alkaline magmas). Upon crystallization, these volatiles probably are bound in hydrous and carbonate minerals and on conductive grain boundaries. We have outlined an important role for basaltic additions to the crust and it may be significant that the volatile content of basalt probably exceeds that in most granulites. In island arcs, there is a strong indication (Anderson 1982) that much of the water is recycled by horizontal tectonics from the surface, brought in by the hydrated oceanic plate to the zone of magma generation under the volcanic line, and removed by the partial melts formed there (the B-level recycling of Fig. 1). A second source of volatiles in the crust also involves horizontal tectonics: underthrusting of clay and carbonate-bearing sediments in continental collision zones (e.g. Cook *et al.* 1979). Newton *et al.* (1980) and Glassley (1983) have postulated that an important source of CO_2 is decarbonation reactions caused by heating of deeply buried sedimentary carbonate. The coincidence of a conductive layer with the position of deeply underthrust clay-rich sediments in the Southern Appalachians (L. Brown, pers. comm. 1984) points directly to introduction of water into mid-crustal levels, beneath higher grade metamorphic rocks. Extrapolation of this result to a general explanation of electrical conductivity anomalies by underthrusting of volatile-rich sediment may be an extreme position, but a tenable one. Certainly, the occurrence of S-type granitoids implies a presence of sediment and volatiles at mid to lower crustal levels, and xenoliths with metasedimentary protoliths indicate deep burial of sediment. A second means—simple burial of volatile rich sediment and oceanic crust—suggests itself in the volcanic arc crustal section outlined by Kay and Kay (1985a, see Fig. 1).

Perhaps the most hotly debated mechanism for introducing volatiles into the crust is the shallow subduction of hydrated oceanic rocks for great distances at low temperatures to positions far inland from oceanic trenches. Helmstaedt's models (e.g. Helmstaedt & Doig 1975) for the Colorado Plateau were the first to postulate this mechanism. The xenoliths in diatremes from the Colorado Plateau (United States) contain blue schists and pervasive hydration of both upper mantle peridotite and lower crustal granulite. Earlier, McGetchin & Silver (1972) had noted the secondary addition of volatiles to many of the xenoliths, and had proposed a regional mantle and crustal hydration event. Smith (1979) has also discussed secondary hydration of mantle assemblages. The regional implications of the hydrated crustal xenoliths have been questioned by Padovani *et al.* (1982) and O'Brien (1983) who emphasize the effect of kimberlite volatiles, the presence of meteoric water in some xenoliths, and the possibly localized (by shear zones) rather than regional extent of the hydration. If the hydration is only local in extent, then McGetchin & Silver's

(1972) conclusions that the lower crust is mafic is called into question: only hydrated mafic rocks will have low enough velocities to match observed refraction velocities.

Bulk composition of the lower crust: an unresolved problem

Study of the composition of the continental crust is plagued by one contradiction: mean seismic velocities of the crust (even allowing for elevated temperature—Christensen 1979) and direct study of exposed deep and shallow crustal sections point to an intermediate 'andesitic' mean crustal composition. But examination of the magmas from island arcs, the supposed site of crustal formation, indicates that the composition of new crust added at arcs is basaltic (high-Al basalt). Basaltic magmas generated in regions of crustal extension, and incorporation of oceanic crust into the continental crust also imply net crustal additions of basaltic composition. Observed deep crustal xenoliths are also frequently 'basaltic'. High volatile content could eliminate the contradiction by lowering seismic velocities of basaltic rocks. Keeping the basaltic composition in a metastable garnet-free mineralogy could also lower the seismic velocity of basaltic rocks to nearly as low a value as observed seismic velocities. By themselves, these two effects would not lower the velocity of basaltic composition rocks to observed values. We are left with two choices: First, the mean composition of new crust (non-recycled additions from the mantle to the crust) was more andesitic in the past, in particular at the time around 2.5 by when large amounts of crust formed (e.g. McCulloch & Wasserburg 1978). Second basaltic continental crust, once formed, is transitory, and returns to the mantle by crustal delamination (Bird 1980). A third process, subduction of siliceous sediment and lack of A-level (see Fig. 1) recycling of the subducted matter in accretionary prisms or in arc magmas, might basify crustal composition (e.g. Kay 1980) but that would only contribute to the problem. The bulk composition of the lower crust remains an unsolved problem.

ACKNOWLEDGEMENTS: This research has been supported by US National Science Foundation grants EAR82-12757 and EAR83-07599 for study in the Aleutian Islands and the central Andes respectively. Cornell Institute for The Study of the Continents (INSTOC) publication number 8.

References

ALBAREDE, F. 1975. The heat flow/heat generation relationship: an interaction model of fluids with cooling intrusions. *Earth Planet. Sci. Lett.* **27**, 73–78.

ANDERSON, A.T. 1982. Parental basalts in subduction zones: implications for continental evolution. *J. Geophys. Res.* **87**, 7047–7060.

ARMSTRONG, R.L. 1981a. Comment on 'Crustal growth and mantle evolution: inferences from models of element transport and Nd and Sr isotopes'. *Geochim. et Cosmochim. Acta*, **45**, 1251.

—— 1981b. Radiogenic isotopes: The case for crustal recycling on a near-steady-state no-continental-growth Earth. *Phil. Trans. R. Soc. London*, **A 301**, 443–472.

ARTH, J.G. & HANSON, G.N. 1972. Quartz diorites derived by partial melting of eclogite or amphibolite at mantle depths. *Contrib. Mineral. Petrol.* **37**, 161–174.

—— & BARKER, F. 1976. Rare-earth partitioning between hornblende and dacitic liquid and implications for the genesis of trondhjemitic-tonalitic magmas. *Geology*, **4**, 534–536.

BARKER, F. & ARTH, J.G. 1976. Generation of trondhjemitic-tonalitic liquids and Archean bimodal trondhjemite-basalt suites. *Geology*, **4**, 596–600.

BICKLE, M.J., BETTENAY, L.F., BARLEY, M.E., CHAPMAN, H.J., GROVES, D.I., CAMPBELL, I.H. & DE

LAETER, J.R., 1983. A 3500 Ma plutonic and volcanic calc-alkaline province in the Archaean East Pilbara block. *Contrib. Mineral. Petrol.* **84**, 25–35.

BIRD, P. 1980. Continental delamination and the Colorado Plateau. *J. Geophys. Res.* **84**, 7561–7571.

BOHLEN, S.R., BOETTCHER, A.L., WALL, V.J. & CLEMENS, J.D. 1983. Stability of phlogopite-quartz and sanidine-quartz: a model for melting in the lower crust. *Contrib. Mineral Petrol.* **83**, 270–277.

BUNTEBARTH, G. 1976. Distribution of uranium in intrusive bodies due to combined migration and diffusion. *Earth Planet. Sci. Lett.* **32**, 84–90.

CHAPPELL, B.W. 1984. Source rocks of I- and S-type granites in the Lachlan fold belt, southeastern Australia. *Phil. Trans. R. Soc. Lond.* **A 310**, 693–707.

COOK, F.A., ALBAUGH, D.S., BROWN, L.D., KAUFMAN, S., OLIVER, J.E. & HATCHER, R.D. JR 1979. Thin-skinned tectonics in the crystalline southern Appalachians: COCORP seismic reflection profiling of the Blue Ridge and Piedmont. *Geology*, **7**, 563–567.

CONRAD, W.K. 1983. *Petrology and geochemistry of igneous rocks from the McDermitt Caldera complex, Nevada-Oregon, and Adak Island, Alaska: Evidence for crustal development.* Cornell University PhD thesis 325 p.

—— 1984. The mineralogy and petrology of compositionally zoned ash flow tuffs, and related silicic volcanic rocks, from the McDermitt Caldera Complex, Nevada-Oregon. *J. Geophys. Res.*, **89**, B10, 8639–8664.

—— & KAY R. 1984. Ultramafic and mafic inclusion from Adak Island: crystallization history and implications for the nature of primary magmas and crustal evolution in the Aleutian arc. *J. Petrol.* **25**, 88–125.

COLLINS, W.J., BEAMS, S.D., WHITE, A.J.R. & CHAPPELL, B.W. 1982. Nature and origin of A-Type granites with particular reference to southeastern Australia. *Contib. Mineral. Petrol.* **80**, 189–200.

CHRISTENSEN, N. 1979. Compressional wave velocities in rocks at high temperatures and pressures, critical thermal gradients and crustal Low-Velocity Zones. *J. Geophys. Res.* **84**, 6849–6857.

—— & FOUNTAIN, D. 1975. Constitution of the lower continental crust based on experimental studies of seismic velocities in granulite. *Bull. Geol. Soc. Am.* **86**, 227–236.

DAWSON, J.B. 1977. Sub-cratonic crust and upper mantle models based on xenolith suites in kimberlite and nephelinitic diatremes. *J. Geol. Soc. Lond.* **134**, 173–184.

DePAOLO, D.J. 1981. Trace elements and isotopic effects of combined wallrock assimilation and fractional crystallization. *Earth Planet. Sci. Lett.* **53**, 189–202.

—— & FARMER, G.L. 1984. Isotopic data bearing on the origin of Mesozoic and Tertiary granitic rocks in the western United States. *Phil. Trans. R. Soc. London*, **A 310**, 743–753.

DEWEY, J.F. & WINDLEY, B.F. 1981. Growth and differentiation of the continental crust. *Phil. Trans. R. Soc. London* **A 301**, 189–206.

FOUNTAIN, D.M. & SALISBURY, M.H. 1981. Exposed cross-sections through the continental crust: implications for crustal structure, petrology, and evolution. *Earth Planet Sci. Lett.* **56**, 263–277.

GLASSLEY, W.E. 1983. Deep crustal carbonates as CO_2 fluid sources: evidence from metasomatic reaction zones. *Contrib. Mineral. Petrol.* **84**, 15–24.

GLIKSON, A.Y. & LAMBERT, I.B. 1976. Vertical zonation and petrogenesis of the early Precambrian crust in western Australia. *Tectonophysics*, **30**, 55–89.

GRAY, C.M. 1977. The geochemistry of Central Australian granulites in relation to the chemical and isotopic effects of granulite-facies metamorphism. *Contrib. Mineral. Petrol.* **65**, 79–89.

HAMILTON, W. 1981. Crustal evolution by arc magmatism. *Phil. Trans. R. Soc. London*, **A 301**, 279–291.

HARMON, R.S., HALLIDAY, A.N., CLAYBURN, J.A.P. & STEPHENS, W.E. 1984. Chemical and isotopic systematics of the Caledonian intrusions of Scotland and Northern England: a guide to magma source region and magma-crust interaction. *Phil. Trans. R. Soc. London*, **A 310**, 709–742.

HARRISON, T.M. & WATSON, E.B. 1983. Kinetics of zircon dissolution and zirconium diffusion in granitic melts of variable water content. *Contrib. Mineral. Petrol.* **84**, 66–72.

HAWKESWORTH, C.J., HAMMILL, M., GLEDHILL, A.R., VAN CALSTEREN, P. & ROGERS, N.W. 1982. Isotope and trace element evidence for late-stage intra-crustal melting in the High Andes. *Earth Planet. Sci. Lett.* **58**, 240–254.

HEIER, K.S. 1979. The movement of uranium during higher grade metamorphic processes. *Phil. Trans. R. Soc. London*, **A 291**, 413–421.

HELMSTAEDT, H. & DOIG, R. 1975. Eclogite nodules from kimberlite pipes of the Colorado Plateau—samples of subducted Franciscan-type oceanic lithosphere. *Phys. Chem. Earth*, **9**, 95–111.

HERZBERG, C.T., FYFE, W.S. & CARR, M.J. 1983. Density constraints on the formation of the continental Moho and crust. *Contrib. Mineral. Petrol.* **84**, 1–5.

HILDRETH, W. 1981. Gradients in silicic magma chambers; implications for lithospheric magmatism. *J. Geophys. Res.* **86**, 10153–10192.

HILL, M., MORRIS, J. & WHELAN, J. 1981. Hybrid granodiorites intruding the accretionary prism Kodiak, Shumagin and Sanak islands southwest Alaska. *J. Geophys. Res.* **86**, 10569–10590.

HOLT, R.W. & WIGHTMAN, R.T. 1983. The role of fluids in the development of a granulite facies transition zone in S India. *J. Geol. Soc. Lond.* **140**, 651–656.

ISACKS, B.L. 1985. Topography and tectonics of the central Andes, *EOS*, **66**, 375.

JAUPART, C., SCLATER, J.G. & SIMMONS, G. 1981. Heat flow studies: constraints on the distribution of uranium, thorium and potassium in the continental crust. *Earth Planet. Sci. Lett.* **52**, 328–344.

JAMES, D.E. 1982. A combined O, Sr. Nd, and Pb isotopic and trace element study of crustal contamination in central Andean lavas, I. Local geochemical variations. *Earth Planet. Sci. Lett.* **57**, 47–62.

KARIG, D.E. 1982. Initiation of subduction zones: implications for arc evolution and ophiolite development. *In*: LEGGETT, J.K. (ed.) *Trench-Forearc Geology* Geological Soc. Spec. Pub. no. 10, 563–576, Blackwell Scientific, Oxford.

—— & KAY, R.W. 1981. Fate of sediments on the descending plate at convergent plate margins. *Phil. Trans. Roy. Soc. Lond.* **A 301**, 233–251.

KAY, R.W. 1980. Volcanic arc magma genesis: Implications for element recycling in the crust—upper mantle system. *J. Geol.* **88**, 497–522.

—— 1985. Island arc processes relevant to crustal and mantle evolution. *Tectonophysics*, **112**, 1–15.

KAY, R.W. & KAY, S.M. 1981. The nature of the lower continental crust: inferences from geophysics, surface geology and crustal xenoliths. *Rev. Geophys. and Space Phys.* **19**, 271–297.

KAY, S.M. & KAY, R.W. 1983. Thermal history of the deep crust inferred from granulite xenoliths, Queensland, Australia. *Amer. Jour. Sci.* **283-A** (Orville volume) 486–513.

—— & —— 1985a. Role of crystal cumulates and the oceanic crust in the formation of the lower crust of the Aleutian arc. *Geology*, **13**, 461–464.

—— & ——1985b. Aleutian tholeiitic and calc-alkaline magma series I: the mafic phenocrysts. *Contrib. Mineral. Petrol.* **90**, 276–290.

——, —— & CITRON, G.P. 1982. Tectonic controls of Aleutian arc tholeiitic and calc-alkaline magmatism. *J. Geophys. Res.* **87**, 4051–4072.

KOZLOVSKY, Y.A. 1984. The world's deepest well. *Sci. Amer.* **251**, 98–104.

LEE, C.D., VINE, F.J. & ROSS, R.G. 1983. Electrical conductivity models for the continental crust based on laboratory measurements on high-grade metamorphic rocks. *Geophys. J. Roy. Astr. Soc.* **72**, 353–371.

LE FORT, P. 1981. Manaslu leucogranite: a collision signature of the Himalaya: a model for its genesis and emplacement. *J. Geol. Res.* **86**, 10545–10568.

MCCULLOCH, M.T. & WASSERBURG, G.J. 1978. Sm-Nd and Rb-Sr chronology of continental crust formation. *Science*, **200**, 1003–1011.

—— ARCULUS, R.J., CHAPPELL, B.W. & FERGUSON, J. 1982. Isotopic and geochemical studies of nodules in kimberlite have implications for the lower continental crust. *Nature*, **300**, 166–169.

MCGETCHIN, R.R. & SILVER, L.T. 1972. A crustal-upper mantle model for the Colorado Plateau based on observations of crystalline rock fragments in the Moses Rock Dike. *J. Geophys. Res.* **77**, 7022–7037.

MCKENZIE, D. 1984, The generation and compaction of partially molten rock. *J. Petrol.* **25**, 713–765.

MILLER, C.F. & MITTLEFEHLDT, D.W. 1982. Depletion of light rare-earth elements in felsic magmas, *Geology*, **10**, 129–133.

MOLNAR, P., CHEN, P.W. & PADOVANI, E. 1983. Calculated temperatures in overthrust terrains and possible combinations of heat sources responsible for the Tertiary granites in the greater Himalaya. *Jour. Geol. Res.* **88**, 6415–6429.

MOORBATH, S., TAYLOR, P.N. & GOODWIN, R. 1981. Origin of granitic magma by crustal remobilisation: Rb-Sr and Pb/Pb geochronology and isotope geochemistry of the late Archaean Qorqut Granite Complex of southern West Greenland. *Geochim. Cosmochim. Acta*, **45**, 1051–1060.

MUELLER, R.P. 1967. Mobility of the elements in metamorphism. *J. Geol.* **74**, 565–582.

NEWTON, R.C., SMITH J.V. & WINDLEY, B.F. 1980. Carbonic metamorphism, granulites and crustal growth. *Nature*, **288**, 45–50.

NICHOLLS, I.A. & HARRIS, K.L. 1980. Experimental rare earth element partition coefficients for garnet, clinopyroxene and amphibole coexisting with andesitic and basaltic liquids. *Geochim. Cosmochim. Acta* **44**, 287–308.

O'BRIEN, T.F. 1983. *Evidence for the nature of the lower crust beneath the central Colorado Plateau as derived from xenoliths in the Buell Park-Green Knobs diatremes.* Ph.D Thesis, Cornell University. 250 p.

OXBURGH, E.R. & MCRAE, T. 1984. Physical constraints on magma contamination in the continental crust: an example, the Adamello complex. *Phil. Trans. Roy. Soc. Lond.* A **310**, 457–472.

PADOVANI, E. & CARTER, J. 1977. Aspects of the deep crustal evolution beneath south central New Mexico. *In*: HEACOCK, J. (ed.) *The Earth's Crust*, Am. Geophys. Union Monograph **20**, 19–55.

——, HALL, J. & SIMMONS, G. 1982. Constraints on crustal hydration below the Colorado Plateau from V_P measurements on crustal xenoliths. *Tectonophysics*, **84**, 313–328.

PIN, C. & SILLS, J.D. 1986. Petrogenesis of layered gabbros and ultramafics from the Ivrea zone, NW Italy: trace element and isotope geochemistry. (This volume).

PITCHER, W.S. 1979. The anatomy of a batholith. *J. Geol. Soc. Lond.* **135**, 157–182.

POLDERVAART, A. 1955. Chemistry of the earth's crust. *Geol. Soc. Am., Spec. Paper*, **62**, 119–182.

PRIDE, C. & MUECKE, G.K. 1981. Rare earth element distributions among coexisting granulite facies minerals, Scourian Complex, NW Scotland. *Contrib. Mineral. Petrol.* **76**, 463–471.

—— & —— 1980. Rare earth element geochemistry of the Scourian Complex N.W. Scotland—evidence for the granite-granulite link. *Contrib. Mineral. Petrol.* **73**, 403–412.

REYMER, A. & SCHUBERT, G. 1984. Phanerozoic addition rates to the continental crust and crustal growth. *Tectonics*, **3**, 63–77.

RINGWOOD, A.E. 1979. *Origin of the Earth and Moon.* Springer Verlag, New York. 259 pp.

ROGERS, N.W. & HAWKESWORTH, C.J. 1982. Proterozoic age and cumulate origin for granulite xenoliths, Lesotho. *Nature*, **299**, 409–413.

RONOV, A.B. & YAROSHEVSKY, A.A. 1969. Chemical composition of the Earth's crust. *In*: HART, P.S. (ed.) *The Earth's Crust and Upper Mantle: Structure, Dynamic Processes, and their Relation to Deep-Seated Geological Phenomena*, Geophysical Monograph **13**, Amer. Geophys. Union, Washington, D.C., 37–57.

SISSON, V.B., CRAWFORD, M.L. & THOMPSON, P.H. 1981. CO_2-brine immiscibility at high temperatures, evidence from calcareous metasedimentary rocks. *Contrib. Mineral. Petrol.* **78**, 371–378.

SMITH, D. 1979. Hydrous minerals and carbonates in peridotite inclusions from the Green Knobs and Buell Park kimberlitic diatremes on the Colorado Plateau. *In*: BOYD, F.R. & MEYER, H.O.A. (eds) *The Mantle Sample: Inclusions in Kimberlites and Other Volcanics*, Amer. Geophys. Union, Washington D.C., 345–356.

SMITHSON, S.B. 1978. Modeling continental crust: structural and chemical constraints. *Geophys. Res. Lett.* **5**, 749–752.

—— & BROWN, S.K. 1977. A model for lower continental crust. *Earth Planet. Sci. Lett.* **35**, 134–144.

—— & HEIER, K.S. 1971. K, U, and Th distribution between normal and charnockitic facies of a deep granitic intrusion. *Earth Planet. Sci. Lett.* **12**, 325–326.

——, JOHNSON, R.A. & WONG, Y.K. 1981. Mean crustal velocity: a critical parameter for interpreting crustal structure and crustal growth. *Earth Planet. Sci. Lett.* **53**, 323–332.

TARNEY, J. & WINDLEY, B.F. 1977. Chemistry, thermal gradients and evolution of the lower continental crust. *J. Geol. Soc. Lond.* **134**, 153–172.

—— & —— 1979. Continental growth, island arc accretion and the nature of the lower crust—a reply to S.R. Taylor and S.M. McLennan. *J. Geol. Soc. Lond.* **136**, 501–504.

TATSUMI, Y. 1982. Origin of high magnesian andesites

in the Setouchi volcanic belt, southwest Japan, II melting phase relations at high pressures, *Earth Planet. Sci. Lett.* **60,** 305–317.

TAYLOR, S.R. 1977. Island arc models and the composition of the continental crust. *Am. Geophys. Union. Maurice Ewing Series* **1,** 229–242.

—— & McLENNAN, S.M. 1979. Discussion on "Chemistry, thermal gradients and evolution of the lower continenal crust" by J. Tarney and B.F. Windley. *J. Geol. Soc. Lond.* **136,** 497–500.

—— & —— 1981. The composition and evolution of the continental crust: rare earth element evidence from sedimentary rocks. *Phil. Trans. Roy. Soc. London,* **301,** 381–399.

THORPE, R.S., FRANCIS, P.W. & O'CALLAGAN, L. 1984. Relative roles of source composition, fractional crystallization and crustal contamination in the petrogenesis of Andean volcanic rocks. *Phil. Trans. R. Soc. London* **A 310,** 675–692.

——, —— & HARMON, R.S. 1981. Andean andesites and crustal growth. *Phil. Trans. R. Soc. Lond.* **A 301,** 305–320.

TOURET, J. & DIETVORST, P. 1983. Fluid inclusions in high-grade anatectic metamorphites. *J. Geol. Soc. London,* **140,** 635–649.

THOMPSON, R.N., DICKIN, A.P., GIBSON, I.L. & MORRISON, M.A. 1982. Elemental fingerprints of isotopic contamination of Hebridean Paleocene mantle-derived magmas by Archaean sial. *Contrib. Mineral. Petrol.* **79,** 159–168.

VOLLMER, R. & NORRY, M.J. 1983. Possible origin of K-rich volcanic rocks from Virunga, East Africa, by metasomatism of continental crustal material: Pb, Nd, and Sr isotopic evidence. *Earth Planet. Sci. Lett.* **64,** 374–386.

WEAVER, B.L. & TARNEY, J. 1980a. Continental crust composition and nature of the lower crust: constraints from mantle Nd-Sr isotope correlation. *Nature,* **286,** 342–346.

—— & —— 1980b. Rare earth geochemistry of Lewisian granulite-facies gneisses, Northwest Scotland: implications for the petrogenesis of the Archaean lower continental crust. *Earth Planet. Sci. Lett.* **51,** 279–296.

WHITE, W.M. & PATCHETT, J. 1984. Hf-Nd-Sr isotopes and incompatible element abundances in island arcs: implications for magma origins and crust-mantle evolution. *Earth Planet. Sci. Lett.* **67,** 167–185.

R.W. KAY & S. MAHLBURG KAY, Institute for the Study of the Continents, Snee Hall, Cornell University, Ithaca, NY 14853.

Fluid inclusions in rocks from the lower continental crust

J. Touret

SUMMARY: In granulites, which constitute a major, if not exclusive rock type of the lower continental crust, fluid inclusions are abundant in many rock-forming minerals. Most contain a characteristic 'carbonic' fluid, high density CO_2 possibly mixed with variable quantities of N_2 and more rarely, $CH_4 \cdot NaCl$ brine inclusions have recently been observed in metasedimentary granulites from southern Norway. P–T estimates from synmetamorphic pure CO_2 inclusions agree with solid mineral thermobarometry in three selected examples of: (a) high-pressure granulites (Fura Complex, Tanzania), (b) intermediate-pressure granulites (southern Karnataka, India) and (c) low-pressure granulites (West Uusimaa Complex, Finland). These data support a model of fluid distribution of the lower crust based on the predominance of density-controlled, mantle-derived CO_2-rich fluids coexisting immiscibly with NaCl brines. In southern Norway, most of the CO_2 was associated with or derived from carbonate-rich melts emplaced with deep seated, intermediate synmetamorphic intrusives.

Many lower crustal rocks now exposed at the surface, either in the roots of deeply eroded old mountain chains or brought as xenoliths in volcanic lavas, are granulites: they are characterized by H_2O-deficient parageneses, notably by the diagnostic occurrence of orthopyroxene and quartz when certain physicochemical conditions are fulfilled (Winkler 1979). The question of the continuity of a granulite lower continental crust remains open, but enough evidence has accumulated from all parts of the world to suggest that granulites are at least a major constituent of the base of the continents between 15 and 30 km (lithostatic pressure from 6 to 10 kb) (Beloussov 1966; Behr 1978; Fountain & Salisbury 1981; Behr & Emmermann 1983).

In the early seventies (Touret 1971, 1974), it was found that granulites contain characteristic fluid inclusions, mostly filled with high density CO_2 ('carbonic' fluid inclusions, Touret 1974), which may give valuable information on the nature and the density of the fluid phase during and after peak metamorphic conditions. Rapid advances in the technology (heating–freezing microscopic stage, laser induced Raman spectroscopy) and in the number of investigated areas (see e.g. Touret 1981; Roedder 1984) have significantly increased our understanding of these deep seated fluids, so that we are now able to propose some general conclusions which are supported by independent observations at the scale of the earth.

The world of fluid inclusions. Specific techniques, potentialities and problems

Fluid inclusions were among the first objects described under the petrographic microscope (Sorby 1858). One may wonder why it has taken more than one hundred years to have them included in the normal scope of petrological investigations. The story is complicated and instructive, and many factors—lack of equipment, inadequate theoretical basis, complexity of fluid inclusion distribution, etc.—are involved. But one point is obvious, which is sufficient to explain almost everything: most observers miss the fluid inclusions in lower crustal rocks because they are convinced that they cannot exist (water deficient mineral assemblages must be vapour absent) and they do not use the appropriate methods of observation: double polished *thick* plates (5 to 10 times thicker than the normal thin section), and, most important, high power (at least ×25) objectives without crossed nicols. In most rock samples, the passage from a ×10 to ×25 objective is a fascinating experience: small dark dots, which could be anything from dust, surface artifacts, holes, etc. suddenly become a world of moving bubbles, daughter minerals, perfect crystallographic shapes, etc.

Once they have been recognized, and repeatedly found in section after section, the inclusions cannot be ignored and the need for a more complete study comes almost by itself: their systematic features soon become as typical for the rock as the 'solid' minerals.

The classical petrographic microscope, combined with simple tools like the crushing stage, is sufficient to locate the fluid inclusions in rock-forming minerals. Their real study, however, needs the determination of the temperature of phase transitions (solid-, liquid-, vapour or fluid). This is done by freezing–heating microscopic stages (microthermometry) which are now routinely used in many laboratories and which were the real breakthrough in the development of fluid

From: DAWSON, J.B., CARSWELL, D.A., HALL, J. & WEDEPOHL, K.H. (eds) 1986, *The Nature of the Lower Continental Crust*, Geological Society Special Publication No. 24, pp. 161–172.

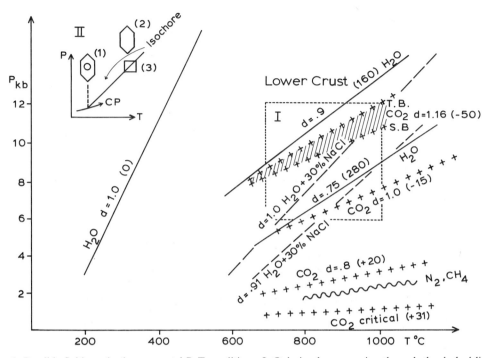

FIG. 1. Possible fluids under lower crustal P–T conditions. I: Only isochores passing through the dashed lined 'box' may represent lower crustal fluids. For each isochore, indication of the density (d) in g/cm³ is followed (in brackets) by the homogenization temperature in °C. Hatched area covers discrepancy for CO_2 at $d = 1.16$ g/cm³ between different authors (T.B.: Touret–Bottinga (1979); S.B.: Swanenberg (1980), Bergmann (1982)) (solid lines: H_2O; dashed lines: $H_2O + 30\%$ wt NaCl; crosses: CO_2; undulating line: maximum densities of N_2 and CH_4 yet recorded in granulites). II (insert): interpretation of fluid inclusion data: for an observed fluid (1) its homogenization temperature (2) defines an isochore along which this inclusion was in the P–T conditions of trapping (3) C.P.: critical point.

inclusion studies. Specific principles and techniques are reviewed in a number of recent publications (Touret 1977; Hollister & Crawford 1981; Roedder 1984) and it will only be noted here that, ideally (i.e. assuming no leakage and neglecting the variation of *solid* volumes at high P and T), fluid inclusions are constant density (isochoric) systems, which may furnish information, not only on the chemical composition, but also—and this is unique—on the density of the enclosed fluid.

On a P–T diagram (Fig. 1), a constant density for a given composition corresponds to a line, or isochore (II, Fig. 1); these have only been determined with some precision for a small number of pure systems (H_2O, CO_2, CH_4, N_2 ...) and, in some cases, their mixtures (Holloway 1981). Fortunately, we now have with the Raman microprobe an extremely powerful and sensitive instrument to check the purity of the gaseous systems, especially for CO_2 which as discussed below is by far the most important fluid at depth.

The interpretation of fluid inclusion data will then be based essentially on a comparison between fluid isochore for the appropriate composition and density and P–T data independently estimated from mineral parageneses. Before giving detailed examples, it might be worthwhile to delimit the problem more precisely by considering which isochores are compatible with the P–T conditions of the lower crust. From metamorphic assemblages, these conditions are loosely—but conservatively—indicated by the box on Fig. 1 (Temperatures 700–1000°C, Pressures 6–12 kb). Only few isochores intersect this P–T 'box' and only those may possibly correspond to fluids of the lower crust. This condition is not sufficient in itself, as fluids may be trapped along the isochore outside of the box, but at least a large majority of all possible isochores are thus eliminated. A closer look at the 'possible' isochores is instructive: CO_2 densities range from 1.0 to 1.16 (approximately). Note the large discrepancy of

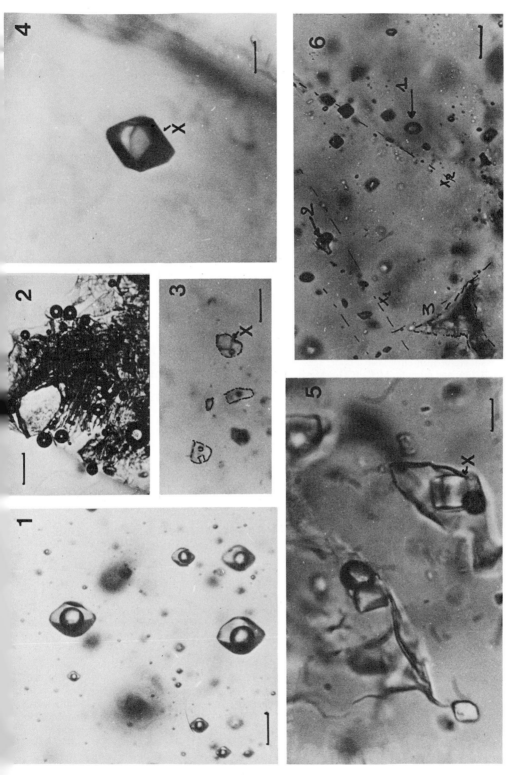

FIG. 2. Fluids in granulites: (1) CO₂ inclusions (gas and liquid) in quartz from a pegmatite, Narestö near Tvedestrand (S. Norway) at about 0°C ($T_{hL} = +28$°C). (2) CO₂ bubbles released by crushing of a quartz grain in glycerine from 'charnockitic gneiss', Tromoy near Arendal (South Norway). (3) Brine in metapelite, Tromoy. X: NaCl cube. (4) CO₂ inclusion in garnet from a metapelite, Tromoy (same rock as 3). X: biotite crystal trapped in the inclusion along with CO₂. (5) NaCl brine inclusions from skarn rock, Barbu, Arendal. X halite cube. (6) Typical fluid inclusions in granulite, Arendal (S. Norway) 1: trail of CO₂ inclusions (trace: x₂), 2: trail of H₂O inclusions (trace: x₁), 3: large, empty cavity. Length of the bar: 1, 3, 4: 5 μm, 2: 0.2 mm, 4: 2 μm, 6: 20 μm.

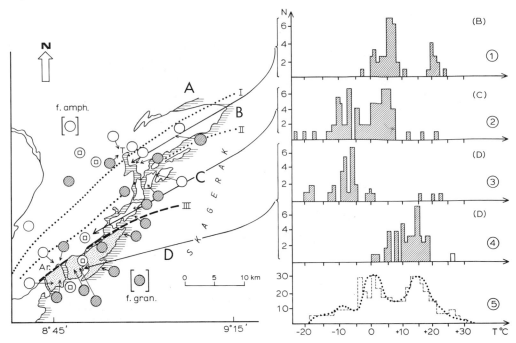

FIG. 3. Fluid inclusion distribution and metamorphic zoning in southern Norway. Left: Early fluids (open circle: H_2O; square in open circle: NaCl brine; shaded circle: CO_2). In square brackets: dominant fluids in amphibolite (H_2O) and granulite facies (CO_2). Metamorphic zones (A–D) and isograds (stippled lines) after Smalley *et al.* 1983: A: amphibolite, B: transitional, C and D: granulite zone. I: Opx in metabasites, II: opx in all lithologies, III: limit of the high-Fe, LIL depleted 'charnockitic gneisses' of Tromoy (dotted area). Ar: Arendal, T: Tvedestrand. Right: Frequencies (N) of homogenization temperatures (T_h) of CO_2 from zones B, C and D. Histogram (1) cordierite bearing gneiss, Tvedestrand (2) orthopyroxene-biotite-plagioclase rich gneiss, Arendal station (3) quartz rich metapelite, Tromoy, Revesand (4) quartz segregation in 3 (5) Composite histogram of the total Bamble area. High density peak: early, synmetamorphic fluids; low density peak: late fluids related to uplift (Touret, 1981).

the calculated position of the high density isochore between different authors (based on the same experimental data and the same equation of state!), a fact which shows the limitations of our knowledge about the most essential fluid systems at high P and T. Corresponding H_2O isochores have a significantly lower density, 0.75–0.9 g/cm^3 (homogenization temperatures 160–280°C), but the isochores of a solution of $H_2O + 30$ wt % NaCl have a density very close to that of CO_2 (0.91 to 1 for $CO_2 = 1$g/cm^3, Fig. 1). Such fluids are also immiscible with CO_2 at high P and T (Bowers & Helgeson 1983) and we will see that they are commonly encountered in lower crustal rocks. Other fluids like CH_4 or N_2 are frequently observed in high grade rocks (Swanenberg 1980; Touret & Dietvorst 1983). But until now only very low densities have been measured (undulating line in Fig. 1), which are not compatible with trapping under peak metamorphic conditions (Isochores grossly outside the 'box' of Fig. 1).

Major fluid inclusion types. Relation with metamorphic grade and nature of the host rock

With the important exception of annealed rocks ('granoblastic' texture with triple mineral junctions at or close to 120°), most rock samples from the lower crust contain abundant (up to several millions in 1 cm^3) but small (a few micrometres) fluid inclusions. Quartz is by far the most frequent host mineral, but also the most complex one by reason of its easy recrystallization and hence remarkable aptitude for trapping fluid at almost any P–T conditions. More rarely, fluid inclusions have been observed in a variety of other minerals: garnet, plagioclase, Al-silicates, pyroxene, etc. These are always very interesting, as they are much more apt to be primary and undisturbed than in quartz. Fortunately enough, the great number of inclusions correspond to only two major types, readily distinguishable at room temperature with a normal microscope: monophase gaseous (high density CO_2, CH_4, N_2), and aqueous (H_2O with dissolved salt, from pure water to NaCl-bearing brines) (Fig. 2).

For reasons discussed in Fig. 1 and as a conclusion of many observations (Touret 1981), the composition of these two types is even more restricted for fluids compatible with peak metamorphic conditions: CO_2 fluids with a density between 1 and 1.16 g/cm^3 (homogenization temperatures between -15 and $-50°C$) and NaCl-bearing brines. CO_2 is frequently mixed with some N_2 or CH_4, but in quantities (less than 10 mole %) which do not seriously affect the microthermometric properties (Touret 1985). All other fluids (low density CO_2, pure N_2 or CH_4, low salinity H_2O) have been trapped after peak metamorphic conditions. They place constraints on the uplift path towards the surface (see e.g. Hollister & Crawford 1981), but they cannot give any indication of the fluid composition in the lower crust.

The widespread occurrence of high density CO_2 fluids at the granulite boundary (orthopyroxene-in isograd) was first documented in southern Norway (Touret 1971, 1974) and then confirmed in many regions of the world (see e.g. Touret 1981; Roedder 1984). The transition is very complicated (Fig. 3): CO_2 rich fluids are common as soon as local melting (anatexis) develops in the rock (Touret & Dietvorst 1983), but they remain restricted to the molten part of the rock (mobilisate) until the orthopyroxene-in isograd is reached. Then above the former conditions, they invade almost all lithologies. An example of this transition is detailed in Fig. 3 (southern Norway). Only the early peak metamorphic fluids have been represented (compare with Fig. 2 in Touret 1985). The overwhelming predominance of CO_2 fluids in the granulite domain is obvious, but there are a few examples of brines, even in the core of the granulite area (LIL depleted zone of Tromoy, Field *et al.* 1980) (Fig. 2). In the transition zone between amphibolite and granulite (B, Fig. 3), CO_2 is more abundant near metagabbros and basic intrusives, now occurring as two-pyroxene granulites. This relation has been recognized since the first observations of CO_2 rich fluids (Touret 1971), and it has been systematically confirmed since. On the other hand, brines are restricted to a few well defined lithologies, all metasedimentary: quartzites, Al-rich metasediments, metapelites and the well known skarns (Arendal) associated with marbles and metacarbonates (Fig. 2). Particularly interesting is the fact that very few CO_2 inclusions are to be found near or in former carbonate rich metasediments.

Pressure and temperature estimates (Furua complex, Tanzania, southern Karnataka, India and West Uusimaa complex, Finland)

While southern Norway is the first place where the relation between deep seated fluids and metamorphic grade has been systematically investigated, it is far from ideal for a precise comparison of P–T data between fluid and solids; the structure is very complicated, the age of the high grade metamorphic episode is debated (Tobi & Touret 1985) and important metamorphic episodes have occurred after the regional granulite metamorphism. Most important, the metamorphic isograds are more or less parallel to the very pronounced regional strike (NE–SW), so that it is impossible to follow a marker horizon across the metamorphic boundaries. This complexity is reflected in the shape of T_h histograms. Many data correspond to relatively low density, late inclusions (Fig. 3, histogram no. 5). Only a handful of measurements define the high density fluids compatible with peak metamorphic conditions (Touret 1971, 1985).

Other areas have proven to be much more favourable with regard to the abundance of early, sometimes typically primary inclusions (Berglund & Touret 1976), the prograde character of the amphibolite–granulite transition, metamorphic isograds clearly transecting the regional structure, and the absence of polymetamorphism and polyphase deformation. Three examples have been selected: Furua granulite complex, Tanzania (Coolen 1980, 1982), southern Karnataka, India (Hansen *et al.* 1984a, b) and West Uusimaa complex, Finland (Schreurs 1985). All have been well studied, both the solid phases and fluid inclusions, and they are complementary, covering a large pressure range from high pressure (Furua complex) to the very low pressure of the West Uusimaa complex in Finland.

Furua Granulite Complex (Coolen 1980, 1982)

The Furua Granulite Complex, about 300 km south-west of Daar es Salaam, is a part of the late Proterozoic Mozambique mobile belt which extends all along the eastern side of the African continent. Geochronological data, including a nearly concordant U-Pb zircon age of 650 Ma (Coolen *et al.* 1982), indicates that granulite facies metamorphism took place during the Pan-African tectonic event. This exceptionally young age for a granulite complex may explain some unique features, notably the preservation of primary

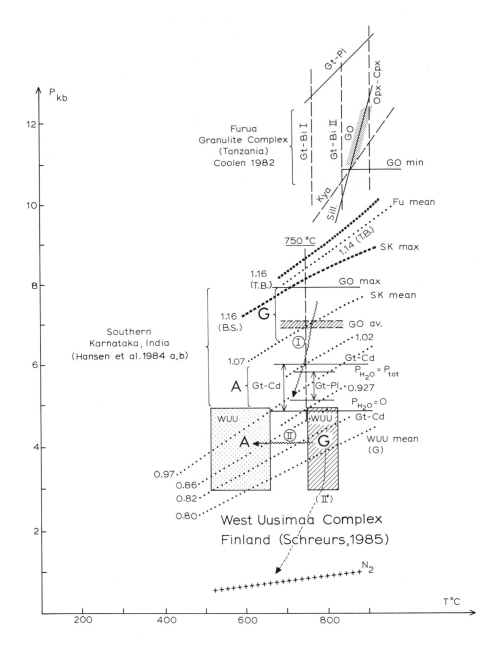

FIG. 4. Comparison of P–T estimates and selected CO_2 isochores, labelled in densities (g/cm³) for 3 well investigated areas (see Figs. 5 and 6). Fu: Furua granulite complex, Sk: southern Karnataka, WUU: West Uusimaa complex. A: amphibolite, G: granulite. (T.B., S.B. as in Fig. 1). Solid phase thermometry: Gt-Bi: garnet-biotite, Opx-Cpx: orthopyroxene-clinopyroxene in basic rocks (Furua: Coolen, 1982); barometry: GO: garnet-orthopyroxene-Al silicate-quartz; Gt-Pl: garnet-plagioclase-Al silicate-quartz; Gt-Cd: cordierite-garnet-sillimanite-quartz; I: granulite-amphibolite transition in southern Karnataka (nearly isothermic, pressure decrease in vertical section of the lower crust); II: same in West Uusimaa Complex (isobaric, thermal dome); (II′): post metamorphic uplift path in West Uusimaa Complex.

inclusions in many rock forming minerals. Similar features are also observed in some granulites of Madagascar (Berglund & Touret 1976).

Two phases of metamorphism have been identified in the area (Coolen 1980). A dominant intermediate to high pressure granulite facies metamorphism (M_1) under T–P conditions of 750–850°C, 7–12 kb was followed by a retrogressive phase (M_2) under amphibolite facies, locally grading into the granulite (600–700°C, 5–7 kb). The Furua granulite complex is one of the few examples of a granulite domain reaching locally the stability field of kyanite (Newton 1983). Many rock types contain exceptional, typically primary fluid inclusions (Coolen 1980), not only in quartz, but also in garnet and plagioclase. Small differences in homogenization temperatures (T_h) exist for the different minerals (see e.g. Fig. 4 in Touret 1981), but as a whole the T_h histograms show well defined maxima at $-50°C$ for plagioclase, $-43°C$ for quartz and $-35°C$ for garnet (but for the latter with less data and a much larger scattering). These results define a representative isochore at $d = 1.16$ g/cm^3 (plagioclase data) or 1.14 g/cm^3 (quartz data); the pressure difference between these two values at a given T is less than the variation for a single isochore according to different authors (T.B. or B.S., Fig. 4).

The straightforward interpretation of T_h in terms of density is only possible for pure CO_2. As other components (N_2, CH_4) are also observed in high grade rocks (e.g. Touret & Dietvorst 1983), the problem of the purity of the CO_2 is certainly not a simple one. Until recently, only microthermometry could be used (Hollister & Crawford 1981) and, as systems like CO_2–N_2 are much more complicated than initially assumed, the risk of error is great. The situation has changed drastically with the development of laser excited Raman microprobes (e.g. Dhamelincourt *et al.* 1979), which enables the analysis *in situ* of a single inclusion. Many analyses have been performed on inclusions from Tanzania and Finland, only a few from India, but on the whole one may be reasonably convinced that, in all investigated cases, major peak histograms are indeed related to a specific CO_2 density and not to a mixing effect.

P–T estimates recorded by the solid phases in the same specimen which have been studied for fluid inclusions are P = 10–11 kb and T = 800–825°C, essentially based on the opx-gar-plag-quartz geobarometer by Perkins & Newton (1981) and on the stability of kyanite. As seen in Fig. 4, the maximum density inclusions record significantly lower pressure than the solid minerals, by about 2 kb for a reference temperature of 800°C. Since the primary character of the inclu-

sions leaves little doubt about the synmetamorphic nature of the enclosed fluid, several reasons may explain this pressure discrepancy, including uncertainty in the position of high density isochores at high P–T and lack of precision in the solid minerals pressure estimates. Partial leakage is unlikely: it would give scattered T_h data, as observed in garnet. The last and most realistic solution to the problem is the possible presence of an additional H_2O partial pressure of about 2 kb, as suggested by the occurrence of stable hydrous phases (hornblende, biotite) in the investigated specimens (Coolen 1982). No water is actually visible in the fluid inclusions, but in small cavities up to 20 mole % H_2O will remain unnoticed in a carbonic fluid (see e.g. Touret 1977)

Southern Karnataka, India (Hansen *et al.* 1984a and b)

At the northern end of the classical late-Archaean granulite province of southern India, southern Karnataka is an illustrative example of a prograde amphibolite/granulite transition in an intermediate pressure granulite terrain, by far the most common type worldwide (Newton & Perkins 1982). The granulite boundary, preceded by a transitional zone of about 10 km, cuts across the regional structure (Fig. 5). For a reference temperature of 750°C, pressures recorded by the solid phases show a significant decrease from granulite to amphibolite facies (7–8 to 5–6 kb, Fig. 4). CO_2 inclusions in samples along a north–south profile (1 to 7, Fig. 5) show a systematic increase in the CO_2 density from amphibolite to granulite, (representative densities selected from histogram peaks in Fig. 5). Note that with the exception of Sample 7, granulite data show less scatter and therefore less perturbation than the amphibolite ones. The selected isochores, indicated in Fig. 4, show an overall agreement with thermobarometry estimates: the best defined isochore at 1.07 g/cm^3 (Sample 6), passes precisely through the average value of the opx-gar-plag-qz geobarometer, the amphibolite isochores (0.97–0.86) correspond to garnet-cordierite and garnet-plagioclase assemblages. There are not enough thermometric data to define the temperature evolution precisely, but the transition path between granulite and amphibolite is evidently similar to I, Fig. 4 (rapid pressure drop, slow temperature decrease).

A similar trend has been observed in southern Norway (Touret 1984) and it seems to be the rule for most intermediate pressure granulites. It may seem at first sight that the agreement of P–T estimates between fluid and solids, repeatedly stressed by many authors (Touret 1981; Rudnick

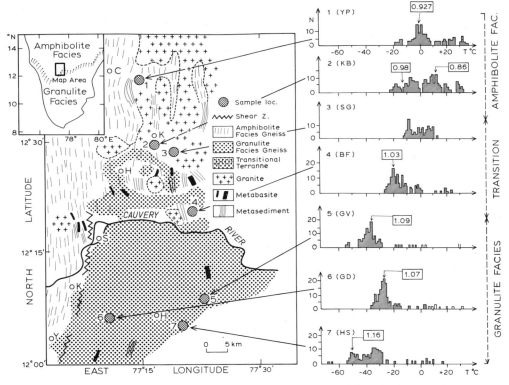

FIG. 5. Geological map and histograms of homogenization temperatures of CO_2 (T_h) from southern Karnataka, India (Hansen *et al.* 1984a). Note the progressive increase in CO_2 density from amphibolite towards granulite. Numbers above histogram: Selected CO_2 densities shown in Fig. 4.

et al. 1984, etc.) is contradicted by the results for Furua, as intermediate granulites also contain stable hydrous phases. A partial H_2O pressure of about 2 kb is most probable in many cases (see e.g. Touret 1971). The problem remains open, but 1–2 kb is certainly within the error margin of any estimate, both for fluids and solids, especially when regional data are compared and not, as at Furua, measurements made on the same thin section.

West Uusimaa Complex, Finland (Westra & Schreurs 1984; Schreurs 1985)

The last example will be taken from the early Proterozoic Svecokarelides in Finland. The West Uusimaa complex, known since the pioneering work of Parras (1958), has been metamorphosed at low pressure (3–5 kb) at temperatures over 800°C in the centre of the dome. The prograde transition amphibolite/granulite, marked by a well defined hypersthene isograd, intersects rock layers of very different composition, each show-

ing specific mineral reactions. The great number of sensitive mineral parageneses allow the use of most geothermobarometers currently available and the results, discussed in detail by Schreurs (1985), support the isobaric character of the amphibolite/granulite transition: P constant at 3–4 kb, temperature from over 800°C to about 650°C in the amphibolite domain (II, Fig. 4). The temperature increase takes place in a zone only a few km wide.

Fluid inclusions are abundant in most of these metamorphic rocks. Two groups have been identified in the whole region (Schreurs 1985):
- 'Early' inclusions (isolated, occurring in undeformed crustal rocks, etc.). They are predominantly carbonic in the granulite domain, aqueous in the amphibolite one.
- 'Late' inclusions, conspicuously trail bound with a prominent preferential direction of trails in many specimens. Their density is less than 0.8 g/cm³ (often much lower with gaseous homogenization as low as $-20°C$) and they occur both in the granulite and in the amphibolite domains. They

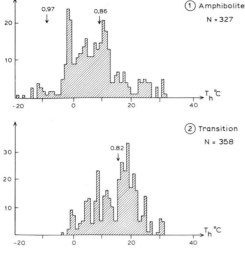

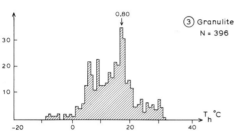

FIG. 6. Histograms of T_h of CO_2 from West Uusimaa Complex, Finland (Schreurs 1985). Note the decrease in density from amphibolite towards granulite. Numbers above the histograms: Selected CO_2 densities shown in Fig. 4.

may be associated with late N_2 inclusions and they define a nearly isothermal uplift trend (II′, Fig. 4, see e.g. Touret 1981), suggesting that the whole area was lifted to the surface by homogeneous uplift (Westra & Schreurs 1984).

T_h histograms (Fig. 6) in marked contrast to the preceding examples, show an increase in the density of the early, high density inclusions from the granulite towards the amphibolite domain. The corresponding isochores agree well with solid mineral estimates (Fig. 4) and the trend of variation corresponds precisely with the isobaric character of the amphibolite/granulite transition.

In conclusion, the data of the three investigated areas are coherent and complementary: CO_2 occurs as a dominant fluid in all granulites. As temperatures remain more or less constant, pressure changes correspond to variations in CO_2 density (Fig. 4). The evolution from low pressure to high pressure granulites leads to an increase in

the CO_2 density from 0.8 to 1.16 g/cm³. Interpreted in terms of depth, this would give an image of a vertical section of the lower crust in which CO_2 fluids are stratified in order of increasing density. However, as shown in southern Norway (Fig. 3), CO_2 fluids are dominant but not unique. They coexist with domains containing NaCl brines, which derive from various types of metasediments and which have a density comparable to that of CO_2.

Origin of CO_2: carbonates and deep synmetamorphic intrusives

Most high grade metamorphic areas are not as favourable for study as the three selected examples. In most cases inclusions trapped at the peak of metamorphism are but a minority among many more late inclusions which can be either less dense, or (hopefully not in too many instances) more dense than the early ones. The late inclusions give useful indications of possible uplift paths (see e.g. Touret 1981), but they confuse the search for well-preserved lower crustal fluids. However, there are sufficient indications to be convinced that southern Norway, the Furua complex, southern Karnataka and West Uusimaa complex are to some extent representative of the lower crust. Note that they cover a large time span, from Archaean to the 650 Ma Pan African event and that the differences are controlled mainly by variation in pressure and to a lesser extent in temperature. We then reach the following conclusions:

● The granulite domain may extend from pressures between 6 and 12 kb, roughly at 18–36 km depth. It is almost ubiquitous at pressures above 7–8 kb.

● In all of this domain the dominant fluid is CO_2, with density increasing from about 1 g/cm³ at 6 kb to 1.16 at 12 kb. But isolated 'pockets' of fluids of different compositions may also exist. These are essentially NaCl-rich brines which seem to be restricted to specific, metasedimentary lithologies.

● Pure water is absent as a free phase. H_2O is present, but in small amounts (less than about 10 mole %) and mixed with the CO_2.

Two aspects of these conclusions must be considered simultaneously: the disappearance of water, and the origin of the observed CO_2. Several indications suggest that the first effect is essentially achieved by preferential dissolution of water in anatectic melts (Touret & Dietvorst 1983), which invariably occur in and near the granulite areas (see e.g., in southern Karnataka, the abundance of granite in and north of the transitional terrain, Fig. 5). The removal of H_2O-bearing anatectic melts from the lower crust has long been an attractive hypothesis to explain the dry character of the granulites and it has been invoked many times (Fyfe 1973; McCarthy 1976; Nesbit 1980, etc.). It is clear, however, that many granulite-facies rocks are themselves migmatites (Touret 1985; Janardhan *et al.* 1982; Rollinson & Windley 1980; Weaver 1980). If the granitic melts are still present, they must have been dehydrated by some other mechanism (Newton 1984).

In any case, the most important and intriguing aspect concerns the origin of CO_2. It might be either sedimentary, from deeply buried carbonate rocks (Glassley 1983), or juvenile, from a degassing mantle (Harris *et al.* 1982) and transported by basaltic (Touret 1971, 1974) or intermediate intrusions (Wells 1979). Newton *et al.* (1980) proposed a pervasive CO_2 'wave' purging the water in front of the streaming carbonic fluids, but large-scale CO_2 flooding is refuted by local mineralogical control of the oxygen isotope distribution (Valley & O'Neil 1984) and by the absence of graphite which should be precipitated if CO_2 was introduced in large quantities in a sufficiently reducing environment (Lamb & Valley 1984).

Clearly more than one mechanism might be operative for a given terrain and each case has to be studied separately. I shall give here some suggestions which concern only southern Norway and which might not be applicable to other cases. In southern Norway a juvenile (mantle derived) origin has been favoured since the discovery of the CO_2 inclusions (Touret 1971, 1974) for a very simple reason: the relative abundance of CO_2 inclusions near synkinematic basic intrusives, compared to their scarcity near carbonate-bearing sediments. Isotopic studies have not been completely conclusive: $\delta^{13}C$ of the inclusions is very low (-15 to $-20‰$ for the densest, synmetamorphic fluids, Hoefs & Touret 1975; Pineau *et al.* 1981). This value eliminates a large scale sedimentary origin, but is also lower than the typical mantle value at $-7‰$. Pineau *et al.* (1981) proposed a derivation from organic matter, which may indeed have been present in some graphite-bearing metapelites (see e.g. Touret 1969). But these rocks do not occur in sufficient volumes and for the most part they still contain graphite. This origin would also not explain the relation between CO_2 and intrusive rocks.

One unexpected result of the isotope studies has been the discovery of large amounts of carbonates in the feldspars of metagabbros (hyperites), with a $\delta^{13}C$ compatible with a mantle origin ($-8.2 \pm 1‰$: Pineau 1977; Pineau *et al.*

1981). These carbonates are visible under the microscope in small cavities and microfissures, not only in metagabbros, but also in most igneous rock types from the granulite facies. At first sight however, they seem to be alteration products which can only be very late, superficial and barely related to peak metamorphic conditions.

But one major argument points to a deep seated origin of some carbonates: in some samples, $\delta^{18}O$ indicates temperatures as high as $700°C$ (Pineau *et al.* 1981). Field *et al.* (1980) and Smalley *et al.* (1983) have shown, that in the core of the granulite area (Zone D, Fig. 3), the so-called 'charnockitic gneisses', rich in iron, conspicuously LIL depleted, are deep seated intrusive rocks which have directly crystallized under granulite facies conditions. They contain indeed perfect idiomorphic minerals, as inclusions in plagioclase (garnet, pyroxene, zircon), which strongly suggest a magmatic origin. The abundance of zircon may be spectacular. These intermediate gneisses contain quartz segregations with many CO_2 inclusions, but with very few carbonates. On the other hand, in the host rock, CO_2 inclusions are rare, but carbonates are abundant. Most occur as late, cryptocrystalline microcrystals, but some have been observed in typical primary magmatic inclusions in feldspars (Touret 1985). The cavities, which are a few microns in diameter, are filled with an aggregate of carbonates (ferroan dolomite and calcite), an unknown isotropic phase, sometimes a squeezed gas bubble and rarely zircons. This association, and the general appearance of the inclusion resemble strongly to some inclusions in carbonatites (e.g. in spinels of the Jacupiranga carbonatite, Brazil) (Touret, unpubl.) and in some olivines (Zabargad, Red Sea: Clocchiatti *et al.* 1981).

These inclusions are believed to be traces of carbonatite-like melts associated with the deep intrusion. They predate the CO_2 inclusions that were either associated with or were formed by the breakdown of the carbonates. Thus, the most recent observations in southern Norway, detailed in another paper (Touret 1985), confirm the initial hypothesis of an essentially juvenile source for the CO_2. It was not introduced in the lower crust as a pervasive flood of gaseous CO_2, but through carbonate melts associated with deep, synkinematic intrusions.

The origin of this association remains conjectural. It is interesting to note that Otto (pers. comm. and 1984), following an independent line of experimental research, has obtained a liquid corresponding to 60 wt % quartz-syenite and 40 wt % carbonatite by partial melting at 25 kb a mixture of quartz-eclogite and carbonate. This was taken as an equivalent of a subducted oceanic crust (basalt + carbonate bearing sediments). With higher geotherms, melting would occur at lower pressure and the resulting liquid, more CO_2 rich, corresponds to trondjhemite (about 80%) and carbonatite (about 20%), an almost ideal parental magma for the 'charnockitic gneisses' of Tromoy. It was suggested by Otto that the liquid produced from quartz-eclogite plus carbonate of the subducted oceanic crust does not rise to the surface, but stays at lower levels (mantle and lower crust) with resulting exsolution of the CO_2 fluids. Many aspects of this scenario need to be investigated in more detail, but it constitutes undoubtedly a very promising working hypothesis to explain the carbonate melts observed in the granulites.

ACKNOWLEDGEMENTS: Thanks are due to J.B. Dawson and H. Wedepohl for organizing a very interesting meeting at London and for much patience and suggestions during the editing of the manuscript. Early drafts of the manuscript were greatly improved by reviews from W.L. Griffin and E. Roedder (but this does not imply that they agree with all ideas expressed by the author). Facilities for electron microprobe analysis and Raman spectrometry were provided by the Free University, Amsterdam, and by WACOM, a working group for analytical geochemistry subsidized by the Netherlands Organization for the Advancement of Pure Research (ZWO).

References

BEHR, H.G. 1978. Subfluenz-Prozesse im Grundgebirgs-Stockwerk Mittel-Europas. *Zeitschr. deutsche Geol. Ges.* **129**, 283–318.

—— & EMMERMANN, R. 1983. Kontinentale Kruste. *In*: Statusbericht des Kont. Tiefbohrprogramms (KTB) der Bundesrepublik Deutschland, Part **1**, 34–38. Alfred Wegener-Stiftung, Bonn 1983.

BELOUSSOV, V.V. 1966. Modern concepts of the structure and development of the Earth's crust and the Upper Mantle of continents. *Quart. J. Geol. Soc. London*, **122**, 239–314.

BERGLUND, L. & TOURET, J. 1976. Garnet-biotite gneiss in 'Système du Graphite', Madagascar: Petrology and fluid inclusions. *Lithos*, **9**, 139–148.

BERGMAN, S.C. 1982. *Petrogenetic aspects of the alkali basaltic lavas and included megacrysts and nodules from the Lunar Crater Volcanic Field, Nevada, USA*. PhD. Thesis, Princeton University.

BOWERS, T.S. & HELGESON, H.C. 1983. Calculations of the thermodynamic and geochemical consequences of non-ideal mixing in the system H_2O–CO_2–NaCl. *Geochim. Cosmochim. Acta.* **47**, 1247–1275.

CLOCCHIATTI, R., MASSARE, D. & JEHANNO, C. 1981. Origine hydrothermale des olivines gemmes de l'île de Zabargad (St. Johns), Mer Rouge, par l'étude de leurs inclusions. *Bull. Miner.* **104**, 354–360.

COOLEN, J.J.M.M.M. 1980. *Chemical petrology of the Furua Granulite Complex, southern Tanzania*. PhD. Thesis, Free University Amsterdam, GUA Ser. (Univ. of Amsterdam), Ser. 1, no. 13, 258 pp.

—— 1982. Carbonic fluid inclusions in granulite from Tanzania—a comparison of geobarometric methods based on fluid density and mineral chemistry. *Chem. Geol.* **37**, 59–77.

——, PRIEM, H.N.A., VERDURMEN, E.A.T. & VERSCHURE, R.H. 1982. Possible zircon U-Pb evidence for Pan-African granulite-facies metamorphism in the Mozambique belt of southern Tanzania. *Precambr. Res.* **17**, 31–40.

DHAMELINCOURT, P., BENY, J.M., DUBESSY, J. & POTY, B. 1979. Analyse d'inclusions fluides à la microsonde M.O.L.E. à effet Raman. *Bull. Soc. Fr. Miner.* **102**, 602–610.

DUBESSY, J., AUDEOUD, D., WILKINS, R. & KOSZTOLANYI, C. 1982. The use of the Raman microprobe Mole in the determination of the electrolytes dissolved in the aqueous phase of fluid inclusions. *Chem. Geol.* **37**, 137–150.

FIELD, D., DRURY, A. & COOPER, D.C. 1980. Rare earth and LIL element fractionation in high grade charnockitic gneisses, South Norway. *Lithos*, **13**, 281–289.

FOUNTAIN, D.M. & SALISBURY, M.H. 1981. Exposed cross-sections through the continental crust: implications for crustal structure, petrology and evolution. *Earth Plan. Sci. Lett.* **56**, 263–277.

FYFE, W.S. 1973. The granulite facies, partial melting and the Archean crust. *Phil. Trans. Geol. Soc. London, A*, **273**, 457–461.

GLASSLEY, W.E. 1983. Deep crustal carbonates as CO_2 fluid sources. Evidence from metasomatic reactions zones. *Contrib. Mineral. Petrol.* **84**, 15–24.

HANSEN, E.C., NEWTON, R.C. & JANARDHAN, A.S. 1984a. Fluid inclusions in rocks from the amphibolite facies gneiss to charnockite progression in southern Karnataka, India: Direct evidence concerning the fluids of granulite metamorphism. *J. Metam. Petrol.* (in press).

HANSEN, E.C., NEWTON, R.C. & JANARDHAN, A.S. 1984b. Pressures, temperatures and metamorphic fluids across an unbroken amphibolite-facies to granulite-facies transition in southern Karnataka, India. *In*: KRÖNER, A., GOODWIN, A.M. & HANSON, G.N. (eds) *Archean Geochemistry*. Amsterdam: Elsevier (in press).

HARRIS, N.B.W., HOLT, R.W. & DRURY, S.A. 1982. Geobarometry, geothermometry and late Archean geotherms from the granulite facies terrain of South India. *J. Geol.* **90**, 509–527.

HOEFS, J. & TOURET, J. 1975. Fluid inclusion and carbon isotope studies from Bamble granulite, southern Norway. A preliminary investigation. *Contrib. Mineral. Petrol.* **52**, 165–174.

HOLLISTER, L.S. & CRAWFORD, M.L. 1981. Fluid inclusions: applications to petrology. Short Course *Mineral. Assoc. Canada*, **6** (Calgary), 304 pp.

HOLLOWAY, J. 1981. Composition and volumes of supercritical fluids in the Earth's crust. Short Course *Mineral. Assoc. Canada*, **6**, 13–38.

JANARDHAN, A.S., NEWTON, R.C. & HANSEN, E.C. 1982. The transformation of amphibolite facies gneiss to charnockite in southern Karnataka and northern Tamil Nadu, India. *Contrib. Mineral. Petrol.* **79**, 130–149.

LAMB, W.M. & VALLEY, J.W. 1985. C-O-H calculations and granulite genesis. *In*: TOBI, A.C. & TOURET, J. (eds) *The Deep Proterozoic Crust in the North Atlantic Provinces*, NATO Adv. Stud. Inst. Ser., C, **158**.

MCCARTHY, T.S. 1976. Chemical interrelationships in a low-pressure granulite terrain in Namaqualand, South Africa, and their bearing on granite genesis and the composition of the lower crust. *Geochim. Cosmochim. Acta*, **40**, 1057–1068.

NESBITT, H.W. 1980. Genesis of the New Quebec and Adirondack granulites: evidence of their production by partial melting. *Contrib. Mineral. Petrol.* **68**, 391–405.

NEWTON, R.C. 1983. Geobarometry of high grade metamorphic rocks. *Am. J. Sci.* **283 A**, 1–28.

—— 1984. Temperature, pressure and metamorphic fluid regimes in the amphibolite facies to granulite facies transition zones. *In*: TOURET, J. & TOBI, A.C. (eds) *The Deep Proterozoic Crust in the North Atlantic Provinces*, NATO Adv. Stud. Inst., Ser. C, **158**.

—— & PERKINS, D. 1982. Thermodynamic calibration of geobarometers based on the assemblages garnet-plagioclase-orthopyroxene-(clinopyroxene)-quartz. *Am. Miner.* **67**, 203–222.

——, SMITH, J.V. & WINDLEY, B.F. 1980. Carbonic metamorphism, granulites and crustal growth. *Nature*, **288**, 45–50.

OTTO, J.W. 1984. *Melting relations in some carbonate-silicate systems: sources and products of CO_2-rich liquids.* PhD. Thesis, University of Chicago.

PARRAS, K. 1958. On the charnockites in the light of a highly metamorphic rock complex in southwestern Finland. *Bull. Comm. Geol. Finlande*, **181**, 137 pp.

PERKINS, D. & NEWTON, R.C. 1981. Charnockite geobarometers based on coexisting garnet-pyroxene-plagioclase-quartz. *Nature*, **292**, 144–146.

PINEAU, F. 1977. *La géochimie isotopique du carbone profond.* Thèse Doctorat des Sciences, Université de Paris 7, env. 300 pp.

——, JAVOY, M., BEHAR, F. & TOURET, J. 1981. La géochimie isotopique du facies granulite du Bamble (Norvège) et l'origine des fluides carbonés dans la croûte profonde. *Bull. Mineral.* **104**, 630–641.

ROEDDER, E. 1984. Fluid inclusions. Reviews in mineral. *Am. Mineral. Soc.* **12**, 644 pp.

ROLLINSON, H.R. & WINDLEY, B.F. 1980. Selective elemental depletion during metamorphism of Archean granulites, Scourie, N.W. Scotland. *Contrib. Mineral. Petrol.* **72**, 257–263.

RUDNICK, R.L., ASHWAL, L.D. & HENRY, D.J. 1984. Fluid inclusions in high grade gneisses of the Kapuskasing structural zone, Ontario: Metamorphic fluids and uplift erosion path. *Contrib. Mineral. Petrol.* (in press).

SCHREURS, J. 1985. *The West Uusimaa Complex, Finland, a Proterozoic low pressure granulite dome.* PhD. Thesis, Free University Amsterdam (in prepar.).

SMALLEY, P.C., FIELD, D., LAMB, R.C. & CLOUGH, P.W.L. 1983. Rare earth, Th-Hf-Ta and large-ion lithophile element variations in metabasites from the Proterozoic amphibolite-granulite transition zone at Arendal, South Norway. *Earth Plan. Sci. Lett.* **63**, 446–458.

SORBY, H.C. 1858. On the microscopic structure of crystals, indicating the origin of minerals and rocks. *Geol. Soc. London, Quarterl. J.* **14** (1), 453–500.

SWANENBERG, H. 1980. Fluid inclusions in high grade metamorphic rocks from S.W. Norway, *Geologica Ultraiectina, Univ. Utrecht*, **25**, 147 pp.

TOBI, A.C. & TOURET, J. 1985. *The Deep Proterozoic Crust in the North Atlantic Province.* NATO Adv. Stud. Inst., Norway 1984, Reidel Publ. Ser. C, **158**, 603 pp.

TOURET, J. 1969. *Le socle précambrien de la Norvège méridionale.* Thèse, Univ. Nancy, 3 vol. 609 pp.

—— 1971. Le facies granulite en Norvège méridionale, I: les associations minérales, II: les inclusions fluides. *Lithos*, **4**, 239–249, 423–436.

—— 1974. Facies granulite et fluides carboniques. *Cent. Soc. Geol. Belgique* (vol. P. Michot, Liège) 267–287.

—— 1977. The significance of fluid inclusions in metamorphic rocks. *In*: FRASER, D.G. (ed.) *Thermodynamics in Geology.* D. Reidel Publ., 203–227.

—— 1981. Fluid inclusions in high grade metamorphic rocks. Short Course. *Miner. Assoc. Canada*, **6**, 182–208.

—— 1984. Fluid regime in southern Norway. *In*: TOBI, A.C. & TOURET, J. (eds). *The Deep Proterozoic Crust in the North Atlantic Provinces.* NATO Adv. Study Inst. Ser. C, **158**, 517–549.

—— & BOTTINGA, J. 1979. Equation d'état pour le CO_2; application aux inclusions carboniques. *Bull. Minéral.* **102**, 577–583.

—— & DIETVORST, P. 1983. Fluid inclusions in high grade anatectic metamorphites. *J. Geol. Soc., London*, **140** (4), 635–649.

VALLEY, J. 1982. Fluid heterogeneity in the lower crust. *Eos*, **63**, 448.

—— & O'NEIL, J.R. 1984. Fluid heterogeneity during granulite facies metamorphism in the Adirondacks: stable isotope evidence. *Contrib. Mineral. Petrol.* **85**, 158–173.

WEAVER, B.L. 1980. Rare earth element geochemistry of Madras granulites. *Contrib. Mineral. Petrol.* **71**, 271–279.

WELLS, P.R.A. 1979. Chemical and thermal evolution of Archean sialic crust, southern West Greenland. *J. Petrol.* **20**, 187–226.

WESTRA, L. & SCHREURS, J. 1984. *In*: TOURET, J. & TOBI, A.C. (eds) *The West Uusimaa Complex, Finland: an early Proterozoic thermal dome.* NATO Adv. Stud. Inst, Norway 1984. Reidel Publ. (in prepar.).

WINKLER, H.G.F. 1979. *Petrogenesis of Metamorphic Rocks.* Springer Verlag, 348 pp.

J. TOURET, Free University, Amsterdam, The Netherlands.

The chemical composition of the Archaean crust

S.R. Taylor & S.M. McLennan

SUMMARY: Three separate observations, from sedimentary rock compositions, subduction zone igneous rocks and heat flow constraints combine to suggest that the composition of the Archaean crust was distinct from that of the present-day continental crust. The sedimentary REE patterns are consistent with derivation of the upper crust from a 1:1 mixture of the Archaean bimodal basic-felsic suite. The andesite model, while adequate to account for present-day continental growth does not provide enough Cr and Ni for the Archaean crust. Average crustal component heat flow values of 14 mW m^{-2} for Archaean terrains are significantly lower than the post-Archaean average of 23 mW m^{-2} for crustally produced heat.

Correlations from the sedimentary trace element data provide Archaean upper crustal values of 1.5% K, 5.7 ppm Th and 1.5 ppm U. Heat flow constraints restrict such compositions to the upper layers. The total crustal values are consistent with a 2:1 basic/felsic composition, and give values for the total crust of 0.75% K, 2.9 ppm Th and 0.75 ppm U, which provide crustal heat flow components of 14 mW m^{-2} for a 30 km thick crust and 19 mW m^{-2} for a 40 km crust.

Crustal evolution

In order to understand and model the thermal and petrological evolution of the earth, it is important to constrain the bulk composition of the continental crust. The processes of crustal growth during the Phanerozoic are generally accepted to be dominated by volcanism and related plutonic activity at island arcs and continental orogenic zones. On the other hand, there is equal acceptance that much of the continental crust is of Archaean age, although the exact amount is under debate. Accordingly, it is of some importance to determine if the bulk composition of the Archaean component of the continental crust differs from younger additions and further, to constrain the bulk composition of the Archaean crust (Taylor & McLennan 1981). In addition, understanding of the vertical distribution of elements within the Archaean crust may provide evidence regarding the processes of crustal fractionation at that time.

There are at least three, entirely independent, lines of evidence which suggest the Archaean crust may have had a different bulk composition from the present-day crust. The best documented line of evidence comes from the sedimentary data which indicate that the composition of the Archaean upper crust differed significantly from that of the present-day upper crust. Secondly, the trace element composition of intermediate volcanic and plutonic rocks in modern arc settings is not entirely consistent with generally accepted constraints on bulk crustal compositions. Finally, there is evidence that the crustal radiogenic component of heat flow is significantly lower for Archaean terrains than for younger terrains suggesting that they contain lower abundances of the heat-producing elements.

In this paper, we will examine briefly these lines of evidence and attempt to place some constraints on the bulk composition of and elemental distribution within the Archaean crust. Many of the arguments presented here have been discussed in much greater detail elsewhere (Taylor & McLennan 1985).

Geochemistry of Archaean sedimentary rocks

Considerable evidence suggests that the composition of the Archaean exposed crust was different to that of the present-day, based mainly on the sedimentary data (Table 1). These suggest that the Archaean upper crust, on average, was less enriched in large ion lithophile elements and lacked the negative Eu-anomaly so characteristic of the post-Archaean. The change from Archaean to Post-Archaean sedimentary REE patterns can be observed at the boundary between Archaean and Proterozoic-style terrains. This change is not isochronous among different cratons, but can be correlated with the widespread intrusion of K-rich granites during the late Archaean (Taylor & McLennan 1985). Caution is warranted in intepreting these observations since sediments sample only the uppermost part of the crust exposed to weathering, erosion and transport and one cannot assume on an *a priori* basis that differences in upper crustal compositions necessarily reflect differences in bulk crustal compositions.

From: DAWSON, J.B., CARSWELL, D.A., HALL, J. & WEDEPOHL, K.H. (eds.) 1986. *The Nature of the Lower Continental Crust*, Geological Society Special Publication No. 24, pp. 173–178.

TABLE 1. *Comparison of trace-element characteristics of Archaean and Post-Archaean fine-grained sedimentary rocks*

	Archaean	Post-Archaean
ΣREE	102 ± 15	185 ± 15
La/Yb	11.5 ± 2.5	15.5 ± 1.3
Eu/Eu*	0.99 ± 0.05	0.65 ± 0.02
La	20 ± 3	39 ± 3
Th	6.7 ± 1.1	14 ± 1
Sc	19 ± 2	16 ± 1
La/Sc	1.3 ± 0.2	2.7 ± 0.3
Th/Sc	0.43 ± 0.07	1.0 ± 0.1
La/Th	3.3 ± 0.3	2.8 ± 0.2
U	1.7 ± 0.3	3.2 ± 0.3
Th/U	3.8 ± 0.3	4.8 ± 0.3
n	45	70

Archaean data from Australia and S. Africa only. Note that Archaean data from the Northern hemisphere, not included here because of minor lithological differences, are consistent with these values (Taylor & McLennan 1985). Post-Archaean data from Australia, New Zealand, Antarctica. Note that when all available data are included, Archaean La/Th = 3.5 ± 0.3.

There is general acceptance that the characteristic post-Archaean sedimentary pattern is representative of the presently exposed upper crust. More debatable, however, is the representativeness of the Archaean data which are derived mainly (but not exclusively) from turbidite sequences in greenstone belts. This question has been reviewed in some detail (McLennan & Taylor 1984; Taylor & McLennan 1985) and it was concluded that all major Archaean lithologies were represented in the Archaean data set. The trace element and petrological characteristics (McLennan 1984; McLennan & Taylor 1984) are best explained by variable mixtures of the Archaean bimodal suite (mafic volcanics and felsic volcanics, tonalites, trondhjemites).

There is a superficial resemblance between the REE patterns of many Archaean sedimentary rocks and those deposited in modern fore-arc settings of island-arcs. However, the latter are clearly derived from andesitic volcaniclastic material whereas the former bear little or no evidence of a dominant andesitic contribution. The depositional site for a number of Archaean turbidite sequences (e.g. Swaziland Supergroup, Pilbara Supergroup) is thought to be at an evolving continental margin during synorogenic

conditions (Bickle & Eriksson 1982; K. Eriksson, *pers. comm.*) in which both contemporary volcanic and pre-existing terrains contribute to the detrital record. Although comparison with modern tectonic environments is hazardous, it is interesting to note that unpublished results from this laboratory indicate that turbidites deposited at modern continental arc environments (e.g. Java Trench, Peru–Chile Trench) commonly have typical post-Archaean REE patterns, thus reinforcing the distinction beween the Archaean and post-Archaean.

The andesite model revisited

A durable model for the composition of the continental crust is that it is equivalent to the average composition of calc-alkaline igneous rocks produced in orogenic regions. The popularity of the model is related to two features: the process is presently observable and it provides large volumes of silica-rich ($SiO_2 \simeq 60\%$) material from mantle sources. Andesitic volcanism is the most obvious product of such orogenic activity, resulting in the label 'andesite model'. In fact, the model includes all aspects of plate margin subduction zone igneous activity (Taylor 1967, 1977). Accordingly, other models which appeal to the intrusion of tonalites, at Andean margins, as the cause of continental growth (e.g. Weaver & Tarney 1984), represent a minor variant of the andesite model.

It is not immediately obvious that the andesite model can be applied to the Archaean. Typical calc-alkaline andesitic rocks are much less common in many Archaean terrains. Tonalites and related plutonic rocks are abundant in the Archaean but an island-arc or continental arc tectonic setting of such rocks is debatable (e.g. Barker *et al.* 1981). Their REE patterns, for example, have HREE depletion indicating equilibration with garnet at mantle depths, in contrast to the flatter patterns typical of modern island-arc rocks. Irrespective of the tectonic setting or average composition of Archaean orogenic material, there are several lines of evidence that modern calc-alkaline andesites and related rocks cannot be represenative of the entire continental crust.

A serious problem for the andesite model is that it apparently fails to provide enough Cr and Ni to account for the abundances observed in the continental crust. These elements are typically low in island-arc rocks with Ni averaging 25 ppm and Cr averaging 55 ppm (Taylor 1977). However, average upper crustal values are not much lower than these values (Ni $\simeq$ 20 ppm; Cr = 35

TABLE 2. *Radioactive element abundances, heat production and heat flow for various crustal compositional models*

	Present-day upper crust (1)	Andesite crust (2)	Average crust (1)	Average crust (3)	Archaean upper crust (1)	Archaen crust (1)
K (%)	2.8	1.25	0.91	1.66	1.50	0.75
Th (ppm)	10.7	4.8	3.5	5.1	5.7	2.9
U (ppm)	2.8	1.25	0.91	1.3	1.5	0.75
Heat production (μWm^{-3})	1.79	0.80	0.58	0.87	0.95	0.48
Heat flow (mWm^{-2})						
40 km Crust	72	32	23	35	38	19
30 km Crust	54	24	17	26	29	14

Sources: (1) Taylor & McLennan 1985.
        (2) Taylor 1977.
        (3) Weaver & Tarney 1984.

ppm) (Taylor & McLennan 1985). We could expect such elements to remain in residual phases in the lower crust during intra-crustal melting which produces upper crustal granitic rocks. Lower crustal xenoliths commonly contain high levels of Cr and Ni (e.g. Arculus *et al.* 1986). Whether xenoliths are statistically sampling the lower crust is unknown, but acidic granulites are rare (see Rudnick & Taylor 1986 and Taylor & McLennan 1985).

An additional problem with the Andesite Model is that the predicted crustal levels of heat-producing elements (K, Th, U) are probably too high. The most recent estimate for present-day island-arc rocks, which assumes Th/U = 3.8, predicts a crustal component of heat flow of about 32 mW m^{-2} for a 40 km crust (Table 2). Recent compilations of the available data, however, suggest that actual value may be significantly below this, perhaps 23 mW m^{-2} (see below).

Constraints from heat flow

Estimates of the average heat production of crustal rocks are the most important geophysical constraints on crustal composition. There is a growing consensus regarding the relative importance of the various components of continental heat flow (e.g. Sclater *et al.* 1980; Vitorello & Pollack 1980; Morgan 1984). Morgan (1984) has provided the most recent estimate; for heat flow provinces that experienced the last tectonothermal event in the post-Archaean, the average crustal radioactive component of heat flow is 23 ± 8 mW m^{-2}. This value excludes only the Central Australia Province which has abnormally high heat flow (Morgan 1984). For a 40 km thick crust, and assuming K/U = 10^4 and Th/U = 3.8, this implies the following bulk crustal abundances: K = 0.9%, Th = 3.4 ppm, U = 0.9 ppm.

On the basis of this evidence, the andesite model and other similar models (e.g. Weaver & Tarney 1984) encounter difficulties because they predict abundances of K, U and Th which generates 40–50% too much heat within the crust.

Of equal importance is that Archaean crustal rocks have a smaller component of radiogenic heat flow, averaging about 14 ± 2 mW m^{-2} out of a total heat flow of 41 mW m^2 (see Morgan 1984, 1985 for details). Earlier models (e.g. Vitorello & Pollack 1980) suggested a fairly gradual change with time, however, Morgan (1984) recently has argued convincingly that there is no compelling evidence to suggest secular changes occur during the post-Archaean and that the Archaean possesses uniquely lower values. This is supported by the uniformly low measured crustal heat productions ($\simeq 0.4\ \mu W\ m^{-3}$) through a 20 km exposed cross section in the Archaean Kapuskasing Structural Zone (Ashwal *et al.* 1984).

The most straightforward interpretation is that the crust stabilized in the Archaean contained only about 60–80% (depending on thickness) of the abundance levels of K, Th and U as crust which stabilized during the post-Archaean. An alternative possibility, suggested recently by Morgan (1985), is that the crust had the same composition but that the low heat production crust was selectively stabilized during the Archaean. The enriched portion of the crust was selectively remobilized and recycled during orogenic activity. We do not favour this alternative because the sedimentary data do not appear to

S.R. Taylor & S.M. McLennan

support it. Clastic sedimentation is generally related to tectonic uplift (e.g. Garrels & Mackenzie 1971) and, as pointed out above, many Archaean sediments are of synorogenic origin. Accordingly, we would expect Archaean sedimentary rocks to bear the signature of this highly enriched crustal component to an extent at least as great as presently observed in sedimentary rocks. However, from Table 1, it was documented that Archaean sedimentary rocks are depleted in heat-producing elements by a factor of two compared to post-Archaean sedimentary rocks.

Nd-isotopic data on Archaean sedimentary rocks pose even greater difficulties for such a recycling model. Nd model ages of sediments date the average age of separation of provenance rocks from the mantle (McCulloch & Wasserberg 1978). Available data on Archaean sedimentary rocks indicate model ages essentially indistinguishable from stratigraphic age thus restricting significant recycling to a minor role. Early Proterozoic sedimentary rocks record some evidence of recycling in the Nd data but probably not on the scale implied by the model of Morgan.

The Archaean upper crust

It is possible to make direct estimates for many elements in the Archaean upper crust directly from the sedimentary data (McLennan *et al.* 1980; McLennan & Taylor 1984; Taylor & McLennan 1985). The procedure involves adopting the average shale REE pattern as equivalent to that of the upper crust. This is analogous to the well-documented case in the post-Archaean. In the post-Archaean, absolute abundances are reduced by 20% to account for carbonates, evaporites, quartzites and so forth. Such lithologies appear to be comparatively scarce in the Archaean and we have refrained from such an adjustment. Thus REE and other incompatible element abundances calculated probably represent upper limits.

The coherence between REE and Th in Archaean sedimentary rocks allows an estimate of Th abundances in the Archaean upper crust (McLennan *et al.* 1980). The most recent estimate for La/Th is 3.5 ± 0.3. For a sedimentary value for La of 20 ppm we calculate Th = 5.7 ppm in the upper crust. Using Th/U = 3.8, K/U = 10^4, K/Rb = 300, Th/Sc = 0.43 ± 0.7, La/Sc = 1.3 ± 0.2, we may also derive U = 1.5 ppm, K = 1.5%, Rb = 50 ppm, Sc = 14 ppm.

The trace element geochemistry of Archaean shales indicates an origin predominantly from a mixture of the end members of the Archaean bimodal suite. By estimating the end-member

TABLE 3. *Comparison of Archaean upper crustal compositions derived (I) from sedimentary rock data, and (II) from a 1:1 mixture of the Archaean bimodal basaltic-felsic igneous suite.*

	I	II
K %	1.5	1.1
Rb ppm	50	37.5
La ppm	20	20.4
Th ppm	5.7	6.8
U ppm	1.5	1.8
Sc ppm	14	22.5

compositions, we can calculate an average upper crustal composition by constraining the mixing proportions with the sedimentary REE pattern. This procedure has been attemped by Taylor & McLennan (1985) and a complete listing of Archaean upper crustal abundances can be found there. In Table 3, we compare the abundances of selected trace elements derived from the mixing model and more directly from the sedimentary data as outlined above. The results agree reasonably well, although for these elements, we consider the sedimentary data to be more accurate.

Such a composition cannot be representative of the entire Archaean crust on account of heat flow constraints. Levels of K, Th, U indicated in Table 3 provide 0.95 μW m^{-3} of heat production. This would result in 28.5 mW m^{-2} and 38 mW m^{-2} of heat flow for a 30 km and 40 km crust, respectively. Such values greatly exceed the crustal radioactive component of 14 mW m^{-2} in the Archaean crust. From simple mass balance constraints, the upper crustal composition can represent no more than 50% of the total Archaean crust. If the uniform heat production measured through the Kapuskasing crustal cross-section is representative (Ashwal *et al.* 1984), then the upper crustal thickness must be much less.

The Archaean total crustal composition

If we assume the bimodal model can be extended to the entire Archaean crust, then we may estimate the bulk composition using the heat flow data to constrain mixing proportions. A 2:1 mixture of mafic and felsic lithologies provides the following abundances: K = 0.75%; Th = 2.9 ppm; U = 0.75 ppm for K/U = 10^4 and Th/U = 3.8. This provides a crustal heat-flow component of 14 mW m^{-2} and 19 mW m^{-2} for a 30 km and 40 km crust, respectively. Considering the many uncertainties, including the small data base for Archaean heat flow provinces (n = 4), uncertainties in estimating end-member compositions, and mean Archaean crustal thickness, the agreement is reasonably good. A complete listing of compositions, deduced by this approach, can be found in Taylor & McLennan (1985). The most important point here is that the Archaean upper crust must be considerably enriched in felsic rocks, compared to the total Archaean crust, in order to satisfy the heat flow constraints.

Geochemical distribution and differentiation of the Archaean crust

In the previous section we noted, from the heat flow constraint, that the Archaean upper crust was more felsic than the bulk Archaean crust. The lack of Eu-depletion in the Archaean sedimentary record indicates that intra-crustal melting and the production of K-rich granites and granodiorites was minor and these rock types did not comprise more than about 10% of the exposed upper crust. Accordingly, the vertical differentiation of the Archaean crust must be due not to intra-crustal melting, but to other effects. In our model, the Archaean crust is dominated by the basic-felsic bimodal suite of igneous rocks of mantle origin. We attribute the vertical difference in crustal composition to the operation of simple density and buoyancy effects, with the less dense felsic magmas becoming preferentially concentrated in the upper layers of the crust. The basic magmas are concentrated in the lower crust. This model for the Archaean crust contrasts with the intra-crustal stratification caused by the production, in the late Archaean, Proterozoic and Phanerozoic, of K-rich granites and granodiorites which dominate the present upper crust. A consequence of these models is that Archaean granulites are less likely to contain evidence of melt extraction than lower crustal rocks of younger ages.

Eu enrichment, trondhjemites and the lower crust

The more acidic members of the Na-rich granitic suites which dominate the Archaean occasionally display enrichment in Eu, relative to chondrite normalized patterns (e.g. Jahn & Zhang 1984). Such enriched Eu abundances are not uncommon in trondhjemites. The REE patterns of Archaean tonalites and trondhjemites are characterized by enrichment in LREE and depletion in HREE. The resulting steep patterns are the reciprocal of those observed in garnet and, accordingly, this evidence is usually cited as indicative of the presence of garnet as a residual phase during partial melting. The source of the tonalites and trondhjemites is thus placed within the stability field of garnet. Modern examples of tonalites from subduction zone environments usually display the flatter patterns typical of modern calc-alkaline rocks, indicating a shallower depth of melting. Garnet discriminates against Eu^{2+} (and Sr^{2+}), and this coupled with the fact that Eu^{2+} partitions strongly into polymerized silicic melts (Möller & Muecke 1984) accounts for the enrichment in Eu observed in the more acidic trondhjemites. (See Rudnick & Taylor (1986) for an extended discussion.) However, the general scarcity of Eu enrichment in Archaean sedimentary rocks shows that such trondhjemites must comprise only a minor (< 10%) fraction of the exposed Archaean crust. The Eu enrichment displayed by the more acidic trondhjemites should not be confused with predicted Eu enrichments in the lower crust, following the extraction of K-rich granitic melts bearing the signature of Eu depletion, which now dominates the exposed upper crust.

References

Arculus, R.J. *et al*, 1984. Eclogites and granulites in the lower continental crust: Examples from eastern Australia and southwestern U.S.A. *Proc. Eclogite Symposium* (in press).

Ashwal, L.D., Morgan, P., Kelley, S.A. & Percival, J.A. 1984. An Archean crustal radioactivity profile: The Kapuskasing structural zone, Ontario (abstract). *Geol. Soc. Amer.* **16**, 433.

Barker, F., Arth, J.G. & Hudson, T. 1981. Tonalites in crustal evolution. *Phil. Trans. Roy. Soc.* A **301**, 293–303.

Bickle, M.J. & Eriksson, K.A. 1982. Evolution and subsidence of early Precambrian sedimentary basins. *Phil. Trans. Roy. Soc.* A **305**, 225–247.

Garrels, R.M. & Mackenzie, F.T. 1971. *Evolution of Sedimentary Rocks*. Norton, London.

Jahn, B-M. & Zhang, Z-Q. 1984. Archean granulite gneisses from eastern Hebei Province, China: rare earth geochemistry and tectonic implications. *Contrib. Min. Pet.* **85**, 224–243.

McCULLOCH, M.T. & WASSERBURG, G.J. 1978. Sm-Nd and Rb-Sr chronology of continental crust formation. *Science*, **200**, 1003–1011.

McLENNAN, S.M. 1984. Petrological characteristics of Archean graywackes. *J. Sed. Pet.* **54**, 889–898.

——, NANCE, W.B. & TAYLOR, S.R. 1980. Rare earth element—thorium correlations in sedimentary rocks and the composition of the continental crust. *Geochim. Cosmochim. Acta*, **44**, 1833–1839.

—— & TAYLOR, S.R. 1984. Archaean sedimentary rocks and their relation to the composition of the Archaean crust. *In:* KRONER, A. HANSON, G.N. & GOODWIN, A.M. (eds) *Archaean Geochemistry*. Springer-Verlag, 47–71.

MÖLLER, P. & MUECKE, G.K. 1984. Significance of europium anomalies in silicate melts and crystal-melt equilibria: a re-evaluation. *Contrib. Min. Pet.* **87**, 242–250.

MORGAN, P. 1984. The thermal structure and thermal evolution of the continental lithosphere. *Phys. Chem. Earth*, **15** 107–193.

—— 1985. Crustal radiogenic heat production and the selective survival of ancient continental crust. *J. Geophys. Res.* **90**, C561–C570.

POLLACK, H.N. 1980. The heat flow from the Earth: A review. *In:* DAVIES, P.A. & RUNCORN, S.K. (eds) *Mechanisms of Continenal Drift and Plate Tectonics.* Academic Press, 183–192.

RUDNICK, R.L. & TAYLOR, S.R. 1986. Geochemical constraints on the origin of Archaean tonalitic-trondhjemitic rocks and implications for lower crustal composition. (This volume).

SCLATER, J.G., JAUPERT, C. & GALSON, D. 1980. The heat flow through oceanic and continental crust an the heat loss from the Earth. *Rev. Geophys. Space Phys.* **18**, 269–311.

TAYLOR, S.R. 1967. The origin and growth of continents. *Tectonophysics*, **4**, 17–34.

—— 1977. Island arc models and the composition of the continental crust. *AGU Ewing Series* **I**, 325–335.

—— & McLENNAN, S.M. 1981. Evidence from rare-earth elements for the chemical composition of the Archean crust. *Geol. Soc. Aust. Spec. Pub.* **7**, 255–261.

—— & —— 1985. *The Continental Crust: Its Composition and Evolution.* Blackwell Scientific, Oxford.

VITORELLO, I. & POLLACK, H.N. 1980. On the variation of continental heat flow with age and the thermal evolution of the continents. *J. Geophys. Res.* **85**, 983–995.

WEAVER, B.L. & TARNEY, J. 1984. Empirical approach to estimating the composition of the continental crust. *Nature*, **310**, 575–577.

S.R. TAYLOR & S.M. McLENNAN, Research School of Earth Sciences, Australian National University, Canberra, ACT, Australia. 2601.

Geochemical constraints on the origin of Archaean tonalitic-trondhjemitic rocks and implications for lower crustal composition

R.L. Rudnick & S.R. Taylor

SUMMARY: The rare earth element (REE) concentrations of high-grade trondhjemitic gneisses from the Archaean Kapuskasing Structural Zone, central Ontario, are similar to REE concentrations of Archaean tonalites and trondhjemites world-wide: REE patterns are steeply fractionated ($(La/Yb)_N = 20-100$), with variable positive Eu anomalies and fractionated HREE ($(Gd/Yb)_N = 2-4$). Calculated models for the melting of a mafic source with a flat REE pattern and variable mineralogy show that amphibole alone is not capable of producing the large amount of REE fractionation observed in Archaean tonalites; garnet amphibolite (with >20% garnet), mafic garnet granulite or eclogite are the most likely sources. The presence of garnet constrains the mafic source to be within the lower crust (deeper than 25 km) or upper mantle. Such crustal or upper mantle melting depths are not geochemically distinguishable, yet have important implications for the composition of the Archaean lower crust.

Introduction

Rocks which have equilibrated at lower crustal conditions are available from: (1) granulite-facies metamorphic terrains and (2) granulite-facies xenoliths in alkali basalts. Rocks from these two sources show marked differences in overall composition, and hence in mineralogical, chemical and physical properties. Granulite-facies metamorphic terrains possess a variety of lithologies, ranging from ultramafic to silicic, including sedimentary, but tend to be dominated by felsic compositions. This is particularly true of Archaean granulite facies terrains, e.g. Scourian gneisses of Scotland (Weaver & Tarney 1980, 1981), south Indian granulites (Condie et al. 1982; Janardhan et al. 1982), Napier complex in Enderby Land, Antarctica (Sheraton & Black 1983), Qianxi Group gneisses, China (Jahn & Zhang 1984), where rocks of the granodiorite-tonalite-trondhjemite suite are pervasive.

In contrast, lower crustal xenoliths are dominantly mafic in composition, for example Lesotho xenoliths, southern Africa (Griffin et al. 1979; Rogers & Hawkesworth 1982), Colorado Plateau xenoliths (Arculus & Smith 1979), Chilean xenoliths (Selverstone 1982), Aleutian xenoliths (Francis 1976), eastern Australian xenoliths (Kay & Kay 1981; Griffin & O'Reilly, this volume; Arculus et al. 1986). Mestasedimentary xenoliths are present only locally, as in the Massif Central xenolith suite (Leyreloup et al. 1977) or Kilbourne Hole, New Mexico (Padovani & Carter 1977). The compositional discrepancy between these two lower crustal rock types leads us to question which is the best representative of the present-day lower continental crust.

Granulite-facies terrains with large amounts of supracrustal rocks (e.g. sediments and K-rich igneous rocks) cannot be representative of the Archaean lower crust because of their high concentrations of heat producing elements, which would produce a higher heat flow than is observed in most Archaean terrains (Morgan 1984). However, granulite terrains dominated by Na-rich felsic rocks are possible lower crustal candidates because of their inherently lower proportions of heat producing elements and their intrusive nature. Yet such high silica rock types are very rare in lower crustal xenolith suites, even in suites from pipes which erupt through Precambrian crust, such as the Palaeozoic alkali basalts from Scotland (Upton et al. 1983) and the Cretaceous kimberlites of southern Africa (Rogers & Hawkesworth 1982), where mafic granulites dominate the lower crustal xenolith suites. This discrepancy suggests that either:

(1) The Archaean lower crust is primarily mafic, but these rocks are not observed in appreciable volumes in most Archaean granulite facies terrains. This may be due to their higher densities, which would tend to inhibit tectonic uplift,

(2) mafic lower crustal xenoliths coming from Archaean crustal segments may represent younger additions to the crust by intrusion of mafic magmas into a dominantly tonalitic lower continental crust, or

(3) the basaltic lavas carrying xenoliths sample selectively. Silicic rocks may be present, but for some reason are not carried to the surface.

It is the purpose of this paper to review trace element characteristics of Archaean tonalite-trondhjemite rocks world-wide, incorporating new data from the Kapuskasing Structural Zone,

From: DAWSON, J.B., CARSWELL, D.A., HALL, J. & WEDEPOHL, K.H. (eds) 1986, *The Nature of the Lower Continental Crust*, Geological Society Special Publication No. 24, pp. 179–191.

central Ontario. By doing this we hope to constrain the possible origins of these volumetrically important rocks. Understanding the origin of Archaean tonalitic rocks may help reconcile the compositional differences observed between granulite-facies terrains and lower crustal xenoliths.

Data sources

The tonalitic gneiss samples investigated here come from areas of granulite to amphibolite facies within the Kapuskasing Structural Zone (KSZ) of central Ontario. The KSZ represents a linear shear zone along which high-grade Archaean gneisses are juxtaposed against greenschist facies rocks of the Abitibi greenstone belt. Because the metamorphic grade decreases gradually to the west within the KSZ, where it forms an indistinct boundary with rocks of the Michipicoten greenstone belt, the KSZ is interpreted to represent an oblique cross section through the Archaean crust (Percival & Card 1983). Maximum metamorphic pressures and temperatures of 8.4 kbars and 800°C occur adjacent to the fault boundary. More detailed information on the geology, metamorphic history and age of the KSZ is found in Percival (1983), Percival & Card (1983), Percival & Krogh (1983) and Rudnick *et al.* (1984). Rocks of the tonalite-trondhjemite suite predominate in the central portion of the KSZ and remain an important rock type in the lower crustal sections. Because of sample heterogeneity, samples >0.5 kg were used, but some heterogeneities are on such a large scale that representative sampling becomes impossible.

Additional REE geochemical data for Archaean tonalite-trondhjemite rocks were obtained from the literature. Specific terrains incorporated in this review include: southern India (3.4 Ga, Condie *et al.* 1982), Finnish Lapland (3.0 Ga, Jahn *et al.* 1984), the Ancient Gneiss Complex, Swaziland (3.0 Ga, Hunter *et al.* 1978, 1984), eastern Finland (2.9 Ga, Martin *et al.* 1983), Scourian granulites, Scotland (2.9 Ga, Weaver & Tarney 1980, 1981; Pride & Muecke 1980), Nuk gneisses, Greenland (2.7–3.0 Ga, Compton 1978) and the Qianxi Group gneisses, China (2.5 Ga, Jahn & Zhang 1984). All of these Archaean terrains possess significant volumes of tonalite-trondhjemite gneisses, of variable metamorphic grade. Many of these gneisses have highly fractionated REE patterns (Fig. 2), which seem to be characteristic of this rock type in the Archaean. Relatively few post-Archaean tonalites possess such highly fractionated REE (notable exceptions are the 1.8 Ga tonalites and trondhjemites of southwest Finland (Arth *et al.* 1978)).

Geochemistry

Major elements were determined from microprobe analyses (Reed & Ware 1973) on rock powders fused and quenched to glass under positive Ar pressure on molybdenum strips. Trace elements were determined using spark source mass spectrometry as outlined in Taylor & Gorton (1977). Precision and accuracy for these elements is better than 5%.

Major and trace element analyses for the KSZ felsic gneisses are presented in Table 1. These

TABLE 1. *Major and trace element geochemistry of KSZ trondhjemites*

	2-2-1	2-3-1	8-17-10
SiO_2	71.46	69.02	68.99
TiO_2	0.20	0.27	0.55
Al_2O_3	15.92	16.62	16.38
FeO*	2.38	2.95	2.91
MnO	—	—	—
MgO	1.00	1.19	1.92
CaO	3.06	4.09	3.31
Na_2O	4.48	4.42	4.27
K_2O	1.38	1.34	1.70
P_2O_5	0.17	0.17	0.22
TOTAL	100.05	100.07	100.25
Rb	46	39	54
Y	2.8	6.2	3.1
Zr	51	142	260
Nb	5.1	4.7	3.5
Cs	3.8	1.2	0.7
Ba	334	687	2140
La	11.4	14.9	44.0
Ce	21.3	29.9	85.8
Pr	2.27	3.32	7.48
Nd	8.20	13.0	26.2
Sm	1.21	2.14	3.31
Eu	0.45	0.71	1.07
Gd	0.83	1.66	1.13
Tb	0.10	0.25	0.17
Dy	—	1.36	—
Ho	0.09	0.24	0.14
Er	0.22	0.53	0.31
Yb	0.21	0.51	0.27
Hf	2.0	3.7	4.4
Pb	12.2	4.0	7.4
Th	3.75	1.46	4.91
U	0.59	0.17	0.31
K/Rb	249	285	261
Th/U	6.3	8.6	15.8
$(La/Yb)_N$	37	20	110
$(Gd/Yb)_N$	3.2	2.6	3.4

* Total Fe as FeO.
— Not detected.

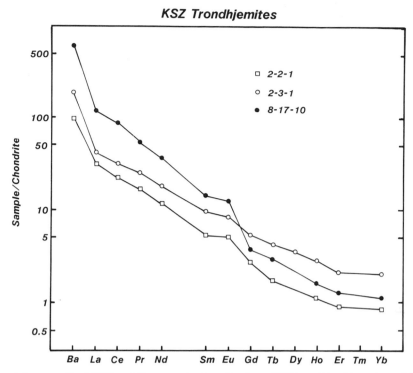

FIG. 1. Chondrite normalized REE and Ba concentrations for three KSZ trondhjemitic gneisses. Normalizing values from Taylor & McLennan (1985).

gneisses are tonalitic to trondhjemitic in composition with SiO_2 between 60 and 71%, and high Al_2O_3 (15.9–16.6%). MgO (1–1.9%), CaO (3—4%), Na_2O (4.4–4.7%) and K_2O (1.3–1.7%) concentrations allow these rocks to be classified as trondhjemites (Barker 1979), yet when the CIPW normative feldspar contents of these rocks are plotted on the triangular silicic igneous rock classification diagram of Barker (1979), all of the KSZ rocks plot within the tonalite field. On an AFM diagram the KSZ felsic gneisses plot near the calc-alkaline trend, similar to tonalitic gneisses from other Archaean high-grade terrains (Sheraton *et al.* 1973; Hunter *et al.* 1978; Jahn & Zhang 1984).

The KSZ tonalitic rocks have variable Rb, Ba, Cs, Th and U concentrations. K/Rb ratios for the tonalites range from 250 to 285. Therefore, the KSZ tonalitic gneisses do not show the significant depletion of Rb relative to K, which has been observed in several granulite-facies terrains (e.g. Scourian granulites of Scotland (Weaver & Tarney 1980) and Qianxi granulites of China (Jahn & Zhang 1984)). Th/U ratios for the tonalites vary from 6.3 to 15.8. As such high Th/U ratios are extreme for unmetamorphosed tonalitic rocks, it

appears U was variably depleted relative to Th in the KSZ gneisses.

Fig. 1 shows the chondrite-normalized REE plus Ba patterns for the KSZ tonalites. These samples have highly fractionated REE with $(La/Yb)_N$ ratios between 21 and 109, slight to large positive Eu anomalies and $(Ba/La)_N$ ratios greater than three. These features are very common in tonalitic rocks of Archaean high-grade terrains world wide (Fig. 2). Because of their similarities with Archaean tonalitic rocks from other areas, the petrogenesis of the highly fractionated KSZ tonalites (2-2-1, 2-3-1 and 8-17-10) will be discussed along with data from other localities in an attempt to define a general model for tonalite origin.

Petrogenesis of felsic gneisses

Partial melting of a mafic rock is the model of tonalite genesis preferred by most workers. Crystal fractionation of a mafic magma is a less preferred model, due to the lack both of significant volumes of intermediate igneous rocks in Archaean terrains (Condie *et al.* 1982) and of any recognizable cumulate material (Weaver & Tar-

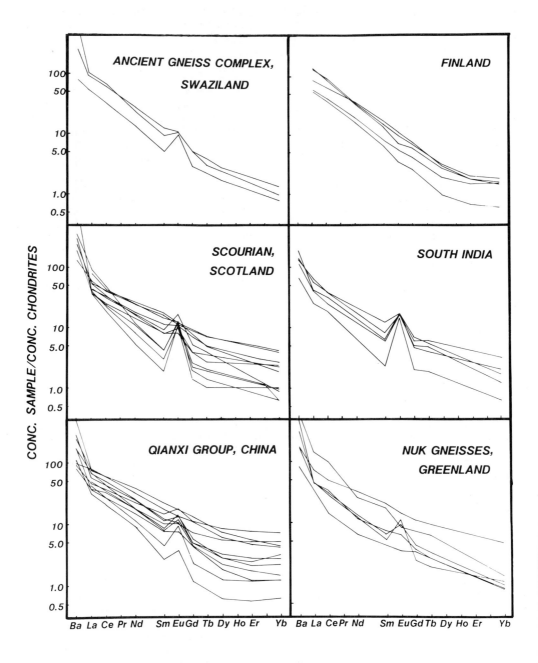

FIG. 2. Chondrite normalized REE and Ba concentrations for Archaean tonalite-trondhjemite rocks world-wide. Data sources: Ancient Gneiss Complex values from Hunter *et al.* (1978, 1984); Finland values from Martin *et al.* (1983); Scourian values from Weaver & Tarney (1980) and Pride & Muecke (1980); south Indian values from Condie *et al.* (1982); Qianxi Group gneiss values from Jahn & Zhang (1984); and Nuk gneiss values from Compton (1978). Chondrite normalizing values are from Taylor & McLennan (1985).

ney 1984). Specifically, three types of mafic source material are invoked:

(1) amphibolite, with or without garnet (Hunter *et al.* 1978; O'Nions & Pankhurst 1978; Jahn *et al.* 1981; Martin *et al.* 1983; Hunter *et al.* 1984; Sheraton & Collerson 1984; Jahn *et al.* 1984);

(2) mafic garnet granulite (Condie & Hunter 1976; Weaver & Tarney, 1980; Gower *et al.* 1982); and

(3) eclogite (Arth & Hanson 1975; Condie & Hunter 1976; Hanson 1978; Weaver & Tarney 1980; Jahn *et al.* 1981; Jahn *et al.* 1984).

In an attempt to distinguish which source material is the most reasonable for the KSZ tonalites, and Archaean tonalites in general, the compositions of liquids derived from 20% melting of each source were calculated. The calculations presented here utilize experimental data and actual modal analyses from the literature to define the mineralogy of mafic rocks at various pressures and temperatures of equilibration and the melt proportions of those phases. Most geochemical modelling of this sort can only be used to place gross constraints on parent rock compositions and mineralogy, and this is all that is attempted here. It is shown that by varying different parameters of the batch melting equation (i.e. REE content of parent material and partition coefficients), all three proposed lithologies are potential sources for Archaean tonalites. However, observations on the geochemical nature of mafic rocks from Archaean terrains are used in conjunction with the partial melting calculations to rule out certain source materials.

In all models detailed below, equilibrium melting is assumed (equation 5 of Hanson 1978) and the REE pattern for the parent rock is taken as twenty times chondritic. Qualitative arguments are provided to show the effects of changing the relative REE fractionation of the source rocks, and the feasibility of these source compositions are evaluated. These calculations are similar to those done by Hanson (1980), and yield similar results, except that we have used garnet in our mafic granulite composition. Garnet is a stable solidus phase in mafic rocks which lie below 25–41 km, depending on the amount of water present (Green 1982).

REE partition coefficients (listed in Table 2) were taken from the literature and span the range for rocks of andesitic to dacitic compositions. In this way, the REE partitioning behaviour in a range of possible melt compositions are evaluated. Because the partition coefficients (K_Ds) exert a strong control on the calculated trace element composition of partial melts, a discussion of the applicability of the available K_Ds is warranted. Many mineral K_D^{REE} show changes in overall values and slight changes in REE pattern corresponding to changes in mineral and melt compositions (see K_D patterns reported in Fujimaki *et al.* 1984 and Irving & Frey 1978). The most important of these for our purposes is the variation observed in K_D of garnets. Pyrope has lower K_D^{REE} and no Eu anomaly compared with almandine, which has higher K_D^{REE} and variable negative Eu anomalies (Irving & Frey 1978). It is not clear whether these differences in K_D are attributable to differences in garnet composition or differences in the equilibrium melt composition. In all likelihood, both exert a control. This leads to the dilemma of which garnet K_Ds to use when modelling partial melting of a mafic source (which will have garnets with compositions lying between the pyrope-almandine end members, with variable amounts of grossular component (Deer *et al.* 1982)) to produce a silicic melt (NB, all available pyrope K_Ds are for pyrope in equilibrium with mafic melts). Since the REE K_D patterns for both pyrope and almandine are nearly parallel, choosing one over the other will not affect the REE *pattern* of the partial melt (except for Eu), but will shift the absolute concentration up or down. The negative Eu anomaly observed in REE K_D patterns for almandine was suggested by Irving & Frey (1978) to be either an artifact of early plagioclase crystallization, or due to the larger size of Eu^{2+} relative to the small $+3$ REE, causing Eu^{2+} to be excluded from the garnet cation sites. The second alternative gains support when one compares the K_Ds of Sr^{2+} (which is similar in size and charge to Eu^{2+}) with those of Sm and Gd (which bracket Eu) in garnet. While K_D^{Sr} is low (0.0154 for almandine in dacite (Philpotts & Schnetzler 1970)), $K_D^{Sm–Gd}$ are large (0.76–14.99 for almandine in andesite-rhyodacite (Irving & Frey 1978)). This partitioning data suggests that Eu^{2+} will partition into garnets less strongly than the $3+$ REE. The apparent correlation of the size of the Eu anomaly with silica content of the coexisting melt suggests that the Eu anomalies in garnet K_Ds may be real. Möller & Muecke (1984) have recently suggested, on a theoretical basis, that Eu^{2+} will partition more strongly into a polymerized melt than Eu^{3+}. Thus, more silicic melts will take in Eu^{2+} relative to $3+$ REE, creating negative Eu anomalies in the coexisting mineral phases. Because we are modelling the evolution of a silicic melt, we will use the almandine/felsic melt K_Ds rather than the pyrope/mafic melt K_Ds of Irving and Frey (1978) in our trace element models.

Amphibolite source

Helz (1976) provides data on stable mineral

TABLE 2. *Parameters used in batch melting calculations*

Partition coefficients

	Andesite					Dacite				
	Pc[1]	Hb[1]	Gar[2]	Cpx[1]	Opx[1]	Pc[1]	Hb[3]	Gar[2]	Cpx[1]	Opx[1]
La	0.302	0.366	0.076	0.047	0.031	0.390	(0.40)	0.37	0.015	0.015
Ce	0.221	0.574	(0.12)	0.084	0.028	0.251	0.90	0.53	0.044	0.016
Nd	0.149	1.012	(0.50)	0.183	0.028	0.189	2.80	0.81	0.166	0.016
Sm	0.102	1.366	1.25	0.377	0.028	0.137	3.99	5.50	0.457	0.017
Eu	1.214	1.212	1.52	(0.500)	0.028	1.113	3.44	1.37	0.411	(0.022)
Gd	0.066	1.490	5.20	0.583	0.039	0.120	5.47	13.6	0.703	0.027
Ho	(0.047)	(1.554)	23.8	(0.741)	(0.114)	(0.117)	(6.06)	31.1	(0.738)	(0.057)
Yb	0.041	1.488	53	0.633	0.254	0.132	4.90	26.0	0.640	0.115

Modal mineralogies (X_i) and melt proportions (P_i)

	Amphibolite				Garnet granulite				Eclogite			
	1		2		1		2		1		2	
	X_i	P_i	X_i	P_i	X_i	P_i	X_i	P_i	X_i	P_i	X_i	P_i
Pc	45	92	34	92	21	85	20	70	—	—	—	—
Hb	55	8	66	8	—	—	—	—	—	—	—	—
Gar	—	—	—	—	29	7.5	46	10	11	11	33	33
Cpx	—	—	—	—	50	7.5	26	10	89	89	67	67
Opx	—	—	—	—	—	—	5	10	—	—	—	—

Amphibolite 1 = Picture Gorge basalt from Helz (1975); Amphibolite 2 = 1921 olivine tholeiite from Helz (1975); Garnet granulite 1 = PHN 1646 from Griffin *et al.* (1979); Garnet granulite 2 = L20 from Griffin *et al.* (1979); Eclogites 1 and 2 represent range of modal mineralogies of natural eclogites from Apted (1981). Modal mineralogies and melt proportions expressed in weight percent.

Source of partition coefficients: [1] Fujimaki *et al.* 1984, [2] Irving & Frey 1978, [3] Arth & Barker 1976.

( ) = interpolated value.

assemblages and mineral melt proportions for various degrees of partial melting of amphibolite (olivine tholeiite and quartz tholeiite compositions) at 5 kbars water pressure. The relevant mineralogies and melt proportions at 20% batch melting are listed for both compositions in Table 2, along with the range and source of partition coefficients used. The REE patterns of the derived melts are slightly LREE enriched, with variable positive and negative Eu anomalies (Fig. 3a). It can be seen that batch melting of an amphibolite with a chondritic REE pattern is not capable of producing either the high $(La/Yb)_N$ ratios observed in many Archaean tonalites, or the fractionation within the HREE observed in Archaean tonalites (best described by the $(Gd/Yb)_N$ ratio which is typically between 2 and 8 for Archaean tonalites and trondhjemites (Fig. 4b)).

If the $(La/Yb)_N$ and $(Gd/Yb)_N$ of the parent amphibolite is increased, then amphibolite could be parental to fractionated tonalites. For example, if the source has $(La/Yb)_N = 4$, the melt can have $(La/Yb)_N$ up to 40 (Jahn *et al.* 1984). However, because of the inability of amphibole to fractionate the HREE relative to one another, the mafic source would have to have the same $(Gd/Yb)_N$ ratio as that observed in the derived melt. Such high $(Gd/Yb)_N$ ratios are generally not found in Archaean mafic rocks (see Sun & Nesbitt 1978; Sun 1984), consequently these fractionated sources are not feasible parents for the massive volumes of tonalite found in Archaean terrains. Therefore, residual amphibole alone cannot produce the HREE depletion in tonalites. The only way in which amphibolite could be parental to fractionated tonalites is if it possesses significant

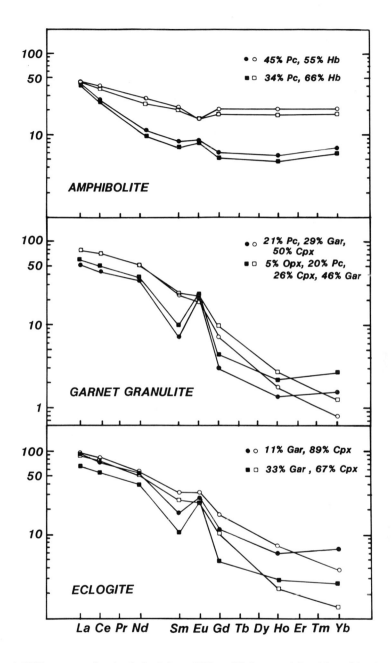

Fig. 3. Calculated REE patterns of melts derived from 20% equilibrium partial melting of (top) amphibolite, (middle) mafic garnet granulite and (bottom) eclogite. Circles represent calculated melts using andesite partition coefficients; squares represent calculated melts using dacite partition coefficients. Parameters used in calculations are given in Table 2.

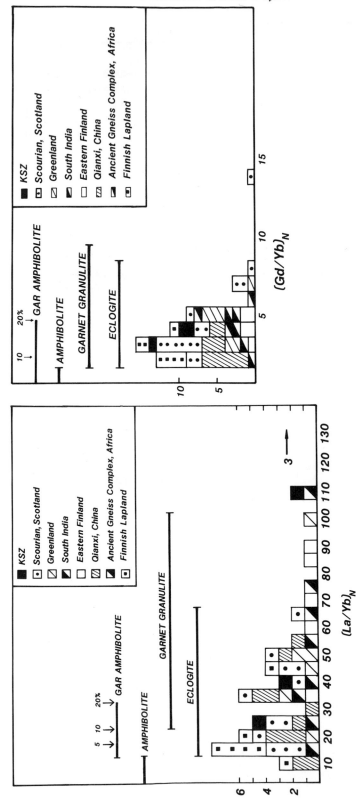

FIG. 4. (A) Range of (La/Yb)$_N$ ratios observed in Archaean tonalites (REE patterns pictured in Fig. 2) compared with (La/Yb)$_N$ ratios produced by 20% partial melting of eclogite, garnet granulite, amphibolite without garnet, and amphibolite with varying proportions of garnet (marked by arrows). The '3' at the right hand side of the figure refers to 3 ratios that plot off the scale of the diagram. (B) Same as (A) but for (Gd/Yb)$_N$ ratios.

amounts of garnet; at least 20% garnet is required to produce the necessary fractionations in a chondritic parent (Fig. 4).

Garnet granulite source

There is a notable lack of experimental data defining the melting behaviour of mafic garnet granulites. Even so, provided the modal proportions used for the parent composition are reasonable, large variations in mineral melt proportions have little effect on the calculated REE compositions of the melt. To model partial melting of garnet granulites, parental mineral modes were taken from the literature. Most rocks of basaltic composition have between 20 and 45% garnet at granulitic pressures and temperatures. The particular mineral modes used in this calculation were taken from Griffin *et al.* (1978) for two mafic lower crustal xenoliths from Lesotho which have REE concentrations similar to basaltic liquids (Roberts & Hawkesworth 1979); i.e. these xenoliths do not represent cumulates or residual material. Table 2 lists the mineral proportions and partition coefficients used in the calculations. The modal proportions of garnet in these two samples span the range for all Lesotho xenoliths reported by Griffin *et al.* (1978), hence, are considered representative of mafic granulites in general. Fig. 3b shows the range in calculated melt compositions produced by varying the parental modes and partition coefficients. Both $(La/Yb)_N$ (22–99) and $(Gd/Yb)_N$ (1.7–9) for these melts overlap most of the range exhibited by Archaean tonalites and trondhjemites (Fig. 4); only the very low $(La/Yb)_N$ ratios (less than 20) are not capable of being produced by melting of this rock type. Like the partial melts of amphibolite, these melt compositions also exhibit marked positive Eu anomalies, a feature common, but not ubiquitous, in Archaean tonalites and trondhjemites (Fig. 2). Therefore, a garnet granulite source with chondritic REE pattern is capable of producing tonalites with all but the smallest degree of LREE fractionation. If the model source is chosen to be LREE enriched, the calculated melt $(La/Yb)_N$ ratios would be even higher, producing less overlap with observed ratios in Archaean tonalites. Alternatively, the model source could be LREE depleted, similar to modern MORB (there is isotopic evidence that at least some Archaean basalts were derived from a depleted mantle source (Carlson *et al.* 1983, and there are some LREE depleted Archaean basalts; Sun & Nesbitt 1978; Sun 1984)). This would cause the $(La/Yb)_N$ ratio of the melt to be lower, producing a total overlap with the observed $(La/Yb)_N$ ratios for Archaean tonalites. Clearly, the $(La/Yb)_N$ ratios of the model sources can be varied to produce a non-unique answer. An important difference between garnet granulite and amphibolite source rocks is the ability of the garnet granulite source to produce higher $(Gd/Yb)_N$ ratios. A garnet granulite source having flat to slightly enriched HREE patterns would be capable of producing the range in HREE fractionation observed in the tonalites, unlike the amphibolite source.

Eclogite source

Apted (1981) reviewed the partial melting behaviour of quartz eclogites and concluded that by 20–30% partial melting, phases such as quartz and kyanite would be completely melted out. He also noted that since both garnet and clinopyroxene are on the liquidus throughout the cooling interval in basaltic melts, partial melting of eclogite can be modelled by modal melting. Using the range of modal proportions of garnet and clinopyroxene cited by Apted (1981) as being representative of natural eclogites, REE contents of equilibrium melts were calculated. The parameters used in the partial melting calculations are listed in Table 2. Fig. 3c shows the range in REE patterns for 20% melting of eclogite with 20 times chondritic REE pattern, varying parental modes and partition coefficients. The range in $(La/Yb)_N$ (13–64) and $(Gd/Yb)_N$ (1.6–7.3) ratios produced by melting of eclogite overlaps the majority of those observed in Archaean tonalites and trondhjemites (Fig. 4). Again, variable positive Eu anomalies are produced. As with the garnet granulite source, the REE pattern of the eclogite source can be changed, with LREE either enriched or depleted relative to HREE to produce similar changes in the melt. Therefore, based on the $(La/Yb)_N$ and $(Gd/Yb)_N$ ratios, eclogite is an equally viable source for Archaean tonalites and trondhjemites.

Partial melting models for the three commonly suggested source rocks for Archaean tonalite-trondhjemites show that garnet-free amphibolite is not a likely source. Garnet-bearing amphibolite, garnet granulite or eclogite remain possible source rocks. Attempts to distinguish between a crustal (garnet amphibolite or garnet granulite) versus mantle (eclogite) source based on the presence of plagioclase, and the effect it would have on Sr and Eu concentrations, have not been successful. The most important conclusion from these calculations is that the dominance of tonalitic rocks with highly fractionated REE in Archaean high-grade terrains requires large volumes of mafic material at depth in the Archaean.

Additional constraints

Petrologic and isotopic data are also useful in defining the depth of melting required to produce tonalitic and trondhjemitic melts. Green (1982) reviewed experimental data on synthetic and natural basaltic compositions and used these to define the depth of melting for modern andesites. He showed that the composition of the phases present during partial melting is strongly dependent on water content. For anhydrous tholeiitic compositions, garnet is stable only at pressures greater than 17 kbars and plagioclase is stable only below 17 kbars, at temperatures 50°C above the solidus. For water undersaturated conditions (5 wt. % H_2O), garnet is stable above 8 kbars, plagioclase does not exist above 15 kbars and amphibole is stable below about 20 kbars at solidus temperatures. Therefore, this implies a minimum melting depth from 25 to 66 km for a garnet bearing basaltic rock (depending upon the presence of water). Thus the source region for tonalite-trondhjemite rocks is confined to the lower crust or upper mantle. The mineralogy of the source material could be granulitic (cpx-gar-plag, at melting depths between 41–82 km) to eclogitic (cpx-gar ± qtz + ky, for melting depths greater than 82 km) for dry conditions, or amphibolitic (amph-cpx-gar-plag or gar-amph-cpx-zo, at depths between 25–50 km) to eclogitic (gar-cpx-ky, at depths greater than 50 km) for water undersaturated conditions.

Isotope studies have been used to establish crustal residence ages and the relative contributions of crust and mantle components to tonalites. Archaean tonalites commonly have low, mantle-like initial $^{87}Sr/^{86}Sr$ ratios (O'Nions & Pankhurst 1978; Peterman 1979). In addition, Sm-Nd model ages (age of LREE enrichment, McCulloch & Wasserburg 1978) are often nearly identical to the crystallization age of Archaean tonalites (e.g. Carlson *et al.* 1983; Martin *et al.* 1983; Jahn & Zhang 1984). These features are thought to reflect derivation of tonalites from mafic rocks with short crustal residence times. An exception to this are some Finnish tonalites which show a long (200–500 Ma) crustal residence time and are interpreted to be derived from an earlier, more differentiated crustal source (Jahn *et al.* 1984). The similarities between crystallization and Nd model ages in most Archaean tonalites suggest the existence of a short-lived mafic crust, perhaps analogous to present-day oceanic crust, which is rapidly recycled by some sort of subduction-like process where it could be transformed to eclogite and then partially melted to produce Archaean tonalites. However, the above interpretations of the short crustal residence times assume

that the mafic precursor to the tonalites was LREE enriched. If the mafic parent had flat or depleted LREE (which is plausible considering the possibility of a depleted mantle during the Archaean (Carlson *et al.* 1983)), then the close coherence between Nd model ages and crystallization ages for tonalites would not reflect a short crustal residence age for the mafic parent, but would instead reflect the age of the melting event that created the tonalite. In these circumstances it is impossible to define how long the mafic parent existed before it partially melted; it may have existed for long periods of time in a crustal environment (presumably continental, if an oceanic crust is likely to be rapidly recycled). The isotope data confirm the likelihood of a mafic parent for most Archaean tonalites, but do not distinguish between a crustal or mantle depth of melting.

Implications for the composition of the Archaean lower crust

The trace element models presented here suggest that the most reasonable source for Archaean tonalite-trondhjemite rocks is a garnet amphibolite, garnet granulite or eclogite. Phase stability studies require this mafic source to be deep-seated, either within the lower crust or upper mantle. The exact depth of this mafic source has important implications for Archaean tectonics and crustal growth.

If the mafic source of tonalites lies within the lower crust, then the Archaean lower crust is dominantly mafic, and this may reconcile the apparent discrepancy between lower crustal compositions as recorded by xenoliths and granulite terrains. This model requires substantial thicknesses of residual mafic or ultramafic material to lie beneath many tonalite-dominated Archaean granulite terrains presently exposed at the earth's surface. Alternatively, the residue may be so dense that it will decouple from the overlying crust and remain at the crust–mantle interface during tectonic uplift, or possibly founder into the mantle (Sun 1984). Such high density material will possess high P-wave velocities and may be interpreted as mantle material during geophysical modelling (NB the debate over the location of the crust–mantle transition zone in southeast Australia (Griffin & O'Reilly, this volume)).

If the mafic source of Archaean tonalites lies within the mantle, then some form of subduction is required to get large volumes of mafic material to mantle depths. This model predicts that the Archaean lower crust is dominated by tonalitic rocks intruded from below, requiring melting of significant volumes (15–20%) of the subducted mafic crust, and little to no interaction between this silicic melt and the overlying mantle wedge. Production of high silica melts by partial melting of the oceanic slab is envisioned for modern arcs in some circumstances (Wyllie 1982), but these melts are generally thought to react with the overlying mantle to produce more intermediate melts. It is not clear why this process would not also occur in Archaean subduction zones. Until more direct evidence is gained regarding the presence or absence of subduction during the Archaean, these conflicting models cannot be reconciled.

ACKNOWLEDGMENTS: We thank W.F. McDonough, S.M. McLennan and S-S. Sun for fruitful discussions on the problem of the origin of Archaean tonalites and trondhjemites. We also thank L.P. Black, W.L. Griffin, R.D. Holmes, J.W. Sheraton, S-S. Sun and K.H. Wedepohl for constructive reviews. R.L.R. was partially supported through a graduate fellowship from the National Science Foundation, U.S.A.

References

APTED, M.J. 1981. Rare earth element systematics of hydrous liquids from partial melting of basaltic eclogite: a re-evaluation. *Earth and Planetary Science Letters*, **52**, 172–182.

ARCULUS, R.J., FERGUSON, J. *et al.* 1986. Eclogites and granulites in the lower continental crust: examples from eastern Australia and southwestern U.S.A. *Proc. Eclogite Symposium* (in press).

—— & SMITH, D. 1979. Eclogite, pyroxenite and amphibolite inclusions in the Sullivan Buttes latite, Chino Valley, Yavapai County, Arizona. *In*: BOYD, F.R. & MEYER, H.O.A. (eds) *The Mantle Sample: Inclusions in Kimberlites and Other Volcanics*, Am. Geophys. Union, 309–317.

ARTH, J.G. & BARKER, F. 1976. Rare-earth partitioning between hornblende and dacitic liquid and implications for the genesis of trondhjemitic-tonalitic magmas. *Geology*, **4**, 534–536.

——, ——, PETERMAN, Z.E. & FRIEDMAN, I. 1978. Geochemistry of the gabbro-diorite-tonalite-trondhjemite suite of southwest Finland and its implications for the origin of tonalitic and trondhjemitic magmas. *J. Petrol.* **19**, 289–316.

—— & HANSON, G.N. 1975. Geochemistry and origin of the early Precambrian crust of northeastern Minnesota. *Geochim. et Cosmochim. Acta*, **39**, 325–362.

BARKER, F. 1979. Trondhjemite: definition, environment and hypothesis of origin. *In*: BARKER, F. (ed.) *Trondhjemites, Dacites and Related Rocks*, Elsevier, 1–12.

CARLSON, R.W., HUNTER, D.R. & BARKER, F. 1983. Sm–Nd age and isotopic systematics of the bimodal suite, Ancient Gneiss Complex, Swaziland. *Nature*, **305**, 701–704.

COMPTON, P. 1978. Rare earth evidence for the origin of the Nuk Gneisses, Buksefjorden region, southern west Greenland. *Contrib. Mineral. Petrol.* **66**, 283–293.

CONDIE, K.C., ALLEN, P. & NARAYANA, B.L. 1982. Geochemistry of the Archean low- to high-grade transition zone, southern India. *Contrib. Mineral. Petrol.* **81**, 157–167.

—— & HUNTER, D.R. 1976. Trace element geochemistry of Archean granitic rocks from the Bar-berton region, South Africa. *Earth and Planetary Sci. Letters*, **29**, 389–400.

DEER, W.A., HOWIE, R.A. & ZUSSMAN, J. 1982. *Orthosilicates*. Second Ed., Longman. 919 pp.

FRANCIS, D.M. 1976. Corona-bearing pyroxene granulite xenoliths and the lower crust beneath Nunivak island, Alaska. *Can. Mineral.* **14**, 291–298.

FUJIMAKI, H., TATSUMOTO, M. & AOKI, K-I. 1984. Partition coefficients of Hf, Zr, and REE between phenocrysts and groundmasses. Proc. Fourteenth Lunar and Planet. Sci. Conf., Part 2. *Jour. Geophys. Res.* **89**, supplement, B662–B672.

GOWER, C.F., PAUL, D.K. & CROCKET, J.H. 1982. Protoliths and petrogenesis of Archean gneisses from the Kenora area, English River subprovince, Northwest Ontaria. *Precambrian Res.* **17**, 245–274.

GREEN, T.H. 1982. Anatexis of mafic crust and high pressure crystallization of andesite. *In*: THORPE, R.S. (ed.) *Andesites: Orogenic Andesites and Related Rocks*, John Wiley & Sons, 465–487.

GRIFFIN, W.L., CARSWELL, D.A. & NIXON, P.H. 1979. Lower-crustal granulites and eclogites from Lesotho, southern Africa. *In*: BOYD, F.R. & MEYER, H.O.A (eds) *The Mantle Sample: Inclusions in Kimberlites and Other Volcanics*. Am. Geophys. Union, 59–86.

—— & O'REILLY, S.Y. 1984. The lower crust in eastern Australia: xenolith evidence. This volume.

HANSON, G.N. 1978. The application of trace elements to the petrogenesis of igneous rocks of granitic composition. *Earth and Planetary Science Letters*, **38**, 26–43.

—— 1980. Rare earth elements in petrogenetic studies of igneous systems. *Ann. Rev. Earth Planet. Sci.* **8**, 371–406.

HELZ, R.T. 1976. Phase relations of basalts in their melting ranges at $P_{H_2O} = 5$ kb. Part II. Melt compositions. *J. Petrol.* **17**, 139–193.

HUNTER, D.R., BARKER, F. & MILLARD, H.T., Jr. 1978. The geochemical nature of the Archean Ancient Gneiss Complex and granodiorite suite, Swaziland: a preliminary study. *Precambrian Res.* **7**, 105–127.

——, —— & —— 1984. Geochemical investigation of Archaean bimodal and Dwalile metamorphic suites,

Ancient Gneiss Complex, Swaziland. *Precambrian Res.* **24**, 131–155.

IRVING, A.J. & FREY, F.A. 1978. Distribution of trace elements between garnet megacrysts and host volcanic liquids of kimberlitic to rhyolitic composition. *Geochim. et Cosmochim. Acta*, **42**, 771–787.

JAHN, B., GLIKSON, A.Y., PEUCAT, J.J. & HICKMAN, A.H. 1981. REE geochemistry and isotopic data of Archean silicic volcanics and granitoids from the Pilbara block, western Australia: implications for the early crustal evolution. *Geochim. et Cosmochim. Acta*, **45**, 1633–1652.

——, VIDAL, P. & KRÖNER, A. 1984. Multi-chronometric ages and origin of Archaean tonalitic gneisses in Finnish Lapland: a case for long crustal residence time. *Contrib. Mineral. Petrol.* **86**, 398–408.

—— & ZHANG, Z. 1984. Archean granulite gneisses from eastern Hebei Province, China: rare earth geochemistry and tectonic implications. *Contrib. Mineral. Petrol.* **85**, 224–243.

JANARDHAN, A.S., NEWTON, R.C. & HANSEN, E.C. 1982. The transformation of amphibolite facies gneiss to charnockite in southern Karnataka and northern Tamil Nadu, India. *Contrib. Mineral. Petrol.* **79**, 130–149.

KAY, R.W. & KAY, S.M. 1981. The nature of the lower continental crust: inferences from geophysics, surface geology and crustal xenoliths. *Rev. Geophys. Space Phys.* **19**, 271–297.

LEYRELOUP, A., DUPUY, C. & ANDRIAMBOLOLONA, R. 1977. Catazonal xenoliths in French Neogene volcanic rocks: constitution of the lower crust. *Contrib. Mineral. Petrol.* **62**, 283–300.

MARTIN, H., CHAUVEL, C. & JAHN, B. 1983. Major and trace element geochemistry and crustal evolution of Archaean granodioritic rocks from eastern Finland. *Precambrian Res.* **21**, 159–180.

MCCULLOCH, M.T. & WASSERBURG, G.J. 1978. Sm-Nd and Rb-Sr chronology of continental crust formation. *Science*, **200**, 1003–1011.

MÖLLER, P. & MUECKE, G.K. 1984. Significance of europium anomalies in silicate melts and crystal-melt equilibria: a re-evaluation. *Contrib. Mineral. Petrol.* **87**, 242–250.

MORGAN, P. 1984. The thermal structure and thermal evolution of the continental lithosphere. *Phys. Chem. Earth*, **15**, 107–193.

O'NIONS, R.K. & PANKHURST, R.J. 1978. Early Archaean rocks and geochemical evolution of the earth's crust. *Earth and Planetary Sci. Letters*, **38**, 211–236.

PADOVANI, E.R. & CARTER, J.L. 1977. Aspects of the deep crustal evolution beneath south central New Mexico. *In*: REACOCK, J.G. (ed.) *The Earth's Crust*. Am. Geophys. Union, 19–55.

PERCIVAL, J.A. 1983. High-grade metamorphism in the Chapleau-Foleyet area, Ontario. *Am. Mineral.* **68**, 667–686.

—— & CARD, K.D. 1983. Archean crust as revealed in the Kapuskasing uplift, Superior province, Canada. *Geology*, **11**, 323–326.

—— & KROGH, T.E. 1983. U-Pb zircon geochronology of the Kapuskasing structural zone and vicinity in

the Chapleau-Foleyet area, Ontario. *Can. J. Earth Sci.* **20**, 830–843.

PETERMAN, Z.E. 1979. Strontium isotope geochemistry of late Archean to late Cretaceous tonalites and trondhjemites. *In*: BARKER, F. (ed.) *Trondhjemites, Dacites and Related Rocks*. Elsevier, 133–147.

PHILPOTTS, J.A. & SCHNETZLER, C.C. 1970. Phenocryst-matrix partition coefficients for K, Rb, Sr and Ba, with applications to anorthosite and basalt genesis. *Geochim. Cosmochim. Acta*, **311**, 307–322.

PRIDE, C. & MUECKE, G.K. 1980. Rare earth element geochemistry of the Scourian Complex N.W. Scotland—evidence for the granite-granulite link. *Contrib. Mineral. Petrol.* **73**, 403–412.

REED, S.J.B. & WARE, N.G. 1973. Quantitative electron microprobe analysis using a lithium drifted silicon detector. *X-Ray Spectrom.* **2**, 69–74.

ROGERS, N.W. & HAWKESWORTH, C.J. 1982. Proterozoic age and cumulate origin for granulite xenoliths, Lesotho. *Nature*, **299**, 409–413.

RUDNICK, R.L., ASHWAL, L.D. & HENRY, D.J. 1984. Fluid inclusions in high-grade gneisses of the Kapuskasing Structural Zone, Ontario: metamorphic fluids and uplift/erosion path. *Contrib. Mineral. Petrol.* **87**, 399–406.

SELVERSTONE, J. 1982. Fluid inclusions as petrogenetic indicators in granulite xenoliths, Pali-Aike volcanic field, Chile. *Contrib. Mineral. Petrol.* **79**, 28–36.

SHERATON, J.W., SKINNER, A.C. & TARNEY, J. 1973. Geochemistry of the Scourian gneisses of Assynt district. *In*: PARK, R.G. & TARNEY, J. (eds) *The Early Precambrian of Scotland and Related Rocks of Greenland*. University of Keele, 31–43.

—— & BLACK, L.P. 1983. Geochemistry of Precambrian gneisses: relevance for the evolution of the east Antarctic shield. *Lithos*, **16**, 273–296.

—— & COLLERSON, K.D. 1984. Geochemical evolution of Archaean granulite-facies gneisses in the Vestfold block and comparisons with other Archaean gneiss complexes in the east Antarctic shield. *Contrib. Mineral. Petrol.* **87**, 51–64.

SUN, S-S. 1984. Geochemical characteristics of Archaean ultramafic and mafic volcanic rocks: implications for mantle composition and evolution. *In*: KRÖNER, A.F., HANSON, G.N. & GOODWIN, A.M. (eds) *Archaean Geochemistry*. Springer-Verlag, Berlin, 25–46.

—— & NESBITT, R.W. 1978. Petrogenesis of Archaean ultrabasic and basic volcanics: evidence from rare earth elements. *Contrib. Mineral. Petrol.* **65**, 301–325.

TAYLOR, S.R. & GORTON, M.P. 1977. Geochemical application of spark source mass spectrography—III. Element sensitivity, precision and accuracy. *Geochim. et Cosmochim. Acta*, **41**, 1375–1380.

—— & MCLENNAN, S.M. 1985. *The Continental Crust: Its Composition and Evolution*. Blackwell Scientific, Oxford. 312 pp.

UPTON, B.J.G., ASPEN, P. & CHAPMAN, N.A. 1983. The upper mantle and deep crust beneath the British Isles: evidence from inclusons in volcanic rocks. *Q. J. geol. Soc. London*, **140**, 105–121.

WEAVER, B.L. & TARNEY, J. 1980. Rare earth geo-

chemistry of Lewisian granulite-facies gneisses, northwest Scotland: implications for the petrogenesis of the Archaean lower continental crust. *Earth and Planetary Sci. Letters*, **51**, 279–296.
—— & —— 1981. Lewisian gneiss geochemistry and Archaean crustal development models. *Earth and Planet. Sci. Letters*, **55**, 171–180.

WYLLIE, P.J. 1982. Subduction products according to experimental prediction. *Geol. Soc. Am. Bull.* **93**, 468–476.

R.L. RUDNICK & S.R. TAYLOR, Research School of Earth Sciences, Australian National University, Canberra, A.C.T., Australia. 2600.

Eclogite facies metamorphism in the lower continental crust

D.A. Carswell & S.J. Cuthbert

SUMMARY: Occurrences of medium-temperature eclogite-facies assemblages in the Norwegian Caledonides, Western Alps and Central European Variscides are reviewed with reference to a generalized tectono-thermal model for their stabilization in continental plate collision zones involving A-type subduction. It is demonstrated that the respective pressure-temperature-time paths for lower continental crust and uppermost mantle during such orogenesis contrast with that for upper crust, and that mineralogical features, such as chemical zoning in garnets, may monitor the tectonic location and thermal evolution of different crustal levels in the resultant thrust-nappe stack. Phanerozoic orogenic belts of this type typically show evidence of the generation of early eclogite-facies assemblages subjected to partial greenschist- or amphibolite-facies overprint some 40–60 Ma later. The survival and ultimate surface exposure of the high-pressure assemblages requires that their exhumation is rapid (mean uplift rate around 1 mm/year) relative to the rate of thermal relaxation in the tectonically thickened continental crust.

Introduction

The minimum lithostatic pressures required for the stability of eclogite-facies rocks (Fig. 1) are best experimentally defined for quartz-normative metabasaltic rocks, with the plagioclase free essentially bimineralic assemblages of pyralspite garnet + omphacitic clinopyroxene diagnostic of eclogites (*sensu stricto*), and for garnetiferous meta-peridotites. Whilst a detailed discussion of the definition of this metamorphic facies is perhaps not appropriate here, it is pertinent to note that in some felsic rocks albitic plagioclase may still be stable relative to jadeite + quartz (Fig. 1) under eclogite-facies P_{LOAD}-T conditions where omphacitic pyroxene and grossular bearing garnet are stable relative to plagioclase in meta-basaltic rocks. Accordingly mineral assemblages of quartz + K feldspar + plagioclase + garnet + orthopyroxene ± clinopyroxene ± kyanite in quartzo-feldspathic gneisses customarily designated as high-P granulites may in fact be cofacial with eclogites, as also may be garnet + phengite + quartz ± kyanite ± zoisite in metapelites and jadeite + garnet + zoisite + K feldspar + quartz in metagranitoids (Compagnoni 1977; Heinrich 1982).

Eclogites are known to have formed in a variety of geological environments, which are reflected in their varying modes of occurrence and mineral chemistry (Eskola 1921; Coleman *et al.* 1965; Smulikowski 1960, 1964, 1968; and Banno 1970). In this contribution we are primarily concerned with those found as layers or lenses in terrains of dominant amphibolite- or granulite-facies continental crust (broadly forming 'Group B' of Coleman *et al.* 1965, and the 'Common Eclogites' of Smulikowski 1968). Mineral thermometers such as that based on $K^{Gnt-Cpx}_{Fe^{2+}-Mg^{2+}}$ partition coefficients, indicate that these eclogites have generally equilibrated at temperatures between 500–900°C (e.g. Carswell *et al.* 1985; Griffin *et al.* 1985) higher than those deduced for eclogites in blueschist-facies terrains (e.g. Brown & Bradshaw 1979) and lower than those for mantle derived xenoliths in alkalic eruptive rocks (e.g. Carswell *et al.* 1981). We consider it appropriate to refer to such rocks as 'medium-T eclogites'.

If Ringwood's (1975) linear (dP/dT = 20 bar/°C) extrapolation of the high-T experimental reaction boundary for plagioclase elimination in a quartz tholeiite composition (Fig. 1) is valid, then eclogite should be stable towards the base of continental crust of average 35 km thickness at temperatures less than about 600°C. Such lower crustal thermal conditions are to be expected in stable continental regions with observed surface heat flows < 65 mW m^{-2} (Chapman, this volume). On the other hand, petrological studies of exposed lower crust gneiss terrains and of lower crustal derived xenolith suites, as well as deep seismic investigations, are generally taken to indicate (as exemplified by several other papers in this volume) that granulite-facies assemblages are dominant in the lower continental crust. In tectonically inactive regions, such lower crustal granulites, most likely to have formed in response to an enhanced thermal flux during some previous tectono-thermal event, may customarily be preserved metastably under P_{LOAD}-T conditions corresponding to the eclogite facies. During establishment of the expected ambient thermal conditions, transformation of lower continental crust rocks to eclogite-facies assemblages along a cooling path under stress-free conditions may well be effectively blocked for kinetic reasons.

From: DAWSON, J.B., CARSWELL, D.A., HALL, J. & WEDEPOHL, K.H. (eds) 1986, *The Nature of the Lower Continental Crust*, Geological Society Special Publication No. 24, pp. 193–209.

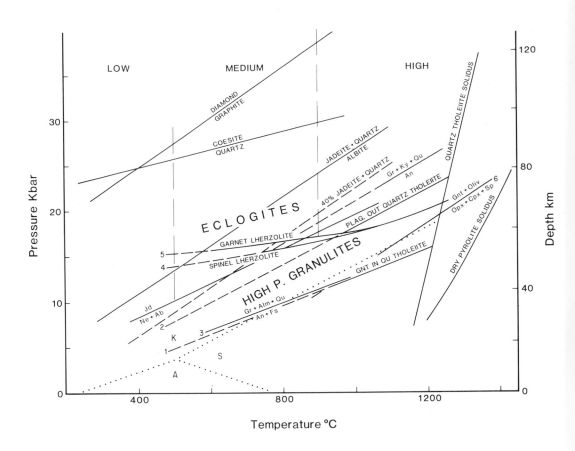

Fɪɢ 1. Pressure (depth)-temperature diagram showing the stability fields for low, medium and high temperature eclogite-facies assemblages relative to the following experimentally determined reaction equilibria: Garnet-in (1) and plagioclase-out (2) reaction curves and extrapolations in quartz tholeiite compositions (Green & Ringwood 1967a; Ringwood 1975). Calculated reaction curve (3) for 3 anorthite + 3 ferrosilite = grossular + 2 almandine + 3 quartz from Bohlen *et al.* (1983); 3 anorthite = grossular + 2 kyanite + quartz: Goldsmith (1980); albite = jadeite + quartz: Holland (1980). Estimated stability curve for 40% jadeite + quartz: Kushiro (1969); nepheline + albite = 2 jadeite: Robertson *et al.* (1957). Univariant reaction curve (4) for spinel lherzolite ⇌ garnet lherzolite reaction in natural garnet peridotite composition: O'Hara (1975). Curve (5) for the same reaction in more magnesian compositions: O'Hara *et al.* (1971). Reaction curve (6) for pyroxenes + spinel = garnet + olivine in the CMAS system: MacGregor (1970). Quartz ⇌ coesite: Mirwald & Massonne (1980); graphite ⇌ diamond: Berman (1979). Stability fields for alumino-silicates (K-Kyanite; S-sillimanite; A-andalusite: Holdaway (1971). Solidus for quartz tholeiite composition (Green & Ringwood 1967a). Dry pyrolite (periodotite) solidus (Green & Ringwood, 1967b). The low pressure boundary conditions for the eclogite-facies are best defined by the reaction curves for spinel lherzolite ⇌ garnet lherzolite and for plagioclase disappearance in quartz tholeiite basalt compositions.

Current exposure of medium-T eclogites within continental crust bears witness to special tectonic processes responsible both for their stabilization and their survival during transport to the surface. In this review we highlight certain features common to occurrences of medium-T eclogites and cofacial rocks in the Norwegian Caledonides, the western Alps and the Central European Variscides. We attempt to deduce their formation within the framework of a generalized model for the tectono-thermal behaviour of continental collision zones.

A tectono-thermal model for eclogite formation in continental crust

Here we examine the general thermal and metamorphic consequences of crustal thickening in a collision zone involving two continental lithospheric plates. A simplistic model showing underthrusting (A-type subduction; Hodges *et al.* 1982) of one continental plate beneath the other and detachment around the Moho in the overthrust plate is illustrated in Fig. 2. In reality an intensely imbricated thrust-nappe stack is likely to evolve in response to prolonged crustal shortening (compression) but the contrasted thermal effects and resultant metamorphic responses in the lower and upper plates can be expected to persist. In the upper plate overthrusting (nappe translation) will bring deep level warmer rocks over cooler rocks, whereas in the lower plate progressive underthrusting (with thrusts stepping down with time) will place original high level colder rocks beneath warmer rocks. The main thermal contrast is still, however, likely to be across the principal plate suture.

Tectonic processes can be expected to operate rapidly compared with the rate at which thermal surfaces re-equilibrate by thermal conduction after disturbance, as emphasized by thermal model calculations (Oxburgh & Turcotte 1974; England & Thompson 1984). Thus the perturbed P–T (depth) profile (Fig. 2 c–d) resulting from the tectonic thickening of continental crust will only gradually decay over a time scale of at least some tens of millions of years. The actual P-T-t (time) paths followed by the rocks will in fact depend on the balance between the rates of resultant thermal relaxation and of surface erosion (exhumation). Whilst the metamorphic temperatures (T_{max}) attained are likely to be largely governed by thermal relaxation towards a steady state geotherm for thickened crust, the maximum pressures recorded by the rocks will be critically dependent upon the erosion rate (England & Thompson 1984).

The predicted forms of the P-T-t paths for original cool upper crustal rocks in the upper part of the underthrust plate and for warmer lower crustal rocks sited in the lower part of the overthrust plate are illustrated in Figs 2e and 2f, respectively. In the former case, the rocks may be expected to enter the eclogite-facies stability field along a prograde (heating) path, whilst in the second case eclogite-facies assemblages may be stabilized along a cooling path. Similarly, any cool oceanic crustal rocks (ophiolites) caught up in the suture zone may be expected to follow a prograde metamorphic path subsequent to collision, whereas any tectonically intercalated subcontinental upper mantle rocks would follow cooling paths. This may help to discriminate between the suggested alternative origins of metaperidotite bodies in such orogenic belts. Contrasted P-T-t paths for the metamorphic evolution of three possible eclogite-facies samples generated in a continental plate collision zone are illustrated in Fig. 3, with reference to a time sequence (a–e) of lithospheric cross-sections.

We now proceed to review the tectonic setting and interpretation of various European occurrences of medium-T eclogites in the light of this generalized model. In particular we examine evidence which might indicate that different rocks have indeed generated their eclogite-facies assemblages along contrasted P-T-t paths, depending on their pre- and post-collision locations in an evolving plate collision zone. In this connection, a consideration of chemical zoning profiles in minerals, notably garnets, is especially important (Spear *et al.* 1984).

High-P metamorphism in the Western Gneiss region, Norwegian Caledonides

Medium-T eclogites occur sporadically over a substantial tract of the Scandinavian Caledonides (see Fig. 1 in Bryhni *et al.* 1977) but are particularly in evidence in the Western (also called Basal) Gneiss Region of southern Norway. Here metabasic eclogite lenses are widely preserved within dominantly amphibolite-facies quartzo-feldspathic gneisses whilst cofacial garnet lherzolite and garnet websterite assemblages and more rarely also eclogites (*sensu stricto*) occur within 'alpine type' peridotite bodies. This region represents a deeply eroded core zone of the Scandinavian Caledonides which structurally underlies a thick pile of Caledonian nappes which now outcrop mostly to the east and south (see, for example, Roberts *et al.* 1981, and Figs. 1 & 2 in Cuthbert *et al.* 1983).

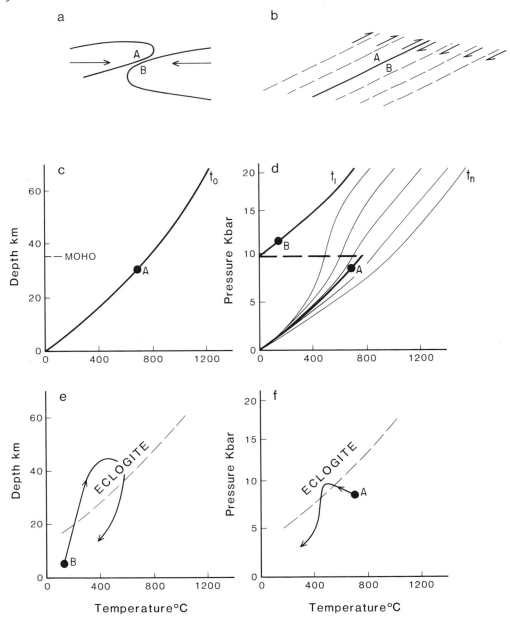

FIG. 2. Tectono-thermal model for the generation of eclogite-facies assemblages in continental crust. (a) A-type subduction as a result of collision between two continental plates. (b) Evolution of thrust-nappe stack in tectonically thickened continental crust. (c) Assumed pre-collisional (time = t_0) 'steady state' continental geotherm of 80 m W/m^2 (Chapman, this volume), appropriate for a partially cooled previous extensional regime with granulite-facies assemblages dominant in the lower continental crust. (d) Perturbed geotherm at t_1 immediately following collision and overthrusting of continental crust slab with assumed detachment at the Moho. Thermal relaxation over some tens of millions of years would eventually lead to a new 'steady state' geotherm at t_n appropriate for the thickened continental crust. (e) Prograde metamorphic path in P-T space leading to the stabilization of eclogite-facies assemblages in original upper crustal rocks (B) of the underthrusting plate. Note that P_{max} is reached before T_{max} and the initial, near isothermal, decompression during exhumation. (f) P-T-t path leading to possible generation on cooling of eclogite-facies assemblages from original granulite-facies lower crustal rocks (A) near the sole of the overthrust slab. An initial (perhaps exaggerated) pressure increase due to thrust-nappe stacking in the overthrust slab is assumed but is not critical to the model. Any tectonically intercalated slices of sub-continental mantle peridotite may be expected to follow a similar path.

K-Ar, Rb-Sr, U-Pb and Sm-Nd mineral ages for various lithologies (including eclogites) mostly fall between 380–450 Ma (e.g. Krogh *et al.* 1973; Gebauer *et al.* 1982; Griffin & Brueckner 1980, 1982; Mearns & Lappin 1982) and confirm the profound influence of the Caledonian orogeny on the rocks of this region. Field differentiation between reworked Precambrian basement and metamorphosed Late Precambrian ('Eocambrian')-Lower Palaeozoic cover rocks has proved extremely difficult (Bryhni 1966; Bryhni & Grimstad 1970; Carswell 1973; Strand 1960) but it seems that the Western Gneiss Region comprises an interfolded and intensely imbricated sequence of both autochthonous and allochthonous cover rocks and pre-Caledonian basement. This is corroborated by recent work in the east of the region (Krill 1985).

Rb-Sr whole rock isochrons and U-Pb zircon 'upper' concordia intersection ages as well as geochemical features (Harvey 1983; Krill 1983; Lappin *et al.* 1979) indicate that substantial volumes of acid-intermediate gneisses within this region represent recrystallized mid-Proterozoic igneous rocks. These and possible genetically related, spatially associated rocks such as anorthosites, titaniferous gabbros-norites and peridotites (Carswell *et al.* 1983; Harvey 1983) appear to have been intruded into the lower continental crust at around 1500 Ma.

High-P/medium-T eclogite facies assemblages are best preserved in the western coastal parts of the region and are generally more thoroughly retrogressed further east. Such high-P relics are most conspicuously preserved in rocks of basic or ultrabasic composition (eclogites and garnetiferous peridotites, respectively) which are typically mantled by shells of later amphibolite. For the most part the dominant acid-intermediate gneisses have biotite and hornblende-bearing amphibolite-facies assemblages but locally almandine rich garnet and clinopyroxenes are preserved in less deformed rocks (Bryhni 1966; Carswell & Harvey 1985; Krill 1983; Lappin *et al.* 1979) indicating the earlier stability of the high-pressure assemblage: perthitic orthoclase + plagioclase + quartz + garnet + clinopyroxene. Scarce pelitic rocks frequently contain relict garnet and kyanite as well as symplektitic intergrowths of biotite and feldspars indicating that the high-P mica phengite may have been initially stable (cf. Heinrich 1982).

Whilst most eclogite lenses have sheared and amphibolitized margins against the encompassing gneisses, a few preserve intrusive contact relationships (Cuthbert & Carswell 1982; Griffin & Carswell 1985) which indicate that they originally formed dykes or larger bodies within the plutonic igneous precursors of the quartzo-feldspathic gneisses. Moreover, some of the large bodies have retained zones of original igneous textured doleritic-gabbroic rocks (Cuthbert & Carswell 1982; Gjelsvik 1952; Griffin & Råheim 1973; Mørk 1982). Indeed it is possible to observe all stages of conversion from the original igneous protoliths through rocks with coronitic development of the high-P garnet and omphacitic clinopyroxene replacing the original low P mineral phases, to thoroughly recrystallized granoblastic textured eclogites.

There is, therefore a growing volume of evidence which suggests that at least a substantial proportion of the exposed Western Gneiss Region witnessed the high-P metamorphism responsible for eclogite formation. It appears that evidence for this 'early' eclogite-facies metamorphic event is generally only retained in the more massive and structurally competent rock masses which have escaped the extensive late deformation and concurrent recrystallization to amphibolite-facies assemblages. Recent evidence provided by Sm-Nd and U-Pb data (Gebauer *et al*; Griffin & Brueckner 1980, 1982; Mearns & Lappin 1982) indicates a Caledonian (400–450 Ma) age for eclogite formation in western Norway. Such observations and the recognition of the development of the high-P eclogite-facies assemblages from a variety of crustal protoliths contradicts certain previous interpretations (O'Hara 1975; Lappin & Smith 1978; Smith 1980, 1981, 1982) that the Norwegian medium-T eclogites invariably represent foreign 'bodies' of deep seated, mostly mantle, origin tectonically emplaced into lower grade continental crust gneisses late in the Caledonian orogenic cycle. An ultimate upper mantle origin does, however, seem likely from general petrological and geochemical considerations for the forsteritic olivine-bearing 'alpine-type' peridotite bodies which occasionally retain high-P garnet lherzolite, garnet websterite and eclogite assemblages (O'Hara & Mercy 1963; Carswell & Gibb 1980; Carswell *et al.* 1983; Medaris 1980, 1984).

Application of mineral thermometers and barometers to the high-P (eclogite-facies) metamorphic relics in the Western Gneiss Region indicate (Krogh 1977; Griffin *et al.* 1985) a regional variation in the maximum T–P conditions attained, which varies from around 775°C and 22 kbar in the coastal regions of Møre and Romsdal to around 550°C and 12.5 kbar further south in Sunnfjord (Carswell 1981; Carswell & Gibb 1980; Carswell *et al.* 1984; Giffin *et al.* 1985; Krogh 1977, 1980a and b).

We have previously presented a tectonic model relating the features described above to northwes-

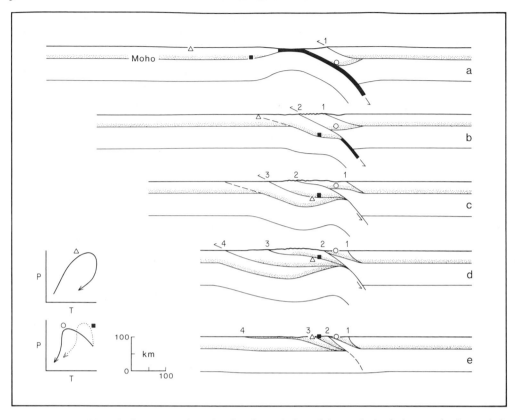

FIG. 3. Schematic lithospheric cross-sections showing the evolution of three eclogite-facies samples (open circle, closed square, triangle) in a continental collision-type orogen. Inset graphs show path of each sample in P-T space (compare with examples in text and Figs. 2 and 4). Lower crust is stippled, oceanic crust black. Numbers indicate successive major thrusts stepping back into the underthrusting plate. Accretion/obduction of oceanic crust (ophiolites) is omitted for clarity.

terly subduction of the margin of the Baltic continental plate beneath the leading edge of the Greenland continental plate on closure of Iapetus Ocean during the Caledonian orogeny (Cuthbert *et al.* 1983). The expected initial thermal perturbation of the regional conductive thermal gradient (Bird *et al.* 1975; England & Thompson 1984) associated with envisaged crustal thickening to about 80 km would have promoted the formation of eclogite-facies assemblages in the lower parts of the thickened prism. However, the survival of such assemblages and their apparent exposure at the surface by the early mid-Devonian, only some 65 Ma later, necessitates their rapid exhumation. Estimates of uplift rates based on considerations of isostacy and fluvial/glacial erosion alone (England & Richardson 1977; Anhert 1971) suggest that the rapid erosional exhumation may have been aided by tectonic stripping near the surface associated with east-

wards translation of the allochthonous Caledonian nappes on to the Baltic foreland. More importantly, crustal imbrication in the Baltic foreland, leading to further northwestwards underthrusting beneath the currently exposed Western Gneiss Region is thought to have resulted in the exposure of the high pressure lithologies above a normal thickness of continental crust.

Whereas in the northwesternmost part of the Western Gneiss Region the eclogite-facies mineral phases are largely chemically homogeneous (due to the high temperatures of formation) further south garnets often display marked compositional prograde growth zonation (as exemplified in Fig. 5A for an eclogite sample from the Dalsfjord area) and may enclose mineral inclusions interpreted as relics of earlier amphibolite- or blueschist-facies assemblages (e.g. Bryhni & Griffin 1971; Krogh 1982). This evidence indicates that such rocks have initially followed a prograde P–T path (as illustrated in Fig. 4) through the blueschist facies into the eclogite facies. This was followed by rapid decompression (reflected in symplektitic development of granulite-facies assemblages) and ultimately, if deformation permitted access to aqueous fluids, the development of lower-T amphibolite-facies assemblages. Such a P-T-t path, characterized by heating during compression on the prograde segment, is compatible with these rocks having originally resided in the lower underthrusting continental plate (cf. Fig., 2e and Fig. 4). By contrast, as outlined by Carswell *et al.* (1983) and Medaris (1980, 1984), the medium-T eclogite-facies assemblages in the alpine-type peridotite bodies show evidence, from diffusion controlled 'retrograde' garnet zoning profiles (e.g. Fig. 5B) and rarely preserved exsolution textures, of equilibration (albeit sometimes arrested) in response to cooling. Such observations are compatible with their emplacement from the upper mantle into thickened continental crust on tectonic imbrication across the crust-mantle interface during the major continental collision event.

High-P metamorphism in the Sesia-Lanzo zone, Western Alps

Important information on the tectonic and metamorphic processes affecting lower continental crust rocks during the collision between the NW European (Penninic) and Austro-alpine (Insubric) continental plates is recorded in the Sesia-Lanzo Zone, the most southeasterly (internal) unit of the Western Alps. The Sesia-Lanzo Zone is now taken to represent a tectonic slice of the Austro-alpine continental crust (Compagnoni *et al.* 1977) which structurally overlies the meta-ophiolites and schistes lustrés of the Piemonte Zone (presumed relicts of the Jurassic Piemonte oceanic crust). The Piemonte Zone (Nappe) has in turn been thrust on to the margin of the Penninic continental plate, portions of which are exposed in the tectonic windows represented by the Monte Rosa, Gran Paradiso and Dora Maira Massifs as well as the Grand Saint Bernard-Briançonnais Nappe further west.

It would appear that the Austro-alpine, Piemonte and Penninic units of the Western Alps have all had a similar metamorphic history during the Alpine orogeny, characterized by an early Eo-Alpine high-P eclogite-/blueschist-facies event and a later (Lepontine) greenschist-facies overprint of variable intensity (Compagnoni *et al.* 1977; Compagnoni 1977; Caby *et al.* 1978; Lardeaux *et al.* 1982; Chopin & Maluski 1980). The latter has been reliably dated at 38–40 Ma but there is considerably more doubt over the date of the early high-P metamorphism. Compagnoni *et al.* (1977) indicated a 70–90 Ma date but, on the basis of the age data provided by Hunziker (1974) and Oberhaensli *et al.* (1982), Rubie (1984) has concluded that the eclogite-/blueschist-facies conditions in the Sesia-Lanzo Zone were initiated 100–130 Ma ago but may have persisted until 60 Ma. Caron (1984) has taken the available data to indicate that the high-P metamorphic conditions were attained at different times in the different structural units—namely at 110–130 Ma in the Sesia-Lanzo Zone, 40–70 Ma in the Piemonte schistes lustrés and at 40–50 Ma in some Briançonnais units. Chopin & Malaski (1980) have dated the early high P event at 60–75 Ma in phengites and paragonites from the Gran Paradiso nappe of the Pennine Zone.

According to Compagnoni *et al.* (1977) the Sesia-Lanzo Zone comprises heterogeneous pre-Alpine continental basement rocks which are subdivided into two main, lithologically contrasted, tectonic units—the lower *Eclogitic Micaschist Complex* (EMS) and the upper *Seconda Zona Dioritico-Kinzigitica* (II DK). The latter comprises mostly amphibolite- and locally granulite-facies gneisses (kinzigites) of probable mainly Hercynian age, and recrystallization to early Alpine high-P assemblages or late greenschist-facies assemblages is restricted to local high strain zones. The II DK has close lithological similarity to the Ivrea Zone, to the east of the Insubric Line (Fault Zone), which appears to have been largely unaffected by the Alpine metamorphism and deformation and exposes similar gneisses structurally underlain by peridotites. The Ivrea Zone is widely taken to represent a tectonic slice through the pre-Alpine lower continental crust into the upper mantle (Mehnert 1975; Rivalenti *et al.* 1980).

The lower EMS unit consists of dominant micaschists and metabasites which only rarely retain high temperature pre-Alpine mineral assemblages as they have suffered near pervasive recrystallization to Early Alpine eclogite-/blueschist-facies assemblages. However, in the western part of the Sesia-Lanzo Zone these rocks have been extensively retrograded to Late Alpine greenschist-facies assemblages (Gneiss Minuti Complex). A further important feature of the EMS unit is the occurrence of Permian granitoid bodies (such as at Mt. Mucrone) in which the original sodic plagioclase has been replaced by intergrowths of jadeite and zoisite and the biotite by phengite and garnet in response to the Early Alpine high-P metamorphism (Compagnoni 1977; Oberhaensli *et al.* 1982).

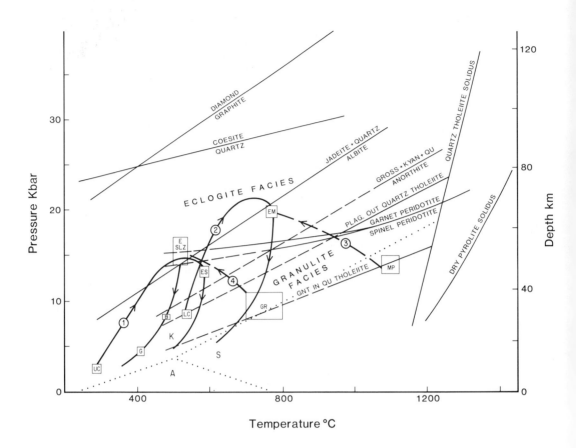

FIG. 4. Pressure (depth)-temperature diagram illustrating contrasted P-T-t paths for the generation and exhumation of eclogite-facies assemblages in protoliths from different structural levels in continental collision zones as follows: (1) along a prograde (heating) P-T-t path for Sunnfjord/Dalsfjord, west Norway [ES] from original upper–middle crustal rocks [UC] in underthrusting lower plate. (2) along a prograde path for the Møre district, west Norway [EM] from a deep crustal level [LC] in underthrusting lower plate. (3) along a cooling path for tectonically intercalated sub-continental mantle peridotite slices [MP] in the Møre district, west Norway. (4) formation by re-equilibration at lower temperature of original lower crustal granulites-amphibolites [GR] in overthrusting slab for the Sesia-Lanzo Zone [ESLZ], western Alps.

The P-T-t path for the Sesia-Lanzo Zone rocks is based on data provided by Lardeaux *et al.* (1982) with [B] and [G] representing indicated conditions for formation of retrograde blueschist- and greenschist-facies assemblages, respectively. However, note that as discussed in the text and also indicated in Carswell *et al.* (1982), the cooling paths (heavy dashed lines) leading to stabilization of medium-T eclogite-facies assemblages from lower crustal [GR] and upper mantle [MP] protoliths may well not be as direct as indicated in this diagram.

In all instances the initial uplift path for the high-P assemblages is characterized by rapid, near isothermal, decompression. Data sources for the various reaction curves as in Fig. 1.

Best estimates of temperatures during the high-P metamorphism in the Sesia-Lanzo rocks are 500°–540°C (Lardeaux *et al.* 1982) with probable pressures of 14–17 kbar—the minimum value being constrained by the observed stability of effectively pure jadeite + quartz (see Fig. 1). By contrast, Lardeaux *et al.* (1982) have indicated the P–T conditions for formation of the pre-Alpine granulite-facies assemblages to have been 700–800°C and 8–11 kbar.

A P-T-t path for the Sesia-Lanzo Zone rocks (Fig. 4) based on these observations might, on first consideration, be taken to be compatible (cf. Fig. 2f) with eclogite formation during Alpine upthrusting and cooling of granulite-facies lower continental crust. However, although the tectonic location within the Alpine belt appears correct, various geological, mineralogical and geochronological observations (Compagnoni *et al.* 1977; Lardeaux *et al.* 1982; Zingg 1983; Kruhl 1984) indicate that the Sesia-Lanzo Zone rocks (like those of the Ivrea and Strona-Ceneri Zones to the east) were already at a fairly high crustal level prior to the Alpine orogeny. Controversy persists over the age of the granulite-facies metamorphism, although Zingg (1983) favours the interpretation that it was Caledonian. However, these continental crust segments are considered to have remained at depth until the Variscan orogeny, which caused limited amphibolite- to greenschist-facies retrogression (Lardeaux *et al.* 1982; Kruhl 1984), but to have been uplifted and cooled to *c.* 300°C by *c.* 180 Ma on the basis of K-Ar mica ages (Zingg 1983). Chemical zoning profiles in garnets from Sesia-Lanzo Zone rocks are unfortunately rather variable and complex (Desmons & Ghent 1977) but suggestive of different episodes of garnet growth and diffusive re-equilibration.

High-P metamorphism in the European Variscan fold belt

Sparse relicts of high-P metamorphic assemblages (mostly medium-T eclogites or high-P granulites) are widely distributed in several basement massifs (Bohemian, Polish Sudetan, Vosges-Schwarzwald, Massif Central, S. Armorican, N.W. Iberian) of the Variscan fold belt of western and central Europe (e.g. Dudek & Fediukova 1974; Kappel 1967; Matthes 1978; Pin & Vielzeuf 1983; Smulikowski & Bakum-Czubarow 1973). Interpretation of the tectonometamorphic evolution of these particular high-P rocks is clouded by general uncertainty over the geochronology of the seemingly complex sequence of Caledonian to Hercynian orogenic events which appears to have been involved in the Palaeozoic consolidation of median Europe (Ziegler 1984)—compounded by the limited, and spatially separated, nature of the exposure.

Early ideas (e.g. Zoubek 1969) of a Precambrian age for these high-P assemblages have been disproved by more recent Rb-Sr and U-Pb dating (Gebauer & Grünenfelder 1979; van Breemen *et al.* 1982; Vidal *et al.* 1980) which indicate that they have formed from protoliths of both Proterozoic and Early Palaeozoic age. However, major divergences of opinion remain over the actual age of the high-P assemblages. Pin & Vielzeuf (1983), in their review, support a Caledonian age (400–450 Ma) for their formation throughout Variscan median Europe, a view supported by the radiometric age determination of Arnold & Scharbert (1973) in Lower Austria and Jäger & Watznauer (1969) in Saxony. However, much younger Hercynian ages of 345 ± 5 Ma and 320^{+29}_{-36} Ma, respectively, have been indicated by van Breeman *et al.* (1982) in the Moldanubian Zone of the Czech part of the Bohemian Massif and by Gebauer *et al.* (1981) in the Massif Central. On the other hand, an intermediate 380^{+14}_{-22} Ma (Acadian-Ligerian) age has been demonstrated by Gebauer & Grünenfelder (1979) for eclogites in the Munchberg Massif of the northern Saxo-Thuringian Zone of the Bohemian Massif. Acadian ages have also been deduced for eclogites in the W. Iberian Massif of Western Galicia (Van Calsteren *et al.* 1979) and for blueschists on the Ile de Groix, S. Armorican Massif (Peucat & Cogné 1977). This situation is not fully resolved and the possibility must be considered that the development of the medium-T eclogite/high-P granulite association was not synchronous across the central European Variscides. We tentatively favour the interpretation put forward by Autran & Cogné (1980), Santallier *et al.* (1978), Matte & Burg (1981) and Zeigler (1984) for a dominant Acadian-Ligerian (i.e. Eo-Hercynian) age of around 380 Ma for the high-P metamorphism, in response to collision of the Gondwana-derived Intra-Alp, Iberian and Avalon allochthonous terrains with the southern margin of the Laurasian craton previously consolidated during the Caledonian orogenic cycle (Ziegler 1984). As in the Norwegian Caledonides and the Western Alps, there is abundant evidence in the Central European Variscides for a later (*c.* 320 Ma) retrograde (lower-P) metamorphic overprint on the early high-P assemblages during the now most apparent tectono-metamorphic phase of the Hercynian orogeny.

As emphasized by Pin & Vielzeuf (1983) a particular feature of the Variscan fold belt is the close spatial association of medium-T eclogites,

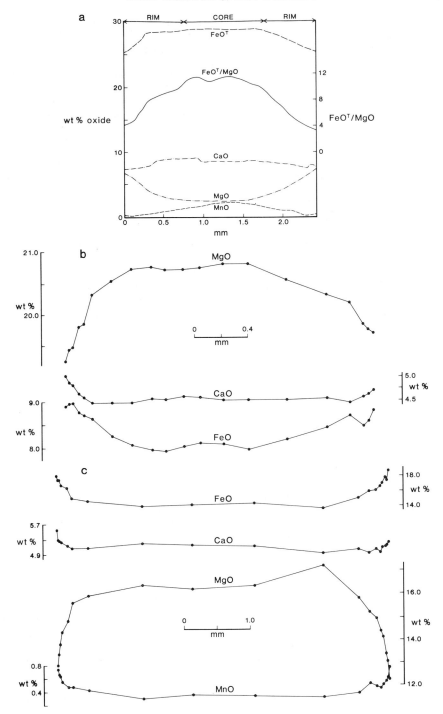

FIG. 5. Chemical zoning profiles across garnet grains in eclogite-facies rocks. Analyses all performed on the Microscan IX electron microprobe at the University of Sheffield. A. Garnet showing prograde growth zoning in eclogite from Dalsfjord, west Norway. Mineral inclusion suites are tschermakitic amphibole + zoisite + ilmenite/rutile in the core zone, and omphacitic clinopyroxene + rutile ± barroisitic amphibole in the rim zones. B. Garnet clast showing extensive diffusion controlled 'retrograde' zoning indicative of a prolonged cooling history—in porphyroclastic textured garnet peridotite from Ugelvik, Otrøy, west Norway. C. Garnet clast again showing extensive 'retrograde' zoning in porphyroclastic garnet pyroxenite associated with peridotite within the St. Leonhard granulite complex at Reitmuhle, Kamptal, lower Austria.

garnet peridotites and garnet websterites with quartzo-feldspathic gneisses. The gneisses have a dominant high-P granulite facies assemblage of quartz-perthite-plagioclase-garnet-kyanite with minor orthopyroxene and/or clinopyroxene in scarcer more mafic variants. This association is particularly in evidence in the Moldanubian Zone of the Bohemian Massif in Southern Czechoslovakia (Dudek & Fediukova 1974; Misar *et al.* 1983), in Lower Austria (Kappel 1967; Scharbert & Carswell 1983) and in the Oberfalzwald of N.E. Bavaria (Busch 1970; Matthes 1978).

Scharbert & Kurat (1974) have estimated minimum P–T conditions for formation of quartz-feldspathic granulites in Lower Austria to have been 11 kbar and 760°C, consistent with formation in the lower continental crust. Czech geologists (e.g. Vesela 1967) have often favoured interpreting the dominant felsic granulites as metamorphosed rhyolites and tuffs stratigraphically associated with more mafic granulites of presumed meta-sedimentary origin. However, recently Vrana & Jakes (1982) have described the deformation induced transformation of relatively pristine unfoliated garnet bearing hypersolvus leucogranites into highly flattened and recrystallized felsic garnet granulites. They view such plutonic igneous rocks to have been the products of relatively dry ($P_{H_2O} < 2$ kbar) anatectic melting in the lower continental crust with crystallization under granulite-facies conditions at $P_{Total} > 7$ kbar and $T > 800°C$. However, van Breeman *et al.* (1982) have interpreted the initial crystallization of the lower crustal rocks of the Moldanubian Zone to have been largely accomplished during the Cadomian orogenic cycle (*c.* 550 Ma) but the actual granulite-facies metamorphism to have occurred much later during the Hercynian orogeny at 345 ± 5 Ma.

Spatially associated medium-T eclogites, garnet websterites and garnet peridotites occur within the quartzo-feldspathic granulites at numerous localities within the Moldanubian Zone of the Bohemian massif. These garnetiferous peridotite bodies and enclosed metabasic rocks, taken to represent the partially recrystallized products of trapped partial melts of olivine tholeiite to picrite compositions (Scharbert & Carswell 1983), have been widely interpreted as tectonic intercalculations of upper mantle rocks (Kappel 1967; Fiala 1966; Scharbert 1973; Matthes 1978; Dudek & Fediukova 1974). Such rocks invariably show evidence of static recrystallization to lower pressure granulite-facies assemblages as well as later deformation induced retrogression to amphibolite-facies assemblages, as also widely observed in the felsic granulites.

Garnets in the eclogite-facies rocks from Lower Austria have conspicuous retrograde chemical zoning profiles (see Fig. 5c) and pyroxenes show extensive exsolution. Comparable mineralogical features are observed in the associated granulites. Unfortunately it is currently only possible to indicate minimum P–T conditions for formation of the quartzo-feldspathic granulites (Scharbert & Kurat 1974) and the extensive metamorphic retrogression suffered by the basic and ultrabasic rocks makes thermobarometry of their relict high P assemblages rather hazardous. However, we suggest that the lower continental crust and upper mantle lithologies were in effective tectonic juxtaposition early in the metamorphic evolution and that their respective high-P/medium-T mineralogies were in effect both coeval and cofacial. We further suggest that the stabilization of these high-P assemblages in the Moldanubian Zone rocks may have been caused by them having been underthrust from the north-west by cooler continental lithosphere in the manner illustrated in Fig. 3. In addition we speculate that the implied earlier high-T assemblages both in the lower crust and uppermost mantle have been generated in response to an enhanced mantle heat flux during a precursive phase of continental lithospheric extension.

Confirmation that the lower continental crust rocks of the Moldanubian Zone were subjected to P–T conditions appropriate for the stability of medium-T eclogites early in the Variscan orogenic cycle requires further isotopic age data and proof that undoubted eclogite-facies assemblages were also developed in crustal protoliths. In the latter connection it is worth noting the reported significant range of eclogite rock compositions (Dudek & Fediukova 1974; Matthes 1978) and the fact that not all eclogite occurrences are directly associated with mantle-derived peridotite bodies. However, clearly more field and analytical data are required to consolidate the interpretation of the high-P rocks in the Moldanubian Zone of the Bohemian Massif and indeed elsewhere in the Variscides.

Discussion

The controversial medium-T eclogite-facies assemblages sporadically preserved in certain amphibolite or granulite-facies terrains are considered to have been generated largely from continental crust protoliths. Mineral-chemical evidence indicates that their formation requires substantial crustal thickening (up to or even in excess of 70 km). A likely tectonic scenario for this is considered to be the collision of two continental lithospheric plates during which the leading edge of one plate undergoes transient subduction beneath the other (Fig. 3). Subsequent further thickening may occur as thrusts progressively step 'down' and back into the underthrust plate, as outlined by Hodges *et al.* (1982).

Continental plate collision would be expected to perturb the steady state temperature-depth profile and result in the stabilization of eclogite-facies assemblages along prograde P-T-t paths in the underthrust plate. In addition the lower crust of the upper plate may be cooled by being brought into contact with underthrust upper crust. However, in this instance kinetic barriers may limit re-equilibration to eclogite-facies assemblages to high strain zones. Eclogite stabilization would be enhanced by any tectonic thickening of the upper plate and may be repeated during continued crustal shortening each time a major new thrust becomes operational in the underthrusting foreland.

Mineral features may monitor the origin and location of rocks during the tectonic events responsible for eclogite stabilization. Garnet zoning and mineral inclusion suites can show an evolution involving increasing T as well as P_{Load} (Fig. 5a) indicating that the rocks probably represent previous upper–middle crust (e.g. Sunnfjord, western Norway). Alternatively zoning may indicate cooling (Figs. 5B & C), with or without an initial pressure increase, indicative of previous lower continental crust (Moldanubian Zone; Sesia-Lanzo Zone?) or, in the case of garnetiferous alpine-type peridotites, tectonically intercalated upper mantle (Moldanubian Zone; western Norway).

The survival and ultimate exposure of medium-T eclogites requires that the rate of exhumation is fast relative to the rate of thermal relaxation following continental underthrusting (England & Richardson 1977; Draper & Bone 1980). Thermobarometry on widely developed retrogressive symplektites and kelyphites with granulite-facies assemblages corroborates this rapid uplift hypothesis, showing that initial uplift and decompression was usually accomplished with little change in temperature. Geochronological evidence for the Caledonides, Alps and Variscides whilst still controversial, indicates that eclogite-facies assemblages formed early in the evolution of each belt and were superimposed upon appreciably older (Proterozoic or Palaeozoic) crustal material. The time interval between the early high-P metamorphism and later extensive lower-P greenschist- or amphibolite-facies assemblages may have been only some 40–60 Ma, indeed our best estimate for complete unroofing of the eclogite-bearing terrain in Western Norway is

some 65 Ma. The necessary average erosion rate of around 1 mm/year is encouragingly in line with estimates of current and recent uplift rates along the Indus–Tsangpo suture zone in the Himalaya (Zeitler *et al.* 1982; Melita 1980).

Buoyant uplift of terrains bearing eclogite-facies assemblages may be aided by continued underthrusting of crustal wedges below them (Fig. 3), allowing eclogites in lower crustal rocks to be exposed above a normal thickness of crust. Thrusts such as the Main Boundary Thrust of the Himalaya may be responsible for such a process. Sequential foreland propagation of such thrusts (Fig. 3) may produce eclogites of different ages at different tectonostratigraphic levels in an orogen (as suggested by Caron (1984) for the western Alps) and may be responsible for the presence of high-P metamorphic rocks at several levels in the Caledonian allochthon of Scandinavia (Bryhni *et al.* 1977). Tilting of the crust as a consequence of underthrusting (Windley 1983) may also aid eventual exposure of high-P rocks from deep levels in the thrust-nappe stack.

Competition between thermal and mechanical effects can have a significant effect on the resulting mineral assemblages. Factors such as the previous thermal structure of the crust, the rate of plate convergence and the rate of uplift will all control the nature of the rocks reaching the surface. It is perhaps significant that in the three areas reviewed the peak metamorphic and retrogressive mineral assemblages are significantly different, probably reflecting different thermal regimes. Rubie (1984) has developed a model for the Sesia-Lanzo zone in which prolonged subduction of oceanic lithosphere beneath the thickened crust retarded relaxation of the isotherms and produced the unusually low equilibration temperatures recorded there. Whilst this model may significantly overestimate the period of oceanic subduction (Tricart, 1984 indicates that ocean closure commenced at *c.* 80 Ma, significantly later than Rubie's (1984) estimate of 130–100 Ma) it serves to illustrate the blanketing effect of thrusting cold material below an evolving eclogite-facies terrain. Retardation of uplift may result from density increases brought about by up-pressure phase transformations (Richardson & England 1979); a reduced rate of plate convergence or even climatic factors affecting erosion rates. Such retardation may allow thermal relaxation in the thickened crust to obliterate the medium-T eclogite-facies assemblages. Thus some exposed apparent 'lower crustal' granulite facies terrains may have originally been at eclogite-facies, particularly those regions containing high P granulites.

Medium-T eclogites appear to be a particular

feature of Phanerozoic continental collision belts, although a recent geochronological study of comparable eclogites in the Eastern Glenelg Lewisian inlier of NW Scotland (Sanders *et al.* 1984) has indicated a Late Proterozoic (Grenville) age. We know of no proven examples of Archaean crustal eclogites. This might be indicative of the fact that only more recently in the earth's history has the continental crust been sufficiently thick, cool and rigid (cf. Tarling 1980) to permit the degree of Type A subduction necessary for the stabilization and survival of medium-T eclogites. On the other hand, England & Bickle (1984) have recently argued that the Archaean continental crust was not greatly differ-ent in thickness and thermal structure from that of the Phanerozoic era.

The Himalayan collision zone currently lacks medium-T eclogites at outcrop, although high-P granulites have been found (Windley 1983). Eclogites might be expected to exist at deeper levels in the thrust-nappe stack and their eventual exposure may be aided by erosion and underthrusting along faults such as the Main Boundary Fault.

ACKNOWLEDGEMENTS: We wish to acknow-ledge research grants (DAC) and a research studentship (SJC) from the Natural Environment Research Council which have supported our studies of eclogite-facies rocks.

References

AHNERT, F. 1970. Functional relationship between denudation, relief and uplift in mid-latitude drainage basins. *Am. J. Sci.* **268**, 243–263.

ARNOLD, A. & SCHARBERT, H.G. 1973. Rb-Sr alterbestimmungen an granuliten der Südlichen Böhmischen masse in Österreich. *Schweiz. Mineral. Petrogr. Mitt.*, **53**, 61–78.

AUTRAN, A. & COGNÉ J. 1980. La zone interne de l'orogène varisque dans l'ouest de la France et sa place dans le developpement de la chaîne hercynienne. *In*: COGNÉ, J. & SLANSKY, M. (eds) *Geologie de l'Europe du Precambrian aux bassins sedimentaries posthercyniens.* Mem. B.R.G.M. **107**, 87–92.

BANNO, S. 1970. Classification of eclogites in terms of physical conditions of their origin. *Phys. Earth Planet. Interiors*, **3**, 405–421.

BERMAN, R. 1979. Thermal properties. *In*: FIELD J.E. (ed.), *The Properties of Diamond.* Academic Press, London.

BEST, M.G. 1982. *Igneous and Metamorphic Petrology.* Freeman, San Francisco.

BIRD, P., TOKSOZ, M.N. & SLEEP, N.H. 1975. Thermal and mechanical models of continent-continent convergence zones. *J. Geophys. Res.*, **80**, 4405–4416.

BOHLEN, S.R., WALL, V.J. & BOETTCHER, A.L. 1983. Geobarometry in granulites. *In*: SAXENA, S.K. (ed.), *Kinetics and Equilibrium in Mineral Reactions.* Advances in Physical Geochemistry, Vol. 3, Springer-Verlag, New York.

BROWN, E.H. & BRADSHAW, J.Y. 1979. Phase relations of pyroxene and amphibole in greenstone, blueschist and eclogite from the Francisan Complex, California. *Contrib. Mineral. Petrol.* **71**, 67–83.

BRYHNI, I. 1966. Reconnaissance studies of gneisses, ultrabasites, eclogites and anorthosites in Outer Nordfjord, Western Norway. *Nor. Geol. Unders.*, **241**, 1–68.

—— & GRIFFIN, W.L. 1981. Zoning in eclogite garnets from Nordfjord, West Norway. *Contrib. Mineral. Petrol.* **32**, 112–125.

—— & GRIMSTAD, E. 1970. Supracrustal and infracrustal rocks in the gneiss region of the Caledonides west of Breimsvatn. *Nor. Geol. Unders.* **266**, 105–140.

——, KROGH, E. & GRIFFIN, W.L. 1977. Crustal derivation of Norwegian eclogites: A review. *Neues Jb. Miner. Abh.* **130**, 49–68.

BUSCH, K. 1970. Die Eklogitvorkommen des Kristallinen Grundgebirges in N.E. Bayern. IV. Die Eklogite der Oberpfalz und ihr metamorpher Abbau. *Neues Jb. Miner. Abh.* **113**, 138–178.

CABY, R., KIENAST, J.-R. & SALIOT, P. 1978. Structure, métamorphisme et modèle d'evolution tectonique des Alps occidentales. *Rev. Geogr. Phys. Geol. Dyn.* **XX**, 307–322.

CARON, J.M. 1984. The diversity of geodynamic regimes leading to high pressure-low temperature metamorphism (Western Alps and Corsica). *Terra Cognita*, **4**, 39–43.

CARSWELL, D.A. 1973. Garnet pyroxenite lens within Ugelvik layered garnet peridotite. *Earth Planet Sci. Lett.* **20**, 347–352.

—— 1980. Mantle derived lherzolite nodules associated with kimberlite, carbonatite and basalt magmatism: A review. *Lithos*, **13**, 121–138.

—— 1981. Clarification of the petrology and occurrence of garnet lherzolites, garnet websterites and eclogite in the vicinity of Rödhaugen, Almklovdalen, West Norway. *Nor. Geol. Tidsskr.* **61**, 249–260.

——, DAWSON, J.B. & GIBB, F.G.F. 1981. Equilibration conditions of upper-mantle eclogites: implications for kyanite-bearing and diamondiferous varieties. *Mineral. Mag.* **44**, 79–89.

—— & GIBB, F.G.F. 1980. The equilibration conditions and petrogenesis of European crustal garnet lherzolites. *Lithos*, **13**, 19–29.

—— & HARVEY, M.A. 1985. The intrusive history and tectono-metamorphic evolution of the Basal Gneiss Complex in the Moldeford area, West Norway. *In*: GEE, D.G. & STURT, B.A. (eds) *The Caledonide Orogen.* John Wiley, London.

——, HARVEY, M.A. & AL-SAMMAN, A. 1983. The petrogenesis of contrasting Fe-Ti and Mg-Cr garnet peridotite types in the high grade gneiss complex of Western Norway. *Bull. Mineral.* **106**, 727–750.

——, KROGH, E. & GRIFFIN, W.L. 1985. Norwegian orthopyroxene eclogites: Calculated equilibration conditions and petrogenetic implications. *In*: GEE, D.G. & STURT, B.A. (eds) *The Caledonide Orogen*. John Wiley, London.

CHOPIN, C. & MULUSKI, H. 1980. ^{40}Ar-^{39}Ar dating of high pressure metamorphic micas from the Gran Paradiso area (Western Alps): Evidence against the blocking temperature concept. *Contrib. Mineral. Petrol.* **74**, 109–122.

COLEMAN, R.G., LEE, D.E., BEATTY, J.B. & BRANNOCK, W.W. 1965. Eclogites and eclogites: Their differences and similarities. *Bull. Geol. Soc. Am.* **76**, 483–508.

COMPAGNONI, R. 1977. The Sesia-Lanzo zone: High pressure-low temperature metamorphism in the Austroalpine continental margin. *Rend. Soc. Ital. Mineral. Petrol.* **33**, 335–374.

——, DAL PIAZ, G.V., HUNZIKER, J.C., GOSSO, G., LOMBARDO, B. & WILLIAMS, P.F. 1977. The Sesio-Lanzo zone, a slice of continental crust with alpine high pressure–low temperature assemblages in the western Italian Alps. *Rend. Soc. Ital. mineral. Petrol.* **33**, 281–334.

CUTHBERT, S.J. & CARSWELL, D.A. 1982. Petrology and tectonic setting of eclogites and related rocks from the Dalsfjord area, Sunnfjord, western Norway. *Terra Cognita*, **2**, 315.

——, HARVEY, M.A. & CARSWELL, D.A. 1983. A tectonic model for the metamorphic evolution of the basal gneiss complex, western South Norway. *J. Metamorphic Geol.* **1**, 63–90.

DAWSON, J.B. 1980. *Kimberlites and their Xenoliths*. Springer-Verlag, Berlin.

DESMONS, J. & GHENT, E.D. 1977. Chemistry, zonation and distribution coefficients of elements in eclogitic minerals from the Eastern Sesia Unit, Italian Western Alps, *Schweiz. Mineral. Petrogr. Mitt.* **57**, 397–411.

DRAPER, G. & BONE, R. 1981. Denudation rates, thermal evolution and preservation of blueschist terrains. *J. Geol.* **89**, 601–613.

DUDEK, A. & FEDIUKOVA, E. 1974. Eclogites of the Bohemian Moldanubian. *Neues Jb. Miner. Abh.* **121**, 127–159.

ELLIS, D.J. & GREEN, D.H. 1979. An experimental study of the effect of Ca upon garnet-clinopyroxene Fe-Mg exchange equilibra. *Contrib. Mineral. Petrol.* **71**, 13–22.

ENGLAND, P. & BICKLE, M. 1984. Continental thermal and tectonic regimes during the Archaean. *J. Geology*, **92**, 353–367.

—— & RICHARDSON, S.W. 1977. The influence of erosion upon the mineral facies of rocks from different metamorphic environments. *J. geol. Soc. London*, **134**, 201–213.

—— & THOMPSON, A.B. 1984. Pressure-temperature-time paths of regional metamorphism. 1. Heat transfer during the evolution of regions of thickened continental crust. *J. Petrol.* **25**, 894–928.

ERNST, W.G. 1977. Tectonics and prograde versus retrograde P-T trajectories of high pressure metamorphic belts. *Rend Soc. Ital. Mineral. Petrol.* **33**, 221–252.

ESKOLA, P. 1921. On the eclogites of Norway. *Sr.* *Norske vidensk-Akad. i Oslo Mat. Natur. Kl.* **8**, 1–118.

FIALA, J. 1966. The distribution of elements in mineral phases of some garnet lherzolites from the Bohemian Massif. *Krystalinikum*, **4**, 31–53.

GJELSVIK, T. 1952. Metamorphosed dolerites in the gneiss area of Sunnmøre on the west coast of southern Norway. *Nor. Geol. Tidsskr.* **31**, 31–134.

GOLDSMITH, T.R. 1980. The melting and breakdown reactions of anorthite at high pressures and temperatures. *Am. Mineral.* **65**, 272–284.

GREEN, D.H. & RINGWOOD, A.E. 1967a. An experimental investigation of the gabbro to eclogite transformation and its petrological applications. *Geochim. Cosmochim. Acta*, **31**, 767–833.

—— & —— 1967b. The stability fields of aluminous pyroxene peridotite and garnet peridotite and their relevance in upper mantle structure. *Earth Planet. Sci. Letters*. **3**, 151–160.

GRIFFIN, W.L. & BRUECKNER, H.K. 1980. Caledonian Sm-Nd ages and a crustal origin for Norwegian eclogites. *Nature*, **285**, 319–321.

—— & 1982. Rb-Sr and Sm-Nd studies of Norwegian eclogites. *Terra Cognita*, **2**, 324.

—— & CARSWELL, D.A. 1985. Geochronological setting of in situ eclogite metamorphism in western Norway. *In*: GEE, D.G. & STURT, B.A. (eds) *The Caledonide Orogen*. John Wiley, London.

——, AUSTRHEIM, H. *et al*. 1985. High pressure metamorphism in the Scandinavian Caledonides. *In*: GEE, D.G. & STURT, B.A. (eds) *The Caledonide Orogen*, John Wiley, London.

—— & RÅHEIM, A. 1973. Convergent metamorphism of eclogites and dolerites, Kristiansund area, Norway. *Lithos*, **6**, 21–40.

GEBAUER, D., BERNARD-GRIFFITHS, J. & GRÜNENFELDER, M. 1981. U-Pb zircon and monazite dating of a mafic-ultramafic complex and its country rocks. Example: Sauviat-sur-Vige, French Central Massiff. *Contrib. Mineral. Petrol.* **76**, 292–300.

—— & GRÜNENFELDER M. 1979. U-Pb zircon and Rb-Sr mineral dating of eclogites and their country rocks. Example: Munchberg Gneiss Massif, northeast Bavaria. *Earth Planet Sci. Lett.* **42**, 35–44.

——, LAPPIN, M.A., GRÜNENFELDER, M., KOESTLER, A. & WYTTENBACH, A. 1982. Age and origin of some Norwegian eclogites. A U-Pb zircon and REE study. *Terra Cognita*, **2**, 323.

HARTE, B. 1983. Mantle peridotites and processes—the kimberlite sample. *In*: HAWKESWORTH C.J. & NORRY, M.J. (eds) *Continental Basalts and Mantle Xenoliths*. Shiva geology series.

HARVEY, M.A. 1983. A geochemical and Rb-Sr study of the Proterozoic augen orthogneisses on the Molde Peninsula, west Norway. *Lithos*, **16**, 325–338.

HEINRICH, C.A. 1982. Kyanite-eclogite to amphibolite facies evolution of hydrous mafic and pelitic rocks, Adula Nappe, central Alps. *Contrib. Mineral. Petrol.* **81**, 30–38.

HELMSTAEDT, H. & GURNEY, J.J. 1984. Kimberlites of southern Africa—are they related to subduction processes? *In*: KORNPROBST, J. (ed.) *Kimberlites 1*:

Kimberlites and Related Rocks. Elsevier, Amsterdam.

HODGES, K.V., BARTLEY, J.M. & BURCHFIEL, B.C. 1982. Structural evolution of an A-type subduction zone, Lofoten-Rombak area, northern Scandinavian Caledonides. *Tectonics,* **1,** 441–462.

HOLDAWAY, M.J. 1971. Stability of andalusite and the aluminium phase diagram. *Am. J. Sci.* **271,** 97–131.

HOLLAND, T.J.B. 1980. The reaction albite = jadeite + quartz determined experimentally in the range 600–1200°C. *Am. Mineral.* **65,** 1129–134.

HUNZIKER, J.C. 1974. Rb-Sr and K-Ar age determination and the alpine tectonic history of the western Alps. *Mem. 1st Geol. min. Univ. Padova,* **31,** 54 pp.

JÄGER, E. & WATZNAUER, A. 1969. Einige Rb/Sr datierungen an granuliten des Sachsischen granulitgebirges. *Monatber. Dtsch. Akad. Wiss.* **11,** 420–426.

KAPPEL, E. 1967. Die Eklogite Meidling im Tal und Mitterbach graben im Niederösterreichischen Moldanubikum südlich der Donau. *Neues Jb. Miner. Abh.* **107,** 266–298.

KRILL, A.G. 1983. Rb-Sr study of rapakivi granite and augen gneiss from the Risberget Nappe, Oppdal, Norway. *Nor. Geol. Unders.* **380,** 51–65.

—— 1985. Relationships between the western gneiss region and the Trondheim region: stockwerk-tectonics reconsidered. *In*: GEE, D.G. & STURT, B.A. (eds) *The Caledonide Orogen* John Wiley, London.

KROGH, E.J. 1977. Evidence of precambrian continent-continent collision in western Norway. *Nature,* **267,** 17–19.

—— 1980a. Geochemistry and petrology of glaucophane-bearing eclogites and associated rocks from Sunnfjord, western Norway. *Lithos,* **13,** 355–380.

—— 1980b. Compatible P–T conditions for eclogites and surrounding gneisses in the Kristiansund area, western Norway. *Contrib. Mineral. Petrol.* **75,** 387–393.

—— 1982. Metamorphic evolution of Norwegian country-rock eclogites, as deduced from mineral inclusions and compositional zoning of garnets. *Lithos,* **15,** 305–321.

KROGH, T.E., MYSEN, B.O. & DAVIS, G.L. 1973. A palaeozoic age for the primary minerals of a Norwegian eclogite. *Carnegie Inst. Wash. Yearb.* **73,** 575–576.

KRUHL, J.H. 1984. Metamorphism and deformation of the N.W. margin of the Ivrea Zone, Val Loana (Italy). *Schweiz Mineral. Petrog. Mitt.* **64,** 151–167.

KISHIRO, I. 1969. Clinopyroxene solid solutions formed by reactions between diopside and plagioclase at high pressure. *Mineral. Soc. Am. Sp. Pap.* **2,** 179–191.

LAPPIN, M.A. 1966. The field relationships of basic and ultrabasic masses in the basal gneiss complex of Stadlandet and Almklovdalen, Nordfjord, S.W. Norway. *Nor. Geol. Tidsskr.* **46,** 439–495.

—— 1977. Crustal and in situ origin of Norwegian eclogites. *Nature,* **269,** 730.

——, PIDGEON, R.T. & VAN BREEMAN, O. 1979. Geochronology of basal gneisses and mangerite syenites of Stadlandet, west Norway. *Nor. Geol. Tidsskr.* **59,** 161–181.

—— & SMITH, D.C. 1978. Mantle equilibrated orthopyroxene eclogite pods from the basal gneisses in the Selje District, western Norway. *J. Petrol.* **19,** 530–584.

LARDEAUX, J.-M., GOSSO, G., KIENAST, J.-R. & LOMBARDO, B. 1982. Relations entre le métamorphisme et la déformation dans la zone Sésia-Lanzo (Alps occidentales) et le problème de l'eclogitisation de la croûte continentale. *Bull. Soc. Geol. France,* **24,** 793–800.

MACGREGOR, I.D. 1970. The effect of CaO, Cr_2O_3, and Al_2O_3 on the stability of spinel and garnet peridotites. *Phys. Earth Planet. Int.* **3,** 372–377.

MATTE, P.L. & BURG, J.P. 1981. Sutures, thrusts and nappes in the Variscan Arc of western Europe: Plate tectonic implications. *In*: MCCLAY, K.R. & PRICE, N.J. (eds.), *Thrust and Nappe Tectonics.* Spec. Publ. Geol. Soc. Lond. **9,** 353–358.

MATTHES, S. 1978. The eclogites of southern Germany. A summary. *Neues Jahrb. Mineral. Monatsch.* **3,** 93–109.

MEARNS, E.W. & LAPPIN, M.A. 1982. A Sm-Nd isotopic study of 'internal' and 'external' eclogites, garnet lherzolites and grey gneiss from Almklovdalen, western Norway. *Terra Cognita,* **2,** 324.

MEDARIS JR., L.G. 1980. Petrogenesis of the Lien peridotite and associated eclogites, Almklovdalen, western Norway. *Lithos,* **13,** 339–353.

—— 1984. A geothermobarometric investigation of garnet peridotites in the western gneiss region of Norway. *Contrib. Mineral. Petrol.* **87,** 72–86.

MEHTA, P.K. 1980. Tectonic significance of the young mineral dates and dates of cooling and uplift in the Himalayas. *Tectonophys.* **62,** 205–217.

MEHNERT, K.R. 1975. The Ivrea Zone. A model of the deep crust. *Neues Jb. Miner. Abh.* **125,** 156–199.

MIRWARD, P.W. & MASSONNE, H.J. 1980. The low-high quartz and quartz-coesite transition to 40 kbar between 600°C and 1600°C and some reconnaissance data on the effect of $NaAlO_2$ component of the low quartz-coesite transition. *J. Geophys. Res.* **85,** 6983–6990.

MISAR, Z., JELINEK, E. & JAKES, P. 1983. Inclusions of peridotite, pyroxenite and eclogite in granulite rocks of pre-Hercynian upper mantle and lower crust in the East Bohemian Massif (Czechoslovakia). *Ann. Sci. Univ. Clermont-Fd. II,* **74,** 85–95.

MIYASHIRO, A. 1973. *Metamorphism and Metamorphic Belts.* John Wiley, New York.

MØRK, M.B.E. 1982. A gabbro-eclogite transition on Flemsøy, Sunnmøre, western Norway. *Terra Cognita,* **2,** 316.

NIXON, P.H. & BOYD, F.R. 1973. Petrogenesis of the granular and sheared ultrabasic nodule suite in kimberlites. *In*: NIXON, P.H. (ed.) *Lesotho Kimberlites* Lesotho Nat. Devel. Corp. Maseru, Lesotho.

O'HARA, M.J. 1975. Pressure required to stabilise garnet-peridotite and eclogite at low temperatures. *Abst. Int. Conf. Geotherm. Geobarom. Penn State University, U.S.A.*

—— & MERCY, E.L.P. 1963. Petrology and petrogenesis of some garnetiferous peridotites. *Trans. Roy. Soc. Edinb.* **65,** 251–314.

——, RICHARDSON, S.W. & WILSON, G. 1971. Garnet peridotite stability and occurrence in crust and mantle. *Contrib. Mineral. Petrol.* **32**, 48–68.

——— & YODER JR., H.S. 1967. Formation and fractionation of basic magmas at high pressures. *Scott. J. Geol.* **3**, 67–117.

OBERHAENSLI, R., HUNZIKER, J.C., MARTINOTTI, G. & STERN, W.B. 1982. Mucronites: An example of Eo-Alpine eclogitisation of Permian granitoids, Italy. *Terra Cognita*, **2**, 325.

OXBURGH, E.R. & TURCOTTE, D.L. 1974. Thermal gradients and regional metamorphism in overthrust terrains with special reference to the Eastern Alps. *Schweiz. Mineral. Petrogr. Mitt.* **54**, 641–662.

PEUCAT, J.J. & COGNÉ, J. 1977. Geochronology of some blueschists from Ile de Groix, France, *Nature*, **268**, 131–132.

PIN, C. & VIELZEUF, D. 1983. Granulites and related rocks in Variscan Median Europe: A dualistic interpretation. *Tectonophysics*, **93**, 47–74.

RINGWOOD, A.E. 1975. *Composition and petrology of the earth's mantle*. McGraw Hill, New York.

RIVALENTI, G., GARUTI, G., ROSSI, A., SIENA, F. & SINIGOI, S. 1981. Existence of different peridotite types and of a layered igneous complex in the Ivrea Zone of the western Alps. *J. Petrol.* **22**, 127–153.

ROBERTS, D., THON, A., GEE, D.G. & STEPHENS, M.B. 1981. Scandinavian Caledonides—Tectonstratigraphy map, scale 1–1,000,000. Uppsala Caledonide Symp.

ROBERTSON, E.C.F., BIRCH, F. & MACDONALD, G.L.F. 1957. Experimental determination of jadeite stability relations to 25,000 bars. *Amer. J. Sci.* **255**, 115–137.

ROBINSON, D.N., GURNEY, J.J. & SHEE, S.R. 1984. Diamond eclogite and graphite eclogite xenoliths from Orapa, Botswana. *In:* KORNPROBST, J. (ed.) *Kimberlites II: The Mantle and Crust-Mantle Relationships*. Elsevier, Amsterdam.

RUBIE, D.C. 1983. A thermal-tectonic model for high-pressure metamorphism and deformation in the Sesia Zone, western Alps. *J. Geol.* **92**, 21–36.

SANDERS, I.S., VAN CALSTEREN, P.W.C. & HAWKESWORTH, C.J. 1984. A Grenville Sm-Nd age for the Glenelg eclogite in north-west Scotland. *Nature*, **312**, 439–440.

SANTALLIER, D., FLOCH, J.P. & GUILLOT, P.L. 1978. Quelques aspects du métamorphisme dévonien en bas-limousin (Massif Central, France). *Bull. Mineral.* **101**, 77–88.

SCHARBERT, H.G. 1973. Pyrope-rich garnet from Moldanubian garnet pyroxenites, Bohemian Massif, Lower Austria, Austria. *Neues Jb. miner. Abh.* **H-2**, 89–93.

——— & CARSWELL, D.A. 1983. Petrology of garnet-clinopyroxene rocks in a granulite facies environment, Bohemian Massif of Lower Austria. *Bull. Mineral.* **106**, 761–774.

——— & KURAT, G. 1974. Distribution of some elements between coexisting ferromagnesian minerals in Moldanubian granulite facies rocks, Lower Austria, Austria. *Tschermaks Mineral. Petrogr. Mitt.* **21**, 110–134.

SMITH, D.C. 1980. A tectonic mélange of foreign eclogites and ultramafites in west Norway. *Nature*, **287**, 366–368.

——— 1981. A reappraisal of factual and mythical evidence concerning the metamorphic and tectonic evolution of eclogite-bearing terrain in the Caledonides. *Terra Cognita*, **1**, 73.

——— 1982. A review of the controversial eclogites in the Caledonides. *Terra Cognita*, **2**, 304.

SMULIKOWSKI, K. 1960. Comments on eclogite facies in regional metamorphism. *Rep. Int. Geol. Congr. XXI Copenhagen* Pt. **XIII**, 372–382.

——— 1964. Chemical differentiation of garnets and clinopyroxenes in eclogites. *Bull. Acad. Pol. Sc. Ser. Geol. Geogr.* **XII**, 11–18.

——— 1968. Differentiation of eclogites and its possible causes. *Lithos*, **1**, 89–1012.

——— & BARUM-CZUBAROW, N. 1973. New data concerning the granulite-eclogite rock series of Stary Gieraltow, East Sudetes, Poland. *Bull. Acad. Pol. Sc. Ser. Geol. Geogr.* **XXI**, 25–34.

SPEAR, F.S., SELVERSTONE, J., HICKMOTT, D., CROWLEY, P. & HODGES, K.V. 1984. P-T paths from garnet zoning: A new technique for deciphering tectonic processes in crystalline terranes. *Geology*, **12**, 87–90.

STRAND, T. 1960. The region with basal gneiss in the N.W. part of S. Norway. *Nor. Geol. Unders.* **208**, 230–245.

TARLING, D.H. 1980. Lithosphere evolution and changing tectonic regimes. *J. geol. Soc. London*, **137**, 459–467.

TRICART, P. 1984. From passive margin to continental collision: A tectonic scenario for the western Alps. *Amer. J. Sci.* **284**, 97–120.

VAN BREEMAN, O., AFTALION, M. *et al.* 1982. Geochronological studies of the Bohemian Massif, Czechoslovakia and their significance in the evolution of central Europe. *Trans. Roy Soc. Edinb. Earth Sci.* **73**, 89–108.

VAN CALSTEREN, P.W.C., BUELRIJK, N.A.I.M., *et al.* 1979. Isotopic dating of older elements (including the Cabo Ortegal Mafic-Ultramafic Complex) in the Hercynian orogen of N.W. Spain. Manifestations of a presumed early palaeozoic mantle-plume. *Chem. Geol.* **24**, 35–56.

VESELA, M. 1967. On the stratigraphical position of granulites in the Moldanubicum. *Krystalinikum*, **5**, 137–152.

VIDAL, PH., PEUCAT, J.J. & LASNIER, B. 1980. Dating of granulites involved in the Hercynian Fold Belt of Europe: An example taken from the granulite-facies orthogneisses at La Picherais, southern Armorican Massif, France. *Contrib. Mineral. Petrol.* **72**, 283–289.

VRANA, S. & JAKES, P. 1982. Orthopyroxene granulites from a segment of charnockitic crust in southern Bohemia. *Vest. Ustr. Ustav. Geol.* **57**, 129–143.

WINDLEY, B.F. 1983. Metamorphism and tectonics of the Himalaya. *J. geol. Soc. London*, **140**, 849–865.

TAHIKRHELI, R.A.K. 1982. Fission-track evidence for Quaternary uplift of the Nanga Parbat region, Pakistan. *Nature*, **298**, 255–257.

ZIEGLER, P.A. 1984. Caledonian and Hercynian crustal

consolidation of western and central Europe—a working hypothesis. *Geol. Mijnbouw*, **63**, 93–108.

ZEITLER, P.K., JOHNSON, N.M., NAESER, C.W. & TAHIRKHELI, R.A.K. 1982. Fission-track evidence for Quaternary uplift of the Nanga Parbat region, Pakistan. *Nature*, **298**, 255–257.

ZINGG, A. 1983. The Ivrea and Strona-Ceneri Zones (southern Alps, Ticino and N-Italy)—A Review. *Schweiz. Mineral. Petrogr. Mitt.* **63**, 361–392.

ZOUBEK, V. 1969. Age relations in metamorphic terrains of the Bohemian Massif: Some methods and results. *Geol. Ass. Can Spec. Pap.* **5**, 73–81.

D.A. CARSWELL, Department of Geology, University of Sheffield, Mappin Street, Sheffield, S1 3JD, UK.

S.J. CUTHBERT, Exploration Division, BRITOIL plc., 150 St. Vincent Street, Glasgow, G2 5LJ, UK.

Geochronology and related isotope geochemistry of high-grade metamorphic rocks from the lower continental crust

S. Moorbath & P.N. Taylor

SUMMARY: We discuss the distribution of Rb, Sr, Sm, Nd, U, Th and Pb in medium-to-high pressure granulites, which are considered representative of the material of the lower crust. The depletion of large-ion-lithophile, heat-producing elements in such granulites is emphasized, together with its consequencs for radiogenic isotope evolution. Geochronological studies of granulites show that sometimes high-grade metamorphism closely follows original crust formation, whilst in other cases a long time interval separates these major events. Isotopic criteria for the recognition of lower crustal involvement in magma genesis are also discussed.

A very important feature of many medium (opx-plag)-to-high (cpx-gt) pressure (~ 7–15 kbar) granulites, particularly in Precambrian terrains, is the low concentration of large-ion-lithophile elements (LILE), usually attributed to depletion during high-grade metamorphism through the production and migration of partial melts and/or fluids. Earlier workers (e.g. Lambert & Heier 1967; Sighinolfi 1971; Heier 1973) particularly noted the depletion of radioactive, heat-producing element, (i.e. U, Th, K, Rb) in granulites. Concurrently, geophysicists demonstrated a linear relation between surface heat flow and surface heat production, which was explained in terms of an exponential decrease of heat production with depth in the continental crust (Roy et al. 1968; Lachenbruch 1970).

The behaviour of U and Rb are particularly important for isotope geochemistry, since these elements are depleted in medium-to-high pressure granulites. This gives rise to terrains with very low U/Pb and Rb/Sr ratios, particularly in felsic granulites. Pb is usually only slightly depleted (if at all) in such rocks compared to average continental crust. Sr is not significantly affected. The effect on radiogenic isotope evolution is profound, because Rb/Sr and U/Pb ratios determine rates of Sr- and Pb-isotopic evolution. Characterization of all components of continental crust and upper mantle in terms of Rb/Sr, U/Pb, Th/Pb, Sm/Nd and associated radiogenic isotopic ratios $^{87}Sr/^{86}Sr$, $^{206}Pb/^{204}Pb$, $^{207}Pb/^{204}Pb$, $^{208}Pb/^{204}Pb$, $^{143}Nd/^{144}Nd$, is essential for studies of crust-mantle evolution as well as interaction between magmas and continental crust.

Here we discuss the distribution and evolution of the above isotopic parameters in exposed terrains of the lower continental crust, and compare and contrast them with higher level continental crust and with underlying upper mantle. Work on lower crustal xenoliths in magmas is summarized elsewhere in this Volume. Of particular interest is the measurement of pre-metamorphic crustal residence time, i.e. the time interval between extraction of the protoliths of a rock unit from the mantle and their metamorphism into granulite facies. Isotopic work has shown that some deep crustal sectors undergo high-grade metamorphism and geochemical depletions within a few tens or hundreds of million years of a major, crust-producing mantle differentiation episode. Thus, medium-to-high pressure granulite-facies mineral assemblages represent the natural culmination of high-P,T, low-P_{H_2O} prograde crustal metamorphism, and their development in the deep crust is closely linked with differentiation and geochemical 'ripening' of juvenile sialic crust (references in Wells 1981). In other cases, high-grade metamorphism and accompanying geochemical differentiation may post-date the primary crust-forming event considerably, and do not form part of that event. Clearly, there are several tectonic situations and/or magmatic processes whereby high-level continental crust can be translated long after original formation to deep crustal levels where it may undergo high-grade metamorphism and geochemical depletions. Conversely, deep crustal granulites may be tectonically transported to higher crustal levels, where they may subsequently suffer retrogression and metasomatic enrichment in LILE. The timing of such events may be elucidated from detailed Rb-Sr, U-Pb and Sm-Nd age and isotope studies.

Several workers (e.g. Tarney & Windley 1977; Weaver & Tarney 1984) have proposed models in which the characteristics of the lower continental crust are represented by Archaean granulite-facies terrains, because the latter record metamorphic conditions appropriate to the deep crust and frequently display severe depletion in heat-producing elements. Undepleted Archaean gra-

From: DAWSON, J.B., CARSWELL, D.A., HALL, J. & WEDEPOHL, K.H. (eds) 1986, *The Nature of the Lower Continental Crust*, Geological Society Special Publication No. 24, pp. 211–220.

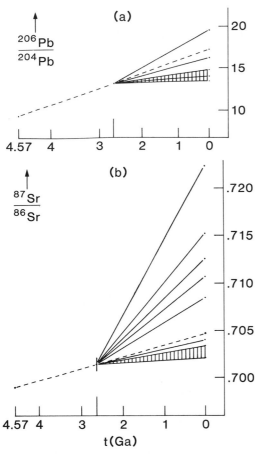

character little modified during metamorphism (Iyer *et al.* 1984), or whether metasomatic, selective enrichment in LILE occurred during retrograde metamorphism (Sighinolfi *et al.* 1981).

Some Proterozoic and Phanerozoic granulites of similar bulk composition to the above show little or no depletion in LILE. Such rocks may have developed in lower-P (high-T) regimes with steep thermal gradients (Heier 1973; Wells 1980, 1981). Depleted medium-to-high pressure granulites covering a large age range probably exist at depth throughout much of the continental crustal area, but the younger ones may not yet have been sufficiently uplifted to reach the surface.

Isotopic characteristics of lower crust generated during Crustal Accretion–Differentiation Superevents (CADS)

Following the demonstration of the depletion phenomenon in granulites, Moorbath *et al.* (1969) discovered that the present-day Pb isotopic composition of late Archaean Scourian granulites of N.W. Scotland is, on average, extremely unradiogenic, and in several samples has remained virtually unchanged since the rocks were metamorphosed some 2.7–2.8 Ga ago. Furthermore, the initial Pb isotopic composition considered in relation to plausible Pb isotopic evolution models for the mantle demonstrated that these granulites could not have had a crustal history prior to about 2.9 Ga. Thus, the time interval between extraction of the igneous precursors of these orthogneisses from the mantle and the high-grade metamorphism with accompanying U-depletion did not exceed about 100–200 Ma. This is illustrated in Fig. 1a, based on data of Chapman & Moorbath (1977). 75% of the analysed granulites fall in the hatched field, indicating grossly retarded Pb isotopic evolution compared with the mantle over the past 2.7 Ga. On a plot of $^{207}Pb/^{204}Pb$ *vs.* $^{206}Pb/^{204}Pb$ (not shown here, but see Chapman & Moorbath 1977) these data yield an age of 2680 ± 60 Ma, agreeing well with other age determinations on Scourian rocks. The model μ_1 ($^{238}U/^{204}Pb$) value for the source region of the gneisses is 7.7, the value for the Pb isotope primary growth line in Fig. 1a. This value is within the range typical of Archaean upper mantle source regions (Moorbath & Taylor 1981). There is also a good correlation (not shown here) between $^{208}Pb/^{204}Pb$ and $^{206}Pb/^{204}Pb$ data, yielding a value of 3.1 for the Th/U ratio of the gneisses themselves. This is in the range of many primary igneous rocks. In general, however, it

FIG. 1. (a) Pb-isotopic evolution for late Archaean granulites which underwent high-grade metamorphism and U-depletion soon after crust formation: the case of the 2.7 Ga Scourian granulites from the Lewisian complex of N.W. Scotland (data from Chapman & Moorbath 1977). Most of the analysed samples have developed to the present with much lower U/Pb ratios than the mantle (dashed line). The hatched area represents the evolution of unradiogenic Pb isotope compositions in fifteen out of twenty analysed samples. (b) Sr-isotopic evolution for Scourian granulite- (hatched area) and amphibolite- (individual growth lines) facies gneisses (data from Chapman 1978 and Moorbath *et al.* 1975). All the granulites underwent severe Rb depletion soon after crust formation. Model mantle evolution shown by dashed line.

nulites have also been reported, for example from Bahia in N.E. Brazil, but there is still debate as to whether the relatively high concentrations of radioactive and other LILE in the Jequié granulites represent the relict primary geochemical

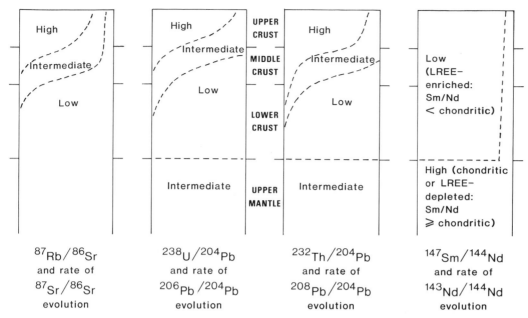

FIG. 2. Isotopic geochemistry of typical, bulk crust and mantle reservoirs. For further discussion see text.

appears that Th is more erratically depleted than U during granulite facies metamorphism (Moorbath *et al*. 1969; Gray & Oversby 1972).

Sr isotope data for Scourian granulites yield analogous information. Because of the extremely low Rb/Sr ratios resulting from Rb-depletion, measured $^{87}Sr/^{86}Sr$ ratios range from 0.7021 to 0.7034 (Chapman 1978), mostly within the same range as modern, sub-oceanic depleted mantle sources. Calculated model initial $^{87}Sr/^{86}Sr$ ratios at 2.7 Ga are close to 0.701, characteristic of average late Archaean mantle (Peterman 1979). This is illustrated in Fig. 1b, based on data from Chapman (1978) and Moorbath *et al*. (1975). All measured Scourian granulites fall in the hatched field, whilst amphibolite-facies Scourian gneisses from the southern Outer Hebrides have much higher Rb/Sr ratios and do not show the severe Rb-depletion of the granulites.

These observations, which are paralleled in other high-P,T granulite terrains, demonstrate one of the great contrasts between U-Pb and Rb-Sr systems. If Scourian granulites are characteristic of the lower continental crust, it follows that the lower continental crust and most upper mantle magma sources have similar average Rb/Sr ratios, but very different *average* U/Pb ratios. This means that whereas Sr isotopes cannot usually discriminate beween magmatic source regions in lower crust and upper mantle. Pb isotopes often can. *Within* the continental crust,

both Rb/Sr and U/Pb ratios vary, depending upon depth in the crust (through pressure, temperature and thence metamorphic grade) and rock composition, but generally increase upwards in the crust. This is shown *qualitatively* in Fig. 2, largely based on the authors' studies of several high-grade gneisses from cratonic areas bordering the N. Atlantic (e.g. N.W. Scotland, N. Norway, W. and E. Greenland). The intracrustal divisions in Fig. 2 cannot be given with greater precision because there are certainly some low-Rb/Sr reservoirs in the upper crust, whilst by no means all lower crustal granulites are felsic. There may be a substantial reservoir of mafic and ultramafic rocks in the lower crust and there are indications (Oxford unpublished data) that such rocks (especially the non-feldspathic ones) not only have very low U and Rb, but also very low Pb and Sr concentrations, resulting in comparatively high U/Pb and Rb/Sr ratios. This may account for the relatively radiogenic Pb and Sr isotope compositions observed in mantle-derived xenoliths. The proportion of mafic and ultramafic rocks in the lower crust, as well as the concentration ratios of incompatible elements in minerals from such rocks, are not yet well understood, but there is little doubt that Pb and Sr, as well as U and Rb, are highly incompatible in the major mafic minerals of the lower crust and mantle.

Upper amphibolite-facies gneisses marginally

qualify for inclusion in any consideration of lower continental crust. They show much greater range and variability in Rb/Sr, U/Pb and Th/Pb than granulites, with corresponding effects on isotopic ratios. Thus the early Archaean Amîtsoq gneisses of W. Greenland fall in the partially-depleted category, forming a geochemical transition between upper and lower crust. Pb-isotopic evolution is extremely retarded in most samples because of early U-depletion, with some present-day $^{206}Pb/^{204}Pb$ and $^{207}Pb/^{204}Pb$ ratios (~ 11.5, ~ 13.1) closely approaching the initial Pb isotopic compositions characteristic of a mantle-source region with $\mu_1 \sim 7.5$ at ~ 3.7 Ga (Black *et al.* 1971; Gancarz & Wasserburg 1977). In contrast, Rb is not depleted in most Amîtsoq gneisses, which have an average Rb/Sr of ~ 0.3, close to the average value for continental crust as a whole (Moorbath *et al.* 1972). Locally, Amîtsoq gneisses are in the granulite facies, where they show moderate Rb-depletion (Griffin *et al.* 1980). Analogous Pb and Sr isotopic relationships are found in granulite- and amphibolite-facies late-Archaean (~ 3.0–2.7 Ga) Nûk gneisses of W. Greenland (Black *et al.* 1973; Moorbath & Pankhurst 1976). The Pb-isotopic picture is complicated in areas of cratonic overlap between early and late Archaean sectors by contamination of Nûk magmas with unradiogenic Pb derived from lower crustal rocks believed to be deep-seated equivalents of the Amîtsoq gneisses (Taylor *et al.* 1980, 1984). Contamination of magmas by isotopically contrasted country rock components during emplacement and differentiation can provide a powerful tool for the detection and geochronological/geochemical characterization of ancient, depleted crust at depth (e.g. Taylor *et al.* 1980; Dickin 1981).

Recently, Sm-Nd studies have been applied to the timing and evolution of continental crust. As indicated in Fig. 2, the average Sm/Nd ratio for continental crust is comparatively homogeneous throughout ($^{147}Sm/^{144}Nd \sim 0.12$), being about 40% lower than for a chondritic mantle, so that the $^{143}Nd/^{144}Nd$ ratio in the continental crust as a whole evolves with time to less radiogenic Nd than most mantle magma sources. It follows that initial and isotopic ratios discriminate well between crust and mantle magma sources, because crust-mantle differentiation is the major cause of Sm/Nd fractionation. Isotopic evolution of Nd in the mantle through time is now well constrained, with the recognition that many crustal rocks were derived from a *depleted* mantle reservoir in which the average Sm/Nd ratio has been greater than chondritic since ~ 3.8 Ga ago (see Jacobsen & Wasserburg 1984).

Scourian gneisses (amphibolite- and granulite-facies) from N.W. Scotland have yielded a Sm-Nd whole rock isochron age of 2920 ± 50 Ma (Hamilton *et al.* 1979), regarded as the time at which their igneous protoliths separated from the mantle. Since most Rb-Sr, U-Pb and Pb/Pb age determinations on other regions of the Scourian Complex fall in the range ~ 2.65–2.75 Ga, Hamilton *et al.* (1979) proposed a thermal model requiring ~ 200 Ma for the newly accreted mantle-derived material to become metamorphically/geochemically differentiated and stabilized as granulite- and amphibolite-facies crust. They conclude that 'to avoid extensive melting, heat-producing elements must have commenced migration from the lower crust < 100 Ma after crustal thickening. The mechanism of this migration is still not clear, but did not involve a transport medium which would significantly affect the Sm-Nd systematics, as would a magmatic phase'. This agrees with the view of Lambert (1976) that unless the radioactive elements were strongly concentrated in the upper part of the crust within 50–100 Ma after an accretion episode, widespread crustal melting by internal radiogenic heat production would be inevitable in the Archaean. Where the redistribution of heat-producing elements is slower, intracrustal melting may be an important factor, eventually producing a lower crust of refractory, depleted granulite restites. Many ancient shield areas at intermediate-to-high crustal levels display progressively more evolved magmatism extending over a period of several hundred million years. In some cases, isotopic measurements clearly identify a succession of intracrustal melts derived from slightly older crust produced during the same CADS, with or without contributions from much older continental crust produced during some earlier CADS. For example, in the late Archaean of W. Greenland, there is a progression from calc-alkaline orthogneisses to granites, with progressively higher initial $^{87}Sr/^{86}Sr$, extending over a period of some 300–400 Ma (Moorbath & Pankhurst 1976; Moorbath *et al.* 1981; Robertson 1983).

These thermal models for individual terrains, including the Scourian granulites, require detailed testing by carrying out Rb-Sr, U-Pb, Pb/Pb, Sm-Nd analyses on the *same* sample suites, whenever possible. Thermal modelling requires precise knowledge of the time interval between crustal accretion (mantle differentiation) and the peak of metamorphic/geochemical differentiation. Wells (1980, 1981) has shown how different magmatic crustal accretion mechanisms (e.g. overaccretion, underaccretion) can influence thermal behaviour and temperature-related processes in juvenile and intruded crust. Whilst concordant, or nearly concordant, whole-rock Rb-Sr and Pb/Pb (as well as zircon U-Pb) dates, with mantle-type initial Sr and Pb isotope compositions, in a given terrain are in themselves strong evidence for a major, short-term CADS (Moorbath & Taylor 1981), the Sm-Nd data can, in principle, yield more precise constraints on the timing of crust-mantle differentiation. However, in some cases the resolution of the combined dating methods may be inadequate for timing the different stages in the development of a short-term CADS, particularly when the total time interval involved is less than about 50–100 Ma.

To illustrate the above points, we mention a few examples of published multi-isotopic studies in granulite terrains which appear to have evolved within a single, short-term CADS. Rb-Sr, U-Pb, Pb/Pb and Sm-Nd age data on mafic granulites in Finnish Lapland are all in agreement within analytical error and show that the emplacement of igneous rocks and the granulite-facies metamorphism occurred at ∼1.9–2.0 Ga ago (Bernard-Griffiths *et al.* 1984). Field observations in this region also indicate that magmatism and granulite-facies metamorphism were almost synchronous. In contrast, in the Suomussalmi gneisses of E. Finland, whole-rock Rb-Sr, Pb/Pb and Sm-Nd dates indicate that crustal accretion and crustal differentiation extended over a period of some 350 Ma (Vidal *et al.* 1980; Martin *et al.* 1983). In China, granulite gneisses of the Qianxi Group from eastern Hebei province yield concordant whole-rock Rb-Sr, Sm-Nd, and zircon U-Pb dates of ∼2.5 Ga (Jahn & Zhang 1984). Initial Sr and Nd isotopic compositions demonstrate that granulite facies metamorphism closely followed emplacement of the mantle-derived protoliths. Previous claims for early Archaean ages (> 3.5 Ga) of the granulites were not substantiated. In W. Greenland, whole-rock Rb-Sr, Pb/Pb and Sm-Nd dates on early Archaean granulite-facies Amîtsoq gneisses are analytically indistinguishable from dates obtained by the same methods (as well as zircon U-Pb) on amphibolite-facies Amîtsoq gneisses, falling in the range 3.55 to 3.70 Ga (Griffin *et al.* 1980). The Sm-Nd whole-rock isochron age of the Amîtsoq gneisses, not previously reported, is 3627 ± 48 Ma, obtained on a suite of eleven amphibolite- and granulite-facies samples. The few examples above contradict the conclusion of Ben Othman *et al.* (1984) that 'internal' differentiation of continental crust (granulite-facies metamorphism) usually follows 'external' differentiation (mantle differentiation) with a very large time interval. This simplistic view was criticized by Moorbath (1984).

Isotopic characteristics of terrains subjected to multi-stage crustal evolution

Isotopic systematics can be severely disturbed and disrupted by tectonism, metamorphism, metasomatism and alteration. Open system behaviour, with partial or complete resetting of ages has been demonstrated in many minerals, as well as in whole-rock Rb-Sr systems (e.g. Field & Råheim 1979, 1980; Cameron *et al.* 1981). We have also observed it in whole-rock Pb/Pb systems in some ancient granitoid gneisses (Oxford unpublished data). A truly multi-isotopic approach (i.e. whole-rock Rb-Sr, Pb/Pb, Sm-Nd, plus zircon U-Pb) can nevertheless yield fundamental information on timing of tectonothermal events, crustal residence age and petrogenesis of rock units, on account of the differential response of these isotopic systems to later events. Here we discuss some simple cases relevant to granulite complexes in the lower continental crust.

First, we consider normal upper crust which undergoes a much later depletion event. Rb-Sr, Pb/Pb and Sm-Nd isotopic studies (Griffin *et al.* 1978; Jacobsen & Wasserburg 1978) on the Vikan granulites of N. Norway demonstrate that the protoliths of these rocks were added from mantle to crust at ∼2.7 Ga ago with a wide range of U/Pb ratios mostly characteristic of upper continental crust. Subsequently they underwent granulite-facies metamorphism with severe U-depletion at ∼1.8 Ga, so that Pb-isotopic evolution at that time became severely retarded. This is illustrated in Fig. 3a for three of the analysed samples. Clearly, Pb-isotopic compositions at 1.8 Ga ago were heterogeneous and mostly much more radiogenic than contemporaneous mantle. There is no suggestion that granulite facies metamorphism has homogenized the Pb-isotopic ratios (see also Gray & Oversby 1972). Originally the linear array of data points in a $^{206}Pb/^{204}Pb$ vs. $^{207}Pb/^{204}Pb$ diagram was interpreted on the basis of a simple secondary-isochron model, yielding an apparent age of 3460 ± 70 Ma, which was taken as the primary age of the rocks (Taylor 1975). However, a ∼2.7 Ga Pb/Pb isochron for Archaean gneisses beyond the area affected by 1.8 Ga granulite facies metamorphism (Griffin *et al.* 1978), as well as Sm-Nd work on Vikan granulites (Jacobsen & Wasserburg 1978) which yielded a t_{CHUR} model (mantle differentiation) age of ∼2.64 Ga, provided the evidence for the currently accepted model of three-stage Pb-isotopic evolution (Fig. 3a, and Moorbath & Taylor 1981). Similarly, a previously reported whole-rock Pb/Pb age of ∼4.0 Ga (Sobotovich *et al.* 1976) from

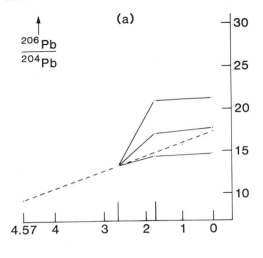

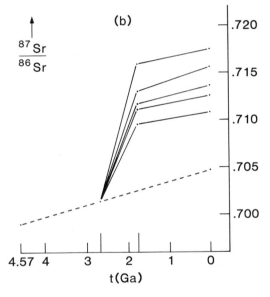

granulite-facies gneisses at Fyfe Hills, Enderby Land, Antarctica, is now regarded as an artefact produced by an erroneous secondary-isochron (two-stage) interpretation of a three-stage Pb-isotopic evolution (Grew & Manton 1979). Subsequent Sm-Nd analyses on these gneisses have yielded model ages of ~ 3.5 Ga (De Paolo *et al.* 1982), whilst two regional episodes of granulite-facies metamorphism have been documented at ~ 3.1 and ~ 2.5 Ga (Black *et al.* 1983).

A similar situation to the above can also occur in the Rb-Sr system. Fig. 3b shows a Sr-isotopic evolution diagram for granulites at Ryggedals-vatn on Langøy in North Norway (P.N. Taylor, unpublished data). The protoliths probably differentiated from the mantle at ~ 2.7 Ga and then developed with a range of normal, upper crustal Rb/Sr ratios, subsequently undergoing Rb-depletion at ~ 1.8 Ga during granulite-facies metamorphism, but without Sr-isotopic homogenization. Consequently, $^{87}Sr/^{86}Sr$ ratios of these granulites at 1.8 Ga were heterogeneous, and much higher than contemporaneous mantle. On a conventional $^{87}Rb/^{86}Sr$ vs. $^{87}Sr/^{86}Sr$ isochron diagram (not shown), the data points scatter well above, and to the left of, any plausible reference isochron line.

Next, we consider depleted crust which undergoes a much later enrichment (metasomatic) event. In Fig. 4a we show the predicted effect of variable Rb addition at time t_2 on a suite of ancient, low-Rb/Sr granulites, such as the Scourian, which have maintained an unradiogenic and relatively homogeneous Sr-isotopic composition. Clearly, because the first (depleted) crustal stage develops in such close proximity to the mantle evolution line, Sr isotopes alone cannot distinguish between a 2.7 Ga and a 1.8 Ga mantle origin for the protoliths. The most likely outcome would be a 1.8 Ga Rb-Sr isochron with mantle-type initial ratio. It is not so much a case of *erasure* of previous crustal history, as lack of resolvable contrast between isotopic evolution of crust and mantle.

In contrast, U-depleted granulite crust has a much more retarded Pb-isotopic evolution than most mantle magma sources (Figs 1a, 2), so that the intial Pb-isotopic composition at the time of late U-addition would be much less radiogenic than in contemporaneous mantle. Figs 4b and 4c show two possible situations involving U-addition at 1.8 Ga to 2.7 Ga-old depleted granulites. Fig. 4b illustrates Pb-isotopic homogenization through crustal anatexis. The homogeneous initial Pb-isotopic composition is much less radiogenic than contemporaneous mantle. Such a situation has been fully documented (Taylor *et al.* 1984) for the ~ 1.74 Ga-old Badcall Quay Red Granite (BQRG) at the boundary between the

FIG. 3. (a) Pb-isotopic evolution for late Archaean (~ 2.7 Ga) rocks subjected to granulite-facies metamorphism and U-depletion at ~ 1.8 Ga: the case of the Vikan gneisses, western Langøy, N. Norway (data from Taylor 1975 and Griffin *et al.* 1978). The three samples illustrated encompass the entire range of Pb isotopic ratios measured in these rocks today. Note that there was no Pb-isotopic homogenization during granulite-facies metamorphism, (b) Sr-isotopic evolution for late Archaean (~ 2.7 Ga) rocks subjected to granulite-facies metamorphism and Rb-depletion at ~ 1.8 Ga; the case of the Ryggedalsvatn granulites, central Langøy, N. Norway (P.N. Taylor, unpublished data). Note that there was no Sr-isotopic homogenization during granulite-facies metamorphism.

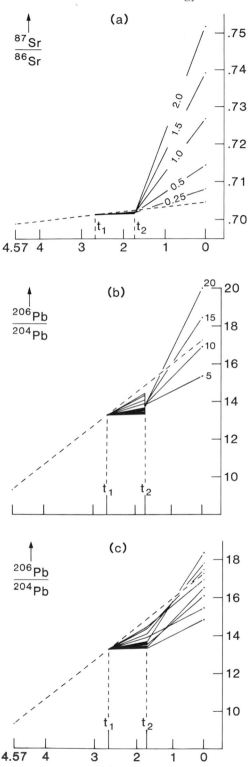

Scourian and Laxfordian sectors of the Lewisian complex of N.W. Scotland. In this granite some samples show evidence for significant U-enrichment at ~ 1.74 Ga. Pb, Nd and Sr initial isotopic compositions demonstrate that this is a crustal melt derived from a basement with the geochemical characteristics of the ~ 2.7 Ga-old Scourian Complex, except that the immediate precursors of the BQRG had slightly higher Rb/Sr ratios than the most depleted exposed Scourian granulites. Most significantly, however, the Sm-Nd model age of the BQRG is ~ 2.75 Ga, agreeing well with the independently determined age for many Scourian rocks.

Fig. 4c illustrates U-addition at time t_2 without Pb-isotopic homogenization, corresponding to locally variable metasomatic retrogression of granulites. It is premature to quote clear-cut examples, but they may well turn up with further work on sheared and retrogressed terrains. In both cases illustrated in Figs 4b and 4c there is no question of erasure of the previous crustal Pb-isotopic record since it is evident that U was added to ancient, low-U/Pb crust at time t_2.

Sm-Nd data for high-grade metamorphic rocks can constrain the timing of crust mantle differentiation, even where subsequent depletion or enrichment events, or any combination of them, has caused isotopic disturbance with partial or complete age resetting in the Rb-Sr and U-Pb systems.

Discussion and conclusions

Isotopic criteria can distinguish juvenile, mantle-derived sialic crust from reworked sialic crust. Partial melting and reworking of ancient continental crust can give rise to rocks with a wide variety of initial Sr, Nd and Pb isotopic composi-

FIG. 4. (a) Effect of variable Rb-addition at time t_2 on a suite of low-Rb/Sr granulites formed at time t_1 which preserved a relatively homogeneous, unradiogenic Sr-isotopic composition between t_1 and t_2. The numbers represent the new Rb/Sr ratios. For further discussion, see text, (b) effect of U-addition at time t_2 to a suite of low-U/Pb granulites, formed at time t_1, with Pb-isotopic homogenization at t_2 during crustal anatexis, (c) ditto, but without Pb-isotopic homogenization at t_2 during metamorphic retrogression.

Note that 4(b) corresponds closely to the petrogenetic history of the Badcall Quay Red Granite, Loch Laxford, N.W. Scotland (Taylor *et al.* 1984), which was derived by anatexis of U-depleted Scourian gneisses.

tions, depending upon the crustal level and composition of the melted or reworked rocks. In general, crustal radiogenic isotopic characteristics differ significantly from the combined radiogenic isotopic characteristics of a contemporaneous depleted mantle magma source region (Fig. 2). Of course, a combination of crustal and mantle isotopic characteristics can be produced in mantle-derived magmas which undergo crustal interaction and contamination, although various petrological, chemical and isotopic criteria can be used to recognize this phenomenon.

Hart *et al.* (1981), Welke & Nicolaysen (1981) and Bridgwater *et al.* (1981) have proposed that some Precambrian rock units have crustal residence ages greatly in excess of concordant dates given by isochrons with mantle-type initial ratios. These workers, together with Bridgwater & Collerson (1976) and Collerson & Fryer (1978), maintain that the radiogenic-isotope memory of a lengthy crustal residence time within a rock unit can be totally erased. The controversy hinges on whether or not isotopic chronometers can be *completely* reset, thus giving the appearance of an ancient rock unit having been newly derived from a mantle source long after the actual time of crust-mantle differentiation and crustal accretion. This implies that isotopic resetting results from one or more of the following processes: (1) complete expulsion of parent and/or daughter isotopes and elements out of a rock, or even out of an entire sector of continental crust, by high-grade metamorphism or by metasomatic 'flushing', (2) massive metasomatic influx of parent and/or daughter isotopes and elements into the rock from a sub-crustal (mantle) source expelling fluids charged with incompatible elements, (3) large-scale isotopic rehomogenization of daughter isotopes and elements, (4) large-scale redistribution of parent elements within the rock unit.

The result of the right combination of these processes could then hypothetically lead to the complete erasure of all previous age and isotope record from an ancient sector of continental crust, yielding concordant 'resetting' dates, accompanied by mantle-type initial isotopic ratios. This situation has actually been proposed for two granulite terrains namely in the Vredefort Dome, South Africa (Hart *et al.* 1981; Welke & Nicolaysen 1981) and at Kangimut sangmissoq, Godthaabsfjord, West Greenland (Bridgwater *et al.* 1981; V.R. McGregor, pers. comm.). In each case, concordant mid-to-late Archaean dates

obtained from several decay schemes, together with mantle-type initial ratios, are interpreted in terms of total age and isotope resetting of early Archaean crustal rocks. We have presented a more detailed criticism elsewhere (Moorbath & Taylor 1985), where we voice our scepticism at the treatment of the continental crust as an efficient chromatographic column in which parent and daughter isotopes and elements (and other LILE) can be quantitatively eluted by metamorphism and/or metasomatic fluids, with subsequent total replacement by juvenile, mantle-derived components of the same type. This view would be more plausible with evidence for unretrogressed granulites (particularly felsic ones) showing gross depletion of such elements as Pb, Sr, Nd which carry the previous isotopic memory. Such evidence has not yet been reported.

Not many individual terrains have yet been studied by all available isotopic methods (particularly by Sm–Nd), but the combined isotopic approach has already shown that every high-grade metamorphic terrain must be treated on its own merits. In particular, crustal residence time varies widely both within a single terrain, and between different terrains, suggesting that no single specific tectonic model, such as that proposed by Ben Othman *et al.* (1984), can possibly apply to all granulite terrains. In some cases P,T-conditions in the deep crust culminate in granulite-facies metamorphism penecontemporaneously with a major crustal accretion event, whilst in other cases granulite-facies metamorphism is superimposed on very much older igneous, sedimentary or metamorphic rocks during a much later tectonothermal episode quite unrelated to the primary CADS. Indeed rocks which have been metamorphosed at high-T, low-P, granulite-facies conditions may never have formed part of the lower continental crust at all. In addition, granulites may be geochemically depleted, or non-depleted, or secondarily enriched in radioactive elements and other LILE by a variety of processes dependent upon their mineralogical and chemical composition, tectonic environment, crustal depth, structural history, etc.

In conclusion, there are evidently 'granulites and granulites', and our main contention here is that each terrain must be examined isotopically (and by other techniques) in detail in order to elucidate its own particular history and evolution, and the part which it played in the overall regional and global scheme.

References

BEN OTHMAN, D., POLVÉ, M.& ALLÈGRE, C.J. 1984. Nd-Sr isotopic composition of granulites and constraints on the evolution of the lower continental crust. *Nature*, **307**, 510–15.

BERNARD-GRIFFITHS, J., PEUCAT, J.J., POSTAIRE, B., VIDAL, P., CONVERT, J. & MOREAU, B. 1984. Isotopic data (U-Pb, Rb-Sr, Pb-Pb, Sm-Nd) on mafic granulites from Finnish Lapland. *Precambrian Res.* **23**, 325–48.

BLACK, L.P., GALE, N.H., MOORBATH, S., PANKHURST, R.J. & McGREGOR, V.R. 1971. Isotopic dating of very early Precambrian amphibolite facies gneisses from the Godthaab district, West Greenland. *Earth Planet. Sci. Lett.* **12**, 245–59.

——, JAMES, P.R. & HARLEY, S.L. 1983. The geochronology, structure and metamorphism of early Archaean rocks at Fyfe Hills, Enderby Land, Antarctica. *Precambrian Res.* **21**, 197–222.

——, MOORBATH, S., PANKHURST, R.J. & WINDLEY, B.F. 1973. $^{207}Pb/^{206}Pb$ whole-rock age of the Archaean granulite facies metamorphic event in West Greenland. *Nature Phys. Sci.* **244**, 50–3.

BRIDGWATER, D. & COLLERSON, K.D. 1976. The major petrological and geochemical characters of the 3600 m.y. Uivak gneisses from Labrador. *Contrib. Mineral. Petrol.* **54**, 43–59.

——, McGREGOR, V.R. & NUTMAN A. 1981. Geologic constraints on isotopic and thermal models for the consolidation of the late Archaean crust, West Greenland (abstract). *Terra Cognita, Spec. Issue,* **93**.

CAMERON, M., COLLERSON, K.D., COMPSTON, W. & MORTON, R. 1981. The statistical analysis and interpretation of imperfectly-fitted Rb-Sr isochrons from polymetamorphic terrains. *Geochim. cosmochim. Acta* **45**, 1087–97.

CHAPMAN, H.J. 1978. *Geochronology and isotope geochemistry of Precambrian rocks from northwest Scotland.* D.Phil. thesis Oxford University.

—— & MOORBATH, S. 1977. Lead isotope measurements from the oldest recognised Lewisian gneisses of northwest Scotland. *Nature*, **268**, 41–2.

COLLERSON, K.D. & FRYER, B.J. 1978. The role of fluids in the formation and subsequent development of early continental crust. *Contrib. Mineral. Petrol.* **67**, 151–67.

DE PAOLO, D.J., MANTON, W.I., GREW, E.S. & HALPERN, M. 1982. Sm-Nd, Rb-Sr, and U-Th-Pb systematics of granulite facies rocks from Fyfe Hills, Enderby Land, Antarctica. *Nature*, **298**, 614–18.

DICKIN, A.P. 1981. Isotope geochemistry of Tertiary igneous rocks from the Isle of Skye, N.W. Scotland. *J. Petrol.* **22**, 155–89.

FIELD, D. & RÅHEIM, A. 1979. Rb-Sr total rock isotope studies on Precambrian charnockitic gneisses from South Norway: evidence for isochron resetting during a low-grade metamorphic-deformational event. *Earth planet. Sci. Lett.* **45**, 32–44.

—— & —— 1980. Secondary geologically meaningless Rb-Sr isochrons, low $^{87}Sr/^{86}Sr$ initial ratios and crustal residence times of high-grade gneisses. *Lithos*, **13**, 295–304.

GANCARZ, A.J. & WASSERBURG, G.J. 1977. Initial Pb of the Amîtsoq gneiss West Greenland, and implications for the age of the earth. *Geochim. cosmochim. Acta*, **41**, 1283–1301.

GRAY, C.M. & OVERSBY, V.M. 1972. The behaviour of lead isotopes during granulite facies metamorphism. *Geochim. cosmochim. Acta*, **36**, 939–52.

GREW, E.S. & MANTON, W.I. 1979. Archaean rocks in Antarctica: 2.5 billion-year uranium-lead ages of pegmatites in Enderby Land. *Science*, **206**, 443–5.

GRIFFIN, W.L., McGREGOR, V.R., NUTMAN, A., TAYLOR, P.N. & BRIDGWATER, D. 1980. Early Archaean granulite-facies metamorphism south of Ameralik, West Greenland. *Earth planet. Sci. Lett.* **50**, 59–74.

——, TAYLOR, P.N. et al. 1978. Archaean and Proterozoic crustal evolution in Lofoten-Vesterålen, N. Norway. *J. geol. Soc. Lond.* **135**, 629–47.

HAMILTON, P.J., EVENSEN, N.M., O'NIONS, R.K. & TARNEY, J. 1979. Sm-Nd systematics of Lewisian gneisses: implications for the origin of granulites. *Nature*, **277**, 25–28.

HART, R.J., WELKE, H.J. & NICOLAYSEN, L.O. 1981. Geochronology of the deep profile through Archaean basement at Vredefort with implications for early crustal evolution. *J. geophys. Res.* **86**, 10663–80.

HEIER, K.S. 1973. Geochemistry of granulite facies rocks and problems of their origin. *Phil. Trans. R. Soc. Lond.* **A273**, 429–42.

IYER, S.S., CHOUDHURI, A., VASCONCELLOS, M.B.A. & CORDANI, U.G. 1984. Radioactive element distribution in the Archaean granulite terrane of Jequié, Bahia, Brazil. *Contrib. Mineral. Petrol.* **85**, 95–101.

JACOBSEN, S.B. & WASSERBURG, G.J. 1978. Interpretation of Nd, Sr and Pb isotope data from Archaean migmatites in Lofoten-Vesteralen, Norway. *Earth Planet. Sci. Lett.* **41**, 245–53.

—— & —— 1984. Sm-Nd isotopic evolution of chondrites and achondrites, II. *Earth planet. Sci. Lett.* **67**, 137–50.

JAHN, B.M. & ZHANG, Z.Q. 1984. Archaean granulite gneisses from eastern Hebei Province, China: rare earth geochemistry and tectonic implications. *Contrib. Mineral. Petrol.* **85**, 224–43.

LACHENBRUCH, A.H. 1970. Crustal temperature and heat production: implications of the linear heat-flow relation. *J. geophys. Res.* **75**, 3291–300.

LAMBERT, R.ST.J. 1976. Archaean thermal regimes, crustal and upper mantle temperatures, and a progressive evolutionary model for the earth. *In*: WINDLEY, B.F. (ed.) *The Early History of the Earth*, Wiley, London, 363–73.

LAMBERT, I.B. & HEIER, K.S. 1967. The vertical distribution of uranium, thorium and potassium in the continental crust. *Geochim. cosmochim. Acta*, **31**, 377–90.

MARTIN, H., CHAUVEL, C., JAHN, B.M. & VIDAL, P. 1983. Rb-Sr and Sm-Nd ages and isotopic geochemistry of Archaean granodiorite gneisses from eastern Finland. *Precambrian Res.* **20**, 79–91.

MOORBATH, S. 1984. Origin of granulites. *Nature*, **312**, 290.

——, O'NIONS, R.K., PANKHURST, R.J. & McGREGOR, V.R. 1972. Further rubidium-strontium age determinations on the very early Precambrian rocks of the Godthaab district, West Greenland. *Nature Phys. Sci.* **240**, 78–82.

—— & PANKHURST, R.J. 1976. Further rubidium-strontium age and isotope evidence for the nature of the late Archaean plutonic event in West Greenland. *Nature*, **262**, 124–6.

——, POWELL, J.L. & TAYLOR, P.N. 1975. Isotopic evidence for the age and origin of the "grey gneiss" complex of the southern Outer Hebrides, Scotland. *J. geol. Soc. Lond.* **131**, 213–22.

—— & TAYLOR, P.N. 1981. Isotopic evidence for continental growth in the Precambrian. *In*: KRÖNER, A. (ed.) *Precambrian Plate Tectonics*, Elsevier, Amsterdam, 491–525.

—— & —— 1985. Precambrian geochronology and the geological record. *In*: SNELLING, N.J. (ed.) *The Chronology of the Geological Record.* Geol. Soc. Lond. Memoir No. 10, 10–28.

——, TAYLOR, P.N. & GOODWIN, R. 1981. Origin of granitic magma by crustal remobilisation: Rb-Sr and Pb/Pb geochronology and isotope geochemistry of late Archaean Qôrqut granite complex of southern West Greenland. *Geochim. et cosmochim. Acta*, **45**, 1051–60.

——, WELKE, H. & GALE, N.H. 1969. The significance of lead isotope studies in ancient, high-grade metamorphic basement complexes, as exemplifed by the Lewisian rocks of northwest Scotland. *Earth planet. Sci. Lett.* **6**, 245–56.

PETERMAN, Z.E. 1979. Strontium isotope geochemistry of late Archaean to late Cretaceous tonalites and trondhjemites. *In*: BARKER, F. (ed.) *Trondhjemites, Dacites and Related Rocks*, Elsevier, Amsterdam, 133–47.

ROBERTSON, S. 1983. Provisional results of isotope investigations into quartzo-feldspathic rocks from Kangiussap nunâ, Ivisârtoq sheet, southern West Greenland. *Geol. Surv. Greenland Report* No. 115, 56–9.

ROY, R.F., BLACKWELL, D.D. & BIRCH, F. 1968. Heat generation of plutonic rocks and continental heat flow provinces. *Earth planet. Sci. Lett.* **5**, 1–12.

SIGHINOLFI, G.P. 1971. Investigations into deep crustal levels: Fractionating effects and geochemical trends related to high-grade metamorphism. *Geochim. cosmochim. Acta*, **35**, 1005–21.

——, FIGUEREDO, M.C.H., FYFE, W.S., KRONBERG, I. & TANNER OLIVEIRA, M.A.F. 1981. Geochemistry and petrology of the Jequié granulitic complex. *Contrib. Mineral. Petrol.* **78**, 263–71.

SOBOTOVICH, E.V., KAMENEV, Y.N., KOMARISTY, A.A. & RUDNICK, V.A. 1976. The oldest rocks of Antarctica (Enderby Land). *Int. Geol. Rev.* **18**, 371–88.

TARNEY, J. & WINDLEY, B.F. 1977. Chemistry, thermal gradients and evolution of the lower continental crust. *J. geol. Soc. Lond.* **134**, 153–72.

TAYLOR, P.N. 1975. An early Precambrian age for migmatitic gneisses from Vikan i Bø, Vesteralen, North Norway. *Earth Planet. Sci. Lett.*, **27**, 35–42.

——, JONES, N.W. & MOORBATH, S. 1984. Isotopic assessment of relative contributions from crust and mantle sources to the magma genesis of Precambrian granitoid rocks. *Phil. Trans. R. Soc. Lond.*, **A310**, 605–25.

——, MOORBATH, S., GOODWIN, R. & PETRYKOWSKI, A.C. 1980. Crustal contamination as an indicator of the extent of early Archaean continental crust: Pb isotopic evidence from late Archaean gneisses of West Greenland. *Geochim. cosmochim. Acta*, **44**, 1437–53.

VIDAL, P., BLAIS, B., JAHN, B.M., CAPDEVILA, R. & TILTON, G.R. 1980. U-Pb and Rb-Sr systematics of the Suomussalmi Archaean greenstone belt (Eastern Finland). *Geochim. cosmochim Acta*, **44**, 2033–44.

WEAVER, B.L. & TARNEY, J. 1984. Empirical approach to estimating the composition of the continental crust. *Nature*, **310**, 575–7.

WELKE, H. & NICOLAYSEN, L.O. 1981. A new interpretative procedure for whole-rock U-Pb systems applied to the Vredefort crustal profile. *J. Geophys. Res.* **86**, 10681–7.

WELLS, P.R. 1980. Thermal models for the magmatic accretion and subsequent metamorphism of continental crust. *Earth planet. Sci. Lett.* **46**, 253–65.

—— 1981. Accretion of continental crust: thermal and geochemical consequences. *Phil. Trans. R. Soc. Lond.* **A301**, 347–57.

S. Moorbath & P.N. Taylor, Dept. of Earth Sciences, Parks Road, Oxford OX1 3PR, UK.

The structural evolution of the lower crust of orogenic belts, present and past

B.F. Windley & J. Tarney

SUMMARY: The lower crust of orogenic belts may be exposed by post-growth uplift in magmatic arcs or by tectonic uplift on post-collisional crustal-scale thrusts. Each tectonic zone of a belt can be expected to have a different type of lower crust, and there is a similar tectonic zonation in collisional belts ranging from the early Proterozoic to the Tertiary. Tonalitic gneisses make up the bulk of the lower crust of Andean-type arcs. Deep sections of sutures may expose nepheline gneisses, blueschists and granulites. Exposed basic dyke swarms represent the feeders to eroded lavas in aulacogens in continental forelands. Post-collisional thrusts uplift earlier lower crust of amphibolite facies gneisses or granulites, of either igneous or sedimentary parentage. The result of the many factors controlling variable uplift in the different tectonic zones of an orogenic belt is that a deep-seated cross-section can only be constructed by compilation of type sections from different belts.

Crustal generation in the Archaean was dominantly magmatic and broadly similar to that seen in the deeper parts of Andean-type arcs, but on a much more extensive scale. Magmatic underplating, associated with strong horizontal thrusting, produced a strongly foliated lower crust with interlayering of mafic and silicic components. Archaean crust probably had a primary vertical zonation with respect to the proportion of these components and of heat-producing elements.

Introduction

The tectonic make-up of the deep continental crust is as problematic as its compositional nature. For instance, the nature and structure of the deep crust may vary considerably, both in space (across a modern orogenic belt) and in time (e.g. between the Archaean and the late Phanerozoic). In modern belts formed by plate tectonic processes, island arcs or Andean-type batholiths represent the primary magmatic stage of continental accretion, and there are only a few exposed sections where their deeper root zones may be examined. It is important to distinguish the structures in these Andean roots from those which form during the collisional and post-collisional stages of Himalayan-type orogenic belts.

Much information is becoming available on the geology of high-level sections of Himalayan-type orogenic belts, but much less is known about the deep structure of such belts simply because uplift and erosion have generally not yet proceeded far enough to expose deep crustal levels. Nevertheless there are several belts, ranging from Proterozoic to Tertiary in age, in which segments have been uplifted or upturned to expose deep crustal levels. These belts comprise a wide variety of internal tectonic zones, each of which surprisingly has a different type of deeper crust component (i.e. segments recording a high P–T history). In order to understand the lower crust of orogenic belts it is necessary to consider the lower crust components of each tectonic zone and then piece them together to arrive at a composite picture of the belt as a whole.

There is much evidence to suggest that modern-style plate tectonic processes have been in operation since the early Proterozoic (cf. Windley 1984). For this reason a similar tectonic zonation may be found across collisional orogenic belts ranging from early Proterozoic to Tertiary in age, and we may expect to find broad similarities in the nature and structure of the lower crust in the separate tectonic zones within these belts. However in the Archaean this zonality is lacking and it is logical to assume that there may be differences in the lower crustal make-up too. Thus we need to consider Archaean belts separately from the early Proterozoic–Tertiary belts. In this paper we aim (1) to describe the deep-seated sections of modern Andean-type root zones (2) to analyse the structure and geology of the lower crust in the different tectonic zones of Phanerozoic orogenic belts (3) to examine the structures in the lower crust of Proterozoic belts from a modern tectonic standpoint and (4) to consider the growth and development of the lower crust in Archaean granulite-gneiss belts.

Modern island arcs and Andean-type cordilleran belts

Island arcs and Andean-type continental margins are currently the major zones of magmatic addi-

From: DAWSON, J.B., CARSWELL, D.A., HALL, J. & WEDEPOHL, K.H. (eds) 1986, *The Nature of the Lower Continental Crust*, Geological Society Special Publication No. 24, pp. 221–230.

tion to the continents, the former predominantly volcanic, the latter dominantly plutonic. Indeed the abundance of andesite in circum-Pacific volcanic arcs has been the basis for the suggestion (Taylor 1977) that the continental crust is essentially andesitic in bulk composition and has grown through time by the lateral accretion and remobilization of island arcs and collisional belts. While island arcs certainly constitute a component of Phanerozoic collisional belts (see below) we suspect that plutonic additions play a far greater role in the generation of continental crust.

Modern island arcs, such as those in the Western Pacific, evolve very rapidly over time scales of the order of a few tens of millions of years and are volumetrically significant. However recent studies of Western Pacific island arcs and remnant arcs through dredging and deep sea drilling (summarized in Natland & Tarney 1982) reveal that they are dominantly basaltic rather than andesitic in composition and may be underlain by significant anorthositic gabbro cumulates. Although not suitable as precursor continental material (i.e. remobilization would produce only minor amounts of silicic magma) they may nevertheless be difficult to subduct and be incorporated in collisional belts or exotic accreted terrains. It is being increasingly recognized in fact that several ophiolites (Oman, Cyprus) may be parts of arc-marginal basin systems obducted onto the upper crust and it is potentially possible that others may be partly subducted and emplaced into the lower crust at continental margins.

Voluminous batholiths are a conspicuous feature of circum-Pacific cordilleran belts and their isotopic ratios demonstrate that they are essentially new additions of sialic material. Plutonism in the Andes extends from the Upper Palaeozoic through to the late Tertiary and has added several hundred kilometres of new crust to the western margin of South America during that time. The uplift of the Andes in the Miocene has meant that much of the batholith beneath the volcano-sedimentary cover is now being unroofed, although in northern Chile and Peru the low rate of erosion has exposed only the upper parts of the plutons. Tonalite forms up to 55% of the exposed plutonic rock types (Pitcher 1978).

In southern Chile however deeper levels of the Patagonian batholith are exposed. This has resulted from two factors. First the rate of erosion has been much greater as a result of higher rainfall and glacial activity. Second, the region was the site of intra-arc extension both in the late Jurassic (Dalziel *et al*. 1974) and in the early Tertiary (Bartholomew & Tarney 1984) as a result of which small volcano-sedimentary basins

formed: subsequent closure of the Tertiary basin for instance was accompanied by major uplift on its eastern margin such that deep levels of the Mesozoic batholith have been up-thrust along the edge of the intra-arc basin. These deep level rocks are mostly tonalites and diorites with subsidiary gabbros, complexly intermixed and variably, but often strongly, foliated such that in places they superficially resemble Precambrian gneisses. Similar rock types are found in the Palaeozoic batholith in central Chile, south of Valparaiso. In both cases there are layers of metamorphosed and deformed sediments within the foliated tonalites. The implication is that whereas the upper crustal component of cordilleran batholiths is dominantly granite, granodiorite and tonalite, the lower crustal rock types are tonalites, diorites and gabbros, frequently foliated, and the more mafic rock types being disrupted and broken up by the more siliceous intrusions and aligned in lensoid fashion parallel to the regional foliation which dips gently toward the continent. It is important to note that the foliation and disruption of the mafic intrusions is synplutonic and was developed penecontemporaneously with the emplacement of the batholith. The implication is that the deep crust might acquire a foliation with associated lithological layering during the processes of crustal generation, and does not necessarily have to be imposed during later collisional belt orogenesis.

Phanerozoic collisional orogenic belts

In this section we shall describe the different tectonic zones of Palaeozoic–Cenozoic orogenic belts in terms of the pre-collisional tectonic setting or environment of the respective components.

(1) Aulacogens

Aulacogens are faulted rift zones which extend from former triple junctions far into the continental borderland of an orogenic belt. They typically contain a considerable thickness of predominantly clastic sediments with lava flows of tholeiitic and/or alkali basalts at several, but particularly at lower stratigraphic levels (Hoffman 1980).

In a deeply eroded aulacogen profile it might be expected that the lowermost early clastic sediments, derived from the adjacent continental platform, and the early basalts might just be preserved. At even lower levels the aulacogen might only be represented by basic dyke swarms, at a high angle to the rift margin, which originally acted as feeders to the lava pile. A pre-collisional example is seen in SE Africa where an intense swarm of Jurassic (Karroo) basic dykes extends southeastwards to the Lebombo Monocline at the Indian continental margin (cf. Fig. 16.7 in Windley, 1984).

(2) The marginal continental shelf of an orogenic belt

A carbonate-clastic sedimentary sequence develops on the shelf of a passive continental margin, but is at most a few kilometres thick and is potentially vulnerable to erosion. However in several orogenic belts it is preserved through being overriden by allochthonous thrust sheets carrying ophiolites. Alternatively as the thrust sheets move over the shelf much or all of the carbonate bank may be transported with it above a décollement. The type example is in the western shelf of the southern Appalachian orogen, the deep structure of which is known from COCORP seismic reflection data (Cook *et al.* 1981). The shelf and its allochthon tends not to be a positive area in an orogenic belt and thus is commonly well preserved. If the shelf were completely eroded to expose the basement, the marginal zone of the belt would be the high-grade and highly deformed continental rise.

Thrusting has deformed the shelf sequence of many orogenic belts together with the basal unconformity and the adjacent basement. This late thrusting may be caused by post-collisional indentation of the continent carrying the shelf sequence on its frontal margin.

The Indian plate has been indenting northwards into the Eurasian/Tibetan at a rate of about 5 cm/yr since 40 Ma ago (Peirce 1978), and as a consequence the Phanerozoic Tethyan sediments on the northern margin of the Indian plate have been thrust southwards to form a thick thrust-nappe stack intercalated with a 10 km thick slice of Precambrian gneissic basement (the Central Crystallines). The thrusting has doubled the thickness of the crust under the High Himalaya and consequently has caused partial melting of the deeper crust to give rise to Miocene potassic leucogranites (with very high initial strontium isotope ratios) which are syn-tectonic with respect to the thrusts (Le Fort 1981). Residual granulites, the complement of these extracted liquids, may be expected in the lower crust of zones overthickened in this way.

In N. Pakistan post-collisional deformation was responsible for major uplift of the northernmost margin of the Indian plate with the result that the Tethyan sediments have been eroded to expose a basement of thrusted high-grade gneisses in a marginal zone extending for up to 100 km south of the Indus Suture. In this zone pelitic sediments of probable late Precambrian age have been metamorphosed to paragneisses and Cambrian porphyritic granites have been refoliated to porphyroclastic augen gneisses with a prominent lineation which resulted from the southward thrusting of the basement of the Indian plate (Coward *et al.* 1982, 1986).

Another example of a Himalayan-type thrust gneiss-granulite basement is in the Hercynian belt of the Massif Central of France (Bard *et al.* 1980a). A hot slab of granulites has been thrust over a cooler slab of amphibolite facies gneisses. The slabs are separated by a ductile shear zone of laminated intensely lineated gneissic rocks which overlies a zone of inverted metamorphic isograds; such zones are typical of Himalayan metamorphism (Windley 1983).

(3) The continental rise

The Piedmont Terrain of the Appalachian orogen provides a classic example of a palaeocontinental rise (Williams & Hatcher 1983). The rocks of this Terrain are mainly pelitic to quartzo-feldspathic continental-derived metasedimentary rocks, and oceanic-derived mafic volcanics and dismembered ophiolites, which are all polydeformed and regionally metamorphosed from upper greenschist to upper amphibolite facies. The intense thrusting that is concentrated at the continental margin thickens and uplifts the ancient continental rise with the result that only deep-seated levels tend to be well exposed today. This uplift is also indicated by seismic reflection data which show that the entire Piedmont Terrain in the southern Appalachians is contained in a major subhorizontal thrust slice that was emplaced above the North American miogeocline.

(4) The suture zone

Ophiolites characterize the upper levels of collisional sutures; but they are easily eroded. There are three groups of rocks in suture zones, which provide evidence of the deep-seated processes that operate in collisional belts.

(a) Nepheline gneisses

Alkaline igneous rocks commonly form in early intra-continental rifts at the start of the Wilson cycle. Following the opening and closure of an ocean, the alkaline rocks may come to be close to or within the suture zone between two collided plates. During high deformation and metamorphism the alkaline rocks are converted to nepheline gneisses. We are not aware of any Phanerozoic nepheline gneisses, but Proterozoic examples are noted below.

(b) Blueschists

In the central Himalaya, where there has been relatively little uplift, there are no blueschists in the Indus suture zone, but they are prominent in the western (Virdi et al. 1977; Shams et al. 1980) and eastern (Ghose & Singh 1980) ends of the belt, where uplift has been greater. The blueschist metamorphism records evidence of high pressure (> 10 kbar) conditions in a trench-subduction zone environment. The $^{39}Ar/^{40}Ar$ mineral age of 75–80 Ma for the blueschists in NW Pakistan suggests that the metamorphism took place in the incipient stage of formation of the Indus suture (Bard 1983; Maluski & Matte 1984; Petterson & Windley 1985).

Pyrope quartzite in blueschists in the western Alps contains coesite, implying a formation pressure of 20–28 Kbar (Chopin 1984). The discovery that crustal material can be subducted to depths of 90 km or more opens up new perspectives for collisional geodynamics.

(c) Granulites

Deep-level exposures of collisional sutures commonly contain tectonic lenses of lower crustal granulites (Windley 1983). These typically are remnants of layered igneous complexes such as dunites, peridotites, gabbros, norites and anorthosites, which have a high pressure granulite facies mineralogy with formation of garnet-clinopyroxene granulites, eclogites, garnet-bearing gabbros, amphibolites, pyroxenites and pyriclasites. Examples occur at Jijal in the Himalayan Indus suture in Pakistan (Jan & Howie 1981), the Ivrea zone of the Italian Alps (Hunziker & Zingg 1980; Rivalenti et al. 1981), and the Hercynian sutures of NW Europe (Matte 1983). The origin of these rocks is controversial; they may be remnants of intracontinental rift-type igneous complexes as in the Betico-Rifean belt in S. Spain and in the northern Pyrenees (Caby et al. 1981), the roots of island arc complexes (Coward et al. 1986), of mantle diapirs (Bard et al. 1980b), or intrusions into the lower crust (Rivalenti et al. 1981).

(5) Island arcs

In the development of suspect terrains it is common for island arcs to be swept into a collisional zone. Late thrusts and indentation may uplift or overturn these arcs to expose deep-level crustal sections; this deformation and associated metamorphism may be very intense and give rise to very complicated geological relationships. We can learn about such problems in the western Himalaya.

In N. Pakistan (Kohistan) and NW India (Ladakh) a Cretaceous island arc has been trapped between the collided Asian and Indian plates (Bard et al. 1980b; Coward et al. 1982, 1986). Deformation has isoclinally folded the arc to expose a remarkably complete crustal section from (top downwards) intra-arc sediments, tholeiitic and calc-alkaline lavas, tonalitic-dioritic plutons, and a stratiform (Chilas) igneous complex of norites and noritic-gabbros which probably formed in the sub-arc magma chamber (Coward et al. 1986). The sediments and lavas are metamorphosed mostly to a greenschist grade, the plutonic rocks are recrystallized to amphibolite facies orthogneisses, and the norites to low pressure granulites. The noritic roots of a comparable Jurassic island arc are exposed in Alaska (Burns et al. 1984).

(6) The fore-arc

The fore-arc of an Andean-type belt typically contains a thick pile of continental- and arc-derived clastic debris. The Xigaze Group in Tibet situated between the Yarlung-Zangbo suture and the Gangdese batholith has been telescoped by collisional deformation by a series of upright anticlines and synclines and metamorphosed to a greenschist grade, but it has not been appreciably uplifted. In more highly deformed orogenic belts the fore-arc sediments may be metamorphosed to amphibolite facies paragneisses and the arc-trench gap in which they formed may be obliterated by thrusting of slices of the meta-sediments over the suture.

(7) Back-arc basins

Back-arc basins may close with or without subduction. In their sutured state they may be represented by low to high grade metamorphic rocks and ophiolites as remnants in the suture or as slices thrust onto the continental foreland.

The Northern Suture in Pakistan separates the Kohistan-Ladakh island arc from the Asian plate, and probably represents a closed back-arc basin (Coward et al. 1986; Petterson & Windley 1985). The suture contains a melange consisting of tectonic lenses up to several kilometres long and tens of metres wide of serpentinite, amphibolite, greenschist, chert, phyllite and marble in a slate matrix (Windley 1983).

(8) Andean-type plutonic arcs

In any orogenic belt where collision has occurred through closure of an intervening ocean by subduction, either margin of the colliding continental masses may have developed a volcano-plutonic zone which may have been active for many tens of hundreds of million years before collision, and indeed volcano-plutonic activity may persist for several tens of millions of years after collision. The British Caledonian late granites are a well known example of the latter. Of course, cordilleran magmatic belts are subject to phases of extension and compression unrelated to collision (Bartholomew & Tarney 1984), and many of the plutonic and metasedimentary rocks within these belts may be highly deformed prior to collision, particularly at deep crustal levels. The alternative phases of extension and compression in such belts may lead to uplift of deep crustal zones. Examples have already been given from southern Chile, but very similar rocks are known from the Coast Range batholith of British Columbia, where 50 Ma granulites are exposed today which have been uplifted from 35 km (Hollister 1982).

Sub-volcanic levels of Andean-type granitoids in collisional orogenic belts are exposed in the Kohistan batholith of N. Pakistan (Coward *et al.* 1986; Petterson & Windley 1985) and in the Hercynian belt of NW Europe (Matte 1983).

(9) Variable uplift across an orogenic belt

In any cross-section of a collisional orogenic belt different tectonic units typically expose rocks at different crustal levels because they have suffered uplift at variable rates. This is caused by several factors:

(a) Variable degrees of magmatic thickening during the island arc and more particularly the Andean-type batholith stages of development.

(b) The differential tectonic history of individual allochthonous terrains during the formation of a collage of mini-plates in a Cordilleran type orogenic belt (Shermar *et al.* 1984). Extinct arcs, aseismic ridges and old continental plateaux already have different crustal thicknesses at the time of docking, accretion may take place at different angles and closing rates, and post-accretion strike-slip faults displace segments of individual terrains. All these processes tend to give rise to a highly variable crustal cross-section of the orogenic belt, individual units of which are exposed at different crustal levels.

(c) Following terminal continent-continent collision, continuous indentation of one plate may lead to variable degrees of post-collisional uplift in different segments of the orogenic belt. For example, in the modern Tibetan-Himalayan belt three regimes have markedly different post-collisional uplift rates.

(i) in the High Himalaya estimates range from 0.7–0.8 mm/yr for the past 25 Ma calculated from mineral ages (Mehta 1980) to 1 mm/yr since the late Pliocene from palaeogeographical investigations (Wang *et al.* 1982).

(ii) Since 17 Ma ago the uplift of the Tibetan plateau has been maintained at a rate of 0.2 mm/ yr (Zhao & Morgan 1983).

(iii) fission track data indicate that the uplift rate of the Nanga Parbat massif in the main syntaxis of the western Himalaya has increased over the past 7 Ma from under 0.5 mm/yr to over several millimetres per year (Zeitler 1985).

The result of these factors controlling variable uplift rates in different segments of an orogenic belt is that deep-seated cross-sections can only be constructed by compilation of type sections from different belts.

Proterozoic lower crust

Most Proterozoic orogenic belts have the same tectonic zonation as Phanerozoic belts (Windley 1984). Examples of lower crustal segments are as follows, starting with the Late Proterozoic.

(a) Pan-African belts are made up predominantly of a collage of island arcs and bordering continental margins, upper crustal sections of which are exposed in NE Africa and Arabia (Kröner 1985). A rare section of lower crust is seen in the Mozambique belt which according to Shackleton (1986) consists of a Himalayan-type thrust stack of high grade gneisses. Individual thrust slabs are separated by ductile shear belts which dip shallowly east towards a palaeo-suture zone.

(b) The Grenville belt of Canada formed in the period 1300–1050 Ma. In Ontario it contains an island arc (the Central Metasedimentary Belt) which was trapped between two collided continental plates, both of which have been eroded down to their gneissic basement (Davidson *et al.* 1984). There are marked similarities in the geology and collisional tectonics between the western Grenville and the western Himalaya (Windley in press).

The Central Metasedimentary Belt began its formation as an island arc shortly before 1300 Ma ago (Bartlett *et al.* 1984). The effects of the 1000 Ma Ottawa collisional orogeny were so intense that most arc volcanics and sediments were metamorphosed to a high grade. However in the Elzevir region there is a window of greenschist grade where the original nature of the arc rocks can be discerned—pillowed tholeiitic basalts followed by calc-alkaline felsic-mafic cycles which overlie the Kalashar complex of gabbros (Condie & Moore 1977). This arc illustrates very well the processes which operate during terminal collision and which convert the arc rocks to a highly deformed amphibolite facies state.

Against the arc the basement gneisses on the western side have been thoroughly reworked in a 3 km wide ductile high strain zone of lineated gneissic tectonites, transposed gneisses and porphyroclastic gneisses (Davidson *et al.* 1984; Thivierge *et al.* 1984). The lineation and foliation in this zone were probably caused by southeastwards post-collisional indentation of the western plate against the island arc (Windley, in press). In the Haliburton-Bancroft area along the suture zone between the arc and the western plate there are nepheline gneisses (Miller 1984).

The western gneissic plate is remarkable because recent mapping has revealed the presence of domains of gneisses and granulites which are separated by 1–2 km wide zones of intense ductile deformation (Davidson *et al.* 1982, 1984; Culshaw *et al.* 1983). Each domain has a different lithology, metamorphic grade (granulite to amphibolite facies) and structural style, and exhibits differences in the proportion of metasupracrustal rocks, anorthosites and gabbros and in their types of gneisses and granulites. This is interpreted as a thrust-nappe stack of allochthonous thrust sheets separated by ductile shear belts, comparable with the basement of a Himalayan-type orogenic belt thrusted as a result of post-collisional indentation.

(c)   The early Proterozoic Wollaston fold belt in central Canada has a tectonic framework similar to that of modern orogenic belts (Ray & Wanless 1980; Lewry *et al.* 1981). An 1865 Ma old Andean-type batholith is situated between an island arc and a telescoped fore-arc and a closed back-arc basin and continental margin. The fold belt has been uplifted to expose highly deformed and metamorphosed lower crustal rocks in most tectonic zones.

(d)   In the main suture zone of the early to mid-Proterozoic Aravalli-Delhi orogenic belt in Rajasthan, NW India (Sugden & Windley 1984), there are nepheline gneisses and an ophiolite suite, and in the sutured back-arc basin there are serpentinites with eclogite lenses and granulites. Most of the orogenic belt is now vertical and has been highly deformed and metamorphosed under mid-to-lower crustal conditions. Nevertheless its

modern-type tectonic framework can still be recognized.

Archaean deep crust

Estimates of the amount of continental crust generated in the Archaean are universally high (e.g. 60–80%, Cook & Turcotte 1981; 50–60%, Moorbath & Taylor 1981; 70–85%, McLennon & Taylor 1982; 85%, Dewey & Windley 1981); thus the nature of the Archaean lower crust must be given appropriate weighting in any generalized models of lower crustal composition (Weaver & Tarney 1984) and crustal structure. The Archaean is made up predominantly of two types of contrasting terrain: the low-grade granite-greenstone belts and the high-grade granulite-gneiss belts. The greenstone belts, with their thick synclinal volcano-sedimentary successions and low metamorphic grade, are best regarded as upper crustal extensional features (Tarney & Windley 1981; Tarney *et al.* 1982), but it is generally assumed that they are underlain by grey gneisses and granulites.

Archaean gneiss-granulite terrains are typically made up of a bimodal suite of meta-igneous tonalitic-trondhjemitic gneisses with abundant mafic lenses, layers and inclusions, layered anorthositic-gabbroic-ultramafic complexes, and metasediments. Many gneiss terrains are characterized by a strong sub-horizontal foliation, or by a complex fold interference pattern produced by refolding of early nappe-like isoclines which formed in a dominantly horizontal tectonic regime (Bridgwater *et al.* 1974). These terrains have suffered extreme deformation which has resulted in tectonic fragmentation of the mafic and exotic components to form the present complex heterogeneous banded assemblage. Geochemical studies of high-grade gneiss terrains (e.g. Weaver & Tarney 1981) have shown that they are a mixture of rock-types, some of which have evolved initially under low-pressure conditions (the mafic gneisses, layered complexes and metasediments) with others which have been generated under high P_{H_2O} conditions (the tonalites and trondhjemites). The problem in understanding high-grade terrains is how to intermix and interlayer these low-pressure and high-pressure components. It is unlikely that the low-pressure component represents foundered and dismembered greenstone belts simply because the metasedimentary component in high-grade terrains is composed of shelf type quartzites, marbles and pelites, whereas that in greenstone belts is dominantly greywacke; nor are the anorthositic components likely to have been part of greenstone belts. The alternative suggestion, proposed initially by Holland & Lambert (1975) and since amplified by Weaver & Tarney (1980, 1981), Weaver *et al.* (1981) and Tarney *et al.* (1982) is that tonalitic magmas, generated through shallow hydrous melting of subducted oceanic crust, are emplaced by magmatic underplating into the deeper continental crust. The controlling factor is that the shallow-generated hydrous magmas cannot rise high in the continental crust but must congeal at deeper levels.

There are several important implications of such a model. The first is that continued magmatic underplating means that crust is continually uplifted and transported from high-grade to low-grade P–T conditions, thus explaining why high-grade gneisses can occur at upper crustal levels over very large areas. The second is that magmas trapped and consolidating deep in the crust will remain ductile for very long periods and thus be very susceptible to deformation transmitted from shallow-dipping subducted slabs to crustal movements: this may account for the fact that Archaean deep crustal terrains have suffered extreme and pervasive deformation well beyond degrees of deformation normally observed in collisional fold belts. The third is that with an initially ductile lower crust magmas will rise to levels consistent with their density, thus ensuring that the crust will be, to a first approximation, both density and compositionally stratified..

The picture of the Archaean lower crust that presents itself from the application of such a model is one where the proportion of mafic and ultramafic components (whether magmatically emplaced or tectonically emplaced), increases downwards relative to that of tonalitic-trondhjemitic rock types, and where there is a strong subhorizontal fabric emphasized by the sheet-like form of the mafic and ultramafic layered bodies and metasediments. The origin of metasediments in the deeper crust is problematic. Relationships in the southern Andes, described above, demonstrate that it is possible to have fairly continuous conformable layers of deformed metasediment within the root zones of the Patagonian batholith. This arises in part because of the emplacement of numerous sheet-like tonalitic, dioritic, or more mafic intrusions within a pre-existing metasedimentary wedge that was already strongly foliated. It is also possible that metasediments and other exotic mafic components may be stripped off during subduction, penecontemporaneously with subduction zone magmatism. A ductile lower crust would however facilitate strong horizontal translational movements with the possibility of a

major overthrusting and incorporation of upper crustal metasediments within deep crustal zones.

In most high-grade terrains, granulite-facies metamorphism follows within perhaps one or two hundred million years after crustal generation (Moorbath 1980). This may result from high p_{CO_2} fluid activity (Newton *et al.* 1980), leading to dehydration of initially hornblendic rock types, but is superimposed upon an already existing crustal make-up. Dehydration is usually thought to be accompanied by loss of radioactive heat-producing elements U, Th, K, Rb, in the expelled fluids, as implied by the low heat flow observed in many high-grade Archaean cratons (Heier 1973), and removal of these elements from the lower crust leads to crustal stabilization. However not all granulite terrains necessarily represent deep crustal material, particularly where CO_2-rich fluids have penetrated to higher crustal levels. The magnitude of this secondary vertical redistribution of heat-producing elements within the Archaean crust is difficult to assess, because the primary levels of these elements in the granulite precursors is not known, and has only been estimated through comparison with equivalent rock types in adjacent amphibolite-facies terrains (Weaver & Tarney 1984). In this connection it is important to note that the levels of heat-producing elements in the deep-level tonalites and trondhjemites in the uplifted sections of the southern Andes (Bartholomew & Tarney 1984) are quite low, suggesting that, although further expulsion of heat-producing elements must occur to account for the abnormally low levels in some Archaean granulites, crust formed by Andean type magmatic processes may actually have a primary vertical zonation with respect to heat-producing elements.

Proponents of the andesite model for the continental crust, which implies intracrustal melting on a major scale leaving a dry granulite facies residuum in the lower crust, now agree that the model is not really applicable to the Archaean crust (Taylor & McLennan 1985), or perhaps 80% of the present crustal volume. Archaean crust would appear to be vertically zoned, with respect to the proportions of mafic and silicic components, as a primary feature of crustal growth processes which are dominantly magmatic and plutonic.

Discussion

Growth of the continental crust has occurred primarily through magmatic additions since about 3800 Ma, and modern counterparts of this magmatic growth are seen in the deeply eroded sections of the Andean batholiths. However, growth of the crust, i.e. its 'ripening' and stabilizing into more rigid continental lithosphere (crust and subcontinental lithosphere), itself promotes changes in the way new crust is made and develops. An increasing crustal area means more sediment, which may be accreted laterally with or without magmatic additions, giving a different crustal structure, rheology and tectonic response. At the same time the declining natural heat production within the Earth and transference of heat-producing elements to the crust has led to changes in the way the Earth convects—one consequence being the increasing size, thickness and ridigity of both continental and oceanic plates with time. This will again lead to differences in the tectonic style of orogenic belts through time, the type of lithological components involved in them and without doubt changes in lower crustal make-up.

Crustal generation in the Archaean, though episodic, was rapid. Archaean crust was generated above a weak, immature, continental lithosphere, and this is reflected in the pervasive strong deformation and sub-horizontal foliation shown by many Archaean high-grade terrains, particularly deep crustal granulites, and much of this deformation was probably syn-plutonic. The weakness of the Archaean lithosphere meant that it was also very susceptible to extensional forces which were able to promote crustal stretching and the development of greenstone belts; possibly the Archaean equivalent of modern back-arc basins.

The Archaean deep crust is largely a mixture of silicic magmatic products derived from the shallow melting of subducted ocean crust, together with more mafic components produced either through melting of the overlying mantle wedge or as a result of tectonic emplacement at subduction zones. Magmatic underplating provides a rheologically weak lower crust which permits strong horizontal thrusting and the intercalation of exotic components and metasediments.

Granulite formation, which normally closely follows crustal generation in the Precambrian, and which is commonly associated with expulsion of radioactive heat-producing elements from the lower crust, is an important process which leads to greater stability and rigidity of the continental crust, and indirectly, of the sub-continental lithosphere. Presumably it is for this reason that many Archaean cratons have remained essentially stable regions for almost 3000 Ma. The increasing rigidity of the continental lithosphere through the Proterozoic is accompanied by a parallel change towards modern style plate tectonics. Thus although modern belts of crustal generation such as the Andes display many similarities with the Archaean, collision belts have a more complex make-up, involving accreted island arcs, fragments of ocean crust and particularly a variety of metasedimentary components, as well as Andean-type crust. The lower crust of such belts is therefore more complex, the degree of complexity of which can be judged from the variety of lithotectonic zones seen in the upper crustal sections of orogenic belts such as the Himalayas.

References

BARD, J.P. 1983. Metamorphism of an obducted island arc: example of the Kohistan sequence (Pakistan) in the Himalayan collided range. *Earth planet. Sci. Lett.* **65,** 133–144.

—— *et al.* 1980a. Le metamorphisme en France. *In*: 26th Int. Geol. Cong., Paris, Coll. C7, *Geologie de la France*, 162–189.

——, MALUSKI H., MATTE P.L. & PROUST, F. 1980b. The Kohistan sequence: crust and mantle of an obducted island arc. *Geol. Bull. Univ. Peshawar, Sp. Issue*, **13,** 87–94.

BARTHOLOMEW, D.S. & TARNEY, J. 1984. Geochemical characteristics of magmatism in the southern Andes (45–46°S). *In*: BARREIRO, B. & HARMON, R.S. (eds) *Andean magmatism*, Shiva, Nantwich, 220–229.

BARTLETT, J.R., BROCK, B.S., MOORE, J.M. JR. & THIVIERGE R.H. 1984. *Grenville traverse A. Cross-sections of parts of the Central Metasedimentary Belt*. Geol. Ass. Canada. Field Trip Guidebook, 1–63.

BRIDGWATER, D., McGREGOR, V.R. & MYERS, J.S.

1974. A horizontal tectonic regime in the Archaean of Greenland and its implications for early crustal thickening. *Precamb. Res.* **1,** 179–198.

BURNS, L.E., PESSEL G.H. & NEWBERRY, R.J. 1984. The Border Ranges complex, S. Alaska: plutonic core of an island arc. *Geol. Ass. Canada*, Prog. with Abst. **9,** 49.

CABY, R., BERTRAND, J.M.L. & BLACK R. 1981. Pan-African closure and continental collision in the Hoggar-Iforas segment, Central Sahara. *In*: KRONER, A. (ed.) *Precambrian Plate Tectonics*. Elsevier, Amsterdam, 407–434.

CHOPIN, C. 1984. Coesite and pure pyrope in high-grade blueschists of the western Alps: a first record and some consequences. *Contrib. Mineral. Petrol.* **86,** 107–118.

CONDIE, K.C. & MOORE, J.M. JR. 1977. Geochemistry of Proterozoic volcanic rocks from the Grenville Province, eastern Ontario. *In*: BARAGER, W.R.A. *et al.* (eds) *Volcanic Regimes in Canada*, Geol. Ass. Canada. Sp. Pap **16,** 149–168.

Cook, F.A. & Turcotte, D.L. 1981. Parametrized convection and the thermal evolution of the earth. *Tectonophysics*, **75**, 1–17.

——, Brown, L.D. Kaufman, S., Oliver, J.E. & Peterson, T.A. 1981. COCORP seismic profiling of the Appalachian orogen beneath the Coastal Plain of Georgia. *Bull. Geol. Soc. Am.*, Pt 1, **92**, 738–748.

Coward, M.P., Jan M.Q., Rex D., Tarney, J., Thirlwall, M., & Windley, B.F. 1982. Geotectonic framework of the Himalaya of N. Pakistan. *J. Geol. Soc. London*, **139**, 299–308.

——, Windley, B.F. *et al.* 1986. Collision tectonics in the NW Himalaya In: Coward, M.P. & Ries, A.C. (eds), *Collision Tectonics*, Geol. Soc. Sp. Pub. Blackwell Scientific, Oxford. 19, 203–219.

Culshaw, N.G., Davidson, A. & Nadean, L. 1983. Structural subdivisions of the Grenville Province in the Parry Sound—Algonquin region, Ontario. *In: Current Research, pt. B, Geol. Surv. Canada, Pap.* 83–1B, 243–252.

Dalziel, I.W.D., DeWit, M.J., & Palmer, K.F. 1974. Fossil marginal basin in the southern Andes. *Nature*, **250**, 291–294.

Davidson, A., Culshaw, N.G. & Nadean, L. 1982. A tectono-metamorphic framework for part of the Grenville Province, Parry Sound region, Ontario. *In: Current Research. Geol. Surv. Canada, Pap.* 82–1A, 175–190.

——, —— & Nadeau, L. 1984. *Grenville traverse B, Cross-section of part of the Central Gneiss Belt*. Geol. Assoc. Canada, Field Trip Guidebook, 1–79.

Dewey, J. & Windley, B.F. 1981. Origin and differentiation of the continental crust. *Phil. Trans. R. Soc. Lond.* **A301**, 189–206.

Ghose, N.L. & Singh, R.N. 1980. Occurrence of blueschist facies in the ophiolite belt of Naga Hills, E. of Kiphire, N.E. India. *Geol. Rund.* **69**, 41–48.

Heier, K.S. 1973. Geochemistry of granulite facies rocks and problems of their origin. *Phil. Trans. R. Soc. Lond.* **A273**, 429–442.

Hoffman, P.F. 1980. Wopmay Orogen: a Wilson-cycle of early Proterozoic age in the northwest of the Canadian Shield. *Geol. Assoc. Can., Sp. Pap.* **20**, 523–549.

Holland, J.G. & Lambert R. St. J. 1975. The chemistry and origin of the Lewisian gneisses of the Scottish mainland: the Scourie and Inver assemblages and sub-crustal accretion. *Precamb. Res.* **2**, 161–188.

Hollister, L.S. 1982. Metamorphic evidence for rapid (2 mm/yr) uplift of a portion of the Central Gneiss Complex, Coast Mountains, B.C., *Can. Mineral.* **20**, 319–332.

Hunziker, J.C. & Zingg, A. 1980. Lower Palaeozoic amphibolite to granulite facies metamorphism in the Ivrea zone (southern Alps, northern Italy). *Schweiz. mineral. petrogs. Mitt.* **60**, 181–213.

Jan, M.Q. & Howie, R.A. 1981. The mineralogy and geochemistry of the metamorphosed basic and ultrabasic rocks of the Jijal Complex, Kohistan, N.W. Pakistan. *J. Petrol.* **22**, 85–126.

Kröner, A. 1985. Ophiolites and the evolution of tectonic boundaries in the late Proterozoic Arabian-Nubian Shield of NE Africa and Arabia. *Precamb. Res.* **27**, 277–300.

LeFort, P. 1981. Manaslu leucogranite: a collision signature of the Himalaya. A model for its genesis and emplacement. *J. Geophys. Res.* **86, B11**, 10545–10568.

Lewry, J.F., Stauffer, M.R. & Fumerton, S. 1981. A cordilleran-type batholithic belt in the Churchill province in northern Saskatchewan. *Precamb. Res.* **14**, 277–314.

Maluski, H. & Matte, P. 1984. Ages of Alpine tectonometamorphic events in the northwestern Himalaya (Northern Pakistan) by $^{39}Ar/^{40}Ar$ method. *Tectonics*, **3**, 1–18.

Matte, P. 1983. Two geotraverses across the Ibero-Armorican Variscan of western Europe. *In: Rast, W. & Delaney, F.M.* (eds) *Profiles of Orogenic Belts.* Geodyn. ser. Vol. **10**, *Geol. Soc. Am.*, 53–82.

McLennan, S.M. & Taylor, S.R. 1982. Geochemical constraints on the growth of the continental crust. *J. Geol.*, **90**, 347–361.

Mehta, P.K. 1980. Tectonic significance of the young mineral dates and the rates of cooling and uplift in the Himalaya. *Tectonophysics*, **62**, 205–217.

Miller, R.R. 1984. Tectonic significance of nepheline-bearing rocks in the Haliburton-Bancroft area of the Grenville Province. *Geol. Ass. Canada. Prog. with Abst.* **9**, 89.

Moorbath, S. & Taylor, P.N. 1981. Isotopic evidence for continental growth in the Precambrian. *In: Kröner, A.* (ed.) *Precambrian Plate Tectonics.* Elsevier, Amsterdam. 491–526.

Natland, J.H. & Tarney, J. 1982. Petrological evolution of the Mariana arc and back-arc basin system: a synthesis of drilling results in the South Philippine Sea. *Init. Repts. Deep Sea Drilling Project*, **60**, 877–908 (U.S. Government Printing Office, Washington).

Peirce, J.W. 1978. The northward motion of India since the late Cretaceous. *R. Astron. Soc. Geophys. J.* **52**, 277–311.

Petterson, M.G. & Windley, B.F. 1985. Rb/Sr dating of the Kohistan arc-batholith in the Trans-Himalaya of N. Pakistan, and tectonic implications. *Earth Planet Sci. Lett.* **74**, 45–57.

Pitcher, W.S. 1978. The anatomy of a batholith. *J. Geol. Soc. Lond.* **135**, 157–182.

Ray, G.E. & Wanless, R.K. 1980. The age and geological history of the Wollaston, Peter Lake, and Rottenstone domains in northern Saskatchewan, *Can. J. Earth Sci.* **17**, 333–347.

Rivalenti, G., Garuti, G., Rossi, A., Siena, F. & Sinigoi, S. 1981. Existence of different peridotite types and of a layered igneous complex in the Ivrea zone of the western Alps. *J. Petrol.* **22**, 127–153.

Shackleton, R.M. 1986. Precambrian collision tectonics in Africa. *In: Coward, M.P. & Ries, A.C.* (eds) *Collision Tectonics.* Geol. Soc. Sp. Publ., Blackwell Scientific, Oxford. 19, 329–349.

Shams, F.A., Jones, G.C. & Kempe, D.R.C. 1980. Blueschists from Topsin, Swat district, N.W. Pakistan. *Mineral. Mag.* **43**, 941–942.

Shermer, E.R., Howell, D.G. & Jones, D.L. 1984. The

origin of allochthonous terrains: perspectives on the growth and shaping of continents. *Ann. Rev. Earth Planet. Sci.*, **12**, 107–132.

SUGDEN, T.J. & WINDLEY, B.F. 1984. Geotectonic framework of the early-mid Proterozoic Aravalli–Delhi orogenic belt, N.W. India. *Geol. Ass. Can. Prog. with Abst.* **9**, 109.

TARNEY, J. & WINDLEY, B.F. 1981. Marginal basins through geological time. *Phil. Trans. R. Soc. Lond.* **A301**, 217–232.

——, WEAVER, B.L. & WINDLEY, B.F. 1982. Geological and geochemical evolution of the Archaean continental crust. *Rev. Bras. Geoc.* **12**, 53–59.

TAYLOR, S.R. 1977. Island arc models and the composition of the continental crust. *In*: TALWANI, M. & PITMAN, W.C. (eds) *Island Arcs, Deep Sea Trenches and Back-Arc Basins.* Amer. Geophys. Un., Maurice Ewing Ser. **1**. 325–335.

—— & MCLENNAN, S.M. 1985. *The Continental Crust: its Composition and Evolution.* Blackwell Scientific, Oxford. 312 pp.

THIVIERGE, R.H., BROCK B.S. & HANMER, S.K. 1984. Structural relationships at the northwestern margin of the Central Metasedimentary Belt, Grenville Province, Ontario. *Geol. Assoc. Canada*, Prog. with Abst. **9**, 110.

VIRDI, N.S., THAKUR, V.C. & KUMAR, S. 1977. Blueschist facies metamorphism from the Indus suture zone of Ladakh and its significance. *Himalayan Geol.* **7**, 479–482.

WANG, C.Y., SHI, Y.L. & ZHOU, W.H. 1982. Dynamic uplift of the Himalaya. *Nature*, **298**, 553–556.

WEAVER, B.L. & TARNEY, J. 1980. Rare earth geochemistry of Lewisian granulite-facies gneisses, N.W. Scotland: implications for the petrogenesis of the Archaean lower continental crust. *Earth Planet. Sci. Lett.* **51**, 279–296.

—— & —— 1981. Lewisian gneiss geochemistry and Archaean crustal development models. *Earth Planet. Sci. Lett.* **55**, 171–180.

—— & —— 1984. Estimating the composition of the continental crust: an empirical approach. *Nature*, **310**, 575–577.

——, —— & WINDLEY, B.F. 1981. Geochemistry and petrogenesis of the Fiskenaesset anorthosite complex, southern West Greenland: nature of the parent magma. *Geochim. Cosmochim. Acta*, **45**, 711–725.

WINDLEY, B.F. 1983. Metamorphism and tectonics of the Himalaya. *J. Geol. Soc. Lond.*, **140**, 849–865.

—— 1984. *The Evolving Continents.* 2nd ed., Wiley, Chichester, 399 pp.

—— in press. Comparative tectonics of the western Grenville and the western Himalaya. *In*: Grenville Symp. Vol. Geol. Ass. Canada.

ZEITLER, P.K. 1985. Cooling history of the NW Himalaya, Pakistan. *Tectonics*, **4**, 127–151.

ZHAO, W.L. & MORGAN, W.J. 1983. The uplift history of the Tibetan plateau from a 'fault block' reconstruction. *Eos. Trans. AGU*, **64**, 860.

B.F. WINDLEY & J. TARNEY, Dept. of Geology. University of Leicester, LE1 7RH, UK.

Petrogenesis of layered gabbros and ultramafic rocks from Val Sesia, the Ivrea Zone, NW Italy: trace element and isotope geochemistry

C. Pin & J.D. Sills

SUMMARY: Layered gabbros and ultramafics, well exposed in Val Sesia, have been interpreted as a layered complex. The lower part (LLG; up to 1 km thick) comprises well layered pyroxenites and gabbros with minor peridotite and harzburgite. The upper part consists of fairly homogeneous gabbros (MG), which grade into diorite (up to 5.5 km thick). This gabbro body intrudes already highly deformed metapelites, but is itself relatively undeformed.

These two groups (LLG and MG-diorite) are distinct on the basis of isotope, trace element and major element chemistry and cannot have been derived from the same liquid. The LLG has ε_{Nd} ranging from $+1.5$ to $+2.5$ and ε_{Sr} from $+7$ to -8 and was derived from a depleted source. The MG-diorite body has ε_{Sr} ranging from $+40$ to $+60$ and a wide range of ε_{Nd} from -2 to -6.

The LLG pyroxenites and gabbros are dominantly cumulates of clinopyroxene with lesser amounts of orthopyroxene. Cumulus plagioclase appears towards the top of the LLG. The LLG is strongly depleted in LREE and LIL elements and was derived from a basaltic liquid with relatively depleted trace element characteristics. It probably fractionated at moderate pressures and was emplaced into the Ivrea Zone along with the adjacent Balmuccia peridotite.

The MG comprises plagioclase $\pm$ olivine cumulates with minor cumulus apatite, ilmenite and pyroxene. The diorites additionally contain cumulus biotite. The whole body has strongly LREE-enriched patterns (Ce_N 4.5–20) and large positive Eu anomalies (Eu/Eu* from 2–12). The MG has calc-alkaline features with high Ba and low Nb. The parental magma was also LREE-enriched, with low Ni, Cr and MgO and high Ba. It was possibly an andesite.

The spread in ε_{Nd} cannot be accounted for by contamination by any of the likely crustal contaminants. The spread in intial isotope ratios either results from a heterogeneous enriched mantle source or from a two stage process involving remelting of an intermediate mafic reservoir with moderate enrichment in Rb and LREE.

The MG-diorite body was emplaced into sediments already undergoing amphibolite-facies metamorphism at pressures of 5–7 kbar. It appears as if the development of magma chambers in the crust was an important process of crustal growth in the Ivrea Zone.

Introduction

The Ivrea Zone is of considerable interest as it is commonly considered to represent a cross-section through the lower continental crust (e.g. Mehnert 1975; Fountain & Salisbury 1981). It consists of a steeply dipping sequence of metasediments and metabasites separated from the Alps proper by the Insubric line (Fig. 1). The metamorphic grade decreases from granulite-facies in the NW to amphibolite-facies in the SE. Pressure estimates of 8–9 kbar, obtained from metasedimentary gneisses intercalated with metagabbros near the base of the Ivrea Zone in Val Sesia (Sills 1984; Fig. 1) show that parts of the Ivrea Zone are derived from the lower crust.

There are a variety of metabasic rocks; metagabbros and associated ultramafic rocks, best developed in Val Sesia, and thinner layers of amphibolite, or more rarely, basic granulite which are interbanded with metasedimentary gneisses and are probably of supracrustal origin

(Sills & Tarney 1984). It is the metagabbros which form the subject of this paper. Rivalenti *et al.* (1973, 1981a) proposed that they formed as a stratiform layered complex emplaced at depth. The metagabbros and ultramafics are less deformed than the metasediments which they clearly intrude in Val Sesia and Val Mastallone. The granulite-facies metamorphism of the metasediments may have been caused by the emplacement of this basic body at depth (Schmid & Wood 1976).

The aim of this paper is to examine the petrogenesis of metagabbros which form a significant proportion of the Ivrea Zone (Fig. 1) in order to understand the processes which generated this section of the lower crust. We present isotope, whole rock and mineral analyses in order to determine whether or not the gabbros and associated ultramafics formed as a stratiform complex. We concentrate on the central part of the Ivrea Zone, near Val Sesia where the metagabbros are best developed. Detailed geochrono-

From: DAWSON, J.B., CARSWELL, D.A., HALL, J. & WEDEPOHL, K.H. (eds) 1986, *The Nature of the Lower Continental Crust*, Geological Society Special Publication No. 24, pp. 231–249.

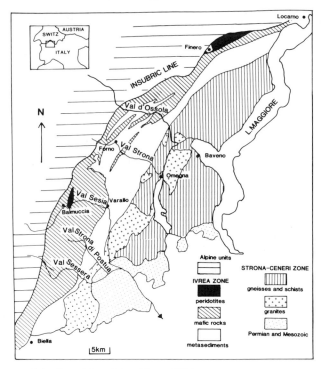

FIG. 1. Generalized map of the Ivrea Zone (after Zingg 1980).

logical implications of the isotope data will be discussed elsewhere (Pin *et al.* in prep.).

Geological setting

The layered gabbro/ultramafic complex in Val Sesia (Rivalenti *et al.* 1975, 1981a; Fig. 2) consists of the following sequence:
(1) The Balmuccia peridotite presumed to be of mantle origin (Shervais 1979; Ernst 1978; Rivalenti *et al.* 1981a).
(2) Deformed and partially recrystallized prominently layered pyroxenites and pyroxene-rich gabbros, up to 600 m thick, with subsidiary ultramafic layers, termed the lower layered group (LLG) by Rivalenti *et al.* (1975).
(3) A zone up to 300 m wide with prominent peridotites and pyroxenites with interbanded gabbros. These have been ascribed either to a cumulate origin (Rivalenti *et al.* 1981a; 1981b) or to a mantle origin (Shervais 1979). These are referred to as the middle layered group (MLG).
(4) A prominent layer, 50–100 m thick, of granulite-facies metasediment which separates the LLG and MLG from more homogeneous gabbros (Ferrario *et al.* 1982).

(5) The upper layered group (ULG) of Rivalenti *et al.* (1975) is up to 500 m thick and consists of gabbros and nickeliferous pyroxenites near the village of Isola, and layered gabbros with anorthosite layers extending as far east as the village of Sassiglioni (Fig. 2). The chemistry shows that the Isola gabbros are anomalous (Sills, unpubl. data) and that the gabbros near Sassiglioni are better grouped with the main gabbro.
(6) The main gabbro(MG)-diorite body which is up to 5.5 km thick. This comprises fairly homogeneous plagioclase-rich gabbros which grade, with increasing proportion of biotite, into rocks which have been termed diorite (Rivalenti *et al.* 1975; Bigioggero *et al.* 1978/79). 'Diorite' is used to cover a range of biotite-bearing rock types, mainly gabbros and monzogabbros, although towards the top of the body true diorites occur. Angular xenoliths of metasediment are found towards the top of the body ranging from a few cm to a few m across. The metasediments are extremely highly deformed (Zingg 1980) whereas igneous textures are preserved in the the main gabbro and diorite.

Rivalenti *et al.* (1975; 1981a) suggested the whole body (LLG, MLG, ULG, MG and diorite) formed as a single stratiform complex but Fer-

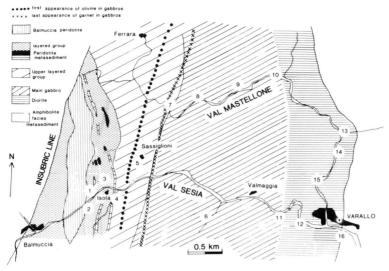

Fig. 2. Generalized map of the layered complex in Val Sesia (after Rivalenti *et al.* 1981a). Sample locations are as follows: 1—IV33, IV35, IV36, IV41, IV301, GZ8, GZ7, GZ10; 2—IV117, IV118; 3—IV176, Fe36, Fe41; 4—GM90, MZ131, IV31; 5—IV305, IV322, IV323; 6—IV46; 7—IV139, IV140, IV142, MAS6, MAS5; 8—MAS4, IV145, IV146, IV147; 9—MAS3; 10—MAS2, IV148, IV104; 11—IV319; 12—IV50; 13—MAS1b; 14—IV107; 15—MAS1a, MAS0; 16—IV313, IV341, IV342.

rario *et al.* (1982) suggested that the MG-diorite body might represent a new influx of magma. Rivalenti *et al.* (1984) suggest that the LLG, MLG and ULG could be derived from a common parental magma, but with the different layers fractionating at different pressures. In contrast to the assumption of Voshage *et al.* (1983), chemical and isotope compositions, considered together, indicate that the LLG and the MG-diorite body are not co-genetic but formed from different magmas, so in the following they will be discussed separately.

Analytical techniques

Isotope analyses were obtained using standard techniques at the Open University and LA10, following Hawkesworth *et al.* (1983). BCR-1 gave a value of 0.51262 ± 2 for $^{143}Nd/^{144}Nd$ during the period of analysis.

Chemical analyses were obtained using a Philips 1400 X-Ray fluorescence spectrometer at the University of Leicester following Weaver *et al.* (1983). Some rare-earth element analyses were performed by instrumental neutron activation analysis at the Open University following Potts *et al.* (1981) and some were performed by inductively coupled plasma emission spectrometry at Kings College, London following Walsh *et al.* (1981). Precision and comparison of the two techniques are given in Sills & Tarney (1984). Mineral analyses were obtained from a Cambridge Instruments Microscan V electron micro-

probe at the University of Leicester. Space allows only very few data to be presented, so full rock and mineral analyses can be obtained from JDS on request.

Samples GZ7, GZ8, GZ10, FE36, FE41, GM90 and MZ131 were supplied by G. Rivalenti; all the remaining samples were collected by the authors. Location of samples is shown in Fig. 2.

Lower layered group (LLG)

Petrography and mineral chemistry

The contact between the LLG and Balmuccia peridotite is generally tectonic, but at one locality on the south bank of the River Sesia near Isola (Fig. 2) an igneous contact occurs. The LLG is quite strongly deformed with a subvertical foliation parallel to the layering. There are several concordant lenses of granulite-facies metasediment within the LLG and these are often adjacent to nickeliferous pyroxenites (Ferrario *et al.* 1982). The significance of these lenses is not known and it is possible that they indicate strong isoclinal folding or thrusting. Detailed descriptions of the stratigraphy, mineralogy and mineral chemistry of the layered series are given in Rivalenti *et al.* (1975, 1981a, 1981b, 1984), Ferrario *et al.* (1982) and Sills (1984). In this work we review the overall chemistry of the layered series and present isotope analyses.

A diagrammatic cross section through the

TABLE 1. *Representative analyses of the lower layered group (IV35–GZ8), the upper layered group (IV322 and 323), the main gabbros (IV146–MAS3) and biotite-rich gabbros and diorites (IV52–IV313). Elements analysed by isotope dilution are in* **bold**. *— means element analysed but not detected*

Rock	IV35	IV33	IV117	IV118	IV36	IV301	IV41	GZ8	IV322	IV323	IV146	IV46	MAS6	MAS5	MAS4	MAS3	IV52	IV319	MAS2	MAS1b	MAS1	IV107	IV313
SiO_2	43.2	41.9	45.3	46.2	43.5	47.0	49.4	44.8	44.4	52.9	52.1	40.6	47.3	50.9	48.5	45.5	48.3	44.4	45.2	47.0	45.8	48.8	52.3
TiO_2	0.44	1.29	1.21	1.07	2.80	0.93	0.37	0.61	2.99	0.11	0.29	3.37	2.59	0.65	0.70	0.73	1.09	2.13	1.70	1.86	1.36	1.29	1.36
Al_2O_3	5.2	15.4	11.0	10.0	13.1	16.8	15.4	7.3	21.9	28.9	29.0	20.2	16.6	26.0	21.4	19.9	16.9	20.0	24.1	22.6	20.7	19.1	21.6
Fe_2O_3	13.54	13.20	11.60	11.48	17.42	8.21	10.94	14.90	12.83	1.07	1.48	14.27	12.78	4.73	8.87	11.24	10.61	12.49	8.49	8.77	9.81	10.14	9.03
MnO	0.20	0.22	0.20	0.20	0.30	0.13	0.18	0.22	0.20	0.02	0.02	0.16	0.20	0.07	0.14	0.18	0.22	0.18	0.09	0.11	0.15	0.19	0.22
MgO	32.78	14.27	14.10	14.98	8.11	13.98	15.95	27.35	5.83	0.23	0.33	6.10	7.00	3.08	5.85	9.81	6.62	4.89	4.32	4.20	6.21	5.91	2.16
CaO	3.80	12.78	15.93	14.85	13.25	11.84	8.25	4.29	10.10	11.65	12.51	10.44	11.44	11.84	12.08	11.66	11.89	10.21	12.14	9.76	11.92	9.72	6.40
Na_2O	0.34	1.22	0.74	0.80	1.83	2.30	0.52	0.23	2.70	4.68	4.43	2.73	2.46	3.40	2.44	1.46	2.52	2.93	2.95	2.56	2.16	2.41	3.56
K_2O	0.01	0.08	0.01	0.01	0.06	0.02	0.02	0.00	0.25	0.38	0.39	0.23	0.23	0.34	0.34	0.22	1.29	1.40	0.99	1.97	0.76	1.81	2.63
P_2O_5	0.01	0.02	0.01	0.01	0.02	0.02	0.02	0.01	0.04	0.06	0.03	1.56	0.38	0.03	0.06	0.03	0.28	0.95	0.70	0.50	0.17	0.14	0.54
Total	99.52	100.36	100.10	99.40	100.39	101.05	101.05	99.71	101.24	100.00	100.58	99.66	100.98	101.04	100.38	100.70	99.72	99.58	100.29	99.72	99.04	99.51	99.80
LOI	0.58	0.72	1.63	1.15	0.36	0.71	0.43	0.59	0.16	0.52	0.62	—	—	—	0.09	0.55	0.37	—	0.20	1.29	1.12	0.77	1.39
mg	84.5	70.9	73.2	74.6	51.2	74.0	76.6	80.5	50.6	32.6	33.4	49.0	55.2	59.4	59.7	66.3	58.4	46.8	53.4	51.9	58.8	56.7	35.0
Trace elements PPM																							
Ni	1586	210	664	187	91	497	78	1067	9	5	6	4	11	7	4	23	28	13	10	11	17	21	3
Cr	4041	366	616	314	112	559	593	4931	27	30	13	16	64	73	81	203	225	23	37	33	115	160	14
V	155	479	448	483	646	214	255	212	239	35	62	388	431	100	190	119	177	287	408	287	294	150	102
Rb	—	—	—	—	**0.4**	—	**0.2**	**0.11**	—	—	**0.7**	**0.8**	**0.6**	**1.0**	**2.2**	**1.7**	21	23	16	31	**6.7**	36	45
Sr	21	68	28	30	112	363	**142**	**11**	525	1085	**928**	**688**	452	**507**	**672**	**442**	400	478	685	**530**	**423**	473	419
Ba	6	36	6	15	2	12	30	9	1004	1629	372	245	290	380	296	175	1600	1596	1718	3798	346	6470	3878
Zr	18	50	54	35	75	50	12	20	76	23	13	40	67	12	62	50	251	391	78	89	194	162	1044
Nb	—	—	—	—	1	—	—	—	2	—	1	2	2	2	9	5	3	4	2	3	4	3	7
Y	9	41	38	33	56	23	6	11	2	—	1	15	16	2	9	5	22	16	15	10	21	12	14
CIPW Norms wt. %																							
qz	—	—	—	—	—	—	—	—	—	—	—	—	—	—	—	—	—	—	—	—	—	—	1.4
cor	—	—	—	—	—	—	—	—	—	—	0.20	—	—	—	—	—	—	—	—	—	—	—	2.9
or	—	0.4	—	0.1	0.4	0.1	0.1	—	1.5	2.3	2.3	1.4	1.4	2.0	2.0	1.1	7.6	8.3	5.9	11.6	4.5	10.7	15.5
ab	2.9	1.1	3.7	6.8	10.9	15.2	4.4	2.0	21.2	37.5	32.9	18.8	20.8	28.8	20.7	12.4	19.0	17.8	16.7	20.4	17.2	20.4	15.5
an	12.7	36.3	26.7	23.7	27.4	35.4	39.6	18.9	46.9	56.7	58.1	41.7	33.6	54.7	46.5	47.2	31.0	37.3	51.4	42.6	44.6	36.0	30.1
ne	—	—	1.4	—	—	2.3	—	—	0.9	1.2	2.5	2.3	—	—	—	—	1.3	3.8	2.7	2.5	0.6	—	—
di	4.9	21.8	42.1	40.1	31.6	18.5	1.0	1.8	2.4	0.6	3.2	—	17.0	2.7	10.8	8.5	21.5	6.0	3.6	2.0	11.1	9.3	—
hy	18.2	—	—	2.6	17.8	—	48.5	34.3	—	—	—	—	9.8	5.8	6.7	7.2	—	—	—	—	—	5.6	—
ol	56.5	29.8	20.9	21.4	3.0	25.7	3.8	37.7	19.2	1.2	0.6	21.7	9.3	4.4	10.1	20.0	13.9	17.0	13.1	13.6	15.7	12.1	—
mgt	2.3	2.3	2.0	2.0	5.3	1.4	1.9	2.6	2.2	0.2	0.3	2.5	2.2	0.8	1.5	1.9	1.8	2.2	1.5	1.5	1.8	1.8	1.6
ilm	0.8	2.5	2.3	2.0	—	1.8	0.7	1.2	5.7	0.2	0.6	6.3	4.9	1.2	1.3	1.4	2.1	4.1	3.3	3.5	2.6	2.5	2.6
ap	—	—	—	—	—	0.1	—	—	0.1	0.1	0.1	3.7	0.9	0.1	0.1	0.1	0.7	2.2	1.6	1.2	0.4	0.3	1.3

LLG (Rivalenti *et al.* 1981b) shows a series of pyroxenites and gabbros, with layers generally 3–5 m thick, with the proportion of gabbros increasing up the section. Within gabbro layers thinner, 5–10 cm pyroxenite layers occur. Subsidiary harzburgite and peridotite layers, 1–2 m thick, occur near the base. The thicker peridotite of the MLG (Rivalenti *et al.* 1981b) is possibly discordant.

In pyroxenites clinopyroxene (cpx) is generally much more abundant than orthopyroxene (opx). Pyroxenes are very aluminous with up to 10 wt. % Al_2O_3 in cpx and 1.5 wt. % TiO_2 (Sills 1984). Where plagioclase is interstitial it is about An_{50}, but where there is cumulus plagioclase (indicated by high Sr and Eu concentrations) it is much more calcic (as high as An_{88}). These gabbros contain more magnesian pyroxenes than some of the pyroxenites. The hornblende is kaersutite (Leake 1978) with TiO_2 up to 5 wt. %, Na_2O up to 4 wt. % but K_2O is below detection (< 0.15 wt. %). The mg nos. (100 Mg/(Mg + Fe)) for pyroxene are quite variable ranging from 85 in harzburgite to 70 in gabbro. Olivine in the basal harzburgite is about Fo_{83} with about 0.4 wt. % NiO. No systematic change in mineral composition has been observed across the LLG (Rivalenti *et al.* 1984).

Chemistry

Representative analyses for a range of rock types are shown in Table 1. The data show clearly that the rocks are cumulates, e.g. many samples have much higher CaO than basalts reflecting cumulus clinopyroxene. The major element chemistry is variable depending on the cumulus mineralogy. Compatible elements such as Ni and Cr are high in ultramafics and decrease rapidly with MgO. V is high in cpx-rich samples. Many incompatible elements, such as P_2O_5 and K_2O are low. TiO_2 ranges up to 3 wt. %. Sr is fairly low in pyroxenites, pyroxene-rich gabbros and ultramafic layers, ranging from 2–100 ppm but is significantly higher in some gabbros (200–400 ppm). There is a reasonable correlation (best held for ultramafic samples) between Ti and Zr (r = 0.89) suggesting both elements acted incompatibly. Magnetite can only have been a minor cumulus phase, otherwise samples would have had much higher Ti/Zr ratios. There is also a reasonable correlation between Zr and Y but Zr/Y ratios are very low (average 1.3) when compared with basaltic liquids (Wood *et al.* 1979), implying a phase which concentrates Y relative to Zr. This could either be clinopyroxene or hornblende (Pearce & Norry 1979). Hornblende, however would also concentrate Ti relative to Y which is

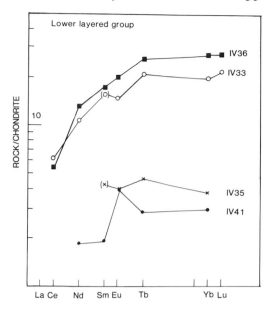

FIG. 3. REE data for the lower layered group, normalized to the chondrite data of Nakamura (1974). IV41—gabbro; IV35—harzburgite; IV33—pyroxenite; IV36—pyroxene-rich gabbro.

not seen as Ti/Y ratios are also quite low. The very low level of K_2O (< 0.10 wt. %) also argues against hornblende accumulation. It is concluded that the main cumulus phase in the pyroxenites and pyroxene-rich gabbros is clinopyroxene, with a lesser amount of orthopyroxene (in the ratio 2:1). The subsidiary ultramafic layers are ortho-pyroxene-olivine cumulates and gabbros towards the top of the section contain cumulus plagioclase. Hornblende, although modally quite abundant, cannot have been a significant cumulus phase.

REE data (Fig. 3, Table 3) are presented for four samples: a harzburgite (IV35), a gabbro (IV41), a pyroxenite (IV33) and a more evolved pyroxene-rich gabbro (IV36). La is below detection in all samples and Ce is below detection in the gabbro and harzburgite. Nd and Sm were measured by isotope dilution for IV41 and IV36; these data suggesting that the INAA Sm analyses were slightly overestimated. A slight negative Eu anomaly for IV36 is removed when the I.D. Sm is plotted, suggesting the slight Eu anomalies in IV33 and IV35 may not be real. The two pyroxenite samples are strongly LREE depleted with Ce_N/Yb_N of 0.2–0.3. IV35 has very low levels of REE, about 4 × chondrite. The gabbro has even lower levels, about 3 × chondrite, but has a positive Eu anomaly, indicating cumulus plagioclase.

TABLE 2. *Trace element and isotope data for the metaigneous rocks from Val Sesia and Val Mastallone. Concentrations are in µg/g. ε values are calculated for 600 Ma and 300 Ma using the parameters of Jacobsen & Wasserburg (1980). Analyses for metasedimentary rocks and leucosomes were of totally spiked samples using a $^{150}Nd-^{149}Sm$ tracer. Two additional samples (ST8 and ST13) are from Ben Othman et al. (1984)*

Sample	rock type	Rb	Sr	Sm	Nd	$^{87}Rb/^{86}Sr$	$^{87}Sr/^{86}Sr$	$^{147}Sm/^{144}Nd$	$^{143}Nd/^{144}Nd$	$\varepsilon_{Nd}600$	$\varepsilon_{Sr}600$	$\varepsilon_{Nd}300$	$\varepsilon_{Sr}300$
GZ10	peridotite	.04	14	.889	2.641	.009	.70416	.1994	.512757 20	+2.1	+4.1	+2.2	−0.4
GZ8	pyroxenite	.11	11	.974	2.628	.029	.70347	.2241	.512826 17	+1.5	−8.1	+2.6	−11
IV41	gabbro	.17	142	.393	1.168	.004	.70964	.2034	.512369 19	−5.8	+83	−5.5	+78
IV36	gabbro	.44	112	3.863	8.33	.011	.7042	.2924	.513105 18	+1.7	+7.3	+5.4	+0.1
GZ7	gabbro	.29	21	1.902	4.96	.04	.70465	.2318	.512857 22	+1.5	−4.4	+2.1	+5.0
FE36	peridotite	—		.144	.374		.70898	.2315	—			—	
FE41	pyroxenite	.37	246	3.481	9.91	.004	.70352	.2123	.512779 30	+1.5	−4.4	+2.1	−9
IV176	gabbro	.61	377	1.044	3.432	.005	.70732	.1839	.512368 38	−4.3	+50	−6.0	+45
GM90	pyroxenite	.20	18	.774	2.433	.032	.70849	.1923	.512325 21	−5.8	+63	−6.0	+60
MZ131	gabbro	2.68	459	1.076	6.712	.017	.70756	.0969	.512245 64	0.0	+52	−3.9	+47
IV146	anorthosite	.66	928	.281	2.561	.002	.70788	.0651	.512082 40	−0.8	+58	−5.8	+53
IV46	gabbro	.83	688	5.17	27.19	.004	.70802	.1127	.512289 18	−0.4	+60	−3.6	+55
IV144	gabbro	2.11	539	1.518	8.26	.011	.70779	.1088	.512212 40	−1.6	+56	−5.0	+51
MAS6	gabbro	.56	452	2.901	13.40	.004	.70807	.1309	.512215 21	−3.2	+60	−5.8	+56
MAS5	gabbro	1.01	507	.445	2.875	.006	.70776	.0936	.512295 38	+1.2	+56	−2.8	+51
MAS4	gabbro	2.18	672	1.668	8.9	.009	.7079	.1147	.512159 46	−3.1	+57	−6.2	+53
MAS3	gabbro	1.71	442	.669	3.221	.011	.70713	.1256	.512359 48	0.0	+46	−2.8	+42
MAS2	diorite	15.90	685	5.18	27.91	.067	.70819	.1122	.512258 20	−1.0	+54	−4.2	+53
MAS1B	diorite	30.80	530	3.431	20.40	.168	.70894	.1017	.512174 24	−1.8	+53	−5.5	+58
MAS1	diorite	6.72	423	5.51	28.08	.036	.7078	.1186	.512210 20	−2.4	+51	−5.4	+49
MAS0	diorite	30.30	428	4.29	26.07	.205	.70928	.0994	.512250 43	−0.1	+53	−3.9	+61
Metasediments													
IV13	granulite	39.8	321	5.26	29.8	.358	.71861	.1066	.512007 50			−8.9	+184
IV27	granulite	38.3	224	4.71	23.1	.494	.71607	.1232	.511967 22			−10.3	+139
IV31	leucosome	25.1	303	2.86	16.6	.315	.7136	.1043	.512016 19			−8.6	+120
IV332	leucosome	85.8	379	9.42	55.0	.656	.7136	.1036	.512092 18			−7.1	+95
IV342	restite?	9.2	259	5.24	43.2	.103	.71702	.0733	.512011 18			−7.5	+177
IV370	acid layer	7.9	492	1.23	10.6	.046	.71247	.0704	.511914 40			−9.3	+115
IV374	mica schist	169	73	11.1	62.4	6.71	.75032	.1071	.511987 26			−9.3	+249
IV384	granulite	42.4	223	6.86	33	.55	.71532	.1257	.512158 47			−6.7	+125
ST8*	granulite	32.9	228	5.17	24.9	.417	.71352	.125	.51237 5			−2.4	+91
ST13*	kinzigite	101	105	3.08	11.8	2.79	.72618	.158	.51237 3			−3.7	+141

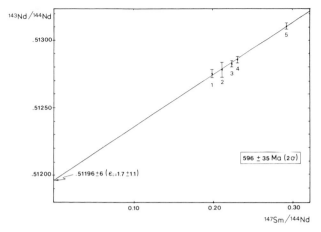

FIG. 4. $^{143}Nd/^{144}Nd$ vs $^{147}Sm/^{144}Nd$ isotope evolution diagram for the lower layered group (whole rock). 1—GM10; 2—Fe41; 3—GZ8; 4—GZ7; 5—IV36.

Isotopic composition

Six samples ranging from peridotite to gabbro have been analysed for Sr and Nd isotopes (Table 2). They are firstly characterized by low present-day $^{87}Sr/^{86}Sr$ (0.7035–0.7047). Extremely low Rb/Sr make these isotope ratios nearly independent of age corrections, whatever the real age of the LLG.

Nd isotopes exhibit fairly radiogenic compositions, corresponding to positive present day ε values (for details on ε notation see DePaolo & Wasserburg 1976). For both Rb–Sr and Sm–Nd systems, sample IV41 appears highly anomalous both with its high $^{87}Sr/^{86}Sr$ (0.70968, duplicate analysis, 0.70964), not supported by a higher Rb/Sr ratio than other samples, and its much lower $^{143}Nd/^{144}Nd$ (0.512369, corresponding to a very negative present day ε value of -5.3). In contrast to the main gabbros (see later) this sample has a LREE depleted pattern as shown by its $^{147}Sm/^{144}Nd$ of 0.212, higher than the chondritic value of 0.1967 (Jacobsen & Wasserburg 1980).

In a Nd isotope evolution diagram (Fig. 4), experimental points plot along a reasonably good linear array (MSWD = 1.99), corresponding to an age of 596 ± 35 Ma (error quoted at the 95% confidence level) with an initial ε value of 1.7 ± 1.1 (see also Voshage *et al.* 1983). The geological significance of this array is uncertain. It could well date the separation of the parental magma to the LLG from a mildly LREE-depleted mantle source. Alternatively, this linear array could be a two component mixing line, without any time significance, but the lack of correlation between $^{143}Nd/^{144}Nd$ and $1/Nd$ does not support this. As

our samples have Rb/Sr and Sm/Nd significantly different from those of the model bulk Earth, their position in the ε_{Nd}–ε_{Sr} diagram depends on their geological age. However for ages in the 600–300 Ma range (assumed to be realistic) the points plot in the time-integrated LREE and LIL-depleted source quadrant; the grouping being better for a 600 Ma age correction of the Sm–Nd isochron (Fig. 5). Whatever the real geological age of the LLG, the Nd isotopes point to an origin from a time-integrated moderately LREE-depleted source (ε_{Nd} c. $+2$ at 600 Ma; c. $+4$ at 300 Ma).

As already mentioned, sample IV41, strongly departs from other LLG samples, its isotopic composition suggesting contamination by a crustal component. Bulk mixing calculations show that it could result from about 25% contamination of material having the Sr–Nd characteristics of granulitic leucosomes such as IV31 or IV370 ($\varepsilon_{Nd} = -8$–9, $\varepsilon_{Sr} = 150$–200, fairly high Sr/Nd = 20–50; Table 2). A further limit to the amount of contamination could be provided by the still LREE depleted pattern (Sm/Nd = 0.336). Thin layers of metasediment occurring within the LLG (Ferrario *et al.* 1982) perhaps lend support to the contamination hypothesis.

In summary, apart from sample IV41, for which some evidence of crustal contamination is found, the LLG is shown to be derived from a time-integrated LREE-depleted reservoir. There is some shift to more radiogenic ε_{Sr} values when compared to typical mantle domains, which might point to either source contamination effects or to late addition(s) of ^{87}Sr, although uncertainty in the age correction makes this

TABLE 3. *Rare-earth analyses for the LLG (IV35, IV33, IV36 and IV41) and for the ULG–MG–diorite body, IV322, 323, 319, 313 and 341 were analysed by ICP, the remaining by INAA. IV322 and 323 are from the ULG; IV139–IV147 are gabbros and IV148– IV341 are biotite-bearing gabbros or diorites*

Rock	IV35	IV33	IV36	IV41	IV322	IV323	IV139	IV140	IV142	IV145	IV147	IV46	IV148	IV319	IV50	IV104	IV107	IV313	IV341
La	—	—	—	—	4.8	6.0	22.3	16.0	6.7	10.7	19.8	17.0	7.3	36.3	31.0	17.3	9.8	31.0	29.7
Ce	—	(5.3)	(5.5)	—	8.7	10.4	57.7	44.4	15.4	20.2	44.1	39.9	14.4	73.9	72.0	39.7	19.6	57.3	60.5
Nd	—	—	8.33	1.17	4.7	4.8	38.9	30.1	8.9	8.6	27.3	27.2	8.5	40.4	38.0	30.0	13.4	30.0	30.9
Sm	0.87	3.2	3.9	0.97	0.8	0.5	5.2	4.4	1.7	1.3	4.5	5.2	1.8	7.1	6.4	8.1	3.4	4.8	5.4
Eu	0.31	1.14	1.57	0.31	2.45	2.56	4.2	3.8	2.25	2.26	4.1	3.7	2.45	3.35	3.2	4.3	4.02	5.08	5.4
Gd	—	—	—	—	0.5	0.2	—	—	—	—	—	—	—	6.5	6.8	6.1	—	4.6	5.2
Tb	0.24	1.1	1.36	0.15	—	0.16	0.71	0.65	0.33	0.18	0.53	0.65	0.26	—	0.78	0.88	0.41	—	—
Dy	—	—	—	—	0.4	—	—	—	—	—	—	—	—	4.1	—	—	—	2.5	3.5
Ho	—	—	—	—	—	—	—	—	—	—	—	—	—	0.79	—	—	—	0.48	0.69
Yb	0.84	4.4	6.16	0.66	0.26	0.11	1.12	1.20	0.77	0.40	0.64	0.76	0.68	1.82	1.70	2.01	1.07	1.07	1.61
Lu	—	0.75	0.95	—	—	0.02	0.14	0.17	—	—	—	—	—	0.30	0.27	0.28	—	—	0.28
Ce_N/Yb_N	—	—	0.21	—	8.7	24.6	13.4	9.6	5.2	13.0	17.9	13.7	5.5	10.6	11.0	5.1	4.7	17.4	9.8

problem difficult to resolve. Sm–Nd systematics favour an age of about 600 Ma for the formation (not emplacement) of the LLG.

Composition of parental liquid

There are considerable difficulties in interpreting cumulates in terms of their equilibrium liquid, particularly when metamorphism and recrystallization have obscured the textural evidence for the cumulus and intercumulus mineralogy. The major element composition is obviously dependent on the cumulus mineralogy and bears no simple relationship to the equilibrium liquid composition and cannot be used for modelling fractionation processes. It is not possible to determine the composition of the magma directly as there are no chilled margins. The high Ni and Cr of ultramafic rocks together with the relatively high MgO contents of pyroxenites and gabbros suggest the magma was basic. Gabbro and pyroxenite dykes cutting the Balmuccia peridotite have similar chemical features to the LLG cumulates (Sinigoi *et al.* 1983), but no dykes have been seen to cross the peridotite–LLG contact. Rivalenti *et al.* (1981b, 1984) proposed that gabbro pods within the mantle peridotite near the contact (e.g. IV301, Table 1) might represent the parental liquid to the LLG cumulates, although these too appear to have a cumulate component. The similarity between the pyroxenite and gabbro dykes to the LLG does suggest however that there may be some genetic relationship.

Harzburgite layers are olivine-orthopyroxene cumulates; both minerals having negligible partition coefficients for Ti, Zr and Y. These elements were presumably located in the intercumulus liquid and the ratios of these element pairs should equal that of the liquid from which they fractionated. The observed correlations give Ti/Zr = 115, Ti/Y = 290 and Zr/Y = 2.2, values which are typical of many basaltic liquids (Wood *et al.* 1979). Similar values are obtained from pyroxenites. REE and other trace element levels have been estimated, making reasonable assumptions as to the original cumulus mineralogy. These calculations yield the following composition for the equilibrium liquid: TiO_2 from 2.0–2.5 wt. %, Y 30–40 ppm, Zr 80–140 ppm, Sm 2–3 ppm, Eu 1–1.7 ppm, Tb 1–1.5 ppm and Yb 3–6 ppm. This wide range results from uncertainties in the proportion of intercumulus liquid. The level of HREE calculated depends on the $K_{HREE}^{cpx/liq}$, for which published values show a considerable range (Henderson 1982). The clinopyroxene is very aluminous, which might favour substitution of the HREE, leading to higher $K_{Yb}^{cpx/liq}$ than used in the calculations. This would slightly reduce the calculated levels of HREE in the equilibrium liquid. K_2O, Rb and LREE are at or below detection limit, implying the equilibrium liquid was depleted in these elements relative to Ti, Zr and HREE.

The trace element ratios calculated above, together with Ni, Cr and MgO contents suggest the parental magma was basaltic. The isotope data, REE and LILE suggest it was derived from a time-integrated depleted mantle source. Rivalenti *et al.* (1981b) proposed that the LLG fractionated at moderate pressure as cpx rather than ol is the main cumulus mineral, supported by the high Al_2O_3 content of the pyroxene (Thompson 1974).

Cumulates like the LLG are only found adjacent to the Balmuccia peridotite in Val Sesia, and to the south of Val Strona near the Insubric line (Ferrario *et al.* 1982; Fig. 1) and are separated from the main gabbro-diorite body by a layer of metasediment. We suggest that the LLG fractionated at greater depth than the main gabbro body (Rivalenti *et al.* 1984) and was emplaced into the Ivrea Zone along with the Balmuccia peridotite. The relative age of the LLG and main gabbro is not yet known.

Main gabbro diorite body

The main gabbro-diorite body reaches a maximum thickness of 5.5 km and is well exposed in Val Sesia and Val Mastallone (Fig. 2). The lower part (ULG of Rivalenti *et al.* 1975; 1984) comprises layered gabbros with minor pyroxenites and anorthosites. It grades into fairly homogeneous plagioclase-rich gabbros, with an average of 65–70% plagioclase and occasional anorthosite layers. Modal biotite increases gradually up the section to form biotite-gabbros, monzogabbros and, at the top, diorites. The lower part of the body has often completely recrystallized to a polygonal equigranular texture but igneous textures are preserved higher up. In Val Sesia and Val Mastallone the diorite is unfoliated and contains numerous sedimentary xenoliths and in this area is clearly intrusive. Further south in Val Strona di Postua, the situation is more complex and the contact between diorite and metasediment is less well defined.

Detailed descriptions of the petrography and mineral chemistry can be found in Rivalenti *et al.* (1975). Unlike the LLG, gabbros contain olivine which is always in reaction relationship with plagioclase (Sills 1984). The main gabbro grades into 'diorite' by the gradual increase in biotite and hornblende. The diorites are olivine free and within a few metres of the top of the body contain

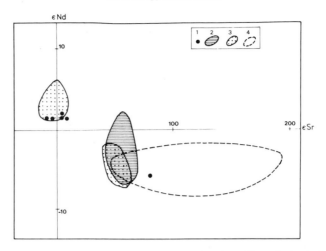

FIG. 5. ε_{Nd}–ε_{Sr} diagram for all samples (lower layered group and main gabbro–diorite body). Due to the uncertainty in the age (see discussion in text), two limiting situations are shown at 600 Ma and 300 Ma respectively. 1—data points for the LLG at 600 Ma (note the discrepant position of IV41 at $\varepsilon_{Sr} \sim 80$). 2—field for the ULG–MG–diorite body at 600 ma. 3—fields for the LLG (top left) and the ULG–MG–diorite body (bottom right) at 300 Ma. 4—field of metasediment at 600 Ma.

very irregular shaped garnet. Biotite occurs as flakes up to 1 cm long and in many cases is clearly cumulus. Some samples near the top of the complex contain quartz and k-feldspar.

Chemistry

Representative analyses are given in Table 1. The AFM plot (Fig. 6) shows that the main gabbro-diorite body broadly follows a calc-alkaline

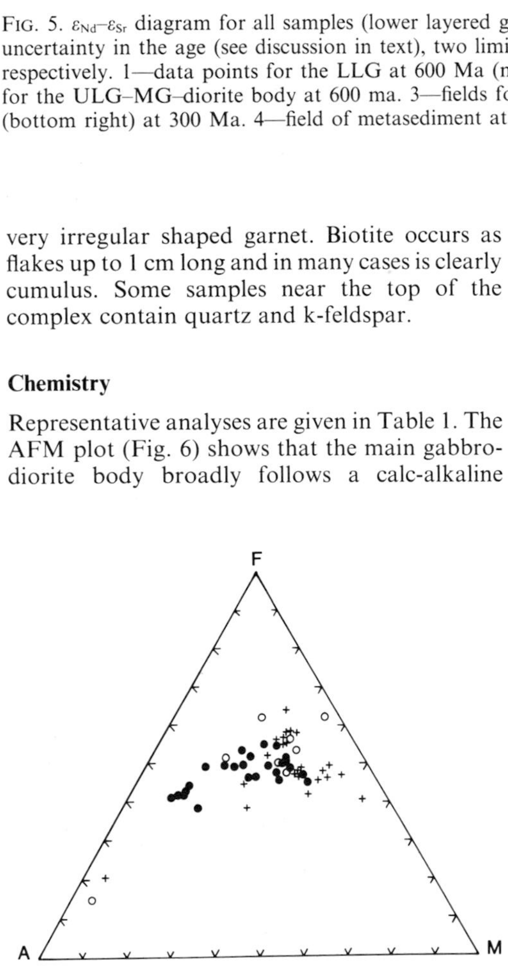

FIG. 6. A($Na_2O + K_2O$)–F(FeO_T)–M(MgO) (wt. % oxide) plot for the ULG-MG-diorite body. Symbols: open circles, ULG; crosses, main gabbro; full circles, diorites.

trend. The major and trace element compositions are consistent with the textural observation that the majority of gabbros are plagioclase cumulates. The REE data (Table 3) show very large Eu anomalies with Eu/Eu* from 2–12, and very steep patterns with Ce_N/Yb_N from 4.5–20. It is impossible to generate such steep patterns from the same liquid that fractionated the LLG without garnet as a major fractionating phase, but the only occurrence of garnet in the LLG is the reaction coronas that developed during slow cooling. The isotope data, REE patterns and many other elements, such as Sr and Ba, indicate that the ULG-MG-diorite body and the LLG had different parental liquids.

A few parameters vary from the base to the top of the body, although none vary smoothly. The diorites show an increase in SiO_2, but SiO_2 is generally low, ranging from 42–57 wt. % (Fig. 7). There is a tendency for mg no. to decrease, but it varies erratically within the gabbro. Pyroxene compositions vary widely (Fig. 8) with mg no. generally decreasing up the section. Al_2O_3 in cpx ranges from 6 wt. % near the base to less than 1 wt. % near the top. CaO decreases while Na_2O increases reflecting the changing plagioclase composition. Plagioclase compositions vary greatly depending on the degree of metamorphic re-equilibration and the stratigraphic position. K_2O (and Rb) increase up the complex, gabbros generally having quite low K_2O from 0.2–0.5 wt. %, but the K_2O, Rb and Ba contents increase

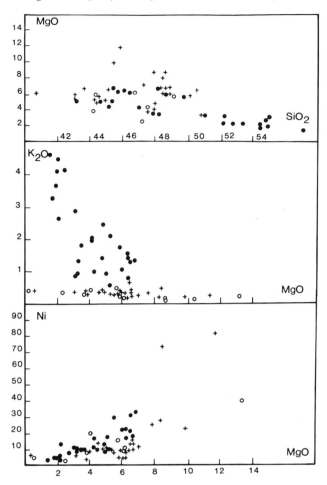

Fig. 7. MgO vs. SiO$_2$ and K$_2$O and Ni vs. MgO for the ULG–MG–diorite body. Symbols as in Fig. 6.

markedly with the presence of biotite which correlates with decreasing mg no. in the diorites (Fig. 7). Ba contents are quite high in all rocks and increase dramatically to 8000 ppm near the top of the body, where biotite may contain as much as 3 wt. % BaO. Zr is also very high in some diorites reflecting cumulus zircon. P$_2$O$_5$ varies within the gabbros depending on whether cumulus apatite is present or not. Ni tends to decrease with mg no. (Fig. 7) but levels are generally low, about 10 ppm. At the same MgO level, hornblende-biotite rich 'diorites' tend to have higher Ni than plagioclase rich gabbros. Hornblende changes from a brown kaersutite near the base to a green pargasitic hornblende near the top and in contrast to the LLG K$_2$O contents are moderate (1–2 wt. %).

The Sr vs. Ba plot (Fig. 9) shows the importance of plagioclase fractionation. In the LLG, Sr and Ba increase together, both being incompatible, whereas in the MG, Sr varies from about 250–900 ppm correlating with the modal percentage of plagioclase. Ba increases up the complex whereas in the diorites Sr is rather invariant, indicating the fractionating phase(s) had a $K_{Sr}^{min/liq}$ of about 1 and a $K_{Ba}^{min/liq}$ of <1. Plagioclase is the most likely phase. Some diorites with very high Ba contain biotite as a cumulus phase for which K_{Ba} is about 3.

In contrast to the LLG, TiO$_2$ vs. Zr shows considerable scatter reflecting the presence of cumulus ilmenite or magnetite. Y contents are generally quite low, reaching a maximum of 35 ppm in some hornblende-rich 'diorites'. Zr/Y

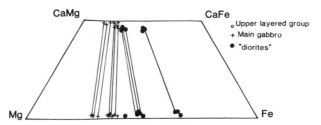

FIG. 8. Pyroxene compositions plotted in part of the Ca–Mg–Fe triangle. Tie lines join grains from the same sample.

ratios are very high: excluding samples with zircon, Zr/Y is an average of about 7.5. The levels of Y and HREE do not increase much up the section. Nb reaches a maximum of 10 ppm, which is low relative to other trace elements such as LREE, Ba and K. This together with the very high Ba is characteristic of calc-alkaline suites (e.g. Pearce 1983).

REE chemistry

REE data for sixteen samples from the ULG, MG and 'diorite' are presented in Fig. 10 a–d and Table 3. All samples have very large positive Eu anomalies, the largest being in plagioclase-rich samples. The ULG samples (IV323 an anorthosite, and IV322 a plagioclase-olivine gabbro) have low total REE abundances which are very similar to some Proterozoic massif anorthosites (Simmons & Hanson 1978) and to plagioclase separated from massif anorthosites (Griffin *et al.* 1974). The smallest Eu anomalies and highest REE abundances occur in biotite-rich gabbros in the transition zone between the MG and true diorites. In general the REE patterns are parallel and increase up the body, which may reflect increasing REE contents in the residual liquids in equilibrium with the more evolved cumulates. IV145 a plagioclase-rich gabbro from the MG has higher REE content than IV322, a gabbro from the ULG with similar plagioclase content.

The textural evidence that the main fractionating phase was plagioclase with olivine, minor pyroxene, apatite and ilmenite and, at a late stage, biotite and zircon, is confirmed by the REE data. Ce correlates quite well with Yb, with the exception of three Yb-rich samples from the transition zone between MG and diorite. There is no systematic change of Ce_N/Yb_N with fractionation, which would be expected if hornblende or garnet were fractionating phases. Zircon, which appears in the biotite-rich gabbros would deplete the evolved liquids in HREE. Gabbros with abundant apatite, tend to have slightly higher REE contents than apatite-free samples.

Isotopic composition

Thirteen samples spanning the 'stratigraphic' section collected in both Val Sesia and Val Mastallone have been analysed (Table 2). All samples have high present-day $^{87}Sr/^{86}Sr$, irrespective of their major element composition. Except for the diorites, Rb/Sr ratios are very low resulting in negligible age correction for *in situ* decay of ^{87}Rb. Thus the ULG–MG–diorite unit is unexpectedly characterized by a fairly radiogenic Sr isotope composition, quite different from 'normal' mantle-derived rocks. In a $^{87}Sr/^{86}Sr$ vs. $^{87}Rb/^{86}Sr$ plot (Fig. 11a), the ultramafic and gabbroic rocks cluster in an elliptical domain elongated parallel to the $^{87}Sr/^{86}Sr$ axis, while diorites define a slightly concave down curve. It is not possible to obtain any age information from the Rb/Sr systematics, but it is worth mentioning that a tie-line between the most evolved (in terms of Rb–Sr) mafic rock (GM90) and the most evolved diorite (Mas0) would correspond to a 'date' of 320 Ma.

Nd measurements exhibit less radiogenic $^{143}Nd/^{144}Nd$ ratios than the chondritic value. In contrast to the LLG they show considerable scatter in an isotope evolution diagram (Fig. 11b) and cannot be fitted to a straight line. This implies either late disturbances of the whole-rock Sm–Nd system or a large initial isotope heterogeneity, or both. As the whole-rock Sm–Nd system is generally considered fairly resistant to resetting, it can be reasonably considered that the ULG–MG–diorite body was highly heterogeneous with respect to Nd isotopes.

Model ages relative to a reservoir with chondritic characteristics (T_{CHUR}, De Paolo & Wasserburg 1976; McCulloch & Wasserburg 1978) range between 0.5 and 1.0 Ga, with the majority clustering between 0.6 and 0.8 Ga. Ages calculated relative to a model depleted mantle source (De Paolo, 1981a) are of course significantly older (1–1.5 Ga). However both calculations assume a simple two-stage history, which is unlikely in this case. The geological age of the ULG–MG–diorite body is not well constrained as yet: three a priori

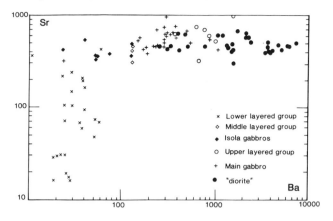

Fig. 9. Sr vs. Ba for all samples analysed.

geological ages have therefore been considered for computation of the initial ε_{Sr} and ε_{Nd}. Firstly, 600 Ma, which is the age suggested for the neighbouring LLG (Fig. 4), although the LLG is different from both the trace element and isotopic points of view. Secondly, 480 Ma, which could be the age of the peak of granulite-facies metamorphism, according to Hunziker & Zingg's (1980) interpretation of Rb–Sr data on large whole-rock samples. Several authors suggest that the granulite-facies metamorphism and emplacement of the mafic complex were contemporaneous (Schmid & Wood 1976). Thirdly, 300 Ma, which is also a possible age, as indicated by U–Pb data from zircons and monazites (Köppel 1974). There is no consensus concerning the interpretation of these data. Hunziker & Zingg (1980) interpret all the data in terms of very slow cooling from the inferred peak of metamorphism at *c.* 480 Ma to about 180 Ma (the closure of K–Ar systems).

In every case the ULG–MG–diorite body has initial $^{143}Nd/^{144}Nd$ and $^{87}Sr/^{86}Sr$ respectively less and more radiogenic than bulk earth at the time considered, pointing to a bulk long term LREE and LILE enrichment of its source material. The only possible exception is sample MAS5, for which a positive ε_{Nd} is calculated at 600 Ma; however, a distinctly positive ε_{Sr} ($+56$) remains, so that this sample would have an unusual position in an ε_{Nd}–ε_{Sr} plot. Owing to its low Sm/ Nd ratio ($^{147}Sm/^{144}Nd$ is very different from the chondritic value), this sample is quite sensitive to age correction, and its discrepant ε_{Nd} could suggest that 600 Ma does not represent a proper age for correction of *in situ* decay of ^{147}Sm. A total spread of seven ε_{Nd} units is found at 600 Ma, against 4.8 at 480 Ma and 3.4 units at 300 Ma.

The major isotope feature of the ULG–MG– diorite body, whatever its age, is the position of the data points in the negative ε_{Nd}, positive ε_{Sr} quadrant of the ε_{Nd}–ε_{Sr} diagram (Fig. 5). Such a position is unusual for mantle derived rocks, and points either to a source enriched in Rb and LREE for a long time before magma generation, or to significant contamination of more 'normal' depleted mantle material by crustal components with high ε_{Sr} and low ε_{Nd} (e.g. Hawkesworth *et al.* 1983). It is worth noting that no clear relationship between rock type, or 'stratigraphic' position, and position on the ε_{Nd}–ε_{Sr} can be seen.

Petrogenesis

The major, trace element and REE data are all consistent with the ULG and MG being plag $\pm$ ol $\pm$ ap $\pm$ ilmt $\pm$ opx $\pm$ cpx cumulates; with pyroxenes being only minor cumulus phases. Hornblende is present as an intercumulus phase. This assemblage is characteristic of low pressure fractionation, in contrast to the LLG, which is dominated by clinopyroxene fractionation. Towards the top of the body, zircon and biotite become fractionating phases. The fractionation of large amounts of plagioclase has restricted the SiO_2 enrichment of more evolved compositions. Late stage cumulates are enriched in Ba, Zr, and K_2O.

The growth of garnet and pyroxene + spinel reaction coronas between plagioclase and olivine and the recrystallization of some gabbros to two pyroxene granulites suggests the lower parts of the body recrystallized at about 750°C and about 7–7.5 kbar (Sills 1984). This either means the body was buried and recrystallized after emplacement, or it was emplaced at these depths and the coronas grew during slow cooling from igneous

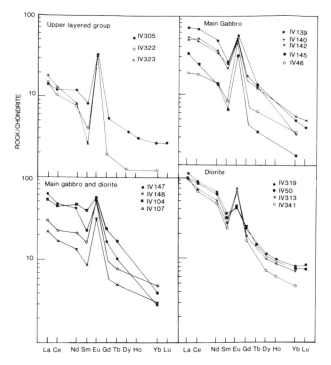

FIG. 10. Chondrite normalized REE plots for the ULG–MG–diorite body.

temperatures. As deformation in the upper part of the MG is localized and igneous textures are commonly preserved, we favour the latter interpretation.

It is interesting to attempt to estimate the composition of the parental liquid. The AFM plot, Ba and Nb suggest calc-alkaline affinities. The very low levels of Ni and Cr, even in olivine bearing samples suggests the liquid had low abundances of these elements. The chemistry suggests that even the most evolved rocks are cumulates. The changing composition of the minerals (e.g. the wide range in mg no., Fig. 8) suggests that the liquid in equilibrium with successive cumulates had evolved.

There are no chilled margins so the composition of the parental liquid must be estimated from the composition of the cumulates. As there is some growth of new minerals during metamorphic re-equilibration, particularly in olivine-bearing samples, the original igneous mineralogy has been inferred from the CIPW norms (Table 1). Trace element concentrations, in particular REE, have been calculated for those samples with low levels of incompatible trace elements, that can reasonably be inferred to have low amounts of intercumulus liquid. Samples with low levels of

P_2O_5 have been chosen as the $K_{REE}^{ap/liq}$ depends critically on the liquid composition (Watson & Green 1981). Partition coefficient data are from Drake & Weill 1975; Arth 1976; Pearce & Norry 1979 and Watson & Green 1981.

There is a fair range of permitted values depending on the proportion of intercumulus liquid assumed and on the value of K_D chosen. However, all the calculations give liquids with LREE-enriched patterns and small positive Eu anomalies. The level of LREE ranges from 60–100 times chondrite and HREE from 5–10 times chondrite. These patterns are very similar to some of the biotite-rich gabbros from the transition zone between gabbros and diorite, such as IV50, IV319 and IV104 (Table 1, Fig. 10). The liquid may have had about 500 ppm Sr and 300–500 ppm Ba. The levels of Ni and Cr in the liquid were probably low, which together with the low mg no. of olivine and the generally low MgO contents of the gabbros, suggest that the liquid was probably intermediate in composition, possibly andesite.

Isotopic constraints

Several occurrences of mafic intrusions contaminated by crustal material are well documented

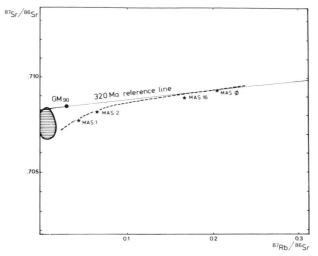

FIG. 11a. Sr isotope evolution diagram for the ULG–MG–diorite body. Data points for the gabbroic cumulates plot in the hatched area, with the exception of GM90. * are diorite samples.

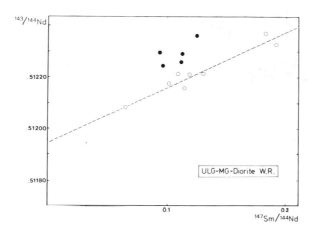

FIG. 11b. Nd isotope evolution diagram for the ULG–MG–diorite body (whole rock). The regression line through 8 data points corresponds to age of 330 Ma.

(Gray *et al.* 1981; Zindler *et al.* 1981). In the present case of a deep seated intrusion in already hot country-rocks affected by high grade metamorphism, contamination by bulk assimilation or by mixing with silicic partial melts would seem likely. Grove & Baker (1984) show large amounts of country rock can be assimilated by a magma undergoing fractional crystallization.

The data points define an elliptical domain in the ε_{Sr-Nd} plot with a steep slope (Fig. 5). This internal variation could possibly be explained as a result of variable degrees of contamination of a mafic, high Sr/Nd melt by crustal components

with a much lower Sr/Nd (De Paolo & Wasserburg 1979). The presence of metasedimentary xenoliths within the diorite attests to the likelihood of this process.

In order to assess the contamination hypothesis more quantitatively, eight samples of metasedimentary rocks and derived mobilizates from the country rocks to the complex have been analysed (Table 2) so that the isotopic range of likely contaminants could be evaluated. It is manifest that these rocks are variable both in their isotope characteristics (ε_{Nd} from -7 to -11 at 300 Ma) and concentrations (Sr 75–500 ppm, Nd 10–60 ppm). More particularly, Sr/Nd ratios, which are an important parameter in mixing models, scatter widely (1–46). With the exception of IV374, an amphibolite facies-metasediment, samples display low Rb/Sr ratios, typical of the granulite-facies. The definition of the isotope characteristics of the local contaminant is dependent on the age chosen. Again ε_{Nd} and ε_{Sr} have been computed for the three possible ages, 600, 480 and 300 Ma; however, mixing calculations have been performed for the 300 Ma situation only because of the best grouping of the ULG–MG–diorite data points obtained at this time, and because leucosomes (some of which are discordant) are not likely to be older than the Hercynian event recorded by the U–Pb system of zircons and monazites. However, if the data for the older ages are used, the conclusions are unaffected.

The large variability of Sr/Nd displayed by analysed metasediments and leucosomes allows a large flexibility in bulk mixing models. However to account for the steep trend of the ULG–MG–diorite data in the ε_{Sr-Nd} diagram, mixing curves characterized by large $R(=(Sr/Nd)_{mafic\,melt}/(Sr/Nd)_{contaminant})$ have to be proposed, viz. $2 \leqslant R \leqslant 10$ for an initial magma plotting near the 'depleted mantle field', or $5 \leqslant R \leqslant 20$ for an initial magma with $\varepsilon_{Nd} \sim 0$, $\varepsilon_{Sr} \sim 40$. If Sr/Nd of 10–20 is taken as a reasonable estimate for the primary mafic melt, the situation with high R requires fairly low Sr/Nd for the contaminants, which is not in good agreement with the likely contaminants (Table 2). Moreover, taking into account the likely absolute Sr and Nd contents of mafic melt and metasedimentary material, a bulk assimilation process would require a very large amount of contamination which is not permitted by major and trace element data (i.e. 30–60% if the starting magma had $\varepsilon_{Nd} = +3$, $\varepsilon_{Sr} = 0$; 10–15% if $\varepsilon_{Nd} = 0$ and $\varepsilon_{Sr} = 40$, with $Nd_{mafic\,melt} = 25$ ppm, $Sr = 400$ ppm, $Nd_{contaminant} = 30–60$ ppm, Sr 100–300 ppm).

These difficulties are even greater if the highly probable mechanism, combined assimilation and fractional crystallization (AFC) (De Paolo 1981b) is considered. In the present case where plagioclase fractionation is the dominant process, the bulk distribution coefficients, D_{Sr} and D_{Nd}, are > 1 and < 1 respectively. This means the shift in ε_{Nd} will be much smaller than the concurrent shift in ε_{Sr} (De Paolo, 1981b), e.g. for a mass assimilated/mass crystallized $M_a/M_c = 0.1$, and for $D_{Sr} = 2$, $D_{Nd} = 0.15$, a 20 ε_{Sr} displacement is correlated to a shift of only 1 ε unit in ε_{Nd}. Thus if the spread in ε_{Sr} of about 20 units exhibited by the ULG–MG–diorite body is ascribed to an AFC process, the concurrent effect on the Nd isotopes is not enough to account for the observed spread of 3.5 ε_{Nd} units. The lack of any obvious relationship between the stratigraphic position and the isotopic composition also argues against a significant amount of contamination. It is however acknowledged that the most obvious effects of crustal contamination, such as enhanced Rb and K_2O contents, are less likely to be observed in cumulates than in the evolved liquids, but Gray et al. (1981) did observe an increase in silica saturation in contaminated gabbros, marked by increased orthopyroxene. In the MG-diorite body there is no correlation between the degree of contamination indicated by the isotope data and the amount of opx, K_2O, Rb or Ba. It is possible that the onset of biotite fractionation was induced by a certain amount of crustal assimilation which increased the K_2O, Rb and H_2O contents of the magma, but the biotite-bearing samples are not obviously more contaminated, in terms of their isotopes, than other samples. The garnet-bearing samples in the top few metres of the diorite may reflect contamination, but unfortunately we have no isotope analyses of these. The more unusual features of the diorites such as high Ba and Zr cannot be due to sediment contamination as these elements are very much higher in the diorites.

In conclusion, although contamination is likely to have affected the ULG–MG–diorite body, this process alone cannot explain all the isotopic features, in particular the large spread in ε_{Nd}. We must therefore consider the possibility of some kind of isotopic heterogeneity and the long-term enrichment of the ULG–MG–diorite source.

Old enriched mantle domains commonly exhibit the isotopic features ascribed to contamination by continental crust (Hawkesworth et al. 1983), and the possibility that the primary magma for the ULG–MG–diorite originated from such a source must be considered. Indeed, enriched phlogopite–peridotite does exist in the Finero massif (Ernst 1978). This type of peridotite has very high Rb/Sr ratios and the same problems are encountered in attempting to explain the wide spread in ε_{Nd} values as for the crustal contamination model. Hawkesworth et al. (1983) show that the mantle underlying some Karoo basalts has very negative ε_{Nd} for only very small positive ε_{Sr}. A liquid derived from this type of mantle which then suffered some crustal contamination might have isotopic characteristics resembling the parental liquid to the ULG–MG–diorite body.

As the contamination hypothesis is not really satisfactory and the enriched mantle hypothesis highly speculative, we have to examine other possible models for the genesis of the ULG–MG–diorite body. A two-stage model, involving two events widely separated in time and an intermediate reservoir between typical mantle and crust, could account for the available chemical and isotopic data. Firstly, consider the generation of a LREE-enriched, rather high Rb/Sr mafic reservoir from an undepleted or slightly depleted mantle, which would evolve to positive ε_{Sr} and negative ε_{Nd} with time. Secondly, melting of this reservoir to produce a calc-alkaline magma which was intruded into the lower crust and fractionated extensively to produce the ULG–MG–diorite body. The time of residence required in the mafic reservoir depends on the degree of LREE and Rb/Sr enrichment which occurred during the first event. It is possible to consider an island arc tholeiite protolith with mildly enriched LREE patterns (Nd c. 20 times chondrite, Sm c. 15 times chondrite, i.e. with a $^{147}Sm/^{144}Nd$ of c. 0.14; cf. Ishizaka & Carlson 1983). Such a material would develop c. -2 ε_{Nd} units per 100 Ma. If it is considered that assimilation during emplacement of the ULG–MG–diorite caused only a minor effect ($\leqslant 1$ ε_{Nd}), the initial values at 300 Ma of -2 to -6 show that the LREE enriched protolith(s) had to exist for at least 100–300 Ma, if it itself had a chondritic source. If this were derived from a depleted mantle these residence times are underestimates. The range of ε_{Nd} could be ascribed either to a variable Sm/Nd in the protolith, or to variable residence times, or both. Similar calculations show that a shift of 20 ε_{Sr} units could have developed in a time of 300 Ma in a reservoir with an Rb/Sr of 0.12 (cf. island arc basalts, Ishizaka & Carlson 1983).

The two stage evolution discussed above has some similarity to Green's (1982) model for the genesis of calc-alkaline andesites. It could involve the generation (and underplating?) of an hydrous basaltic parent, enriched in trace elements, 600–700 Ma ago, the later partial melting of which c. 300 Ma ago would produce a low Ni, low Nb, high Al, high Ba liquid with enriched isotopic characteristics. Other models such as the melting of a mixed amphibolite–metasedimentary pile as proposed for Hercynian granitoids elsewhere (Michard-Vitrac *et al.* 1980) might be appropriate, but we lack isotope data on the amphibolites in order to test this hypothesis. It is worth noting that the high-K, high Rb/Sr, c. 280 Ma old granites of the Strona Ceneri zone (e.g. Baveno, Mte Orfano; Fig. 1) have initial $^{87}Sr/^{86}Sr$ ratios of about 0.708 (Hunziker & Zingg 1980), essentially similar to that of the ULG–MG–diorite, and it is tempting to infer some genetic links between them. Preliminary Sm–Nd data suggests a similar spread of negative ε_{Nd} values. If such a hypothesis is viable, these high level granites could represent the residual liquids which escaped after extensive fractionation in the lower crust to form the ULG–MG–diorite body. This hypothesis still requires further work, in particular precise dating of the

ULG–MG–diorite body and a geochemical study of the granites.

Discussion and conclusions

Despite the many problems still unresolved, we can say that the ULG–MG–diorite is a cumulate body derived from a LREE and Ba-enriched calc-alkaline melt, probably of andesitic composition, which was emplaced and fractionated in the mid to lower continental crust. Isotope data show that the source was heterogeneous, but the present data are unable to distinguish between a variably enriched mantle source or a two stage model involving the remelting of an intermediate mafic reservoir. The extent of crustal contamination is also uncertain. It is hoped that O and Pb isotope data will help the resolution of these uncertainties.

The sequence of events in the Ivrea Zone seems to be firstly the deposition of a series of metasediments, now extremely highly deformed and intercalated with amphibolites, possibly in an accretionary wedge (Sills & Tarney 1984). These sediments were then intruded by (and possibly partially assimilated by) an andesitic melt which fractionated extensively to form the ULG–MG–diorite body. The lower layered group and Balmuccia peridotite were emplaced together, but we do not know whether they originally underlay the metasedimentary sequence or whether they were emplaced at a later stage. One of the main agents of crustal growth in the Ivrea Zone is thus seen to be the addition of a crustal magma chamber (which may possibly be related to the Hercynian granitoids?).

The main conclusions are:

(1) The lower layered group consists of clinopyroxene dominated cumulates, which fractionated at high pressure. The parental magma was basaltic and derived from a time-integrated depleted mantle source at about 600 Ma.

(2) The ULG–MG–diorite body is completely independent of the LLG and is a plagioclase dominated cumulate body, fractionated from a LREE and Ba enriched calc-alkaline magma. The isotope data point to a heterogeneous source with long term enrichment in Rb and LREE.

ACKNOWLEDGEMENTS: CP would like to thank C.J. Hawkesworth for access to isotope facilities at the Open University and P. van Calsteren and A. Gledhill for instruction and guidance during the Nd-analyses. JDS would like to thank P.J. Potts and O.W. Thorpe at the Open University for INAA data and J.N. Walsh at

Kings College for ICP REE data. We would like to
thank G. Rivalenti for supplying some of the samples
and JDS is very grateful for the assistance of G.

Rivalenti, S. Sinigoi and F. Siena in the field. This study
was undertaken whilst JDS was in receipt of a NERC
research fellowship which is gratefully acknowledged.

References

ARTH, J.G. 1976. Behaviour of trace elements during
magmatic processes—a summary of theoretical
models and their applications. *J. Res. U.S. Geol.
Surv.* **4**, 41–7.
BEN OTHMAN, D., POLVE, M. & ALLEGRE, C.J. 1984.
Nd–Sr isotopic composition of granulites and constraints on the evolution of the lower crust. *Nature*,
307, 510–5.
BIGIOGGERO, B., BORIANI, A., COLOMBO, A. & GREGNA-
NIN, A. 1978–79. The diorites of the Ivrea basic
complex, Central Alps, Italy. *Mem. degli Inst. de
Geol. e Min. Univ. Padova*, **33**, 71–85.
DE PAOLO, D.J. 1981a. Neodymium isotopes in the
Columbia front Range and crust–mantle evolution
in the Proterozoic. *Nature*, **291**, 193–6.
—— 1981b. Trace element and isotopic effects of
combined wallrock assimilation and fractional crystallisation. *Earth Planet. Sci. Lett.* **53**, 189–202.
—— & WASSERBURG, G.J. 1976. Nd isotopic variations
and petrogenetic models. *Geophys. Res. Lett.* **3**, 249–
52.
—— & —— 1979. Petrogenetic mixing models and Nd–
Sr isotope patterns. *Geochim. Cosmochim. Acta*, **43**,
615–27.
DRAKE, M.J. & WEILL, D.F. 1975. Partition of Sr, Ba,
Ca, Y, Eu^{2+}, Eu^{3+} and other REE between plagioclase feldspar and magmatic liquid; an experimental
study. *Geochim. Cosmochim Acta*, **39**, 689–712.
ERNST, W.G. 1978. Petrochemical study of lherzolite
rocks from the western Alps. *J. Petrol.* **19**, 341–92.
FERRARIO, A., GARUTI, G. & SIGHINOLFI, G.P. 1982.
Platinum and palladium in the Ivrea–Verbano basic
complex, western Alps, Italy. *Econ. Geol.* **77**, 1548–
55.
FOUNTAIN, D.M. & SALISBURY, M.H. 1981. Exposed
cross sections through the continental crust: implications for crustal structure, petrology and evolution. *Earth Planet. Sci. Lett.* **56**, 263–77.
GRAY, C.M., CLIFF, R.A. & GOODE, A.D.T. 1981.
Neodymium-strontium isotope evidence for
extreme contamination in a layered basic intrusion.
Earth Planet. Sci. Lett. **56**, 189–98.
GREEN, T.H. 1982. Anatexis of mafic crust and high
pressure crystallisation of andesite. *In*: THORPE, R.S.
(ed) *Andesites* (John Wiley), **4**, 465–87.
GRIFFIN, W.L., SUNDVOLL, B. & KRISTMANNSDOTTIR,
H. 1974. Trace element composition of anorthositic
plagioclase. *Earth Planet. Sci. Lett.* **24**, 213–223.
GROVE, T.L. & BAKER, M.B. 1984. Phase equilibrium
controls on the tholeiitic versus calc-alkaline differ-
entiation trends. *J. Geophys, Res.* **89**, 3253–74.
HAWKESWORTH, C.J. & VAN CALSTEREN, P.W.C. 1984.
Radiogenic isotopes—some geological applications.
In: HENDERSON, P. (ed) *Rare Earth Geochemistry*.

Developments in Geochemistry 2. Elsevier, Amsterdam, 375–421.
——, ERLANK, A.J., MARSH, J.S., MENZIES, M.A. &
VAN CALSTEREN, P.W.C. 1983. Evolution of the
continental lithosphere: evidence from volcanics
and xenoliths in Southern Africa. *In*: HAWKES-
WORTH, C.J. & NORRY, M.J. (eds) *Continental
Basalts and Mantle Xenoliths*. Shiva Press, Nantwich, 111–138.
——, HAMMIL, M., GLEDHILL, A.R., VAN CALSTEREN,
P.W.C. & ROGERS, G. 1982. Isotope and trace
element evidence for late-stage intra-crustal melting
in the high Andes. *Earth Planet. Sci. Lett.* **58**, 240–
54.
HENDERSON, P. 1982. *Inorganic Geochemistry*. Pergamon, Oxford. 353 pp.
HUNZIKER, J.C. & ZINGG, A. 1980. Lower Palaeozoic
granulite facies metamorphism in the Ivrea Zone
(southern Alps, northern Italy). *Schweiz. mineral.
petrogr. Mitt.* **60**, 181–213.
ISHIZAKA, K. & CARLSON, R.W. 1983. Nd–Sr systematics of the Setouchi volcanic rocks, SW Japan: a clue
to the origin of orogenic andesite. *Earth Planet. Sci.
Lett.* **64**, 327–40.
JACOBSEN, S.B. & WASSERBURG, G.J. 1980. Sm–Nd
isotopic evolution of chondrites. *Earth Planet. Sci.
Lett.* **50**, 139–55.
KÖPPEL, V. 1974. Isotopic U-Pb ages of monazites and
zircons from the crust–mantle transition and adjacent units of the Ivrea and Ceneri Zones (S. Alps,
Italy). *Contrib. Mineral. Petrol.* **43**, 55–70.
McCULLOCH, M.T. & WASSERBURG, G.J. 1978. Sm–Nd
and Rb–Sr geochronology of continental crust
formation. *Science*, **200**, 1002–11.
MEHNERT, K.R. 1975. The Ivrea Zone. A model for the
deep crust. *N. Jb. Miner. Abh.* **125**, 156–99.
MICHARD-VITRAC, A., ALBAREDE, A., DUPUIS, C. &
TAYLOR, H.P. 1980. The genesis of Variscan (Hercynian) plutonic rocks: inferences from Sr, Pb and O
studies on the Malateda igneous complex. Central
Pyrenees (Spain). *Contrib. Mineral. Petrol.* **72**, 57–
72.
NAKAMURA, N. 1974. Determination of REE, Ba, Fe,
Mg, Na and K in carbonaceous and ordinary
chondrites. *Geochim. Cosmochim. Acta.* **38**, 757–75.
PEARCE, J.A. 1983. Role of the sub-continental lithosphere in magma genesis at active continental
margins. *In*: HAWKESWORTH, C.J. & NORRY, M.J.
(eds) *Continental Basalts and Mantle Xenoliths*.
Shiva Press, Nantwich, 230–49.
PEARCE, J.A. & NORRY, M.J. 1979. Petrogenetic implications of Ti, Zr, Y and Nb variations in volcanic
rocks. *Contrib. Mineral. Petrol.* **69**, 33–47.
POTTS, P.J., THORPE, O.W. & WATSON, J.S. 1981.

Determination of the rare-earth element abundances in 29 international rock standards by instrumental neutron activation analysis: a critical appraisal of calibration errors. *Chem. Geol.* **34**, 331–52.

RIVALENTI, G., GARUTI, G. & ROSSI, A. 1975. The origin of the Ivrea–Verbano basic formation (western Italian Alps)—whole rock geochemistry. *Boll. Soc. Geol. It.* **94**, 1149–86.

——, ——, —— & SIENA, F. 1981a. Existence of different peridotite types and of a layered igneous complex in the Ivrea Zone of the western Alps. *J. Petrol.* **22**, 127–53.

——, ——, ——, —— & SINIGOI, S. 1981b. Chromian spinel in the Ivrea–Verbano basic complex, western Alps, Italy. *Tscherm. mineral. petrogr. Mitt.* **29**, 33–53.

——, ——, SIENA, F. & SINIGOI, S. 1984. The layered series of the Ivrea–Verbano igneous complex, western Alps, Italy. *Tscherm. mineral. petrogr. Mitt.* **33**, 77–99.

SCHMID, R. & WOOD, B.J. 1976. Phase relationships in granulite metapelites from the Ivrea–Verbano Zone (northern Italy). *Contrib. Mineral. Petrol.* **54**, 255–79.

SHERVAIS, J.W. 1979. Thermal emplacement model for the Alpine lherzolite massif at Balmuccia, Italy. *J. Petrol.* **20**, 795–820.

SILLS, J.D. 1984. Granulite facies metamorphism in the Ivrea Zone. N.W. Italy. *Schweiz. mineral. petrogr. Mitt.* **64**, 169–91.

—— & TARNEY, J. 1984. Petrogenesis and tectonic significance of amphibolites interlayered with the metasedimentary gneisses in the Ivrea Zone, southern Alps, Northwest Italy. *Tectonophysics*, **107**, 187–206.

SIMMONS, E.C. & HANSON, G.N. 1978. Geochemistry and origin of massif-type anorthosites. *Contrib. Mineral. Petrol.* **66**, 119–135.

SINIGOI, S., COMIN-CHIARAMONTI, P., DEMARCHI, G. & SIENA, F. 1983. Differentiation of partial melts in the mantle: evidence from the Balmuccia peridotite, Italy. *Contrib. Mineral. Petrol.* **82**, 251–9.

THOMPSON, R.N. 1974. Some high pressure pyroxenes. *Mineral. Mag.* **39**, 768–87.

VOSHAGE, H., HUNZIKER, J. & HOFMANN, A.W. 1983. Sm–Nd and Rd–Sr evidence for the evolution of the Ivrea Zone (southern Alps). *Terra Cognita*, **3**, 205.

WALSH, J.N., BUCKLEY, F. & BARKER, J. 1981. The simultaneous determination of the rare-earth elements in rocks using inductively coupled plasma source spectrometry. *Chem. Geol.* **33**, 141–53.

WATSON, E.B. & GREEN, T.H. 1981. Apatite/liquid partition coefficients for the REE and Sr. *Earth Planet. Sci. Lett.* **56**, 405–21.

WEAVER, B.L., MARSH, N.G. & TARNEY, J. 1983. Trace element geochemistry of basaltic rocks recovered at site 516, Rio Grande Rise, D.S.D.P. Leg 72. *In*: Initial Repts. of the Deep Sea Drilling Project. U.S. Govt. Printing Office, Washington, D.C., **72**, 451–55.

WOOD, D.A., JORON, J-L., MARSH, N.G., TARNEY, J. & TREUIL, M. 1979. Major and trace element variations in basalts from the north Philippine Sea drilled during Deep Sea Drilling Project. Leg 58. A comparative study of back-arc basin basalts from Japan and mid-ocean ridges. *In*: Initial Repts. of the Deep Sea Drilling Project. U.S. Govt. Printing Office, Washington, D.C. **58**, 873–94.

ZINDLER, A., HART, S.R. & BROOKS, C. 1981. The Shabogamo Intrusive Suite: Sr and Nd isotopic evidence for contaminated mafic magmas in the Proterozoic. *Earth Planet. Sci. Lett.* **54**, 217–35.

ZINGG, A. 1980. Regional metamorphism in the Ivrea Zone (southern Alps, N-Italy): field and microscopic investigations. *Schweiz. mineral. petrogr. Mitt.* **60**, 153–70.

C. PIN, Lab. de Chron. des Terrains Crist. et Volc. (LA10), Université de Clermont-Ferrand, 5 Rue Kessler, 63068 Clermont-Ferrand CEDEX, France.

JANE D. SILLS, Department of Geology, University of Leicester, Leicester, LE1 7RH, UK.

Evolution of the late Archaean lower continental crust in southern West Greenland

S. Robertson

SUMMARY: Accretion of new continental crust onto the base of pre-existing sialic crust is indicated by field, geochemical and isotopic data from two suites of Archaean gneisses from the Inner Godthåbsfjord region of southern West Greenland. Crustal accretion combined with horizontal tectonic interleaving of existing rock suites thickened the crust resulting in high-grade metamorphism. Enrichment in LIL elements resulted from the depletion in these elements of granulite-facies gneisses below the present level of erosion. Related fluid movements were significant in leading to the anatexis of rocks enriched in LIL elements. The present-day distribution of rock suites is at least in part a function of the juxtaposition of different levels of the late Archaean crust.

Introduction

Rock suites which have been metamorphosed at temperatures and pressures thought to be typical of the lower continental crust are widespread in Archaean terrains. New data are presented from the Ivisârtoq region of southern West Greenland (Fig. 1) which throw light on the nature of, and the processes operating in, the late Archaean lower continental crust.

The data are based on field mapping in 1981–3 of the Ivisârtoq region by a team from the University of Exeter for the Geological Survey of Greenland (GGU), with geochemical and isotopic investigations at the Universities of Exeter and Oxford, respectively.

Regional geology

The Archaean geology of the Godthåb area in southern West Greenland has been described by McGregor (1973, 1979), Bridgwater et al. (1976), Brown et al. (1981) and Chadwick & Coe (1983). Preliminary descriptions of the Ivisârtoq region were given by Hall (1981), Chadwick (1985), Coe & Robertson (1984) and Brewer et al. (1984). The chronology established in the Ivisârtoq region (Table 1) is the same as that described by McGregor (1973, 1979) for the Godthåb region except for a more continuous record of late Archaean plutonic activity. Two major periods of generation of continental crust produced the Amîtsoq gneisses (c. 3700 Ma) and the Nûk gneisses (c. 3000 Ma). Both suites of gneisses include tonalites, trondhjemites, granodiorites and rare granites. All are thought to be derived by partial melting of basic rocks (Compton 1978; McGregor 1979). Syntectonic granites for which the term Ivisârtoq granite gneiss is proposed were intruded

immediately after the emplacement of the Nûk gneisses.

Subsequent syn- to post-tectonic granites showing evidence of formation by anatexis at the present level of erosion, have yielded Rb-Sr ages of 2640 to 2500 Ma (Robertson 1985). These culminated in the emplacement of the Qôrqut granite complex, a sheeted granite complex which outcrops in a linear belt extending SW from the Inland Ice in the Ivisârtoq region to the Buksefjorden region (Fig. 1).

The structure of the NE part of the Ivisârtoq region has been outlined by Brewer et al. (1984) and Chadwick (1985). Amîtsoq gneisses and Malene supracrustal rocks (Table 1) were intercalated by movements on horizontal tectonic slides which preceded and accompanied the emplacement of the Nûk gneisses. Domes and synforms, some cuspate, were superimposed on this broadly horizontal complex. Isotopic ages of syntectonic granites suggest that the domes developed c. 2650 to 2600 Ma ago (Robertson 1985).

Age and isotope data

Analytical methods

Rb-Sr and Pb/Pb isotopic compositions were measured at the University of Oxford. Rb/Sr ratios were determined on pressed powder pellets with a Philips 1410 XRF spectrometer using a method similar to that of Pankhurst & O'Nions (1973). Sr was separated from powdered whole-rock samples by conventional dissolution and cation exchange techniques and isotopically analysed either on a V.G. Micromass 30 or an Isomass 54E mass spectrometer. Pb was separated from rock powders by the double electrodeposition method of Arden & Gale (1974) and

From: DAWSON, J.B., CARSWELL, D.A., HALL, J. & WEDEPOHL, K.H. (eds) 1986, *The Nature of the Lower Continental Crust*, Geological Society Special Publication No. 24, pp. 251–260.

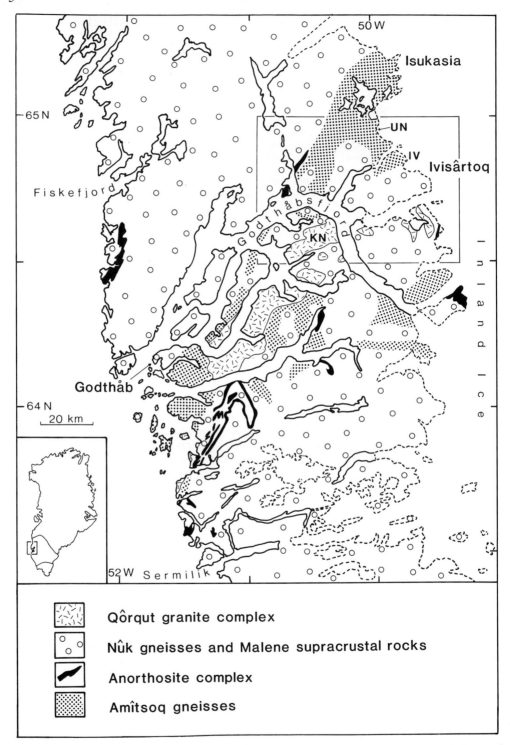

FIG. 1. Simplified geological map of part of southern West Greenland based on 1:500 000 Geological map of Greenland, Sheet 2 (1982), Garde *et al.* (1983, Fig. 7) and results of recent mapping in the Ivisârtoq region. Inset map shows the location of the area within the Archaean craton. The Ivisârtoq region is shown at the head of Godthåbsfjord together with the sample localities: KN = Kangiussap nunâ, UN = Ujaragssuit nunât, IV = NE Ivisârtoq.

TABLE 1. *Summary of the Archaean chronology of the Ivisârtoq region*

Qôrqut granite and pegmatite complex	Rb-Sr 2490 ± 20 Ma	$Sr_o = 0.7058 \pm 0.0005$
	Pb-Pb 2625 ± 105 Ma	
Foliated granite sheets	Rb-Sr 2610 ± 55 Ma	$Sr_o = 0.7038 \pm 0.0016$
Dome-synform phase of deformation	c. 2650 Ma	
Trondhjemites and granodiorites	Rb-Sr 2680 ± 55 Ma	$Sr_o = 0.7030 \pm 0.0004$
Ivisârtoq granite gneiss	Rb-Sr 2750 ± 40 Ma	$Sr_o = 0.7031 \pm 0.0003$
Amphibolite-facies metamorphism and		
enrichment in LIL elements	Rb-Sr 2770 ± 40 Ma	$Sr_o = 0.7022 \pm 0.0006$
Nûk gneisses	Pb/Pb 2960 ± 70 Ma	
Tectonic interleaving of Amîtsoq		
gneisses and Malene supracrustal rocks		
Anorthosite-leucogabbro complex		
Malene supracrustal association		
Ameralik dykes		
Amîtsoq gneisses	c. 3700 Ma	
Akilia supracrustal association		

loaded on single rhenium filaments with H_3PO_4 and silica gel. Pb isotopic ratios were measured on a V.G. Isomass 54E mass spectrometer. The sample preparation blank was 6–8 ng which has a negligible effect on the reported analyses. Pb concentrations were determined on a Philips PW1220 XRF at the University of Exeter.

The decay constants and the model parameters used are:

$\lambda_{87} = 1.42 \times 10^{-11}$ y^{-1}, $\lambda_{238} = 0.155125 \times 10^{-9}$ y^{-1}, $\lambda_{235} = 0.98465 \times 10^{-9}$ y^{-1}; $^{238}U/^{235}U = 137.66$, age of the earth = 4570 Ma; primordial Pb isotopic compositions: $^{206}Pb/^{204}Pb = 9.307$, $^{207}Pb/^{204}Pb = 10.294$. All errors are given at the 2σ level.

Amîtsoq gneisses

Rb-Sr data for Amîtsoq gneisses (Table 2, Fig. 2a) plot above the reference line typical of Nûk gneisses affected by late Archaean metamorphism (Moorbath & Pankhurst 1976; Robertson 1983) and scatter above and below the reference line typical of rocks affected by early Archaean high-grade metamorphism (Griffin et al. 1980). These relations suggest an increase and decrease respectively in the Rb/Sr of individual rock samples at sometime after they were intruded. The possibility that $^{87}Sr/^{86}Sr$ was also changed by introduction of Sr cannot be discounted. Element mobility occurred at or before 2770 Ma since well-fitted isochrons demonstrating no significant subsequent metasomatism are obtained from all younger rocks (Robertson 1983). Assuming that the scatter of data points resulted from the fractionation of Rb/Sr which occurred at or before c. 2800 Ma, sample 291544 has gained c. 150% Rb relative to Sr while 291560 has lost c. 60% Rb relative to Sr.

Whilst Pb/Pb isotopic analyses of the Amîtsoq gneisses (Table 3, Fig. 2b) from NE Ivisârtoq (Fig. 1) plot on the 3750 Ma reference line typical of the Amîtsoq gneisses at Isukasia (Fig. 1) (Moorbath et al. 1975), the remainder of the samples from elsewhere in the Ivisârtoq region produce an errorchron (MSWD = 4.21) which yields an age of 3886 ± 110 Ma with an apparent μ_1 of 9.3. The age and μ_1 are significantly higher than the values of 3750 ± 110 Ma with $\mu_1 = 8.0$ measured on the Amîtsoq gneisses at Isukasia (Moorbath et al. 1975). The data are consistent with significant mobility of U and/or Pb before 2600 Ma since a well fitted Pb/Pb isochron with an age of 2625 Ma is obtained for the Qôrqut granite complex from Kangiussap nunâ (Robertson 1985).

Nûk gneisses

Nûk gneisses from Kangiussap nunâ which are considered to be equivalent to the 'type' Nûk gneisses of the Godthâb region (see McGregor et al. 1983 for discussion of terminology) have yielded an Rb-Sr age of 2767 ± 40 Ma with initial $^{87}Sr/^{86}Sr$ of 0.7022 ± 0.0006 (Robertson 1983). This agrees within error with the age obtained for late Archaean high-grade metamorphism (Black et al. 1973; Pidgeon & Kalsbeek 1978). Sr model ages for the gneisses assuming an initial $^{87}Sr/^{86}Sr$ of 0.701 are little older than the measured isochron age indicating that if these gneisses were intruded at c. 3000 Ma as is indicated for equivalent Nûk gneisses elsewhere in the Godthâbsfjord region (Baadsgaard & McGregor 1981; McGregor et al. 1983), Rb/Sr increased significantly at or shortly before the measured age. The Nûk gneisses from Kangiussap nunâ are enriched in

TABLE 2. *Rb-Sr analytical data for Amîtsoq greisses, Ivisârtoq region*

Sample no.	Rb ppm	Sr ppm	$^{87}Rb/^{86}Sr$	$^{87}Sr/^{86}Sr$
291527	67	267	0.718	0.7484
291529	89	265	0.962	0.7482
291530	100	164	1.764	0.7891
291531	71	224	0.908	0.7512
291543	87	303	0.823	0.7398
291544	78	334	0.671	0.7314
291552	75	248	0.873	0.7467
291553	76	257	0.851	0.7426
291559	44	564	0.221	0.7141
291560	45	416	0.305	0.7214
291568	53	312	0.486	0.7318
291660	91	258	1.020	0.7516
291661	111	277	1.150	0.7549
291663	55	288	0.552	0.7348

Precision on $^{87}Sr/^{86}Sr \pm 1\%$. Average 2σ error on $^{87}Sr/^{86}Sr$ (normalized to $^{86}Sr/^{88}Sr = 0.1194) = 0.01\%$.

Rb (40–200 ppm), Pb (10–45 ppm) and to some extent K_2O (1.3–5 wt%) (Robertson 1985) compared with values of c. 50 ppm, 20 ppm and 2 wt% measured for Nûk gneisses elsewhere in southern West Greenland (McGregor 1979). These data, combined with the fact that Amîtsoq gneisses gained Rb relative to Sr in the late Archaean, suggest that Rb was introduced into the Ivisârtoq region soon after the emplacement of the Nûk gneisses. The possibility that Sr was also introduced from an external reservoir containing unradiogenic Sr cannot be discounted (see Grant & Hickman 1984). This would require that the Sr in the Nûk gneisses equilibrated isotopically with the metasomatic fluids (Grant & Hickman 1984). It seems likely that the Pb and K contents of the gneisses were also increased at this time.

Pb/Pb isotope data for fourteen out of seventeen samples of the Nûk gneisses from Kangiussap nunâ (Table 3, Fig. 2c) yield an age of 2962 ± 73 Ma with an apparent single-stage μ value of 6.64 ± 0.05 and an MSWD of 4.33. The age is within error of the intrusive age of the Nûk gneisses. The three samples omitted from the regression (291454, K-feldspar mineral separates from 291454 & 291455) were collected in northern Kangiussap nunâ where Nûk and Amîtsoq gneisses are intersheeted. Amîtsoq gneisses have not been recognized elsewhere in Kangiussap nunâ. Interpretations of the structure before the dome-synform phase of deformation reveal that

the three samples also represented a higher tectonic level than the remainder of the samples 3000–2600 Ma ago.

The apparent single stage μ value of 6.6 is lower than the value of 7.5 measured from areas to the north and south of the Godthåbsfjord region (Taylor *et al.* 1980) indicating that the gneisses contain Pb derived from a source with Pb isotopic evolution retarded with respect to the late Archaean mantle. Possible interpretations of the data include:

(1) The gneisses were derived from a source with single-stage μ of 7.5 and were contaminated with unradiogenic Pb either during the emplacement of the gneisses (Taylor *et al.* 1980) or as a result of subsequent metasomatism (Fig. 2c).
(2) Granitic partial melts with unradiogenic Pb isotopic compositions derived from the Amîtsoq gneisses contributed to the igneous precursors to the Nûk gneisses (Fig. 2c).
(3) The Nûk gneisses were generated from a mantle source region with a low μ value (Fig. 2d).
(4) The basic precursors to the Amîtsoq gneisses were also the source rocks for the Nûk gneisses (Fig. 2d).

Interpretation of data

The Rb-Sr and Pb/Pb isotope data for both the Amîtsoq and the Nûk gneisses are consistent with significant movement of LIL elements after the emplacement of the Nûk gneisses at c. 3000 Ma but before 2770 Ma. Changes in Rb/Sr can be quantified for the Amîtsoq gneisses such that some samples have gained 150% Rb relative to Sr while others have gained up to 60% Rb relative to Sr. The Nûk gneisses from Kangiussap nunâ are enriched in Rb, Pb and K_2O compared with similar gneisses elsewhere in the Godthåbsfjord region. These elements were introduced into the gneisses exposed at present erosion level and were presumably transported by fluids.

Late Archaean metamorphism in the Ivisârtoq region attained upper amphibolite-facies while granulite-facies conditions existed over large areas of southern West Greenland. Granulite-facies rocks outcrop 40 km to the south of the Ivisârtoq region and it is reasonable to assume that similar rocks occur at a relatively shallow depth below the present erosion level. The preferred hypothesis is that the introduction of metasomatic fluids transporting LIL elements was related to the dehydration of granulite-facies rocks with associated depletions of LIL elements (Wells 1976; Tarney & Windley 1977).

Metamorphic ages of c. 2800 Ma have been measured from many parts of southern West

TABLE 3. *Pb/Pb analytical data for Amîtsoq and Nûk gneisses, Ivisârtoq region. Kf = K-feldspar mineral separates. Pb isotope ratios include mass fractionation corrections assessed from standards NBS 981 and 982. Average precision for $^{206}Pb/^{204}Pb$, $^{207}Pb/^{204}Pb$ and $^{208}Pb/^{204}Pb \pm 0.15\%$.*

Sample no.	$^{206}Pb/^{204}Pb$	$^{207}Pb/^{204}Pb$	$^{208}Pb/^{204}Pb$	Pb ppm
Amitsoq gneisses				
291527	12.690	13.891	33.795	32
291530	12.779	13.879	33.777	34
291544	14.128	14.415	35.158	
291552	12.686	13.863	32.451	25
291559	11.977	13.582	32.053	20
291560	12.246	13.648	32.456	28
291568	12.592	13.801	32.453	29
291568Kf	12.023	13.658	32.395	71
291660	13.625	14.029	33.923	32
291660Kf	12.634	13.878	32.499	
291661	13.657	14.037	33.764	48
291663	13.408	14.024	35.286	27
Nuk gneisses				
291357	15.439	14.433	36.294	22
291369	15.903	14.554	38.821	44
291382	13.109	13.888	35.133	32
291404	13.356	13.970	34.979	28
291404Kf	12.402	13.796	32.448	92
291418	13.706	13.995	34.116	38
291449	13.889	14.115	36.089	20
291450	13.410	14.013	36.106	27
291453	17.921	14.960	38.912	17
291454	12.952	14.030	32.951	42
291454Kf	12.593	13.962	32.398	111
291455	12.994	14.062	33.063	29
291573	13.370	13.950	34.519	26
291578	13.321	13.969	37.112	25
291582	14.195	14.158	35.706	19
291594	13.085	13.922	34.242	21
291595	12.564	13.816	33.381	33
291596	14.658	14.194	35.990	18

Greenland. However, the Pb/Pb age of 2960 Ma for the Nûk gneisses from Kangiussap nunâ indicates that the introduction of metasomatic fluids may have followed shortly after intrusion of the gneisses. This is at variance with the Rb-Sr data. However, measured Rb-Sr ages are commonly younger than Pb/Pb ages for the same suite of rocks. Rb-Sr and Pb/Pb ages for the Qôrqut granite complex from Kangiussap nunâ are 2490 Ma and 2625 Ma (Robertson 1985). This is thought to have resulted from partial open system behaviour perhaps associated with late or post-magmatic fluid movements. Radiogenic Sr was selectively removed thus maintaining the low

initial $^{87}Sr/^{86}Sr$ (P.N. Taylor *pers. comm.* 1983). Biotite may be responsible for this process since it has high mineral/melt Kd values for Rb (*c.* 3) and very low values for Sr (*c.* 0.1) in granitic rocks (Hanson 1978). Ages of *c.* 2800 Ma recorded the closure of the isotopic system and termination of the metamorphic event.

The Pb isotopic data for the Nûk gneisses could be explained by any of the four hypotheses presented above. However, the second and third hypotheses are unrealistic when other factors are considered. The Amîtsoq and Nûk gneisses have similar major element compositions and therefore either very small or very large amounts of

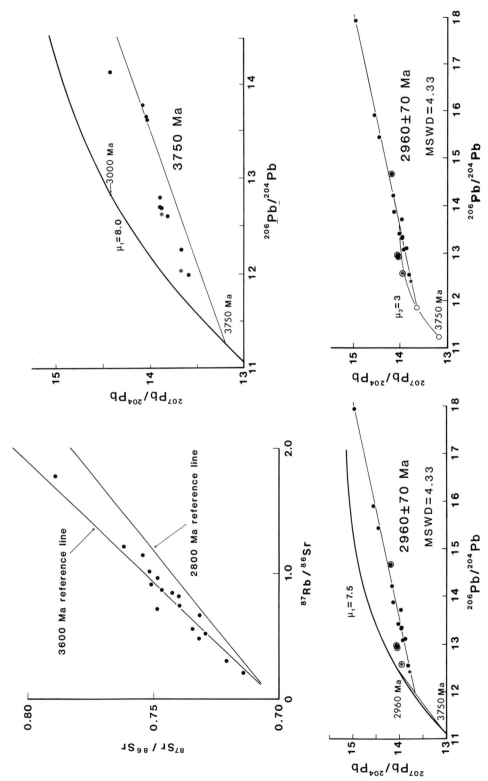

Fig. 2. (a) **Rb-Sr** whole rock isotope diagram for Amîtsoq gneisses from the Ivisârtoq region. 3600 Ma and 2800 Ma reference lines are representative of early and late Archaean high-grade metamorphic rocks respectively. (b) Pb/Pb isotope diagram for Amîtsoq gneisses from the Ivisârtoq region. 3750 Ma reference line is representative of Amîtsoq gneisses at Isukasia. Asterisks are K-feldspar mineral separates. (c) Pb/Pb isochron diagram for Nûk gneisses from the Ivisârtoq region. Chord is shown connecting early and late Archaean isotopic compositions on the single-stage growth curve. $\mu_1 = 7.5$. Asterisks are K-feldspar mineral separates. Four samples are omitted from the isochron regression (see text). (d) Pb/Pb isochron diagram for the Nûk gneisses as above. μ_1 to 3750 Ma $= 8.0$ and μ_2 from 3750 to 3000 Ma $= 3.0$.

partial melting would be required to satisfy the constraints of the geochemical data. The low initial $^{87}Sr/^{86}Sr$ also suggests either very small amounts of partial melting or a very low Rb/Sr in the contributing gneisses. A second process producing the element mobility in the Amîtsoq gneisses would also have to be envisaged.

The Amîtsòq gneisses in the Ivisârtoq region were derived from a source with $\mu_1 = 7.5$–8.0 (Robertson 1985). The apparent single-stage μ value for Nûk gneisses is 6.6. To produce this from mantle sources would require a μ_2 ratio between 3750 and 3000 Ma of 3 (Fig. 2d). This would imply both a marked depletion of U with respect to Pb and preservation of significant heterogeneities in the mantle for 800 Ma at a period when heat production and rates of convection were significantly higher than at present. The fourth hypothesis implies that basic rocks with a significant crustal residence time were the parents to the Nûk gneisses. The Sr contents of the Nûk gneisses are significantly higher ($c.$ 500 ppm) than the Amîtsoq gneisses ($c.$ 250 ppm) (McGregor 1979) perhaps because the basic source rocks to the latter contained less plagioclase. The Nûk gneisses from Kangiussap nunâ have Sr contents ($c.$ 250–300 ppm) similar to the Amîtsoq gneisses perhaps as a result of their derivation from early Archaean basic rocks. However, it has been shown that Rb and presumably Pb and K were introduced into the Ivisârtoq region after the emplacement of the Nûk gneisses. The Rb-Sr and Pb isotopic data are consistent with the derivation of the Nûk gneisses in Kangiussap nunâ from early Archaean basic rocks with subsequent contamination with Pb derived from underlying Nûk and/or Amîtsoq gneisses. The isotopic data would also be consistent with derivation of the Nûk gneisses from a mantle source region with $\mu = 7.5$ and subsequent isotopic equilibration with underlying Archaean basic rocks or granulite-facies Amîtsoq gneisses (compare with Grant & Hickman 1984).

The three samples which plot above the Pb/Pb errorchron are fundamentally different from the remaining samples in that they represented a higher tectonic level 3000 to 2800 Ma ago where Nûk and Amîtsoq gneisses were intersheeted. The igneous precursors to these magmas may have been derived from late Archaean basic rocks and subsequently contaminated with Amîtsoq-type Pb. These samples have Sr contents of $c.$ 500 ppm typical of Nûk gneisses elsewhere in the Godthåbsfjord region. The remainder of the samples have been enriched in Pb derived from Nûk gneisses below the present erosion level.

The isotopic data therefore do not require the presence of Amîtsoq gneisses below the present

erosion level in Kangiussap nunâ. The fourth hypothesis is as plausible as the first. Field evidence shows that Nûk gneisses outcrop within the domal structures and in Kangiussap nunâ Amîtsoq gneisses outcrop only in the north in the core of a synformal cusp. It seems evident that Nûk gneisses occupied a lower tectonic level before the dome-synform phase of deformation.

The isotopic data for the late Archaean granites (Table 1) from the Ivisârtoq region are consistent with their derivation from Nûk gneisses (Robertson 1985) indicating the absence of Amîtsoq gneisses below the present erosion level.

Regional implications

The interpretation of Pb isotope data obtained from the Nûk gneisses by Taylor *et al.* (1980) indicates that Amîtsoq gneisses occur only in limited areas below the present level of erosion up to 30 km to the south of the tract where Amîtsoq gneisses have been recognized (Fig. 1). Late Archaean metamorphism in the area of outcrop of the Amîtsoq gneisses attained amphibolite-facies whereas the Nûk gneisses both to the north and south of this region have been metamorphosed to granulite-facies over extensive areas. Pressures of 5.0–7.5 kbar during the late Archaean amphibolite-facies metamorphism in the Godthåbsfjord region have been calculated by Dymek (1978) and Wells (1976), and Wells (1976) calculated values of 9–11 kbar for granulite-facies rocks in the Buksefjorden region. These values suggest that at that time the Godthåbsfjord region was at a higher crustal level than the adjacent areas.

The interpretation of the data presented for the Ivisârtoq region suggests that the Nûk gneisses occur structurally beneath the Amîtsoq gneisses and were accreted onto the base of the Amîtsoq crust. This interpretation extrapolated to the remainder of the Archaean terrain of West Greenland requires that the area of outcrop of Amîtsoq gneisses was once much more extensive. However, the original extent of Amîtsoq gneisses is limited by the occurrence of Malene supracrustal enclaves in the granulite-facies Nûk gneisses. The old gneisses have subsequently been removed by erosion. The area of granulite-facies Nûk gneisses represented a late Archaean crustal level lower than that of the base of the Amîtsoq gneisses. Nûk gneisses could not be contaminated with Amîtsoq-type Pb.

Granites were intruded at a late stage in each of the crustal accretion events in the Godthåbsfjord region. The granites also appear to have been emplaced at the highest crustal levels recognized. Granites are rare in the adjacent granulite-facies terrains perhaps because subsequent erosion revealed deeper (sub-granite) levels of the crust. Three main periods of granite intrusion have been recognized:

(1) Early Archaean (*c.* 3600 Ma) granites are abundant at Isukasia (Nutman *et al.* 1983). The Isukasia region is thought to have been the highest crustal level exposed in the Godthåbsfjord region although it was metamorphosed to amphibolite-facies.

(2) Granites in north-east Ivisârtoq have yielded an Rb-Sr age of 2750 Ma.

(3) Syn- to post-tectonic granites yielding ages of 2650–2500 Ma are abundant throughout the Godthåbsfjord region (Moorbath *et al.* 1981; McGregor *et al.* 1983).

The distribution of granites and relative abundance increasing through time is consistent with the postulated crustal levels in the late Archaean.

Discussion and speculation

Two crustal accretion-differentiation super-events (CADS) (Moorbath 1977) have been recognized at the present erosion level in the Archaean of southern West Greenland. Each commenced with mainly basaltic volcanism and the deposition of supracrustal rocks followed by the emplacement of large volumes of granitic magmas. The late Archaean magmas underplated the pre-existing sialic crust, although with inter-sheeting of the two rock suites together with intense modification of the older gneisses in the contact zones. Both suites of gneisses were emplaced during horizontal tectonic regimes. The combination of horizontal tectonic interleaving and accretion of new crustal material thickened the crust resulting in granulite-facies metamorphism with associated depletions of LIL elements in the lower parts of the crust (Wells 1976; Tarney & Windley 1977). These elements accumulated at the crustal level represented by the Ivisârtoq region.

Granite intrusion was contemporaneous with the high-grade metamorphism at *c.* 3600 Ma and 2800 Ma. The migration of volatiles associated with the granulite-facies metamorphism was a significant factor leading to crustal anatexis.

The dome-synform structures which developed in the Ivisârtoq region at *c.* 2650 Ma with granite emplacement at *c.* 2600 Ma have similarities with granite-greenstone terrains described in other Archaean cratons (Goodwin 1981; Martin *et al.* 1983; Percival & Krogh 1983). It is conceivable that greenstone belt-type supracrustal rocks were located in the synformal cusps of the Ivisârtoq

region although at a higher tectonic level than that exposed at the present day. Possibly a third CADS occurred at *c.* 2700 to 2600 Ma. Granite (s.l.) gneisses would have been accreted onto the base of the Nûk gneisses *c.* 2700 Ma ago. The associated heat flow produced a third high-grade metamorphic event. The products may be represented at the present erosion level by granodiorites and trondhjemites which outcrop in restricted areas of the Ivisârtoq region and have yielded an Rb-Sr age of 2680 Ma (Robertson 1985).

High-grade metamorphism in the lower crust at *c.* 2600 Ma led to the migration of LIL element bearing metasomatic fluids to higher levels in the crust resulting in widespread anatexis producing the Qôrqut granite complex. Metasomatism at *c.* 2600 Ma has not been recognized istopically in the Ivisârtoq region. However, Pb/Pb and Rb-Sr isotopic rehomogenization occurred on scales of more than 10 cm and 1–5 cm, respectively, as a result of metamorphism at 2600 Ma. Feldspar blastesis in the domal structures (Chadwick 1985) is attributed to this event. Another interesting point is the uniformity of the initial Sr and Pb isotopic compositions of the Qôrqut granite complex throughout the Godthåbsfjord region (Moorbath & Taylor 1981; Taylor *et al.* 1984; Robertson 1985) compared with the inhomogeneity of the potential source rocks. Metasomatism may have 'smeared out' regional variations in isotopic compositions.

It is suggested that the Archaean continental crust was thickened from beneath by successive periods of intrusion of quartzo-feldspathic gneisses. This accounts for the apparently 'normal' thicknesses of continental crust beneath deeply eroded Archaean terrains.

A significant thickness of basic rocks must have been emplaced into the lower crust to provide the source rocks for the quartzo-feldspathic gneisses, and the existence of basic intrusions is indicated by anorthosite and leucogabbro which intruded the Malene supracrustal rocks (Fig. 1). This rock suite which is thought to have been derived by fractional crystallization of basic magmas (Weaver *et al.* 1981) accreted onto the base of the pre-existing crust to be intruded by the Nûk gneisses.

Basic rocks were also intruded during the CADS as suggested by the Amîtsoq iron-rich suite of rocks derived by mixing basic magmas with sialic melts late in the early Archaean CADS (Nutman *et al.* 1984) and by the occurrence of basic dykes such as the Intra-Nûk dykes (Chadwick & Coe 1983). A significant thickness of basic rocks with partial melting in the lower parts is suggested by the lack of partial melting of sialic crust. The basic rocks are not recognized in outcrop in Archaean terrains but are recognized as enclaves in younger igneous rocks.

The lower continental crust was therefore stratified with basic rocks, quartzo-feldspathic gneisses and granites occupying three overlapping zones. The crust may also be stratified on the basis of tectonic style. The upright domes and synforms are the youngest structures in the Ivisârtoq region and may therefore represent the highest crustal level. Flat-lying structures associated with the emplacement of the Nûk gneisses represent a lower crustal level. The domes and synforms in the Ivisârtoq region were superimposed on the older flat-lying structures as the crustal level exposed was raised by underplating 2700 Ma ago. During a CADS, flat-lying structures may dominate the lower crust while upright structures characterize the higher crustal levels where lateral thermal gradients in the crust produced differential buoyancy of some regions facilitating the development of the domes.

ACKNOWLEDGEMENTS: This work forms part of the Ivisârtoq project at the University of Exeter which is jointly supported by the Natural Environment Research Council and the Geological Survey of Greenland. The author was in receipt of a Shell Research Studentship which is gratefully acknowledged. Special thanks go to S. Moorbath, P.N. Taylor and R. Goodwin at the Department of Earth Sciences, University of Oxford for technical guidance and discussions on the age and isotope data. I would also like to thank my colleagues at Exeter for useful discussions. This paper is published by kind permission of the Director, Geological Survey of Greenland.

References

ARDEN, J. & GALE, N.H. 1974. New electrochemical technique for the separation of lead at trace levels from natural silicates. *Anal. Chem.* **46**, 2–9.

BAADSGAARD, H. & McGREGOR, V.R. 1981. The U-Th-Pb systematics of zircons from the type Nûk gneisses, Godthåbsfjord, West Greenland. *Geochim. et Cosmochim. Acta*, **45**, 1099–109.

BLACK, L.P., MOORBATH, S., PANKHURST, R.J. & WINDLEY, B.F. 1973. $^{207}Pb/^{206}Pb$ whole rock age of Archaean granulite facies metamorphic event in West Greenland. *Nature Phys. Sci.* **244**, 50–3.

BREWER, M., CHADWICK, B., COE, K. & PARK, J.F.W. 1984. Further field observations in the Ivisârtoq region of southern West Greenland. *Rapp. Grønlands geol. Unders.* **120**, 55–67.

BRIDGWATER, D., KETO, L., McGREGOR, V.R. & MYERS, J.S. 1976. Archaean gneiss complex of Greenland. *In*: ESCHER, A. & WATT, W.S. (eds) *Geology of Greenland*, 19–25. Geol. Surv. of Greenland.

BROWN, M., FRIEND, C.R.L., McGREGOR, V.R. & PERKINS, W.T. 1981. The late Archaean Qôrqut granite complex of southern West Greenland. *J. geophys. Res.* **86**, B11, 10617–32.

CHADWICK, B. & COE, K. 1983. Buksefjorden 63v1 Nord; the regional geology of a segment of the Archaean block of southern West Greenland. *Grønlands geol. Unders.* 70 pp.

—— 1985. Contrasting styles of tectonism and magmatism in the late Archaean crustal evolution of the Ivisârtoq region, Inner Godthåbsfjord, southern West Greenland. *Precamb. Res.* **27**, 215–238.

COE, K. & ROBERTSON, S. 1984. Contrasting styles of Archaean crustal evolution in parts of southern West Greenland. *J. Geodynamics*, **1**, 301–311.

COMPTON, P.M. 1978. *A study of the Nûk gneisses and Qôrqut granite in the Buksefjorden region, southern West Greenland*. Unpubl. PhD thesis, Univ. Exeter.

DYMEK, R.F. 1978. Metamorphism of the Archaean Malene supracrustals, Godthåb district, West Greenland. *Proc. 1978 Archaean Geochm. Conf.* 339–43. Toronto U.P.

GARDE, A.A., HALL, R.P., HUGHES, D.J., JENSEN, S.B., NUTMAN, A.P. & STECHER, O. 1983. Mapping of the Isukasia sheet, southern West Greenland. *Rapp. Grønlands geol. Unders.* **115**, 20–9.

GOODWIN, A.M. 1981. Archaean plates and greenstone belts. *In*: KRONER, A. (ed) *Precambrian Plate Tectonics*, 105–35. Elsevier.

GRANT, N.K. & HICKMAN, M.H. 1984. Rb-Sr isotope systematics and contrasting histories of late Archaean gneisses, West Greenland. *Geology*, **12**, 599–601.

GRIFFIN, W.L., McGREGOR, V.R., NUTMAN, A.P., TAYLOR, P.N. & BRIDGWATER, D. 1980. Early Archaean granulite-facies metamorphism south of Ameralik, West Greenland. *Earth Planet. Sci. Lett.* **50**, 59–74.

HALL, R.P. 1981. *The Archaean geology of Ivisârtoq, inner Godthåbsfjord, southern West Greenland*. Unpubl. PhD thesis, Committee for National Academic Awards, UK.

HANSON, G.N. 1978. The application of trace elements to the petrogenesis of igneous rocks of granitic composition. *Earth Planet. Sci. Lett.* **38**, 26–43.

MARTIN, H., CHAUVEL, C., JAHN, B.M. & VIDAL, P. 1983. Rb-Sr and Sm-Nd ages and isotopic geochemistry of Archaean granodioritic gneisses from Eastern Finland. *Precamb. Res.* **20**, 79–91.

McGREGOR, V.R. 1973. The early Precambrian gneisses of the Godthåb district, West Greenland. *Phil. Trans. R. Soc. Lond. Ser. A*, **273**, 343–58.

—— 1979. Archaean gray gneisses and the origin of the continental crust; evidence from the Godthåb region, West Greenland. *In*: BARKER, F. (ed) *Trondhjemites, Dacites and Related Rocks*, 168–204. Elsevier.

——, BRIDGWATER, D. & NUTMAN, A.P. 1983. The Qârusuk dykes: post-Nûk, pre-Qôrqut granitoid magmatism in the Godthåb region, southern West

Greenland. *Rapp. Grønlands geol. Unders.* **112**, 101–12.

MOORBATH, S. 1977. Ages, isotopes and evolution of Precambrian continental crust. *Chem. geol.* **20**, 151–67.

——, O'NIONS, R.K. & PANKHURST, R.K. 1975. The evolution of early Precambrian crustal rocks at Isua, West Greenland—geochemical and isotopic evidence. *Earth Planet. Sci. Lett.* **27**, 229–39.

—— & PANKHURST, R.J. 1976. Further rubidium-strontium age and isotope evidence for the nature of the late Archaean plutonic event in West Greenland. *Nature*, **262**, 124–6.

—— & TAYLOR, P.N. 1981. Isotopic evidence for continental growth in the Precambrian. *In*: KRONER, A. (ed) *Precambrian Plate Tectonics*, 491–525. Elsevier.

NUTMAN, A.P., BRIDGWATER, D., DIMROTH, E., GILL, R.C.O. & ROSING, M. 1983. Early (3700 Ma) Archaean rocks of the Isua supracrustal belt and adjacent gneisses. *Rapp. Grønlands geol. Unders.* **112**, 5–22.

——, —— & FRYER, B.J. 1984. The iron-rich suite from the Amîtsoq gneisses of southern West Greenland: early Archaean plutonic rocks of mixed crust and mantle origin. *Contrib. Mineral. Petrol.* **87**, 24–34.

PANKHURST, R.J. & O'NIONS, R.K. 1973. Determination of Rb/Sr and $^{87}Sr/^{86}Sr$ ratios of some standard rocks and evaluation of X-ray fluorescence spectrometry in Rb-Sr geochemistry. *Chem. Geol.* **12**, 127–36.

PERCIVAL, J.A. & KROGH, T.E. 1983. U-Pb zircon geochronology of the Kapuskasing structural zone and vicinity in the Chapleau-Foleyet area, Ontario. *Can. J. Earth Sci.* **20**, 830–43.

PIDGEON, R.T. & KALSBEEK, F. 1978. Dating of igneous and metamorphic events in the Fiskenaesset region of southern West Greenland. *Can. J. Earth Sci.* **15**, 1021–5.

ROBERTSON, S. 1983. Provisional results of isotope investigations into quartzo-feldspathic rocks from Kangiussap nunâ, Ivisârtoq sheet, southern West Greenland. *Rapp. Grønlands geol. Unders.* **115**, 56–9.

—— 1985. *Late Archaean crustal evolution in the Ivisârtoq region, southern West Greenland.* Unpubl. PhD thesis, Univ. Exeter.

TARNEY, J. & WINDLEY, B.F. 1977. Chemistry, thermal gradients and evolution of the lower continental crust. *J. geol. Soc. London*, **134**, 153–72.

TAYLOR, P.N., MOORBATH, S., GOODWIN, R. & PETRYKOWSKI, A.C. 1980. Crustal contamination as an indicator of the extent of early Archaean continental crust: Pb isotopic evidence from the late Archaean gneisses of West Greenland. *Geochim. et Cosmochim. Acta*, **44**, 1437–53.

WEAVER, B.L., TARNEY, J. & WINDLEY, B.F. 1981. Geochemistry and petrogenesis of the Fiskenaesset anorthosite complex, southern West Greenland: nature of the parent magma. *Geochim. et Cosmochim. Acta*, **45**, 711–25.

WELLS, P.R.A. 1976. *The metamorphic petrology of high grade Archaean rocks, Buksefjorden, southern West Greenland.* PhD thesis, Univ. of Exeter.

S. ROBERTSON, Dept of Geology, University of Exeter. Present address: British Geological Survey, Murchison House, West Mains Road, Edinburgh, EH9 3LA.

Chemical and isotopic effects of late Archaean high-grade metamorphism and granite injection on early Archaean gneisses, Saglek–Hebron, northern Labrador

L. Schiøtte, D. Bridgwater, K.D. Collerson, A.P. Nutman & A.B. Ryan

SUMMARY: Early Archaean (3700–3800 Ma) layered tonalitic to granodioritic Uivak gneisses form approximately 60–70% of the Archaean gneiss complex between Saglek and Hebron fiords, northern Labrador. Effects of late Archaean metamorphic and igneous activity can be recognized on the geochemistry and isotope chemistry of the Uivak gneisses: depletion in rocks affected by 2900 Ma granulite facies metamorphism, with a loss of K, Rb and redistribution of REE and a corresponding enrichment of LIL elements in the amphibolite facies area. LIL enrichment is at least in part by addition from intrusive late Archaean granitoid sheets. The depleted granulites (K/Rb 500–2000) show a scatter of data points on a Rb/Sr– $^{87}Sr/^{86}Sr$ diagram to the left of a 3560 Ma reference isochron (ISr 0.700) consistent with marked Rb loss and some homogenization of Sr. Amphibolite facies gneisses from a broad transitional zone into granulite facies have K/Rb values of 200–300 and plot on a *c.* 3560 Ma isochron with a high ISr (0.7038) consistent with a slight loss of Rb at 2900 Ma but without marked Sr homogenization. Amphibolite facies gneisses away from the transition zone show marked enrichment in LIL elements with unusually low K/Rb ratios (100–200). Rb/Sr data plot close to a 3560 Ma reference isochron, εSr_{3560} values are strongly negative, interpreted as due to Rb addition at 2900 Ma rather than derivation from an unusually depleted mantle source. Pb-Pb isotope data from the depleted granulite show a cluster of unradiogenic data points. Samples from a monolith of gneisses in a transitional zone between amphibolite and granulite facies yields a 3808 Ma isochron (μ_1 8.5) suggesting there was locally little disturbance of the Pb/U system at 2900 Ma. Pb-Pb data from the Rb-enriched amphibolite facies area show a markedly higher proportion of radiogenic Pb compared to the granulite facies rocks which cannot be explained by U movement alone. Local galena mineralization suggests that Pb as well as U was mobile during the late Archaean and we interpret the results as mixtures of > 3600 crustal Pb with younger material. Sm-Nd data points (Collerson & McCulloch 1982) show a scatter about a 3665 Ma isochron with an average εNd_{3665} of $+ 2.48$. Marked LIL + LREE enrichment is suggested and we interpret the scatter of εNd_{3665} values as being due to late movement of LREE.

The late Archaean granitoid sheets have a chemistry and petrology suggesting derivation largely by partial melting of earlier sial with modifications by fluid redistribution of alkalis. New minerals such as allanite and apatite grow in older rocks adjacent to young veins. Sr, Nd and Pb data all suggest mixing between early Archaean crust and a component with mantle-like isotopic character at *c.* 2900 Ma. The mixing process was complex and took place by ingestion of older rocks in younger granite, mixing of partial melts of different origin and fluid transport of components from younger intrusions into older gneisses.

Introduction

This paper summarizes and interprets the current field and geochemical results from Archaean quartzofeldspathic gneisses from the Saglek–Hebron fiord area, northern Labrador. Good coastal exposure together with detailed mapping observations from field seasons between 1975 and 1983 has allowed the establishment of a lithological chronology. The same units of early Archaean gneiss can be recognized across a late Archaean amphibolite-granulite boundary and in different degrees of migmatization by late Archaean granite sheets. The period of *c.* 800 Ma separating the main period of rock formation and latest high-grade metamorphism allows the effects of super-imposed depletion, enrichment and mixing processes on the isotopic systems to be recognized.

Discussion is concentrated on a single unit, the early Archaean tonalitic to grandioritic Uivak 1 gneisses which were emplaced at *c.* 3800 Ma (ion probe data on the centres of zircon grains, Collerson 1983a) affected by migmatization and metamorphism between 3800 and 3600 Ma (Rb-Sr, Sm-Nd errorchrons and comparisons with grey tonalitic-granodioritic Amitsoq gneisses of West Greenland regarded as correlatives) and affected by a complex series of metamorphic events and migmatization between 3000 Ma and 2500 Ma (U-Pb ion probe data, Collerson 1983a; conventional U-Pb zircon data, Baadsgaard *et al.* 1979; Hurst & Tilton 1976; Rb-Sr errorchrons,

From: DAWSON, J.B., CARSWELL, D.A., HALL, J. & WEDEPOHL, K.H. (eds) 1986, *The Nature of the Lower Continental Crust*, Geological Society Special Publication No. 24, pp. 261–273.

TABLE 1. *Major events in the Saglek fiord–Hebron fiord area simplified from Bridgwater* et al. (in press)

Intermittent movement on faults	Proterozoic to present
Proterozoic metamorphism granulite facies to west, greenschist facies in area studied	*c.* 1600–1800 Ma
Deposition of Proterozoic cover rocks (Ramah and Mugford sequences)	Early Proterozoic
Intrusion of dykes	Early Proterozoic
Late Archaean granites Widespread syntectonic granites	2500–2800 Ma
High grade metamorphism	2900 ± 100 Ma
Upernavik supracrustals	> 3000
Saglek dykes	< 3500
Early Archaean gneiss complex comprising a) Nulliak supracrustal rocks, b) migmatized Uivak grey gneisses, c) Fe-rich granites and accompanying basic rocks	*c.* 3500–3800 Ma

Collerson *et al.* 1981). Both geochemical and isotopic results show considerable secondary scatter within one unit, indicating that we are not dealing with closed systems. Rather than forcing errorchrons through the data points to try to obtain information about the primary nature of the early rocks our approach is to look for reasons for the scatter and to link these with field observations of metamorphic grade and other mappable parameters. A comparatively simple pattern emerges which could provide a model for the interpretation of data from other polymetamorphic areas.

Regional setting

Early Archaean gneisses form the major component in the high grade gneiss complex between Saglek and Hebron fiords, northern Labrador. A simplified table of events is given in Table 1, detailed description and references in Bridgwater *et al.* (in press). Three main early Archaean units can be recognized, the Nulliak assemblage, a suite of supracrustal rocks and dominantly basic intrusives (Bridgwater & Collerson 1977; Collerson & Bridgwater 1979; Bridgwater *et al.* in press); the Uivak 1 gneisses, a suite of layered tonalitic and granodioritic gneisses; and the Uivak 2 gneisses a younger suite of Fe-rich quartz monzonites and

associated gabbroic rocks. The three units can be recognized throughout the Saglek–Hebron fiord area.

The early Archaean gneisses were intruded by mafic and less commonly ultramafic dykes (the Saglek dykes) and interlayered with a second group of Archaean sediments and volcanic rocks, the Upernavik supracrustals. These form the most prominent mappable units (black units on Fig. 1). They contain a large component of quartz-rich sediments and are interpreted as deposited in the period 3000–3600—at least in part—on an earlier sialic (Uivak) basement. The present outcrop pattern is due to the thrusting and folding together of the basement and cover rocks during several late Archaean episodes of deformation. Interlayering of the Uivak gneisses and the Upernavik supracrustals was followed by the injection of late-Archaean granitoids ranging from an early suite of tonalites and granodiorites through regional swarms of trondhjemites and pegmatitic granites emplaced under high grade metamorphic conditions at approximately 2900 Ma to younger granites derived from partial melts of older crust during the latest stages of Archaean metamorphism in the area (2700–2500 Ma). The youngest granitic activity was associated with widespread retrogression and shearing. The best estimates for the age of high grade

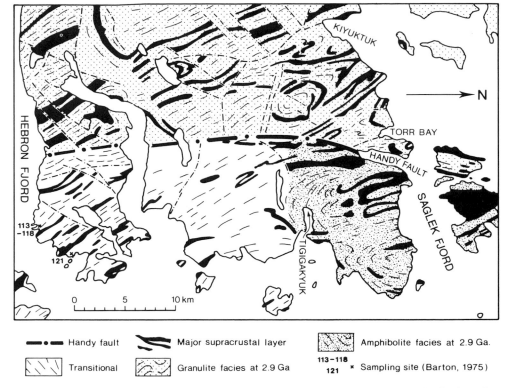

FIG. 1. Sketch map to show main structural pattern in the Saglek–Hebron fiord area and distribution of metamorphic facies at 2900 Ma.

metamorphism and the accompanying granite sheeting is given by conventional U-Pb dating of totally recrystallized zircons (Hurst & Tilton, 1976) and by ion probe dating on the rims of zoned zircons (Collerson 1983a). These give a general age of 2900 ± 100 Ma for the peak of late Archaean metamorphism and granite activity.

The Archaean gneiss complex was intruded by a series of Proterozoic dykes and affected by faulting which began in the early Proterozoic and continued to be active (probably intermittently) into the Phanerozoic. The western edge of the area shown in Fig. 1 was affected by Poterozoic deformation and metamorphism. Over the eastern part of the area Proterozoic metamorphism reached greenschist facies and caused some resetting of whole rock Rb-Sr systems on a scale of a few centimetres (Collerson 1983b). These Proterozoic effects are a contributory cause to the scatter seen on whole rock isochrons, but do not show a marked correlation with regional changes in petrology and are therefore not regarded as the cause of bulk changes in the character of the Archaean gneiss complex. The most prominent

fault mapped in the area, the Handy fault, extends approximately N–S from Handy island through St. John's Harbour to Hebron fiord (Fig. 1). In the Saglek fiord area the Handy fault separates rocks of markedly different metamorphic facies so that to the west the gneiss complex is dominated by rocks which recrystallized under granulite facies conditions at approximately 2900 Ma while to the east the rocks show no evidence of late Archaean granulite metamorphism.

Traced southwards the gneisses of the eastern block between Saglek and Hebron fiords show a transitional zone between a regional amphibolite facies area in the north and a regional granulite facies area in the south. In this transitional zone different units respond to the increase in metamorphism to the south differently so that basic supracrustal rocks and the more iron-rich varieties of gneiss tend to contain orthopyroxene further north than the grey gneiss units with which they are interlayered. The quartzofeldspathic gneisses within the transitional zone vary considerably in their response to the increasing grade of metamorphism from north to south probably depending on local structural control of partial melting and the movement of fluids.

Uivak 1 gneisses from areas with contrasting degrees of late Archaean metamorphism and migmatization

Uivak 1 gneisses form an estimated 60–70% of the total gneiss complex between Saglek and Hebron fiords. They are heterogeneous on a scale of tens of metres due to varying amounts of the two main phases: an early medium to fine-grained biotite gneiss of tonalitic to granodioritic composition, and a slightly younger (pre-Uivak 2) leucocratic migmatite phase of granitic and trondhjemitic sheets and veins emplaced into the grey gneisses. This early Archaean migmatization has not been shown to vary systematically on a regional scale.

We describe Uivak 1 gneisses from four different settings within the area Saglek to Hebron fiord.

(a) Amphibolite facies gneisses from outer Saglek fiord

Uivak gneisses from the Big Island—St. John's harbour—Cape Uivak area (Fig. 1) show distinct early Archaean migmatite structures cut by Saglek dykes. Late Archaean granite veins occur on most outcrops but were avoided during sampling. The early veins vary from 1–10 cm and form from a few per cent to nearly fifty per cent of individual outcrops.

During the late Archaean the originally discordant veins in the gneisses and the intersecting Saglek dykes were rotated so that the structures are now essentially parallel. Recrystallization and presumed metasomatic introduction of material from late Archaean pegmatites and granite sheets caused a further loss of structure. Local sulphide mineralization both in the gneisses and in the interlayered supracrustal rocks occurred and galena has been sampled from two localities. Apatite grew in the marginal selvage of basic and ultramafic enclaves and allanite crystals grew on the interface between leucocratic and biotite-bearing septa in the gneisses.

Petrographically the amphibolite facies tonalitic and granodioritic Uivak 1 gneisses are simple with quartz-biotite-oligoclase ± hornblende ± microcline and accessory Fe-Ti oxides, apatite, allanite, epidote, zircon and sphene. Many gneisses show biotite growing at the expense of hornblende and the late growth of microcline. Zircons are frequently zoned with clear rims around an older core. In a few instances the central core of older material has been dissolved.

(b) Amphibolite facies Uivak 1 gneisses from the amphibolite/granulite facies transitional zone

Traced southwards from Tigigakyuk Inlet to Hebron fiord (Fig. 1) the quartzofeldspathic gneisses show changes associated with the regional increase in metamorphism seen in the development of orthopyroxene in the more mafic rocks. There is a general further loss of early Archaean structures such as well-defined separation into granodioritic and granitic layers. The layered gneisses become plastic with further disruption of basic enclaves and remnants of Saglek dykes. We ascribe these changes to tectonism and an increase in incipient melting under high grade conditions. Ryan (1977) introduced the term Iterungnek gneiss for the modified Uivak gneiss between Tigigakyuk and Hebron fiord. Within the Iterungnek gneisses the effects of late Archaean metamorphism vary from outcrop to outcrop. Enclaves of relatively well-preserved amphibolite facies layered gneisses similar in the field to the Uivak 1 gneisses of outer Saglek are preserved as far south as Hebron fiord enclosed in more recrystallized and migmatized gneisses with orthopyroxene, comparable to the rocks described below from west of the Handy fault. Barton (1975) took a single monolith from one of the best preserved enclaves of early Archaean gneiss 7.5 km southwest of Hebron township (Fig. 1) which he then sliced into 10 cm slabs (B 113a–B 118). A second site sampled by Barton (1975) from Hebron village (B 121) is interpreted by us as more thoroughly affected by the 2900 Ma metamorphic event. Both the field description and the mineralogy of the monolith correspond closely to the Uivak 1 gneisses further north. Isotopically there are marked differences which are discussed later.

(c) Partially remobilized granulite facies Uivak 1 gneisses

West of the Handy fault all pre-3000 Ma units in the gneiss complex recrystallized under granulite facies conditions. We distinguish between two main varieties of granulite facies Uivak 1 gneiss: units in which there has been partial melting on an outcrop scale but no introduction of foreign material and units where there has been considerable migmatization. The unmigmatized granulite facies Uivak 1 gneisses are rather monotonous brown-weathering rocks which locally develop pegmatitic partial melts. These melts intrude basic Saglek dyke remnants and break up trains of supracrustal material. There is a marked reaction between the quartzofeldspathic gneisses and mafic and ultramafic inclusions, which are

FIG. 2. Kiyuktok gneiss showing palaeosome (dark) and neosome with garnet and orthopyroxene together with a pegmatite sheet. Outcrop illustrated is approximately 0·5 m wide.

overgrown by coarse-grained orthopyroxene mantles and biotite rims. The essential mineralogy of granulite-facies Uivak gneisses is quartz, plagioclase, orthopyroxene, orthoamphibole, and hornblende with minor amounts of clinopyroxene, microcline biotite and Fe-Ti oxides. The plagioclase shows mesoperthite textures. Orthopyroxene is partly, or in some cases nearly completely, replaced by fibrous minerals (? Capoor amphibole) and rimmed by biotite. Zircons are subhedral, zoned with clear margins and cloudy cores and are larger than most in amphibolite facies gneisses. Both the dominant medium-grained gneiss forming the ground mass and the pegmatitic material are more mafic compared to the amphibolite facies Uivak gneisses.

(d) Migmatized granulite-facies Uivak gneisses (the Kiyuktok gneisses)

The granulite facies areas both east and west of the Handy fault contain large areas of migmatitic gneisses which contain relic Saglek dykes and Nulliak supracrustal units in a medium to coarse-grained strongly veined migmatitic matrix (Fig. 2). The gneisses are characterized in the field by a 'blebby' texture in which aggregates of mafic minerals are surrounded by quartz and feldspar. The migmatized high-grade Uivak gneisses have been named Kiyuktok gneisses (Kerr 1980; Collerson *et al.* 1981a, 1982a). In detail the relations within the Kiyuktok gneisses are complex. The leucocratic component (neosome) ranges from pegmatitic sheets tens of metres wide subparallel

to the regional foliation (but cross cutting early migmatite veins) through subconcordant migmatite sheets with sharp contacts to closely set 1–2 cm veins and leucocratic areas with diffuse contacts against a finer-grained more mafic component (palaeosome). The palaeosome varies from wavy septa a few millimetres wide between pegmatite sheets to layers of medium- to fine-grained biotite gneiss free of pegmatite a few tens of centimetres wide. The finer-grained darker matrix is frequently recrystallized close to pegmatitic layers, with a loss of the earlier fabric. Orthopyroxene grains up to 10 cm long are found in the margins of pegmatites. The majority of 'blebby' gneisses show transitional metamorphic assemblages between granulite and amphibolite facies while supracrustal rocks on the same outcrops contain granulite facies assemblages. Orthopyroxenes and garnet which form the main component of the mafic aggregates break down respectively to cummingtonite-hornblende-biotite aggregates and to green, Ti-poor biotite and Mg chlorite. The centres of the wider pegmatite sheets contain amphibolite facies assemblages.

Geochemistry and isotope chemistry of the Uivak 1 gneisses

(a) Amphibolite facies gneisses from outer Saglek fiord (Fig. 1)

The type amphibolite facies Uivak 1 gneisses are dominated by granodiorites with subordinate tonalitic units. The leucocratic layers range

TABLE 2. *Selected analyses to illustrate processes described in text*

	1	2	3	4	5	6	7
SiO_2	53.51	70.50	66.82	67.65	70.84	70.90	73.70
TiO_2	1.13	0.04	0.52	0.47	0.25	0.33	0.02
Al_2O_3	12.18	14.90	15.64	16.39	15.37	14.40	14.30
FeO^*	11.00	2.15	4.11	3.27	2.13	1.87	0.71
MnO	0.21	0.03	0.06	0.05	0.03	0.02	0.03
MgO	8.50	0.73	1.19	1.51	1.28	0.65	0.15
CaO	7.55	2.10	3.46	3.87	3.69	2.21	1.18
Na_2O	1.27	4.37	4.43	5.22	4.64	4.37	4.33
K_2O	1.79	2.89	1.67	0.96	1.10	3.37	4.24
P_2O_5	0.09	0.14	0.17	0.13	0.12	0.10	0.05
LOI	1.38	0.61	0.82	0.38	0.31	0.71	0.25
Total	98.61	98.46	98.89	99.90	99.76	98.93	98.96
Rb	92	119	106	41	10	73	190
Sr	779	509	262	517	534	393	90
Ba	247	354	183	380	431	767	318
Pb	13	19	19	17	5	15	30
Cs	2.0	ND	5.0	3.6	ND	0.9	ND
Y	26	17	14	8	2.87	14	24
Zr	108	207	201	125	80	112	56
Nb	10	5	9	4	1.1	7	0
Cr	428	8	20	35	36	5	4
Ni	192	>5	5	12	19	4	0
V	147	26	36	ND	33	26	6
La	22	73.5	35	27.2	20.76	46	9.24
Ce	48	144.4	60	57.6	33.01	79	16.6
Nd	23	65.4	27	17.16	10.95	20	6.26
Sm	5.1	11.99	4.4	4.08	1.69	3.5	1.29
Eu	1.4	2.35	0.97	1.13	0.83	0.6	0.333
Gd	ND	7.34	ND	ND	1.20	ND	1.05
Tb	0.4	ND	0.32	0.374	ND	0.09	ND
Dy	ND	ND	ND	ND	0.54	ND	0.97
Er	ND	ND	ND	ND	0.39	ND	0.462
Yb	1.5	ND	0.8	0.869	0.27	ND	0.43
Lu	0.4	ND	0.13	0.153	0.08	ND	ND
U	1.8	ND	ND	3.8	ND	ND	ND
Th	3.6	ND	7	6.04	ND	9.9	ND

1 75-32 e. Alkali metasomatized basic inclusion in migmatized Uivak gneiss to show increase in K_2O, LREE, Rb, Ba, compared to normal basalt.

2 75-32 y. Uivak grey gneiss from same outcrop as 1. Veined by both early and late Archaean granite sheets. Chemistry interpreted as modified by metasomatic addition of LIL and LREE. (Analysis from Collerson & Bridgwater, 1979.) (See Fig. 3 for REE patterns from less enriched Uivak samples from same outcrop).

3 GGU 236969. Amitsoq gneiss from Isua with ε Nd_{3650} +2 (Hamilton *et al.* 1983). Th, Rb, LREE, Cs, Pb, values regarded as secondary due to enrichment from adjacent granitic sheets.

4 GGU 158528. Non-riched Amitsoq grey tonalite gneiss preserved as sheet in Isua supracrusal rocks and protected from migmatization. Comparable non-enriched Amitsoq gneiss samples have ε Nd_{3650} values close to zero (H. Baadsgaard, personal communication, 1984).

5 82-21 c. Partly depleted granulite facies Uivak gneiss, Torr Bay, west of Handy fault. Note low K_2O, Rb, Pb characteristic of these gneisses. Nd isotopes on this outcrop have equilibrated at circa 2900 Ma (L. Schiøtte, unpublished).

6 74-53 a. A partly enriched paleosome from the mixed Kiyuktok gneisses.

7 75-208 n. Discordant partly retrogressed garnet pegmatite, the neosome in the Kiyuktok gneisses. Analysis from Vocke (1983). Pb and Sr data from Collerson *et al.* (1981). Now interpreted as showing addition of radiogenic Pb at circa 2900 Ma (Fig. 5) but equilibration of Sr with earlier crust to give high ISr_{2900}.

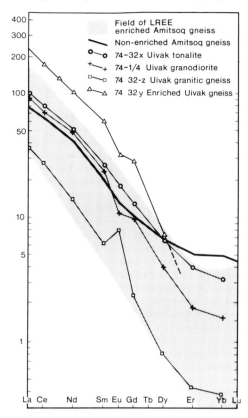

FIG. 3. REE patterns from less migmatized and strongly migmatized Uivak grey gneisses compared to unmigmatized and migmatized Amitsoq gneisses. We interpret the steep REE patterns seen in the migmatized gneisses as due to LREE enrichment from the adjacent granitic sheets. Data from Vocke (1983), Nutman & Bridgwater (in press) and Collerson & Bridgwater, 1979).

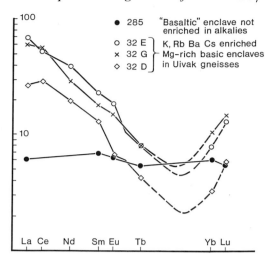

FIG. 4. REE patterns from basic enclaves in Uivak gneisses from same locality as gneisses in Fig. 3 (32) compared to other enclaves apparently not affected by LREE addition.

the local concentration of accessory phases such as allanite.

Bridgwater & Collerson (1976, 1977) interpreted the K_2O and Rb enriched nature of the Uivak 1 gneisses as due to the introduction of LIL elements into original tonalites during migmatization by granitic sheets. This process is not provable in the Saglek Hebron area since there are no unmigmatized outcrops. Recent work in the Isukasia area, West Greenland (Nutman & Bridgwater in press) has however shown that in the least deformed Amitsoq gneisses there is a progression from unmigmatized tonalites through to granodiorites which are enriched in LIL elements, where they are in contact with granitic sheets chemically identical to the leucocratic veins in the Uivak gneisses. This process of LIL enrichment is even more marked where early Archaean Amitsoq veined gneisses are sheeted by late Archaean granite. The Uivak gneisses studied here show close physical and isotopic similarities to the contemporaneous Amitsoq gneisses; the most important difference between the two areas being a higher degree of late Archaean migmatization in the Saglek area. We therefore suggest that the more extreme LIL and LREE enrichment seen in the migmatized Uivak gneisses is a result of similar but more extensive post-magmatic addition from the granitic veins into the grey gneisses forming the original complex. Support for secondary addition of LREE is seen in the late growth of LREE-bearing phases such as allanite and apatite. Loss of HREE into fluids circulating during migmatization may be controlled by break

between silica-rich granites (ss) and trondhjemite. Simple major and minor element plots against Si show considerable scatter, particularly for LIL elements. Average values for K_2O (2.5%), Rb (110 ppm) and K/Rb (170) are higher than most Archaean grey gneisses (Bridgwater & Collerson 1976, 1977) even when all leucocratic veins are removed. REE patterns (Table 2, Fig. 3) are variable, with a few tonalitic samples with chondrite normalized La_N values of about 60 and $(La/Yb)_N$ between 40–60 while the majority of granodiorites have La_N values of 60–400 and $(La/Yb)_N$ ratios betwen 60 and 250 (data extrapolated from Collerson & Bridgwater 1979, Vocke 1983, and D.B. unpublished results). Rare earth patterns from the leucocratic veins are steep with large positive Eu anomalies. The variable LREE contents of the veins are thought to be controlled by down of hornblende and zircon. Independent evidence for REE movement is seen in the margins of Mg-rich basic inclusions in the Uivak migmatites. These are enriched in Rb, Cs and K_2O (Fig. 4, Table 2).

Considerable isotopic work has been carried out on the Uivak gneisses since the first laboratory demonstration of early Archaean rocks in the area (Hurst *et al.* 1975). None of the whole rock systems define true isochrons. Conventional and ion probe U/Pb data on zircons confirm the field and petrological evidence that the rocks are highly disturbed (Hurst & Tilton 1976; Baadsgaard *et al.* 1979; Collerson 1983a). Late Archaean overgrowths and new zircon crystals developed in rocks with an early Archaean history. In view of the scatter on whole rock isochrons we consider the ion-probe data of Collerson (1983a) as the best indication of the age of formation of the Uivak 1 gneiss. This gives a scatter of points close to concordia at 3800 Ma, that is within error of the conventional U/Pb zircon age of the least disturbed Amitsoq gneisses which we regard as correlatives of the Uivak gneisses (Baadsgaard 1983, and pers. comm., 1984).

Rb-Sr data are published by Hurst *et al.* (1975) and in an abstract by Collerson & McCulloch (1982). The original material used by Hurst *et al.* (1975) has been redetermined by P.N. Taylor (pers. comm., 1980) and gives a Rb-Sr errorchron age of 3570 ± 160 Ma, ISr 0.70035 ± 80. The larger sample suite determined by Collerson & McCulloch yields an age of $3714 \pm ^{400}_{291}$, ISr $0.69938 \pm ^{252}_{334}$. (All Rb-Sr calculations use $\lambda_{Rb} = 1.42 \times 10^{-11}$ y^{-1x}.)

Sm-Nd determinations are only reported in an abstract (Collerson & McCulloch 1982). They show a scatter about a 3665 Ma isochron with an average positive εNd_{3665} of 2.48.

Pb-Pb whole rock data show a marked scatter (Baadsgaard *et al.* 1979; this paper; P.N. Taylor, pers. comm.) (Table 3, Fig. 5). The system is clearly highly disturbed and no convincing isochron can be drawn through the data points although there is a marked concentration within a belt with an approximately 2900 Ma slope. Compared to other suites of grey gneisses from the North Atlantic craton the amphibolite facies Uivak gneisses show an abnormally high proportion of radiogenic Pb (Fig. 5).

(b) Amphibolite facies Uivak 1 gneisses from the transitional zone between amphibolite and granulite facies

The monolith from the mouth of Hebron fiord studied by Barton (1975) is one of the few

TABLE 3. *Pb isotope compositions*

Sample no.	206/204	207/204	
Uivak gneisses at Saglek:			
74–161A	15.495	14.532	(1)
74–161C	13.711	14.205	(1)
74–161F	13.962	14.117	(1)
74–40A	13.891	14.290	(1)
75–260C	16.005	14.469	(1)
74–161B	14.307	14.290	(1)
75–260B	13.986	14.188	(1)
75–320	15.986	14.593	(1)
75–296A	16.300	14.890	(1)
74–298	16.216	15.272	(1)
75–291F	16.379	14.976	(1)
27SJ19	16.830	15.034	(2)
27SJ20	13.750	14.422	(2)
3SB3	13.862	14.153	(2)
32SB42	14.060	14.363	(2)
1SBI27	15.481	14.886	(2)
42MI32	14.606	14.321	(2)
43AMI33	12.877	13.811	(2)
LSD136	14.416	14.416	(2)
LSD138KM	14.297	14.228	(2)
LSD138P	13.922	14.150	(2)
3SB1	13.045	13.820	(2)
32SB43	16.031	14.638	(2)
4BI6	13.032	14.019	(2)
DB82.74.1B	14.31	14.66	(5)
Uivak gneisses at Hebron:			
113A	12.78	13.72	(3)
113B	12.47	13.58	(3)
114	12.99	13.79	(3)
115	13.16	13.85	(3)
116	13.11	13.85	(3)
117	13.93	14.16	(3)
118	13.58	13.96	(3)
121	12.95	14.08	(3)
Kiyuktok gneiss:			
238A	12.44	13.56	(4)
238K	13.37	13.90	(4)
238L	13.80	14.03	(4)
238M	15.97	14.77	(4)
238N	16.65	14.77	(4)
DB82.53D	13.02	14.05	(5)
Depleted granulites:			
DB82.21A	12.81	13.84	(5)
DB82.21B	12.79	13.83	(5)
DB82.21C	12.42	13.79	(5)
DB82.21D	12.60	13.82	(5)
DB82.21E	12.43	13.69	(5)
DB82.21G	12.33	13.68	(5)
DB82.21M	12.37	13.71	(5)
Galena mineralization:			
	12.22	13.88	(6)

(1) Baadsgaard *et al.* 1979; (2) Unpublished data by P. Taylor; (3) Barton *et al.* 1983; (4) Collerson *et al.* 1981; (5) New data by L. Schiøtte; (6) Unpublished data by G. Hanson.

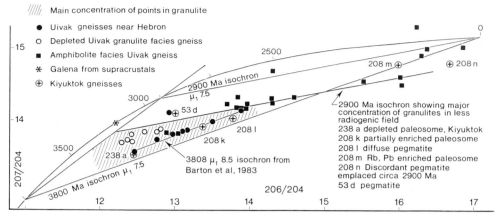

FIG. 5. Pb-Pb whole rock data to show contrasting behaviour of Pb and U in rocks affected by different degrees of metamorphism at 2900 Ma. The granulite facies and transitional granulite facies gneisses were markedly less radiogenic already at 2900 Ma compared to their amphibolite facies counterparts (a feature which thus cannot be ascribed to U-depletion during granulite facies metamorphism). The Kiyuktok gneisses show mixed characteristics at 2900 Ma, consistent with addition of more radiogenic Pb during migmatization. Galena data, G.Harrison, unpublished.

outcrops studied from the area south of Tigigak-yuk which has a mineralogy and whole rock chemistry close to that of the amphibolite facies Uivak 1 gneisses further north. Other outcrops in the same area are either highly depleted granulites or migmatized rocks comparable to rocks west of the Handy Fault described below. The only significant difference which we note between the Hebron fiord monolith and the Saglek fiord Uivak gneisses is that it shows lower Rb contents and consequently higher K/Rb ratios (average respectively 231 and 187). The Hebron village site sampled by Barton (B 121) has lower Rb (76 ppm) with K/Rb (331) and a very low U content of 0.169. Rb/Sr results from all the samples analysed by Barton (1975, 1982) yield an isochron of 3556 ± 83 Ma, ISr 0.7038 ± 8. Exclusion of the one sample from Hebron village does not change the isochron. Additional Rb-Sr data are given by Collerson *et al.* (1982b) and lie within error of the results given by Barton. The results give an age which is identical to that obtained from the Uivak gneisses but a higher ISr. Pb-Pb whole rock data (Table 3) from the monolith (Barton *et al.* 1983) plot on a Pb-Pb 3808^{+112}_{-123} isochron (μ_1 8.5) with little scatter and markedly less radiogenic Pb than seen in the Uivak gneisses at Saglek (Fig. 5), confirming Barton's (1982) contention that the gneisses at Hebron differ isotopically from those further north. The U-poor sample (121) from Hebron village falls markedly to the left of the Pb-Pb isochron from the monolith. Sm-Nd and Rb-Sr results are published from a single sample from

the Hebron village site (Hebron 1, Collerson *et al.* 1981b). This has a higher εNd_{3560} and higher εSr_{3560} than Uivak 1 gneisses from Saglek.

(c) Geochemistry of the granulite facies Uivak gneisses

Unmigmatized orthopyroxene-bearing gneisses from west of the Handy fault contain on average higher Fe, Mg, Ca and lower Na and Si than gneisses east of the fault (Table 2). The difference between the two suites is even more marked in LIL trace elements, in particular Rb (average 17 ppm against 110 ppm) and Pb (average 7 ppm against 18 ppm). REE patterns overlap those of the least LREE enriched amphibolite facies gneisses with $(La/Yb)_N$ ratios between 30 and 60. The high enrichment in LREE recorded from the amphibolite facies granodiorites is not seen.

Rb-Sr isotopic results from a single outcrop of unmigmatized granulite facies Uivak 1 gneisses do not define an isochron but lie within a Rb-poor field to the left of the errorchrons from the amphibolite facies gneisses (Fig. 6).

Pb-Pb data (Table 3, Fig. 5) also fall within a comparatively restricted field with a markedly lower content of radiogenic Pb at 2900 Ma than is seen in the amphibolite facies gneisses. Preliminary Nd-Sm results (L. Schiøtte) are consistent with redistribution of REE between pegmatitic and non-pegmatitic material at approximately 2900 Ma.

(d) Geochemistry of migmatized granulite facies gneisses

The Kiyuktok gneisses show a range from palaeosomes comparable to depleted granulites (c, above) to alkali-rich pegmatitic sheets and veins. The average composition (excluding discordant sheets) is enriched in alkalis compared both to the unmigmatized granulite facies Uivak gneisses and to the amphibolite facies Uivak gneisses (Bridgwater & Collerson 1976 and Table 2) precluding formation of these rocks by separation of a leucocratic melt *in situ*. This is also shown by REE studies (Vocke 1983; Collerson *et al.* 1982a) which yield patterns which cannot be modelled by the simple extraction of a quartzo-feldspathic melt phase. Both field relationships and geochemical studies point to the introduction of a fluid-rich alkali melt represented by the leucocratic veins which has mixed with depleted granulite Uivak 1 gneisses on a scale down to individual centimetre layers. The formation of orthopyroxene in the veins shows that migmatization started under granulite-facies conditions.

Rb-Sr data from the Kiyuktok gneisses (Collerson *et al.* 1981a) scatter about a 2753^{+121}_{-110} Ma ISr 0.7086 isochron consistent with a high proportion of older crustal material in these rocks and partial Sr homogenization during the late Archaean. Published Pb-Pb data (Collerson *et al.* 1981a) show a range in composition. The most depleted palaeosomes have a Pb-Pb isotopic composition close to that of the unmigmatized granulite facies gneisses while discordant pegmatites have a high proportion of radiogenic Pb, comparable to that seen in the amphibolite facies gneisses. There is however no strict correlation between rock type and Pb isotopic composition (Fig. 5). Sm-Nd data (Collerson & McCulloch 1982) show a spread in values from data points on or close to the 3665 Ma isochron from the amphibolite facies gneisses to more radiogenic compositions suggesting addition of material with Nd isotopic character similar to mantle in the late Archaean.

Interpretation

The four sets of geochemical and isotopic data from Uivak 1 gneisses show marked differences. We reject the possibility that these reflect primary variation in the Uivak gneisses from one area to the next since this would require an accidental correlation between pre-3600 Ma characters with post-2900 Ma metamorphic grade and geography. The complex folding and tectonic layering which gives the outcrop patterns on Fig. 1 makes such a correlation very unlikely. We therefore argue that the differences are due to element mobility during high-grade metamorphism.

Effects on the Rb-Sr system

There is a direct correlation between the present isotopic composition of Sr in the rocks, their metamorphic state at 2900 Ma and their Rb contents. This is best seen by plotting histograms of K/Rb ratios and ε Sr_t diagrams for each group of gneisses. The amphibolite facies gneisses from Saglek show average negative ε Sr_{3560} values, whereas those from Hebron show average positive values (data from Hurst *et al.* 1975; Barton 1982; Collerson *et al.* 1982b) while the Rb-poor granulite facies gneisses have high positive values (Fig. 6). The variation between the groups can be explained in terms of Rb depletion and addition at 2900 Ma with Sr remaining stable on a scale of a few centimetres. Acceptable errorchrons are obtained from the relatively Rb-rich amphibolite facies gneisses because the amount of radiogenic Sr developed since 2900 Ma effectively masks the scatter produced by accidental mixing (Bridgwater & Collerson 1976, 1977). In contrast the Rb-depleted rocks do not define an early Archaean isochron, reflecting both the small amount of radiogenic Sr developed since 2900 Ma and possibly a greater degree of Sr homogenization at 2900 Ma. Partial Sr homogenization is seen in the Kiyuktok migmatitic gneisses from which a late Archaean errorchron is obtained, but even in these, Rb-depleted samples lie to the left of the isochron showing that homogenization of Sr with the migmatizing phase was not perfect. High ISr values for the pegmatitic parts of the Kiyuktok gneisses suggest a component of crustal Sr.

The Pb trace element and Pb-Pb isotopic results show clear differences between the four groups. The amphibolite facies gneisses contain at least twice as much total Pb on average as the depleted granulites. The isotopic composition of the Pb from the unmigmatized granulite facies gneisses is dominated by an unradiogenic component and data points cluster just above a 3800 Ma reference isochron with a mantle-like μ_1 of 7.5 (Fig. 5). In contrast the Pb from the amphibolite facies gneisses contains a much higher proportion of radiogenic Pb. A 2900 Ma reference isochron on Fig. 5 separates the majority of data points from the amphibolite and granulite gneisses showing that this difference was already present in the late Archaean. This implies that the Pb in the amphibolite facies area has a component derived from a source which was comparatively rich in U before 2900 Ma, for example either the

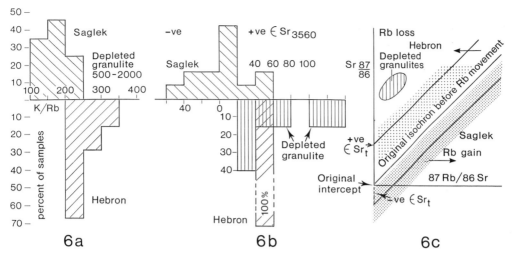

Fig. 6. Effects of movement of Rb at 2900 Ma. a Contrasting K/Rb ratios. b ε Sr_{3560} values (assuming present day bulk values of $^{87}Rb/^{86}Sr = 0.085$ and $^{87}Sr/^{86}Sr = 0.7048$). If a more realistic age of formation of the Uivak 1 gneisses is used (> 3600 Ma), ε Sr_t values change accordingly. When extrapolating to 3800 Ma (the U/Pb ion probe age) with the present Rb/Sr ratios, ε Sr $_{3800}$ for the Uivak gneisses at Saglek averages −29.5 ($Sr_1 = 0.698022$ that is below BABI). For the Uivak gneisses at Hebron ε Sr_{3800} averages +11.0. c Theoretical effects of statistically random loss or gain of Rb from the Uivak gneisses. Slight loss or gain (±25% Rb) will result in a scatter of points in zones respectively to the left and right of the isochron at time t_1. Provided that the time t_1 to present is considerably larger than t_0 to t_1 this scatter will be seen as a slight displacement of the present day isochron to give respectively a high initial ratio in the case of Rb loss and a low initial ratio in the case of Rb gain. (See the decrease in MSWD values as a function of time, Vidal *et al.* in press.)

late Archaean mantle or possibly the Upernavik metasediments. Pb from the monolith of amphibolite facies gneisses in the transitional zone (Fig. 1) falls on an early Archaean isochron suggesting very little disturbance during the late Archaean. The high μ_1 value of 8.5 could possibly indicate a component of older crustal material (for example Pb from pre-3800 Ma Nulliak supracrustal rocks) or slight U loss at 2900 Ma. The Pb results from the Kiyuktok gneisses are interpreted as a mixture of old Archaean Pb such as found in the depleted granulites and a late Archaean Pb more radiogenic component comparable to that in the amphibolite facies gneisses, presumably introduced during migmatization. Some mixing took place so that samples from thinner neosome veins contain comparatively unradiogenic Pb while LIL-enriched palaeosome contains a larger component of radiogenic Pb.

Sm-Nd system: the average εNd_{3665} of +2.48 of the amphibolite facies gneisses could indicate derivation from an Archaean mantle which was already depleted in crustal forming components before 3800 Ma and correlated with a primary low ISr (Collerson & McCulloch 1982). However the correlation between positive εNd_t and negative εSr_t in individual samples is not marked

(Collerson *et al.* 1981b). Evidence for Rb movement is strong. Evidence for LREE movement is more circumstantial but we argue that the large spread in εNd_{3665} shown by Collerson & McCulloch (1982) supports the mineralogical and geochemical model suggested. Direct evidence of mixing Nd with different compositions is seen in the Kiyuktok gneisses (Collerson *et al.* 1982a). A comparable less extreme addition of LREE containing a component of more radiogenic Nd during metasomatism of the regional gneisses near to pegmatite and granite veins would raise the average εNd_{3665} with a small decrease of the slope of the isochron (Rosing 1983). We do not regard the εNd_{3665} value of +2.48 as an accurate measure of the isotopic composition of the mantle at the time of formation of the Uivak gneisses, although it remains possible that the Uivak gneisses could have been derived from a +ve ε_{Nd} source.

Conclusions

The geochemical and isotopic characters of a single unit of early Archaean gneisses are shown to vary with the effects of metamorphism and migmatization. Mixing between old crustal material and material derived from a mantle-like source during, or shortly before, the mixing event is seen in the isotopic systems and is related to the introduction of fluid-rich magmas which at the present levels of exposure partially equilibrated with the crust during their emplacement. Comparatively radiogenic Pb and Nd were added to rocks which retain their geological evidence of an early history. Sr in the younger migmatitic material partially equilibrated with the regional > 3600 Ma crust. Regional changes in Sr initial ratios and the proportion of radiogenic Pb present can be explained in terms of Rb, U and Pb mobility. All the rocks studied must be regarded as mixtures of an older crustal component modified to varying degrees by younger material which is itself isotopically a mixture of mobilized old crust and new material derived from a mantle-like source not long before 2900 Ma. In some samples the Pb-Pb, Rb-Sr and Nd-Sm systems are uncoupled. Taking the Saglek–Hebron area as typical of polymetamorphic gneiss terrains, we suggest that whole-rock isochrons and initial Sr and Nd intercepts can at best be regarded as guides to the original age and origin of the protoliths. Pb isotopes can be a better guide to the metamorphic history of the rocks rather than indicating their primary origin. In polymetamorphic areas ε Nd_t and ε Sr_t values may be sensitive indicators of elemental redistribution rather than showing differences in source rocks and the degree of separation of continental crust from the Archaean mantle.

ACKNOWLEDGEMENTS: Field work in Labrador was supported by the Geological Survey of Canada, the Department of Mines and Energy, Newfoundland and the Canadian and Danish Scientific Research Councils and the Royal Society (grants to D. Bridgwater, K.D. Collerson, T. Rivers & A.P. Nutman). We thank P.N. Taylor for unpublished data (Pb-isotopes) and discussion. New Pb-Pb, Rb-Sr and Sm-Nd determinations which have provided the basis to reassess earlier work are being carried out by L. Schiøtte at CNRS Clermont-Ferrand under the guidance of P. Vidal. We thank M. Rosing for discussions on the Sm-Nd system and B. Fryer for comments on this contribution.

References

BAADSGAARD, H. 1983. U-Pb isotope systematics on minerals from the gneiss complex at Isukasia, West Greenland. *Rapp. Grønlands Geol. Unders.*, **112**, 35–42.

——, COLLERSON, K.D. & BRIDGWATER, D. 1979. The Archean gneiss complex of northern Labrador: 1. Preliminary U-Th-Pb Geochronology. *Can. J. Earth Sci.*, **16**, 951–961.

BARTON, J.M., JR. 1975. Rb-Sr isotopic characteristics and chemistry of the 3.6 B.Y. Hebron gneiss, Labrador. *Earth Planet. Sci. Lett.*, **27**, 427–435.

—— 1982. 'A reappraisal of the Rb-Sr systematics of the early Archaean gneisses from Hebron, Labrador' by K.D. Collerson *et al.*—a reply. *Earth Planet. Sci. Lett.*, **60**, 337–338.

——, RYAN, B. & FRIPP, R.E.P. 1983. Rb-Sr and U-Th-Pb isotopic studies of the Sand River gneisses, Central zone, Limpopo mobile belt. *Spec. Publ. Geol. Soc. South Africa*, **8**, 9–18.

BLACK, L.P., GALE, N.H., MOORBATH, S., PANKHURST, R.J. & McGREGOR, V.R. 1971. Isotopic dating of very early Precambrian amphibolite facies gneisses from the Godthaab district, West Greenland. *Earth Planet. Sci. Lett.*, **12**, 245–259.

BRIDGWATER, D. 1979. Chemical and isotopic redistribution in zones of ductile deformation in a deeply eroded mobile belt. Part 1. Chemical redistribution. *U.S. Geol. Surv. open file report* 70–1239, 505–526.

—— & COLLERSON, K.D. 1976. The major petrological and geochemical characters of the 3600 m.y. Uivak gneisses from Labrador. *Contrib. Mineral. Petrol.* **54**, 43–59.

——, —— 1977. On the origin of Early Archaean gneisses. *Contrib. Mineral. Petrol.* **62**, 179–192.

——, COLLERSON, K.D., HURST, R.W. & JESSEAU, C.W. 1975. Field characters of the early Precambrian rocks from Saglek, coast of Labrador. *Geol. Surv. Can.*, Pap 74-1a, 287–296.

——, NUTMAN, A.P., SCHIØTTE, L. & RYAN, A.B. 1985. The Archean rocks of the outer Saglek—Hebron fiord area, northern Labrador: Progress report, 1983. Submitted to the Canadian Geological Survey.

COLLERSON, K.D. 1983a. Ion microprobe zircon geochronology of the Uivak gneisses: implications for the evolution of early terrestrial crust in the North Atlantic craton. *In:* ASHWAL, L.D. & CARD, K.D. (eds), *Workshop on Cross Section of Archean Crust*, Lunar and Planetary Inst., Rep. 83–03, 28–33.

—— 1983b. The Archaean gneiss complex of northern Labrador 2. Mineral ages, secondary isochrons and diffusion of strontium during polymetamorphism of the Uivak gneisses. *Can. J. Earth Sci.*, **20**, 707–718.

—— & BRIDGWATER, D. 1979. Metamorphic development of early Archean tonalitic and trondhjemitic gneisses: Saglek area, Labrador. *In:* BARKER, F. (ed.) *Trondhjemites, Dacites and Related Rocks*, Elsevier, 205–273.

——, KERR, A. & COMPSTON, W. 1981a. Geochronology and evolution of late Archaean gneisses in Northern Labrador: An example of reworked sialic crust. *Spec. Publ. Geol. Soc. Australia*, **7**, 205–222.

——, McCULLOCH, M.T. & MILLAR, D. 1981b. Nd and Sr isotopic relationships in the early Archaean Uivak and Amitsoq gneisses. *Ann. Rep. Res. School Earth Sci., Austr. Natl. Univ.*

——, KERR, A., VOCKE, R.D. & HANSON, G.N. 1982a. Reworking of sialic crust as represented in late

Archean-age gneisses, northern Labrador. *Geology*, **10**, 202–208.

——, BROOKS, C., RYAN, A.B. & COMPSTON, W. 1982b. A reappraisal of the Rb-Sr systematics of the early Archaean gneisses from Hebron, Labrador. *Earth Planet. Sci. Lett.* **60**, 325–336.

—— & McCULLOCH, M.T. 1982. The origin and evolution of Archaean crust as inferred from Nd, Sr and Pb isotopic studies in Labrador. *5th Intern. Conf. Geochronology, Cosmochronology and Isotope Geology, Nikko, Japan*, Extended Abstracts, 61–62.

HAMILTON, P.J., O'NIONS, R.K., BRIDGWATER, D. & NUTMAN, A. 1983. Sm-Nd studies of Archaean metasediments and metavolcanics from West Greenland and their implications for the Earth's early history. *Earth Planet. Sci. Lett.*, **62**, 263–272.

HURST, R.W., BRIDGWATER, D., COLLERSON, K.D. & WETHERILL, G.W. 1975. 3600 m.y. Rb-Sr ages from very early Archaean gneisses from Saglek Bay, Labrador. *Earth Planet. Sci. Lett.* **27**, 393–403.

—— & TILTON, G.R. 1976. Isotopic investigations of the 3.62 B.Y. old iron-rich Uivak gneiss, Saglek Bay, Labrador. *Geol. Soc. Am. Abstr. with programs* **8**, section **6**, 933.

KERR, A. 1980. *Late Archaean igneous, metamorphic and structural evolution of the Nain province at Saglek Bay, Labrador.* MSc thesis, Memorial University, Newfoundland (unpublished).

NUTMAN, A.P. & BRIDGWATER, D. 1986. Early Archaean Amitsoq tonalites and granites of the Isukasia area, southern West Greenland. Development of the oldest sial. *Contrib. Mineral. Petrol.* (in press).

ROSING, M. 1983. *A metamorphic and isotopic study of the Isua supracrustals, West Greenland.* Cand. scient. thesis, Univ. of Copenhagen (unpublished).

RYAN, A.B. 1977. *Progressive structural reworking of the Uivak gneisses, Jerusalem harbour, northern Labrador.* M.Sc. thesis, Memorial University, Newfoundland (unpublished).

——, MARTINEAU, Y., BRIDGWATER, D., SCHIØTTE, L. & LEWRY, J. 1983. The Archean-Proterozoic boundary in the Saglek Fiord area, Labrador. Report 1. *Geol. Surv. Can. Pap.*, **83**-1A, 297–304.

——, ——, KORSTGAARD, J. & LEE, D. 1984. The Archean/Proterozoic boundary in northern Labrador, Report 2. *Geol. Surv. Can. Report* **84**-1A, 545–551.

VIDAL, P., BERNARD-GRIFFITHS, J., COCHERIE, A., LeFORT, P., PEUCAT, J.J. & SHEPPARD, S., in press. Geochemical comparison between Himalayan and Hercynian leucogranites. MS submitted to Physics and Chemistry of the Earth.

VOCKE, R.D., JR. 1983. *Petrogenetic modelling in an Archean gneiss terrain, Saglek, northern Labrador.* Ph.D., State University of New York, Stony Brook. Unpublished.

L. SCHIØTTE & D. BRIDGWATER, Geological Museum, Øster Voldgade 5–7, DK-1350, Copenhagen K. Denmark.

K.D. COLLERSON, Department of Geology, University of Regina, Canada.

A.P. NUTMAN, Department of Geology, Memorial University, Newfoundland, Canada.

A.B. RYAN, Department of Mines and Energy, St. John's Newfoundland, Canada.

Composition of the Canadian Precambrian shield and the continental crust of the earth

D.M. Shaw, J.J. Cramer, M.D. Higgins & M.G. Truscott

SUMMARY: Following a summary of the pre-Archaean history of the earth's crust, the present compositions of the upper and lower continental crust are appraised.

There is no sure evidence that the Archaean crust differed in bulk composition from the present day: estimates that it was richer in ferrides and poorer in incompatible elements are difficult to distinguish from regional variations.

Estimates of lower crustal composition vary considerably according to the manner in which they were determined. A new estimate is nevertheless presented.

The abundances of heat-producing elements in both the upper and lower crust are in accord with measured values of heat flow at the surface.

The early evolution of any Precambrian shield is a matter of conjecture and, since the evidence in the rocks has been obscured by later events, must be reconstructed indirectly by any available means. Before considering the present composition of the upper and lower continental crust in the Canadian shield an attempt will be made to describe how the earth's crust developed in pre-Archaean times, summarized from previous papers (Shaw 1980, 1981).

Early crustal evolution

The recent history of the earth is dominated by two well-established facts, namely that much continental crust is old and all oceanic crust is young. These observations are benchmarks in any interpretation of crustal history, and have so been used by many authors. However they are temporal, comparative statements and can not be put to profitable use when the early Archaean earth is being considered. But any model of the evolution of the young earth must include an explanation of how subsequent events would give rise to these circumstances.

It is now accepted that earth accreted from cold nebular condensates consisting of metallic and chondritic materials. Accretion disengages a considerable amount of heat, to which may be added a second heat source by gravitational sinking of metal to form the core. It is therefore accepted widely that the early earth underwent a high temperature phase when much, perhaps all, of its material was molten and most of the original volatile components were lost. However the continued presence of noble gases and some volatile compounds of H, O, C, N, S, halogens, render the molten earth hypothesis controversial and unacceptable to some authors.

The earth's mantle consists largely of silicates and its melting would be followed by fractional crystallization, from the bottom up, as heat was lost by conduction from the interior, aided by convection, and by surface radiation. The fractionation concentrated less refractory elements such as K, Na, Ca, Al, Si, REE, U, Th etc. in upper parts of the mantle. At some point the surface temperature of the earth began to permit solidification of a solid crust (1100–1300°C) consisting of anorthositic, basaltic or dioritic material forming rafts or 'ice-floes' floating on a more basic magma ocean, as has been inferred for the moon. Eventually the crust would extend to cover the whole earth's surface. From then on the release of magma to the surface took place through volcanic action.

During this time, perhaps 4.5 to 4.1×10^9 yr BP, the earth presumably was exposed to the extensive flux of impacting asteroidal fragments which sculptured the surfaces of all the inner planets. The effect of this was to brecciate and remelt the early crust perhaps destroying all the primary solid rock, leaving a cratered landscape over much of the earth's surface, much as the moon is today.

Throughout this period the solid earth was surrounded by a hot water-rich proto-atmosphere. As the crust continued to cool a point was reached when water condensed, along with acid halides and other volatile species. The initial acid rains would re-evaporate on reaching the hot surface, but eventually liquid water accumulated on the surface, drowning much of the exposed rock while altering and leaching it extensively. If all the water of the present hydrosphere was available a universal ocean of some 2 km depth would have formed. In the absence of present-day mountain-building forces the crustal relief is unlikely to have been extensive and most of the

From: DAWSON, J.B., CARSWELL, D.A., HALL, J. & WEDEPOHL, K.H. (eds), 1986, *The Nature of the Lower Continental Crust*, Geological Society Special Publication No. 24, pp. 275–282.

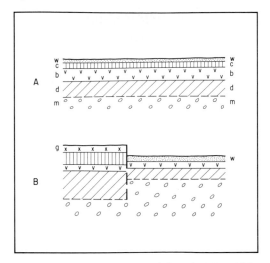

FIG. 1. A schematic summary of the principles which governed the redistribution of rock masses near the earth's surface to form continental nuclei from the primitive continuous crust. A The primitive crust (c) is intermediate in overall composition, overlain by the ocean (w), and is underlain successively by basalt (b), depleted mantle (d), fertile mantle (m). B Generation of sialic rock (g) in a continental nucleus (see text) heightens the relief, exposing rocks to sub-aerial erosion. The volume of depleted mantle associated with the continental mass increases and, in effect, provides it with roots, thereby giving it more rigidity, relative to the thinner oceanic lithosphere,

TABLE 1. *Canadian Precambrian shield surface composition estimate*

	Wt.%		ppm[+]		ppm[+]
SiO_2	64.93	Be	1.3	Y	21
TiO_2	0.52	B†	9.2	Ir	ppb 0.024
Al_2O_3	14.63	Ga	14	Au	ppb 1.81
Fe_2O_3	1.36	Li	22	La	32.3
FeO	2.75	V	53	Ce	65.6
MnO	0.068	Cr	35	Nd	25.9
MgO	2.24	Co	12	Sm	4.51
CaO	4.12	Ni	19	Eu	0.937
Na_2O	3.46	Cu	14	Gd	2.79
K_2O	3.10	Zn	52	Tb	0.481
P_2O_5	0.15	Zn*	60	Ho	0.623
H_2O^+	0.79	Cd*	ppb 75	Yb	1.47
H_2O^-	0.13	Hg§	ppb 96	Lu	0.233
CO_2	1.28	Bi*	ppb 35	Pb	17
S	0.06	Nb	26	Pb*	18
Cl	0.01	Ta‡	5.7	Sr	316
F	0.05	Sc	7.0	Ba	1070
C	0.02	Zr	237	Rb	110
Less O	0.04	Hf	5.8	Rb*	109
		Th	10.3	Tl	ppb 524
		U	2.45	Tl*	ppb 499

Sum	99.63

[+] except where ppb indicated.
Data sources:
† analyst: M.D. Higgins.
‡ analyst: K. Speranzini.
§ analyst: Barringer Research Inc. (Bradshaw *et al.* 1970).
* weighted mean values calculated from Heinrichs *et al.* (1980).
Other data reported from Shaw *et al.* (1967, 1976), Shaw (1967, 1968).

crust was below sea-level (see Fig. 1). There was no distinction between continental and oceanic crust.

The generation of continental nuclei was favoured by four main processes:
a) build-up of sub-aerial volcanic edifices which initiated weathering and sedimentation;
b) generation of light-density, more evolved, sialic rocks by i) magmatic differentiation at volcanic centres; ii) weathering of basic rocks; iii) anatexis of any pre-existing rocks, particularly sediments, in the presence of water (see Campbell & Taylor 1983);
c) isostatic rising of lighter-density rock complexes;
d) generation of low density depleted mantle (by melting out of sialic materials) beneath the growing continental nuclei, forming coherent root-like underplating (Jordan 1978).
In areas where the tectonic forces did not favour development of such nuclei more basic basaltic material remained as incipient oceanic crust.

These processes are sufficient to explain the growing size of continental nuclei and ocean basins, accompanied by decreasing numbers of each. The early 'jostling' of lithospheric mini-plates, during the Prearchaean and Archaean, produced the alternation of cratonic areas fringed by greenstone belts characteristic of older Precambrian terrains: modern plate tectonics was embryonic or absent, because a) plates were much smaller than now; b) there was little contrast in thickness between continental and oceanic plates; c) higher heat generation resulted in faster convective overturn. But as time went on the continental nuclei coalesced, developed roots, grew in mean elevation and thus began to achieve stability, which permitted greater chance of survival during collision with thinner, newly-developed ocean lithosphere (Fig. 1). It became ever more true that continental crust is old and oceanic crust is young.

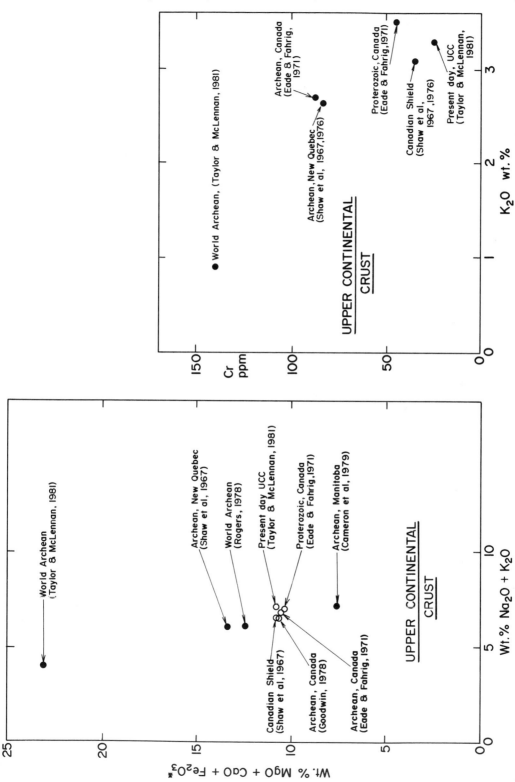

FIG. 2. Composition of UCC according to several authors. Note the variability in Archaean estimates: (a) cafemic vs. alkali oxides; (b) Cr vs. K_2O.

Table 2. *Various Precambrian shield composition estimates*

	A	B	C	D	E	F	G	H	I
SiO_2	65.3	66.50	65.1	65.0	64.0	61.3	57.4	64.93	66.0
TiO_2	0.49	0.42	0.50	0.62	0.5	0.46	0.9	0.52	0.6
Al_2O_3	15.9	16.34	16.0	16.0	17.0	14.5	15.6	14.63	16.0
Fe_2O_3	1.4	3.48*	1.5	1.2	5.4*	1.4	—	1.36	—
FeO	3.0	—	3.0	3.4	—	3.0	9.5	2.75	4.5
MnO	0.08	0.06	0.08	0.09		0.075		0.068	
MgO	2.2	1.38	2.3	2.1	3.1	3.5	5.2	2.24	2.3
CaO	3.8	2.75	3.4	3.3	3.9	5.1	7.3	4.12	3.5
Na_2O	4.0	4.12	4.1	3.5	4.6	3.4	3.1	3.46	3.8
K_2O	2.53	3.06	2.70	3.51	1.5	2.64	0.9	3.10	3.3
P_2O_5	0.13	0.15	0.15	0.19		0.14		0.15	
H_2O	0.80		0.7	0.7		1.1		0.92	
CO_2	0.16		0.20	0.10		2.3		1.28	
Ba ppm			790	810		1280	240	1070	700
Sr ppm			410	310		376	300	316	350
U ppm			1.2	2.2		2.0	0.75	2.45	2.5
Th ppm			9.7	13.6		6.8	2.9	10.3	10.5
Ni ppm			23	12		40	90	18.6	20
Cr ppm			88	45		84	140	35.0	25

A   Archaean upper crust, Canada. Goodwin, 1978, Table III.
B   Archaean upper crust, SE Manitoba, Canada. Cameron *et al.* 1979, Table V.
C   Archaean upper crust, Canada. Eade & Fahrig 1971, Table 15.
D   Proterozoic crust, Canada. Eade & Fahrig 1971, Table 15.
E   Archaean crust (several regions). Rogers 1978, Table 6.
F   Archaean crust, New Quebec. Shaw *et al.* 1967, Tables XII, XV; 1968, Table 3; 1976, Table 4.
G   Archaean upper crust. Taylor & McLennan 1981, Tables 1, 2.
H   Canadian shield, from Table 1.
I   Present-day upper crust. Taylor & McLennan 1981, Tables 1, 2.
* Total Fe as Fe_2O_3

The foregoing account has been greatly abbreviated, and supporting evidence has been largely omitted, since a fuller account has already been published in the references given. It would seem, moreover, that the generation of crustal material was episodic rather than continental. McCullough and Wasserburg (1978) for example found Nd/Sm evidence of a great period of crustal growth in the period 2.7–2.5 × 10⁹ yr BP. There is difficulty in reconciling this for the Canadian Superior province, where the existence of 2.6 × 10⁹ yr Matachewan diabase dykes indicates that cratonization had already been achieved. Compositional evidence will however be examined in the next section.

Surface composition of the Precambrian shields

Estimates of the surface composition of the Canadian shield have been made by a number of workers. Since the last publication of our group (Shaw *et al.* 1976) a number of additional measurements have been made on the composite samples which we first reported in 1967. Table 1 assembles all the data available on these samples, including some additional high-quality analyses from Heinrichs *et al.* (1980).

A number of other recent estimates of upper continental crust (UCC) composition are in Table 2. Most of these show rather similar abundances, although some are Archaean and others are not. However the Archaean crustal estimate in column G differs considerably from all the other (Fig. 2a,b) in being poorer in Si, Na, K, Ba, U, Th and richer in Ti, Fe, Mg, Ca, Ni, Cr: this probably is a consequence of the fact that it is based on volcanic rocks to a large degree (Taylor & McLennan 1981, p. 389). Estimates H of the overall Canadian shield, and I of the present-day UCC, are not dissimilar: moreover they do not differ markedly from the Archaean estimates in columns A, B, C, E and F. Comparing columns H and I on a closer basis, our figures show a little less SiO_2 and Al_2O_3, but the differences are reduced if our figures are recalculated on a volatile-free basis; our CaO is somewhat higher than Taylor and McLennan's. Our trace element data are also very close to those of Taylor and McLennan except for Ba, Cr (Table 2) and Cu, HREE (not shown here). Taylor and McLennan do not list values for Be, B, Ga, Li, Hg, Ta, Ir, Au.

The Proterozoic crust estimate (column D, Table 2) by Eade and Fahrig contains more K, U, Th and less Ni and Cr than their Archaean estimate, and they have discussed this extensively. In the context of the other estimates in Table 2, their documented abundance differences in K, U, Th, although real, may represent regional rather than metamorphic effects.

It must be concluded that there is no strong evidence that the UCC changed in major chemistry from Archaean to younger times, although U, Th, Ni, Cr data are ambiguous. If there was a major growth of the continental crust during later Archaean times, then that growth did not lead to a change in composition which can be confidently distinguished from the variability shown by different regional estimates.

Lower continental crust (LCC) composition

Many workers have recognized the problem posed by the fact that the heat flow through the continental crust is lower than it should be if the surface abundance of U, Th and K were to hold throughout the crustal thickness: this suggests that these elements, and perhaps others too, have different abundances in the lower crust. Seismic evidence indicates that wave velocities increase in the lower crust, and this also suggests compositional differences compared with the surface. Since we cannot yet sample the lower parts of the crust, most approaches to estimating its composition have proceeded by analogical reasoning: this accepts the axiom that the lower crust consists of granulite surface rocks and proceeds on the principle that surface granulites can be used as models.

In some cases, such as the Kapuskasing structural zone in Ontario, the argument is assisted by geological evidence (Percival & Card 1983) that a section through the former crustal thickness is now exposed by tilting and thrusting; elsewhere such evidence is lacking.

An extensive literature has accumulated on the use of granulites to model the LCC, and it is not feasible to review it here. Earlier work was reviewed by Holland and Lambert (1972) and more recent developments are documented by Weaver and Tarney (1984).

TABLE 3. *Granulite averages, oxides in wt.%, elements in ppm*

	Wt.% SiO_2 class				
	<45	45–55	55–65	>65	Weighted mean*
n	14	25	26	23	88
SiO_2	40.14	49.18	58.61	70.13	61.54
TiO_2	1.33	1.51	0.83	0.39	0.81
Al_2O_3	19.23	14.39	16.66	14.65	14.92
Fe_2O_3	13.25	14.50	8.24	4.06	8.02
MnO	0.14	0.21	0.14	0.06	0.12
MgO	15.53	7.79	3.18	1.71	4.06
CaO	5.90	8.91	6.15	2.81	5.20
Na_2O	1.25	2.23	3.75	3.55	3.12
K_2O	1.51	0.79	1.57	2.04	1.58
P_2O_5	0.17	0.20	0.20	0.08	0.13
L.O.I.	1.56	0.32	0.60	0.50	0.48
Sum	100.01	100.03	99.93	99.98	99.98
B	6	15	21	3	9.3
V	219	307	121	43	139
Cr	289	317	102	50	144
Co	61	58	33	28	39
Ni	275	173	58	41	89
Cu	82	65	23	24	38
Zn	133	136	94	59	89
Nb	14	10	13	16	14
Ta	2.4	1.6	1.6	2.9	2.3
Sc	36	36	20	8	19
Zr	161	113	143	207	169
Hf	4.3	3.2	3.3	5.0	4.2
Th	16	2.5	5.2	11	7.7
U	3.8	0.7	1.1	1.3	1.1
Y	46	47	32	19	30
Pb	14	12	18	30	23
Sr	212	223	550	318	317
Ba	671	355	536	491	458
Rb	78	21	37	52	41

* Weighting factors (from left) 2, 31, 13, 54: These factors are the mean values of estimates of the regional abundances of high metamorphic grade Precambrian rock types (U, B, I, S) taken from: Ronov & Yaroshevskiy 1969; Cameron *et al.* 1979 (SE Manitoba); Lambert & Heier 1968 (four granulite areas in Australia).
n: number of samples in each set, except for B, for which the numbers are (from left) 7, 18, 14, 16.

A more direct approach has been used by Leyreloup (e.g. Dupuy *et al.* 1979) and others, in using analyses of granulite xenoliths presumed to be lower crust samples, carried to the surface in basaltic pipes.

Yet another approach has been used by Taylor and McLennan (1981), which consists essentially of accepting that the whole crust may be measured by an andesite model and then subtracting from this a mass with the composition of the observed upper crust.

A major difficulty with any direct estimate of lower crustal composition is to know what proportions of different granulitic rock types are appropriate. For example the relative abundance of mafic granulites (as reported by Cameron *et al.* (1979), Lambert & Heier (1968) and others) varies from nearly zero in some regions to sixty per cent in others. There is therefore no easy way to put together the vast arrays of analytical data on granulite rocks so as to obtain a lower crustal composition in which one might have confidence. Nevertheless average proportions of different rock types have been calculated, and are included in Table 3.

We have recently embarked on this stormy ocean, with the philosophy that it is preferable to use representative suites of granulites from a variety of regions, rather than to concentrate on any particular one. Analytical data were measured using XRF, INAA and PGNAA (prompt gamma neutron activation analysis) for the most part. Accuracy was controlled by analysis of international standard reference samples and precision by replicate analysis. The sample suite was assembled by J.J. Cramer and comprises 89 samples from: Pikwitonei, Manitoba; Adirondacks; Montana (xenoliths); Parry Sound, Ontario; southern India.

The complete data matrix, including the REE which will be reported elsewhere, is too large to be included here, so the results (Table 3) were averaged in four silica classes. This presentation does not permit discussion of the regional differences in composition, which are by no means negligible and for which discussion is again deferred.

Comparison of the abundances in the first columns of Table 3 shows the following trends accompanying the increase from left to right in silica:

Increase: Si, (Na), (K), Pb;

Decrease: Ti, Fe, Mn, Mg (Ca), (P), V, Cr, Co, Ni, Cu, Zn, Sc, Y.

Elements in parentheses do not show a unidirectional trend, which arises from the unusual composition of some of the samples containing < 45 wt% SiO_2.

Since a number of the elements analysed show no clear trend, those elements will be insensitive to the manner in which an overall average is calculated. Accordingly it was decided to use average rock type abundances to weight the four class means, noting that the averages thus

TABLE 4. *Estimates of lower crustal composition*

	A	B	C	D	CC
SiO_2	54.0	61.2	56.3	61.5	63.2
TiO_2	0.9	0.5	1.1	0.8	0.7
Al_2O_3	19.0	15.6	17.1	14.9	14.8
Fe_2O_3	—	—	8.8	8.0	6.22
FeO	9.0	5.3	—	—	—
MnO	0.17	0.08	0.108	0.12	0.09
MgO	4.1	3.4	5.0	4.1	3.15
CaO	9.5	5.6	5.5	5.2	4.66
Na_2O	3.43	4.4	23.1	3.1	3.29
K_2O	0.6	1.0	1.42	1.6	2.34
P_2O_5	—	0.18	0.16	0.13	0.14
L.O.I.	—	—	2.2	0.5	1.41
Sum	100.7	97.26	99.79	99.95	100
B	—	—	—	9.3	9.3
Li	—	—	7	—	—
V	230	—	167	139	96
Cr	65	88	220	144	90
Co	33	—	29	39	26
Ni	35	58	61	89	54
Cu	78	—	22	38	26
Zn	—	—	85	89	71
Nb	4	5		14	20
Ta	—	—	0.8	2	4
Sc	40	—	25	19	13
Zr	30	202	174	169	203
Hf	1.6	3.6	4.6	4	5.0
Th	1.95	0.42	5.8	7.7	9.0
U	0.63	0.05	0.57	1.1	1.8
Y	22	7		30	26
Pb	7.5	13		23	20
Sr	425	569	266	317	317
Ba	175	757	545	458	764
Rb	8	11	27	41	76

A   Present-day lower crust. Taylor & McLennan 1981, Tables 1, 2.
B   Archaean lower crust. Weaver & Tarney, 1984, Table 2.
C   Lower crust below Bournac Pipe. Dupuy *et al.* 1979.
D   Lower crust, from Table 3.
CC Proposed composition of continental crust, obtained as the average of column D and Table 1.

obtained (last column of Table 3) can have no more confidence than the weighting factors.

With that caution the average may be taken as a rough average composition of the granulite terrains represented, and by extension as a tentative representation of the LCC. It is compared with three others in Table 4 and Fig. 3. The

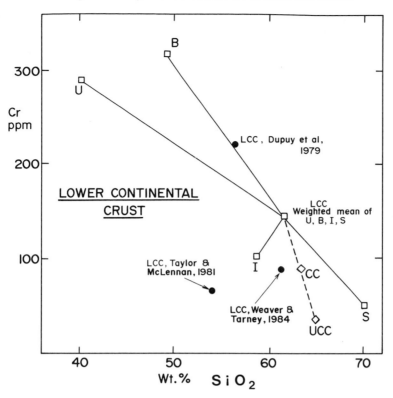

FIG. 3. Cr and SiO_2 values of the four SiO_2 class granulite averages (U, B, I, S) are shown with the weighted mean (Table 3). The position of the mean is evidently very sensitive to the weighting factors, which are not known with high confidence. The position of the UCC and the estimate of the overall continental crust composition (CC) are included from Table 4. Also shown are three other estimates of LCC composition.

differences are notable, both for major elements (Si, Al, K) and for trace constituents (V, Cr, Cu, Nb, Sc, Zr, Th, U, Y, Pb, Ba, Rb). It may be concluded that although the composition of the upper continental crust, as exemplified by the Canadian Precambrian shield, is reasonably well-known, the lower crust is not: in the words of Taylor and McLennan (1981, p. 386) 'The general problem of arriving at estimates of lower-crustal composition is severe'.

Conclusion

After a review of the probable early course of development of the earth's continental and oceanic crusts, new analytical data on the surface composition of the Canadian Precambrian shield have been presented. The composition of the UCC appears from this and other estimates to be reasonably well-known. There is no strong evi-

dence that the crust has changed in overall composition since Archaean times.

Estimation of the composition of the LCC is much more difficult, owing to several kinds of uncertainty as to the best way to model an inaccessible region. A new estimate nevertheless has been made, based on the analysis of granulites from various areas: this estimate and three others, when examined together, show a lack of congruency which highlights the problems.

Our estimates of the upper and lower crust (Table 1 and column D, Table 4) differ in some of the ways which other writers have found. The LCC has less Si, K, Rb, Th, U, as stated by Holland and Lambert (1972) and others: it also appears to have less Zr, Nb and Ta. It however is more basic than the UCC, as evidenced by higher Ti, Fe, Mg, V, Cr, Co, Ni, Cu, Zn, Sc, Y, and has higher ratios K/Rb, K/U.

Using the figures in Table 4, with constants given in Shaw (1968), the heat production of the

UCC and the LCC are 0.567 and 0.336 in units of microcal cm^{-2} s^{-1}, or 0.903 for the whole CC. The measured surface heat flow is about 1.0 unit in stable areas, away from tectonically active zones, which would thus allow about 0.1 unit for the mantle contribution. This is not at variance with upper mantle abundances of K, U, Th.

ACKNOWLEDGEMENTS: We are indebted to the following individuals who kindly supplied representative samples of the granulite terrains in which they worked: Carter Hearn, US Geological Survey, Reston; Jaap Hubregtse, Manitoba Department of Mines; Sisir Sen, Indian Institute of Technology, Kharagpur; Danny L. Thompson, Princeton University. Other samples we collected ourselves. We thank K.H. Wedepohl for a helpful review.

Funding support is gratefully acknowledged through NSERC grants A0155 and E6626 to D.M. Shaw. Difficulites in transcribing the manuscript were valiantly overcome by Linda Hillier.

This is paper number 142 of the McMaster Isotope and Nuclear Geochemical Group.

References

BRADSHAW, P.M.D., CLEWS, D.R. & WALKER, J.L. 1970. Exploration geochemistry. *Mining in Canada*, June 1970, 24–31.

CAMERON, E.M., ERMANOVICS, I.F. & GOSS, T.I. 1979. Sampling methods and geochemical composition of Archean rocks in southeastern Manitoba, Canada. *Precambrian Res.* **9**, 35–55.

CAMPBELL, I.H. & TAYLOR, S.R. 1983. No water, no granites—no oceans, no continents. *Geophys. Res. Lett.*, **10**, 1061–1064.

DUPUY, C., LEYRELOUP, A. & VERNIÈRES, J. 1979. The lower continental crust of the Massif Central (Bournac, France) with special references to REE, U and Th composition, evolution, heat-flow production. *Phys. Chem. Earth*, **11**, 401–415.

EADE, K.E. & FAHRIG, W.F. 1971. Geochemical evolutionary trends of continental plates—a preliminary study of the Canadian shield. *Geol. Surv. Canada. Bull.* **179**, 51 pp.

GOODWIN, A.M. 1978. The nature of Archean crust in the Canadian Shield. *In*: TARLING, D.H. (ed.) *Evolution of the Earth's Crust*, Academic Press. 175–218.

HEINRICHS, H., SCHULZ-DOBRICK, B. & WEDEPOHL, K.H. 1980. Terrestrial geochemistry of Cd, Bi, Tl, Pb, Zn and Rb. *Geochim. Cosmochim. Acta*, **44**, 1519–1534.

HOLLAND, J.G. & LAMBERT R. ST. J. 1972. Major element chemical composition of shields and the continental crust. *Geochim. Cosmochim. Acta*, **36**, 673–683.

JORDAN, T.H. 1978. Composition and development of the continental tectosphere. *Nature*, **274**, 544–548.

LAMBERT, I.B. & HEIER, K.S. 1968. Geochemical investigations of deep-seated rocks in the Australian Shield. *Lithos*, **1**, 30–53.

MCCULLOCH, M.T. & WASSERBURG, G.J. 1978. Sm-Nd and Rb-Sr chronology of continental crust formation. *Science*, **200**, 1003–1011.

PERCIVAL, J.A. & CARD, K.D. 1983. Archean crust as revealed in the Kapuskasing uplift Superior province, Canada. *Geology*, **11**, 323–326.

ROGERS, J.J.W. 1978. Inferred composition of early Archean crust and variation in crustal composition through time. *In*: WINDLEY, B.F. & NAQVI, S.M. (eds) *Archean Geochemistry*, Elsevier. 35–40.

RONOV, A.B. & YAROSHEVSKY, A.A. 1969. Chemical composition of the earth's crust. *In*: HART, P.J. (ed.) *The Earth's Crust and Upper Mantle*, Geophysical Monograph **13**, 1–735, Amer. Geophys. Union, 37–57.

SHAW, D.M. 1967. U, Th and K in the Canadian Precambrian shield and possible mantle compositions. *Geochim. Cosmochim. Acta*, **31**, 1111–1113.

—— 1968. Radioactive elements in the Canadian Precambrian shield and the interior of the earth. *In*: AHRENS, L.H. (ed.) *Origin and Distribution of the Elements* 855–870. Pergamon Press.

—— 1980. Evolutionary tectonics of the Earth in the light of early crustal structure. *In*: STRANGWAY, D.W. (ed.) *The Continental Crust and its Mineral Deposits*, Geol. Assoc. Can. Spec. Pap. **20**, 65–73.

—— 1981. The earth's evolution: an interdisciplinary synthesis. *Trans. Roy. Soc. Canada*, Series IV, Vol. XIX, 233–254.

——, REILLY, G.A., MUYSSON, J.R., PATTENDEN, G.E. & CAMPBELL, F.E. 1967. The chemical composition of the Canadian Precambrian shield. *Can. J. Earth Sci.*, **4**, 829–854.

——, DOSTAL, J. & KEAYS, R.R. 1976. Additional estimates of continental surface Precambrian shield composition in Canada. *Geochim. Cosmochim. Acta*, **40**, 73–84.

TAYLOR, S.R. & MCLENNAN, S.M. 1981. The composition and evolution of the continental crust. REE evidence from sedimentary rocks. *Phil. Trans. Roy. London A.*, **301**, 381–399.

WEAVER, B.L. & TARNEY, J. 1984. Empirical approach to estimating the composition of the continental crust. *Nature*, **310**, 575–577.

D.M. SHAW, J.J. CRAMER, M.D. HIGGINS & M.G. TRUSCOTT, Department of Geology, McMaster University, Hamilton, Ontario, Canada, L85 4MI.

High grade metamorphism in the granulite belt of Finnish Lapland

M. Raith & P. Raase

SUMMARY: The granulite complex of Finnish Lapland is located in the prominent collisional suture which separates the Archaean crustal blocks of the South Lapland and Inari cratons. The protoliths of the granulite complex represented a geosynclinal sequence of Proterozoic flysch-type sediments with intercalations of pyroclastic, effusive and intrusive rocks of tholeiitic to rhyolitic composition and calc-alkaline affinity. Deformation and progressive metamorphism set in about 2.2 b.y. ago by eastward subduction of the volcano-sedimentary sequence and culminated in the granulite facies event during the collision of the Archaean crustal blocks. The granulite complex thereby was thrust westward into higher crustal levels onto the cooler basement unit of the South Lapland craton and was overthrust by the Inari craton. High-grade metamorphic equilibration of the rocks continued until about 1.9 b.y. during the isostatic uplift of the thickened crust.

The strongly sheared lower part of the granulite complex consists of sillimanite-garnet granulites, garnet-biotite gneisses, and hypersthene-plagioclase rocks. Its intensely migmatized middle and upper parts are dominated by medium to coarse-grained sillimanite-garnet-biotite $\pm$ cordierite metatexites and diatexites. An up-dated evaluation of $P-T-X(CO_2,H_2O)$ conditions indicates that the post-kinematic assemblages in the granulite complex equilibrated at 830 to 760°C and 7.2 to 6.2 kbars, in the presence of fluids rich in carbon dioxide ($X_{CO_2} = 0.9 - 0.7$). Increased activity of water in the middle and upper parts of the complex resulted in widespread anatexis and advanced re-equilibration of the rocks. The high heat flow in the granulite complex is explained by intense upward motion of hot fluids in the strongly sheared and foliated granulites. The rock series of the adjacent South Lapland and Inari cratons equilibrated at conditions of the upper amphibolite facies (720–750°C and high activities of water). The changes in the mineral assemblages and $P-T-X(CO_2,H_2O)$ data across the units indicate that heat transfer and flux of CO_2-rich fluids from the upthrusted hot granulite complex into the cooler basement unit was essentially confined to the proximity of the narrow thrust zone. The cooling path of the granulite complex passed near the Al_2SiO_5 triple point ($\sim 500°C/3.5$ kbars) thus demonstrating the slow thermal relaxation of this rock unit.

Introduction

The granulite complex of Finnish Lapland represents one of the classic exposures of former lower continental crust metamorphosed to the granulite grade (Eskola 1952). The geology and structure of the belt, its lithology and metamorphic history have been studied in great detail by Meriläinen (1976), Hörmann et al. (1980), Barbey et al. (1980, 1982), Barbey & Cuney (1982) and Bernard-Griffiths et al. (1983). The results of these investigations and studies of the adjacent rock units have allowed the establishment of a refined geodynamic model of the crustal evolution of this part of the Baltic Shield (Barbey et al. 1984). This paper essentially summarizes and up-dates research on the high-grade metamorphic development of the granulite complex of Finnish Lapland carried out at the Mineralogical Institute of the University of Kiel (cf. Hörmann et al. 1980; Raith et al. 1982).

Geology

The granulites of Finnish Lapland form a prominent thrust belt of Proterozoic supracrustal rocks compressed between two blocks of Archaean continental crust, the Inari-Kola craton and the South Lapland-Karelia craton which both represent typical granite-gneiss-greenstone domains (Fig. 1). This thrust belt possibly extends into the White Sea granulite complex in the SW part of the Kola Peninsula (Kratz 1978).

In a recent synthesis of the geological evolution Barbey et al. (1984) interpret the granulite belt as part of a collisional orogenic belt, the Belomorian fold belt, and suggest the following evolutionary stages: In the first tensional stage (2.4–2.2 b.y.) stretching, rifting and spreading of the Archaean continental crust gave rise to a geosyncline floored by oceanic crust and bordered by passive continental margins. A sequence of terrigenous sediments and volcanic rocks was deposited

From: DAWSON, J.B., CARSWELL, D.A., HALL, J. & WEDEPOHL, K.H. (eds) 1986, *The Nature of the Lower Continental Crust*, Geological Society Special Publication No. 24, pp. 283–295.

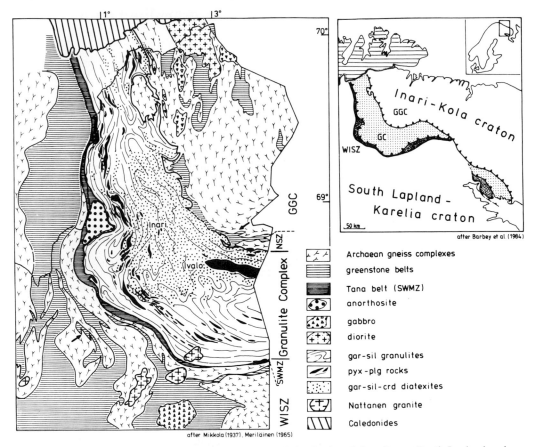

FIG. 1. Geological map of the granulite complex of Finnish Lapland and the adjacent South Lapland and Inari cratons. GGC—granite gneiss complex; NSZ—Nellimö schist zone; SWMZ—SW marginal zone (Tana belt); WISZ—West Inari schist zone. The position of the investigated traverse given in Figs 4 to 8 is indicated by arrows.

forming the parental rocks of the granulite complex. The subsequent compressional stage resulted in the closure of the geosyncline, subduction of the oceanic crust and continental collision. The volcano-sedimentary sequence was thereby subjected to intense deformation and progressive metamorphism which culminated in the granulite facies event about 1.9 b.y. ago.

Lithology and metamorphic evolution

The granulite complex, at its northeastern border, is welded by anatexis to the Inari craton and, at its southwestern margin, is thrust onto the South Lapland craton (Hörmann et al. 1980). Here, a thin unit is distinguished by its different lithology and metamorphic grade (SW marginal zone, Meriläinen 1976) and is interpreted as a separate thrust sheet (Tana belt, Barbey et al. 1980).

The **SW marginal zone** consists of an intensely sheared layered sequence of amphibolites, hornblende-clinopyroxene gneisses and quartz-feldspar gneisses. Geochemical data indicate that this series of rocks originally formed a differentiated volcanic suite ranging from basaltic to andesitic and plagidacitic in composition (Hörmann et al. 1980; Barbey et al., 1984). The series also contains few lenses and bodies of metaperidotites, pyroxenites, hornblendites, troctolitic gabbros and anorthosites. The textural relations and mineral assemblages of the metabasites give evidence of two distinct stages (MI and MII) of high-grade metamorphism. First, garnet-clinopyroxene

TABLE 1. *Major and trace element compositions of the main rock types of the granulite complex (data from Hörmann et al. 1980)*

	(1)	(2)	(3)	(4)	(5)	(6)	(7)	(8)	(9)	(10)	(11)
SiO_2	48.45	52.19	59.23	61.43	66.37	69.98	75.50	74.17	62.25	67.59	82.09
Al_2O_3	15.61	20.00	19.29	19.86	16.71	16.49	16.02	13.62	16.62	19.74	9.06
TiO_2	1.19	0.46	0.39	0.78	0.57	0.53	0.03	0.06	0.56	0.07	0.31
Fe_2O_3	3.48	2.69	0.64	1.32	0.86	1.07	0.23	0.14	2.33	0.99	1.12
FeO	8.71	7.04	4.90	5.71	3.49	4.33	0.02	0.70	7.33	1.95	1.80
MnO	0.23	0.33	0.08	0.04	0.06	0.12	0.005	0.01	0.14	0.01	0.04
MgO	7.84	5.23	3.92	2.93	1.76	2.19	0.08	0.33	3.38	4.05	0.97
CaO	11.28	7.49	5.95	0.37	3.90	1.77	1.74	0.93	1.65	0.45	1.25
Na_2O	1.85	3.26	3.36	0.99	3.23	2.35	3.16	2.19	1.53	0.85	2.10
K_2O	0.43	0.51	0.94	4.68	1.01	0.78	3.00	6.63	3.58	2.75	0.73
P_2O_5	0.10	0.05	0.13	0.03	0.23	0.06	0.08	0.08	0.03	0.03	0.05
H_2O+CO_2	0.53	0.04	0.62	1.28	1.60	0.67	0.56	0.80	0.70	1.68	0.57
Li	8.7	4.8	10.7	7.7	7.2	11.0	6.2	4.3	7.7	11.7	3.1
Rb	6.5	10.7	42.5	143	32.3	98	58.7	149	88	55	18
Ba	142	293	475	1309	265	2004	1540	1750	1396	895	250
Sr	171	161	313	106	358	197	95	267	137	166	132
Ni	76	55	42	71	14	25	1	2	54	21	14
Cr	203	118	109	127	35	86	1	7	121	10	58
V	291	145	68	83	73	57	48	4	103	60	66
Th	1.3	1.4	1.7	12.0	3.6	13.0	7.5	5.2	3.9	15.1	7.0
U	0.9	0.8	0.8	0.3	1.1	1.1	1.3	0.8	0.8	1.8	1.3
La			19.8	56.9			3.7	12.9	38.6	62.3	35.9
Yb			0.09	3.22			0.75	0.56	7.75	2.33	2.01
Y			1	51.0			3.7	6.1		24.3	17.9

(1) Hypersthene-clinopyroxene-plagioclase rock; Kuttura (70I) plag.opx.cpx.hbl (ilm,mt,pyr,ap).
(2) Hypersthene-plagioclase rock; Kultala (97II) plag,opx,bio (ilm-hem,mt,pyr,ap).
(3) Garnet-hypersthene-plagioclase rock; Kuttura (72I) plag,gar,opx,bio,qtz (ilm,mt,pyr,pyrr,ap).
(4) Sillimanite-garnet granulite; Kuttura (62I) qtz,kf,gar,plag,bio,sill (ore,rut,zirc).
(5) Sillimanite-garnet granulite; NE' Kuttura (J5II) qtz,plag,kf,gar,bio,sill (ore,rut,ap).
(6) Garnet granulite; W' Kultala (93I) qtz,kf,plag,gar,bio (ore,ap).
(7) Sillimanite-garnet granulite; Sotajoki (154I) kf,qtz,plag,sill,gar.
(8) Sillimanite-garnet granulite; NE' Kuttura (76I) kf,qtz,plag,gar,bio,sill (zirc).
(9) Garnet-cordierite metatexite; W' Kultala (89I) kf,qtz,cord,plag,gar,bio,sill (ilm,pyr,pyrr).
(10) Garnet-cordierite diatexite; Ivalo (158I) kf,qtz,cord,gar,sill,bio,plag (ilm,pyr,pyrr,rut,zirc).
(11) Garnet-feldspar quartzite; Mukkapalo (65III) qtz,kf,plag,gar,bio (rut,sill).

hypersthene–plagioclase rocks

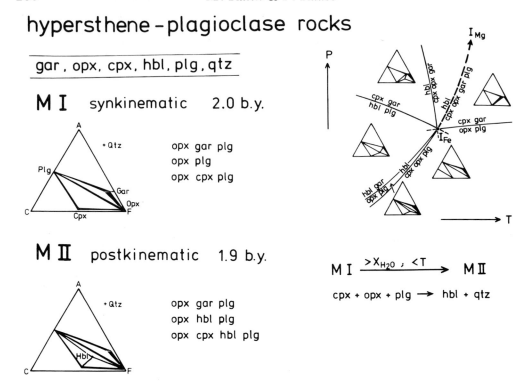

FIG. 2. Schematic P-T diagram showing the phase relations for amphibolites and mafic granulites in the model system CaO-FeO-Al$_2$O$_3$ with excess quartz and at constant fluid pressure. The high-grade metamorphic development of the assemblages in the hypersthene-plagioclase rocks of the granulite complex is illustrated by ACF diagrams.

assemblages transitional to the eclogite facies formed at high pressures and moderate temperatures. Subsequently, due to decreasing pressure, typical amphibolite facies assemblages developed through the breakdown of garnet+clinopyroxene to hornblende and plagioclase. Increased temperatures near the boundary of the upthrust hot granulite complex, however, gave rise to garnet-two pyroxene-plagioclase-bearing assemblages of the hornblende granulite facies.

The **granulite complex** represents an intensely folded and sheared sequence of clastic sediments and differentiated igneous rocks metamorphosed at conditions of the granulite facies.

The *lower part* of the complex is built up of strongly foliated felsic sillimanite-garnet granulites and garnet-biotite flaser gneisses with bands and lenses of basic to intermediate hypersthene-plagioclase rocks. According to the geochemical data (Hörmann *et al.* 1980; Barbey *et al.* 1984) (Table 1) these rocks represent a sequence of metamorphosed greywackes and shales with abundant intercalations of pyroclastic, effusive

and intrusive rocks of basaltic to rhyolitic composition and of calc-alkaline affinity.

This part of the belt has not been affected by anatexis during the progressive metamorphic evolution. Therefore, the structures and assemblages of the synkinematic stage MI of granulite facies metamorphism are largely preserved.

In the *middle and upper parts* of the granulite complex, medium to coarse-grained sillimanite-garnet-biotite metatexites and diatexites with or without cordierite are the predominant rock types (Table 1). These rocks obviously represent a thick metasedimentary sequence of psammitic and pyroclastic greywackes and shales with only minor intercalations of igneous rocks. Like the lower part of the belt, the metasedimentary sequence was strongly folded and sheared during the peak stage of metamorphism. But subsequently widespread anatexis occurred and largely destroyed the typical granulite structures.

The metamorphic development of the granulite complex has been described in detail by Hörmann *et al.* (1980) and only the main results are

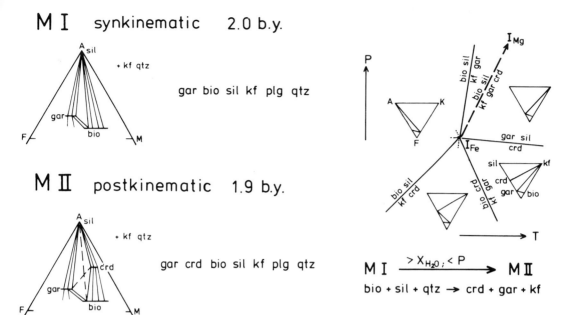

FIG. 3. Schematic P-T diagram showing the phase relations for high-grade gneisses and granulites in the model system K_2O-FeO-Al_2O_3 with excess quartz and K-feldspar and at constant fluid pressure. The high-grade metamorphic development of the assemblages in the granulites and diatexites is illustrated by AFM diagrams.

discussed in the following sections. The mineral assemblages can be assigned to two chemically distinct groups of rocks, i.e. basic to intermediate pyroxene-plagioclase rocks and pelitic to psammitic granulites and diatexites.

Pyroxene-plagioclase rocks

The phase relations of the pyroxene-plagioclase rocks involve the minerals orthopyroxene, clinopyroxene, garnet, amphibole, plagioclase and quartz, and can be portrayed schematically in the model system CaO-FeO-Al_2O_3-SiO_2-H_2O with SiO_2 as an excess component. At constant fluid pressure the compatibility relations are defined by five univariant reactions which are located around an invariant point (Fig. 2). In the extended Fe-Mg system which may serve as a first approximation to the natural rocks, the formerly invariant assemblage hbl-plg-opx-cpx-gar becomes stable along a univariant reaction curve and the univariant reactions become continuous. Decreasing water fugacity will shift the reactions to lower temperatures (cf. Wells 1979).

The synkinematic high-grade stage MI was characterized by the development of dry assemblages of the medium pressure hypersthene granulite facies, i.e. plagioclase + orthopyroxene; plagioclase + orthopyroxene + clinopyroxene; plagioclase + orthopyroxene + garnet. Garnet-clinopyroxene assemblages which would indicate a higher pressure regime were never formed in the granulite complex. They occurred, however, in the Fe-rich metabasites of the SW-marginal zone and the adjacent part of the South Lapland craton.

During the postkinematic stage MII, after the granulite complex was thrusted upward into higher crustal levels, amphibole was formed in the pyroxene-plagioclase rocks at the expense of pyroxenes and plagioclase (Fig. 2). This reaction was initiated essentially by a decrease in temperature and/or an increase of water fugacity. It was during this metamorphic stage that the garnet-clinopyroxene assemblages in the adjacent units of the South Lapland craton were largely replaced by the assemblage hornblende + plagioclase ± quartz.

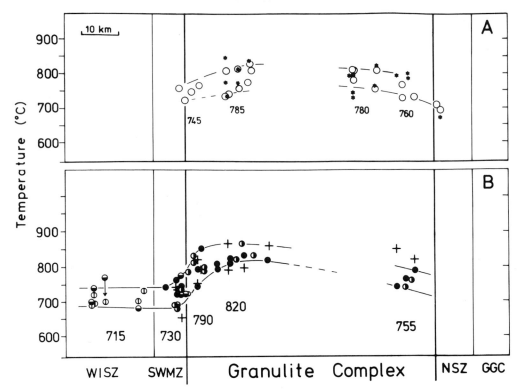

FIG. 4. Results of geothermometry for amphibolites and pyroxene-plagioclase rocks (B) and garnet-sillimanite granulites and cordierite diatexites (A) along the traverse shown in Fig. 1. Fe-Mg exchange equilibria: * gar-crd Thompson (1976), ◒ gar-cpx Ellis & Green (1979), ● gar-opx Lal & Raith (1984), Φ gar-hbl Graham & Powell (1984); Two pyroxene thermometry: ◑ Wells (1977) data corrected by −60°C; + gar-opx thermometry Harley & Green (1982); ○ gar-sill-plg-qtz thermometry (Newton & Haselton 1981). In the model calculations the pressure data shown in Fig. 6B were taken into account. The estimates are based on the core compositions of the coexisting minerals.

Granulites and diatexites

The phase relations in these rocks involve the minerals garnet, biotite, cordierite, sillimanite, feldspars and quartz. They can be illustrated schematically in the model system K_2O-FeO-(MgO)-Al_2O_3-SiO_2-H_2O. In the presence of quartz and K-feldspar as excess phases the phase relations are defined by four univariant reactions which are located around an invariant point (Fig. 3). In the extended Fe-Mg system the invariant six phase assemblage gar-bio-cord-sill-kf-qtz will be stable along a univariant reaction curve and the univariant reactions will become continuous (cf. Vielzeuf & Boivin 1984).

In the synkinematic progressive stage of granulite facies metamorphism (MI) garnet-sillimanite-biotite assemblages were formed at low fugacities of water. The quartz-feldspar rocks were not affected by anatexis. During the subsequent post-kinematic stage MII, however, increased fugacities of water and/or a decrease of pressure initiated widespread anatexis and formation of garnet-sillimanite-biotite-cordierite assemblages in the middle and upper part of the granulite complex (Fig. 3).

Evaluation of P-T-X_{fl} conditions of high-grade metamorphism

The metamorphic conditions in the granulite complex and its adjacent rock units, since the pioneering petrographic work of Eskola (1952) were studied first by Hörmann et al. (1980). Since then development of well calibrated petrological thermometers and barometers applicable to the granulite facies rocks has made considerable progress (cf. Newton & Perkins 1982; Bohlen et al. 1983; Harley 1984). Thereafter, an up-dated reevaluation of the P-T-X_{fl} conditions in the

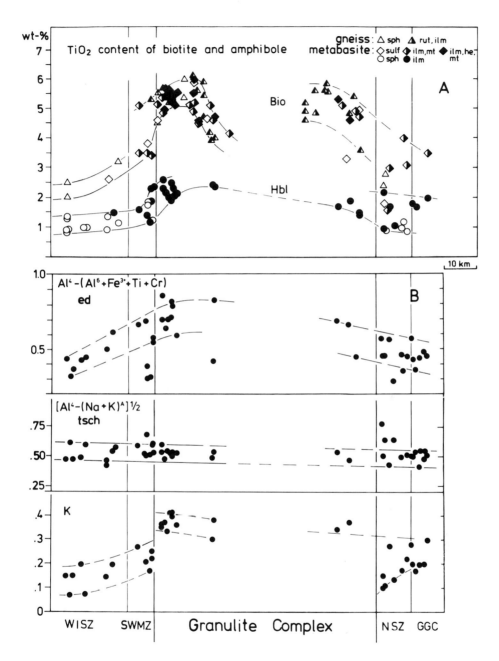

FIG. 5. Variation of compositional parameters of biotite and amphibole along the investigated traverse. (A) Variation of TiO_2 content of amphiboles and biotites. Sph—sphene, ilm—ilmenite, rut—rutile, mt—magnetite, he—hematite, sulf—pyrite, pyrrhotite; (B) Variation of edenite and tschermakite components and potassium content of amphiboles. The edenite component is expressed as $Al^4 - (Al^6 + Fe^{3+} + Cr^{3+} + Ti^{4+})$ neglecting the presence of a minor glaucophane component; the tschermakite component corresponds to $(Al^4 - (Na + K)^A)1/2$.

290 M. *Raith & P. Raase*

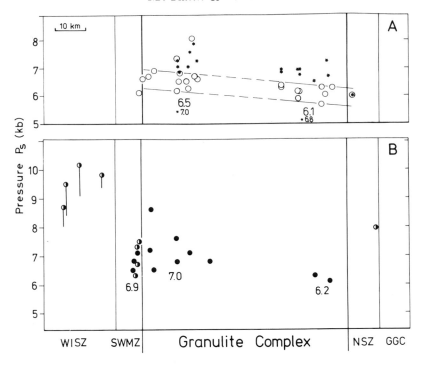

FIG. 6. Results of geobarometry for amphibolites and pyroxene-plagioclase rocks (B) and garnet-sillimanite granulites and cordierite diatexites (A) along the traverse shown in Fig. 1. ◑ gar-cpx-plg-qtz barometry Newton & Perkins (1982); ● gar-opx-plg-qtz barometry Newton & Perkins (1982); ○ gar-sill-plg-qtz barometry Perchuk et al. (1981); * crd-gar-sil-qtz barometry Mirwald (pers. comm.). In the model calculations the temperature data shown in Fig. 4 were taken into account. The estimates are based on the core compositions of the coexisting minerals.

granulite complex was undertaken on the basis of the mineral data published by Hörmann et al. (1980) and Klatt (1980) as well as an extensive set of unpublished own microprobe analyses.

In this context we are faced with the problem that the granulite complex has been subjected to high-grade polyphase metamorphism. There is evidence, however, that the rock, re-equilibrated during the high-grade postkinematic stage MII of granulite facies metamorphism. The assemblages are in complete textural equilibrium and compositional zoning in the minerals is absent or weak, even in the refractory phases such as garnet. The compositions of the coexisting phases will therefore essentially record the P-T-X_{fl} conditions of this stage. The conditions of the previous synkinematic stage MI can only be inferred from reconstructed mineral assemblages and the pertinent experimental stability data. Mineral assemblages which formed during the late retrogressive stages constrain the uplift and cooling path of the granulite complex.

Temperature

The temperature data calculated for the assemblages of the amphibolites and pyroxene-plagioclase rocks with a variety of thermometers are in excellent agreement (Fig. 4B). Estimates obtained from the garnet-sillimanite granulites and the cordierite-bearing diatexites are similar in the NE-part but slightly lower in the SW-part of the granulite complex (Fig. 4A). It is assumed that the temperature estimates which are based exclusively on the core compositions of coexisting phases are close to the thermal peak of the metamorphic stage MII.

Within the granulite complex an increase in temperature towards its base from about 755°C to 820°C is indicated by the model data for the metabasite assemblages (Fig. 4B). A marked decrease of temperature occurs towards the SW marginal zone and the adjacent South Lapland craton where lower and rather uniform temperature values of about 715°C are recorded. This

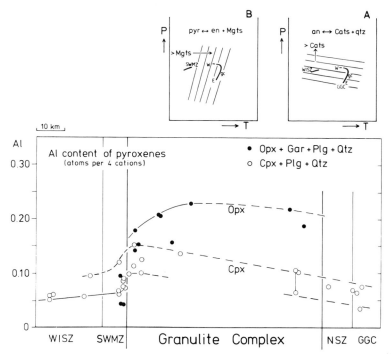

FIG. 7. Variation of the Al contents of pyroxenes along the investigated traverse. The P-T diagrams A and B show schematically the pressure and temperature dependence of Al contents of clinopyroxene in the assemblage with plagioclase and quartz and of orthopyroxene coexisting with garnet. The Al contents in both phases are controlled by the equilibria anorthite $\rightleftharpoons$ Ca tschermakite + quartz and pyrope $\rightleftharpoons$ enstatite + Mg tschermakite respectively.

points to a thermal adjustment between the upthrust hot granulite complex and the cooler basement complex in the west. In the narrow thermal transition zone the strongly foliated rocks of the granulite complex cooled down to about 750°C and those of the SW marginal zone heated up by about 20°C. This regional thermal pattern is also evident from the compositional parameters of minerals which have not yet been calibrated as thermometers (Fig. 5).

The regional variation in the TiO_2 content of the amphiboles coexisting with sphene or ilmenite reproduces perfectly the thermal trend which was derived from geothermometry (Fig. 5A). The positive correlation of the TiO_2 content with temperature is in agreement with the findings from other prograde metamorphic terrains (cf. Raase 1974). A positive correlation with metamorphic grade is also shown by the edenite component and potassium content of amphibole (Fig. 5B). That these parameters are essentially temperature controlled has been confirmed by experimental studies (Spear 1981).

The TiO_2 content of biotites from metabasites,

gneisses, granulites and diatexites which represent a wide range of bulk compositions, irrespective of the changes in the opaque mineral assemblage, show a similar temperature-controlled regional pattern (Fig. 5A). In some parts of the granulite complex, however, systematically lower TiO_2 values indicate that re-equilibration of biotite in contrast to amphibole obviously continued to lower grade. It should be emphasized that the dramatic increase of TiO_2 in biotite and amphibole when passing into the granulite complex may not only be attributed to the temperature increase but could in part be due to the pronounced decrease in the water activity, because of the higher thermal stability of the TiO_2-rich biotite and amphibole solid solutions.

Pressure

The pressure data derived from the mafic granulites could indicate a slight regional gradient from about 7 kbars in the SW marginal zone and the lower part of the granulite complex to about 6.2 kbars in the uppermost part of the complex (Fig.

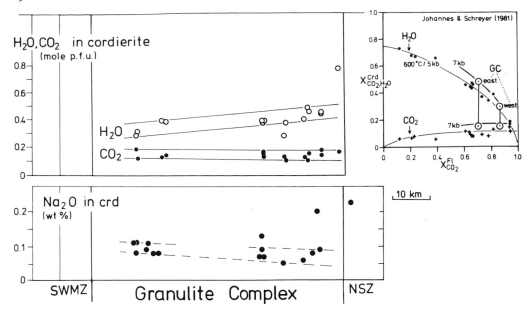

FIG. 8. Variation of H_2O, CO_2 and Na_2O content of cordierite along the investigated traverse. The $X^{crd}_{H_2O,CO_2}$ versus $X^{fl}_{CO_2}$ diagram shows the results of an experimental study of the CO_2–H_2O partitioning between cordierite and fluids of varying composition by Johannes and Schreyer (1981). The partitioning data at 600°C and 5 kbars are shown by dots (H_2O) and crosses (CO_2), their displacement with increase in pressure (7 kbars) is also indicated. According to the experimental data the metamorphic fluids in the granulite complex were characterized by a continuous decrease in carbon dioxide towards the eastern and upper part of the complex.

6B). Higher pressure values of 8 to 10 kbars in the South Lapland and Inari cratons probably are not realistic because garnet and clinopyroxene are no longer in textural equilibrium. Moreover, these high pressure values are not confirmed by the regional variation of the tschermakite component of clinopyroxene in the assemblage with plagioclase and quartz (Fig. 7). The Al content of the clinopyroxenes increases continuously from the Inari craton (GGC and NSZ) towards the base of the granulite complex, but decreases rather abruptly when passing into the SW marginal zone and the South Lapland craton. The tschermakite components of clinopyroxene in the assemblage with plagioclase and quartz is controlled by the strongly pressure dependent equilibrium anorthite $\rightleftharpoons$ Ca tschermakite + quartz (cf. Ellis 1980) (Fig. 7A). The data indicate that pressure and also temperature increased steadily towards the base of the granulite complex but that temperature decreased suddenly at the transition to the SW marginal zone, whereas pressures probably did not change much. This interpretation is corroborated by the regional variation of the Al content of orthopyroxenes coexisting with

garnet which is controlled by the temperature and pressure dependent equilibrium pyrope $\rightleftharpoons$ enstatite + Mg tschermakite (cf. Harley 1984) (Fig. 7B). From the compositional trend of the Al content of clinopyroxenes along the investigated traverse it is also suggested that the Inari craton and the granulite complex were thrusted as welded units onto the cooler basement units in the west. The tschermakite component of amphibole, on the other hand, is nearly constant throughout the entire investigated profile which indicates only minor regional variation in pressure (cf. Raase 1974). This is in accordance with the observation that growth of amphibole occurred after the main phases of deformation and thrusting when the units were approximately at the same crustal level.

Pressure estimates for the granulite complex were also derived for the garnet-sillimanite-plagioclase-quartz assemblage of the granulites and cordierite diatexites (Fig. 6). These data show a gradual decrease in pressure from about 6.5 kbars at the base of the granulite complex to about 6 kbars at its eastern margin. Somewhat higher pressure estimates which compare well with the metabasite estimates were obtained from a recent calibration of the garnet-cordierite-sillimanite-quartz geobarometer (Mirwald, pers. comm.).

The regional variation of the P-T data in the granulite complex could also indicate that the re-equilibration of the rocks during retrogressive metamorphism was more advanced in the eastern migmatized part of the complex. This might also be supported by the results of two feldspar thermometer and the structural state of K-feldspar (Hörmann *et al.* 1980). The Al-Si order of K-feldspar is lowest in the 'dry' lower part of the granulite complex where no anatexis and retrogression has affected the granulites, but highest in the 'wet' amphibolite facies terrains of the adjacent Archaean craton. These observations underline the role of the fluid phase composition in the re-equilibration of the mineral assemblages.

Composition of the intergranular fluid phase

The intergranular fluid phase in the granulite complex was extremely CO_2-rich during the high-grade stages of metamorphism. This is substantiated by the study of fluid inclusions in quartz (Klatt 1980; Klatt & Schoch 1984). High density CO_2 inclusions (0.8–0.9 gcm^{-3}) with 4–10% methane equivalent predominate in the western part. Towards the eastern part of the complex brine rich aqueous inclusions of high density (5–25% NaCl equivalent; 0.95 gcm^{-3}) become more and more important. In the adjacent Archaean cratons only briny aqueous inclusions occur, thus demonstrating the hydrous environment of these amphibolite facies terrains and the absence of any significant fluid exchange across the boundaries of the rock units.

Further evidence of the CO_2-rich fluid regime during granulite facies metamorphism comes from the fluid phase trapped in cordierite (Fig. 8). The H_2O contents of cordierite when applied to the experimental data of Johannes and Schreyer (1981) indicate a regional gradient in fluid composition, X_{CO_2} ranging from 0.9 in the western part of the granulite complex to 0.7 in its eastern migmatized part. Since anatexis in the granulite complex set in only in connection with the uplift and high-grade re-equilibration of the rocks (stage MII), a CO_2-rich environment must have prevailed already during the preceding progressive stages of high-grade metamorphism. The source of the carbonic fluid still has to be ascertained, the occurrence of graphite in the metasediments, however, precludes the necessity of an external source. The P-T data discussed in the previous section indicate a high heat flow in the granulite complex as compared to the adjacent Archaean cratons. This could be due to an intense upward motion of hot fluids in the highly sheared and foliated rocks of the granulite complex.

Conclusions

The evolution of the granulite belt and the adjacent rock units in the eastern part of the Baltic Shield is best explained by plate tectonic processes (Hörmann *et al.* 1980; Barbey *et al.* 1980; Raith *et al.* 1982; Barbey *et al.* 1984). In the geodynamic model favoured by Barbey *et al.* (1984) stretching and rifting of an extended Archaean sialic crust (Inari and South Lapland cratons) led to the formation of a geosynclinal basin floored by oceanic crust and filled with a series of volcanic rocks and flysch type sediments (Tana belt and granulite complex). The deformation and progressive metamorphism (MI) of these supracrustal series set in about 2.2 b.y. ago by eastward subduction and culminated with the collision of the Archaean crustal blocks. During collision the granulite complex was thrust westward into higher crustal levels onto the cooler basement unit of the South Lapland craton and was overthrust by the Inari craton.

The present-day metamorphic features of the rock units essentially reflect the P-T-X_{fl} conditions of the late collisional stage (MI) and of the subsequent high-grade postkinematic stage (MII) which marks the onset of isostatic uplift. The development of mineral assemblages in the granulite complex and petrological considerations indicate that peak conditions of granulite facies metamorphism (MI) did not exceed 850°C and 8 kbars. The intergranular fluid phase was extremely rich in CO_2 ($X_{CO_2} \approx 0.9$). The temperature and pressure model data decrease slightly from $\sim 820°$C/7 kbars in the lower part of the complex to $\sim 755°$C/6.2 kbars in its upper (eastern) part. This variation could reflect the vertical P-T gradient during the peak of metamorphism MII, but it most likely resulted from the more advanced re-equilibration of the rocks in the medium and upper parts of the complex in connection with increasing activity of water ($X_{H_2O} \approx 0.3$) and widespread anatexis during the metamorphic stage MII. The continuous changes in the composition of minerals (clinopyroxene, amphibole) across the boundary between the granulite complex and the Inari craton indicates a continuous metamorphic gradient which suggests that the granulite belt and the Inari craton were welded together during the metamorphic stage MII.

From the results of geothermometry it is concluded that the Tana belt did not form a part of the granulite belt as stated by Barbey *et al.* (1980, 1984). Temperatures of metamorphism (MI) were significantly lower than in the granulite complex ($\sim 720°C$) and the pressures were higher as indicated by the occurrence of relic garnet-clinopyroxene assemblages in the amphibolites. The metamorphic development of the Tana belt series is thus comparable to that of the adjacent South Lapland craton.

The temperature data further show that heat transfer and flux of CO_2-rich fluids from the overthrust hot granulite complex into the cooler basement units was restricted to the uppermost part of the Tana belt series and gave rise to mineral assemblages of the hornblende granulite facies.

Subsequent to deformation and thrusting, with the beginning of isostatic uplift about 1.9 b.y. ago, the entire terrain was subjected to thorough re-equilibration at static conditions. Due to the decrease of pressure, hydrous mineral assemblages of the high-grade amphibolite facies ($\sim 720°C$) were formed in the Tana belt and the adjacent South Lapland craton, whereas in the granulite complex slightly hydrated assemblages of the hornblende granulite facies developed (820–755°C/7–6 kbars; $X_{CO_2} \approx 0.8$–0.6). The decrease of pressure and gradual increase of water

activity towards the upper part of the granulite complex gave rise to widespread anatexis of the granulites and the formation of cordierite-bearing diatexites. From the regional variation of the tschermakite component in amphibole it may be concluded that the postkinematic re-equilibration of the rocks in the granulite complex and the adjacent units took place at rather uniform pressures. The temperature estimates derived for the re-equilibrated assemblages as well as compositional data of minerals (TiO_2 content of amphibole), on the other hand, demonstrate that the pre-existing regional thermal gradient was still effective. An isotope and fission track study of mineral cooling ages is needed to establish the uplift and cooling history of the granulite complex. Late formation of andalusite and kyanite in the cordierite-bearing anatexites (Hörmann *et al.* 1980) suggests that the P-T path of retrogressive metamorphism passed near the Al_2SiO_5 triple point ($\sim 500°C/\sim 3$ kbars) and closely followed an Archaean continental geotherm (cf. Condie 1981).

ACKNOWLEDGEMENTS: Reviews of the manuscript by R.C. Newton and H.A. Seck are gratefully acknowledged.

References

BARBEY, P., CAPDEVILA, R., HAMEURT, J. 1982. Major and transition trace element abundances in the khondalite suite of the granulite belt of Lapland (Fennoscandia): evidence for an early Proterozoic flysch-belt. *Precambrian Res.*, **16**, 273–290.

——, CONVERT, J., MARTIN, H., CAPDEVILA, R. & HAMEURT, J. 1980. Relationships between granite-gneiss terrains, greenstone belts and granulite belts in the Archean crust of Lapland (Fennoscandia). *Geol. Rundschau*, **69**, 648–658.

——, CONVERT, J., MOREAU, B., CAPDEVILA, R. & HAMEURT, J. 1984. Petrogenesis and evolution of an early Proterozoic collisional orogenic belt: the granulite belt of Lapland and the Belomorides (Fennoscandia). *Geol. Surv. Finland*, **56**.

—— & CUNEY, M. 1982. K, Sr, Ba, U and Th geochemistry of the Lapland granulites. LILE fractionation controlling factors. *Contrib. Mineral. Petrol.*, **81**, 304–316.

BERNARD-GRIFFITHS, J., PEUCAT, J.J., POSTAIRE, B., VIDAL, P., CONVERT, J. & MOREAU, B. 1983. Isotopic data (U-Pb, Rb-Sr, Pb-Pb and Sm-Nd) of mafic granulites from Finnish Lapland. *Precambrian Research*, **23**, 325–348.

BOHLEN, S.R., WALL, V.J. & BOETTCHER, A.L. 1983. Experimental investigation and application of garnet granulite equilibria. *Contrib. Mineral. Petrol.*, **83**, 52–61.

BROWN, W.L. & PARSONS, I. 1981. Towards a more practical two-feldspar geothermometer. *Contrib. Mineral. Petrol.*, **76**, 369–377.

CONDIE, K.C. 1981. *Archean Greenstone Belts*. Developments in Precambrian Geology, 3, Elsevier, Amsterdam.

ELLIS, D.J. 1980. Osumilite-sapphirine-quartz granulites from Enderby Land, Antarctica: P-T conditions of metamorphism, implications for garnet-cordierite equilibria and the evolution of the deep crust. *Contrib. Mineral. Petrol.*, **74**, 201–210.

—— & GREEN, D.H. 1979. An experimental study of the effect of Ca upon garnet-clinopyroxene Fe-Mg exchange equilibria. *Contrib. Mineral. Petrol.*, **71**, 13–22.

ESKOLA, P. 1952. On the granulites of Lapland. *Amer. J. Sci. Bowen Volume*, Part **1**, 133–171.

GRAHAM, C.M. & POWELL, R. 1984. A garnet-hornblende geothermometer: calibration, testing, and application to the Pelona Schist, Southern California. *J. metamorphic Geol.* **2**, 13–21.

HARLEY, S.L. 1984. Comparison of the garnet-orthopyroxene geobarometer with recent experimental studies and applications to natural assemblages. *J. Petrol.*, **25**, 697–712.

—— & GREEN, D.H. 1982. Garnet-orthopyroxene barometry for granulites and garnet peridotites. *Nature*, **300**, 697–700.

HÖRMANN, P.K., RAITH, M., RAASE, P., ACKERMAND, D. & SEIFERT, F. 1980. The granulite complex of Finnish Lapland: petrology and metamorphic conditions in the Ivalojoki-Inarijärvi area. *Bull. Geol. Surv. Finland*, **308**, 100 pp.

JOHANNES, W. & SCHREYER, W. 1981. Experimental introduction of CO_2 and H_2O into Mg-cordierite. *Amer. J. Sci.*, **281**, 299–317.

KLATT, E. 1980. *Seriengliederung, Mineralfazies und Zusammensetzung der Flüssigkeitseinschlüsse in den präkambrischen Gesteinsserien Nordlapplands.* Doktor. Dissert., Universität Kiel, 125 pp.

—— & SCHOCH, A.E. 1984. Comparison of fluid inclusion characteristics for high-grade metamorphic rocks from Finnish Lapland with medium grade rocks from north-western Cape Province, South Africa. *27th Intern. Geol. Congr. Moscow*, Vol. V, Sec. 10, 73.

KRATZ, K.O. (Ed.) 1978. *The Continental Crust of the Eastern Part of the Baltic Shield* (in Russian). Nauka Publ., Leningrad, 231 pp.

LAL, R.K. & RAITH, M. 1984. Calibration of Fe-Mg partitioning between garnet and orthopyroxene as a geothermometer. In prep.

MERILÄINEN, K. 1976. The granulite complex and adjacent rocks in Lapland, northern Finland. *Bull. Geol. Surv. Finland*, **281**, 129 pp.

MIKKOLA, E. 1941. *The General Geological Map of Finland.* Sheets B-7, C-7, D-7, Munio, Sodankylä, Tuntsjoki. Explanation to the map of rocks. Suomen geol. Toimikunta, Helsinki.

NEWTON, R.C. & HASELTON, H.J. 1981. Thermodynamics of the garnet-plagioclase-Al_2SiO_5-quartz geobarometer. In: NEWTON, R.C., NAVROTSKY, A. & WOOD, B.J. (eds) *Thermodynamics of Minerals and Melts*, 131–147, Springer Verlag, Berlin.

—— & PERKINS, D., III. 1982. Thermodynamic calib-ration of geobarometers based on the assemblage garnet-plagioclase-orthopyroxene-(clinopyroxene)-quartz. *Amer. Mineral.*, **67**, 203–222.

PERCHUK, L.L., PODLESSKII, K.K. & ARANOVICH, L. YA. 1981. Calculation of thermodynamic properties of endmember minerals from natural parageneses. In: *Advances in Physical Geochemistry*, pp. 111–129, Springer Verlag, Berlin.

RAASE, P. 1974. Al and Ti contents of hornblende, indicators of pressure and temperature of regional metamorphism. *Contrib. Mineral. Petrol.*, **45**, 231–236.

RAITH, M., RAASE, P. & HÖRMANN, P.K. 1982. The Precambrian of Finnish Lapland: evolution and regime of metamorphism. *Geol. Rundschau*, **71**, 230–244.

SECK, H.A. 1971. Koexistierende Alkalifeldspäte und Plagioklase im System $NaAlSi_3O_8$-$KAlSi_3O_8$-$CaAl_2Si_2O_8$-H_2O bei Temperaturen von 650°C bis 900°C. *Neues Jahrb. Mineral. Abh.*, **115**, 315–345.

SPEAR, F.S. 1981. An experimental study of hornblende stability and compositional variability in amphibolite. *Amer. J. Sci.*, **281**, 697–734.

THOMPSON, A.B. 1976. Mineral reactions in pelitic rocks: II. Calculations of some P-T-X(Fe-Mg) phase relations. *Amer. J. Sci.*, **276**, 425–454.

VIELZEUF, D. & BOIVIN, P. 1984. An algorithm for the construction of petrogenetic grids: application to some equilibria in granulite paragneisses. *Amer. J. Sci.*, **284**, 760–791.

WELLS, P.R.A. 1977. Pyroxene thermometry in simple and complex systems. *Contrib. Mineral. Petrol.*, **62**, 129–139.

—— 1979. Chemical and thermal evolution of Archaean sialic crust, southern West Greenland. *J. Petrol.*, **20**, 187–226.

M. RAITH, Mineralogisch-Petrologisches Institut, Universität Bonn, Poppelsdorfer Schloß D 5300 Bonn, West Germany.

P. RAASE, Mineralogisch-Petrographisches Institut, Universität Kiel, Olshausenstraße 40–60, D 2300 Kiel, West Germany.

The South India–Sri Lanka high-grade terrain as a possible deep-crust section

R.C. Newton & E.C. Hansen

SUMMARY: Precambrian granulite-facies terrains are regarded as possible models for the deep crust. In the South India–Sri Lanka late Archaean–early Proterozoic terrain, there is a regional gradation from greenschist and amphibolite facies south through an orthopyroxene isograd to the Granulite Massifs of S. India and of Sri Lanka. The metamorphic gradation is unusually complete and continuous and may represent a partial Archaean crustal profile. Large areas of low-grade grey gneiss may have been formed after charnockite in a retrogression event or may have somehow escaped late Archaean granulite facies metamorphism.

Palaeopressures indicate an increase from 5 kbar to more than 7 kbar in the Transition Zone, though various fault-bounded blocks in the high-grade massif area show different pressure ranges, indicative of different amounts of uplift. The highest pressures of above 9 kbar are for Rb- and U-depleted intermediate charnockites. Undepleted acid charnockites show pressures of 6.5–7.5 kbar, and supracrustal sequences show still lower pressures. The supracrustal-rich Central Highlands of Sri Lanka show pressures of 6 to 7.5 kbar, and most or all rocks are undepleted in Rb. Associated temperatures of all the Massif rocks are in the range 700°–800°C.

This grossly-stratified Archaean crustal section is only the upper half of a thickened crust: there is no indication of what lies below the deepest exposed section, and no compelling reason that depleted intermediate charnockite continues down to the Mohorovicic Discontinuity. Another possibility is deep thrust-bounded repetitions of the section which is exposed at the surface.

Granulite terrains as deep-crustal models

Numerous authors have advanced specific granulite terrains as plausible deep-crustal models. A discussion of petrological and geophysical features of this model is given in the introductory paper of this Symposium by J.B. Dawson *et al.*

The granulite lower crustal model is so prevalent that further implications of the observed lithologies and structures of individual granulite-facies terrains for the deep crust merit study in greater detail. The present paper examines the South India–Sri Lanka granulite-facies terrain as a model for the deep crust. A number of advances in the study of this terrain have been made in recent years, including description of the tectonic framework of South India (Drury & Holt 1980; Drury *et al.* 1984), of palaeotemperatures and palaeopressures over portions of it (Raith *et al.* 1982; Harris *et al.* 1982), of the petrology and geochemistry of the amphibolite-facies to granulite-facies transition zone (Condie *et al.* 1982; Janardhan *et al.* 1982; Hansen *et al.* 1984), and appearance of the Sri Lanka Geological Survey Department geologic map of the island (1982) at a scale of 8 miles to the inch. Many features of the terrain are still poorly known. In particular, geochronological data are sparse. Nevertheless, the existing data allow some assessment of the notion that the South Indian terrain is an exhumed section of the late Archaean lower crust.

The South India–Sri Lanka high-grade terrain

Figure 1 shows the important geologic units. *The Karnataka Craton*, occupying the Mysore Plateau, is a classical gneiss-greenstone terrain containing low- to middle-grade metasedimentary-volcanic troughs, the Dharwars, set in tonalitic to granitic biotite- and amphibole-bearing gneisses, the *Peninsular Gneiss*. The Karnataka Craton has an eastern block, richer in discrete granite bodies, including the long, linear Closepet Granite, and a western block, with most of the supracrustal belts (Swamy Nath *et al.* 1976; Drury *et al.* 1984).

A metamorphic gradation exists in acid gneisses and supracrustal units southwards in the Karnataka Craton (Pichamuthu 1965), from greenschist facies to granulite facies at the southern borders of Karnataka. The onset of the granulite facies is defined by a diffuse orthopyroxene isograd for acid gneisses (Subramaniam 1967) which extends across southern India. A *Transition Zone* follows to the south, in which orthopyroxene becomes more prominent and amphibole

From: DAWSON, J.B., CARSWELL, D.A., HALL, J. & WEDEPOHL, K.H. (eds) 1986, *The Nature of the Lower Continental Crust*, Geological Society Special Publication No. 24, pp. 297–307.

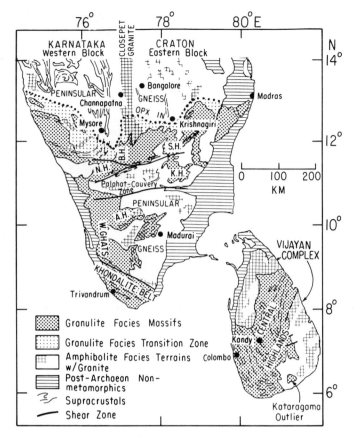

FIG. 1. Geologic units of South India–Sri Lanka high-grade terrain. High-grade massif areas: AH = Annamalai Hills: BH = Biligirirangan Hills: KH = Kollaimalai Hills: NH = Nilgiri Hills; SH = Shevaroy Hills.

becomes rare in the gneisses. The culmination of this progression is the *Charnockite-Khondalite Massif* area entirely in the granulite facies. Massive charnockites (orthopyroxene gneisses) dominate the Hill Ranges and concentrations of supracrustal lithologies, including khondalites (sillimanite-garnet metapelites), occupy the flanks of some upland areas. The Sri Lanka Central Highlands and the Khondalite Belt of southernmost India are dominantly metasedimentary granulites. The highland massif blocks are separated by several long shear belts, which may be surfaces of differential uplift (Drury & Holt, 1980).

Large tracts of lower-grade biotite and amphibole gneiss occupy the lowlands to the south of the orthopyroxene isograd. These are cut in numerous places by granite plutons and are otherwise similar to the Peninsular Gneiss of Karnataka. They are termed Peninsular Gneiss on the state geological maps of Tamil Nadu and Kerala (Karunakaran 1974; Varadan 1975). The

correlative gneiss-granite terrain of the Sri Lanka coastal lowlands is the Vijayan Gneiss Complex.

Age relations of the south Indian shield are poorly known. Some of the Peninsular Gneiss of the Karnataka Craton is as old as 3.4 Ga (Beckinsale *et al.* 1980). Most of the reliable dates for the granulite-facies metamorphism in southern India are close to 2.6 Ga (see Ramiengar *et al.* 1978, for a summary). One Rb-Sr whole-rock isochron and a few model ages exist for the Sri Lanka Highlands, giving a minimum age of 2.1 Ga (Crawford & Oliver 1969). Model ages for the Khondalite Belt of southern India range from 2.15–3.10 Ga (Crawford 1969). A generalization is that much of the high-grade area results from late-Archaean to early-Proterozoic granulite-facies metamorphism.

The granulite-facies transition zone

The transition in southern Karnataka and northern Tamil Nadu from dominant amphibolite-facies grey gneiss to dominant charnockite takes place over 25–50 km. Fig. 2 shows the transition zone south of Krishnagiri, Tamil Nadu. Charnockite increases from incipient patches near the south end of Krishnagiri Reservoir to the dominant country rock in the vicinity of Karimangalam. According to Condie *et al.* (1982), migmatization and granite bodies are especially prominent in the transition zone, decreasing rapidly in amount north and south of this zone. Textural and structural relations indicate that migmatization preceded charnockite emplacement and that charnockite formed by metamorphic conversion of gneisses, migmatites and granites. Charnockites and granites have been dated at 2.55 Ma by zircons (personal communication from A.S. Janardhan, quoting P. Vidal). Metamorphic conversion from acid gneiss to charnockite was almost exactly isochemical (analyses 1, 2, Table 1), both in major and minor elements. The advent of orthopyroxene does not correspond to sudden depletion in the large-ion lithophile (LIL) elements such as Rb and U.

The regional transition zone relations are very similar in the section south of Channapatna, Karnataka. Dark charnockite in short veins and irregular patches appears in migmatic grey gneiss at about latitude 12°30′N. The gneissic foliation is disrupted by coarse recrystallization (Fig. 3). Chemical analyses of adjacent gneissic and charnockitic portions (analyses 3, 4, Table 1) again show nearly isochemical conversion, which precludes that igneous processes are responsible for the charnockitization of gneiss. Rather, conversion was accomplished by a metamorphic low-$P(H_2O)$ vapour phase whose access was controlled by deformation.

Janardhan *et al.* (1982) and Condie *et al.* (1982) both concluded that copious CO_2-rich vapours converted amphibole to orthopyroxene in gneisses, without much chemical change other than dehydration, and that the CO_2-rich fluids were preceded by a wave of H_2O-richer solutions which effected K-metasomatism and anatexis. A similar conclusion was reached by Friend (1981) for the Kabbal locality, southern Karnataka, and for the type charnockite locality at Madras by Weaver (1980).

Hansen *et al.* (1984) determined an average mineralogical palaeotemperature of $750° \pm 50°C$ for the transition zone south of Channapatna, but were unable to discern a definite temperature increase over the interval. Fig. 4 shows their palaeopressure profile normal to the orthopyroxene isograd, determined from nearly pressure-independent mineralogical geobarometers. There is a steady increase from about 5 kbar where incipient charnockite first appears, to 7.5 kbar in

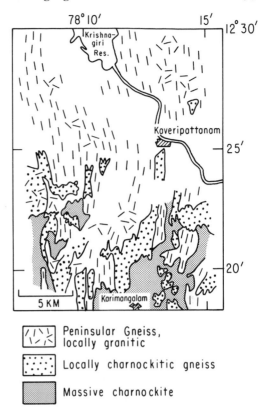

FIG. 2. Section of the granulite facies Transition Zone, south of Krishnagiri, Tamil Nadu. Incipient charnockite appears at about 12°30′N and becomes the dominant rock at about 12°15′N. After Condie *et al.* (1982), Fig. 2.

the Biligirirangan Hills charnockite massif. The plot includes, in addition to standard mineralogical geobarometers, pressures determined from the densities of CO_2 inclusions in quartz by the P-V-T equation of state of Touret and Bottinga (1979). The presence of abundant high-density CO_2 inclusions in charnockites confirms that CO_2 was the dominant vapour species. The lack of apparent breaks in the palaeopressure profile encourages the view that an Archaean boundary between the amphibolite facies upper crust and the granulite facies lower crust is exposed in the region.

Lack of pronounced LIL depletion seems to be generally true of granulite-facies gneisses near to an orthopyroxene isograd, as in the West Uusimaa region of southern Finland (Schreurs 1985). Many of the charnockites at Madras are undepleted and some show only moderate depletion (Weaver 1980). The Hansen *et al.* (1984) traverse

R.C. Newton & E.C. Hansen

TABLE 1. *Analyses of charnockites and gneisses from South India and Sri Lanka*

Wt%	Transition zone		Transition zone		S. India massifs		Sri Lanka		Khondalite belt	
	1) gneiss	2) charn.	3) gneiss	4) charn.	5) charn.	6) charn.	7) charn.	8) charn.	9) gneiss	10) charn.
SiO_2	71.25	70.43	68.4	71.4	70.33	62.10	66.5	70.8	65.23	66.07
TiO_2	0.31	0.33	0.62	0.49	0.34	0.56	0.47	0.42	0.88	0.89
Al_2O_3	14.45	15.70	13.6	13.7	15.00	15.95	14.7	14.0	15.53	15.02
$Fe_2O_{3(T)}$	2.15	2.68	4.43	3.48	2.58	5.27	3.82	2.46	7.29	6.44
MgO	0.63	0.77	0.77	.79	1.03	3.10	1.39	0.63	1.12	1.07
MnO			0.06	.06	0.03	0.08	0.08	0.04	0.09	0.07
CaO	2.28	3.40	2.44	1.95	3.00	5.60	4.03	2.21	2.17	2.24
Na_2O	4.88	4.97	3.65	3.76	4.88	4.80	3.86	3.01	2.32	2.58
K_2O	1.49	1.28	3.31	3.67	1.53	0.96	2.26	4.70	4.55	5.01
ppm										
Rb	46	38	70	60	6.7	5.0	50	110	227	207
Cs	0.11	0.20								
Ba	437	368	500	550						
Sr			130	110	347	595	570	400	154	175
Zr	134	178	310	230	140	85	140	190		
Cr	10.8	3.5								
Th	3.1	2.6	15.0	15.0	5	<1				
U			1.4	1.0	0.2	0.1				
K/Rb	267	281	390	450	930	1600	377	356	167	202

Transition zone (anal. 1,2) Condie *et al.* 1982, S. of Krishnagiri, Tamil Nadu. Avg. of grey gneisses (4), charnockites (4), SiO_2 68–73%.
Anal. 3,4. Janardhan *et al.* 1982, Kabbal, Karnataka. Grey gneiss (3) 3–1B and immediately adjacent charnockite (4) 3–1A.
S. India massifs (anal. 5) Janardhan *et al.* 1982, BR Hills, Karnataka. Avg. of 3 depleted charnockites.
Anal. 6. Janardhan *et al.* 1982, Nilgiri Hills North Slopes, Tamil Nadu. Avg. of 2 depleted intermediate charnockites.
Sri Lanka Central Highlands. Garnitiferous charnockite (anal. 7) SR 13–7 and acid charnockite (anal. 8) SR 13–3 (present authors, unpublished data).
Khondalite belt (anal. 9,10) Ravindra Kumar *et al.* 1985, Ponmudi, Kerala. Garnet-biotite-graphite paragneiss (avg. of 2) and immediately adjacent charnockite (avg. of 2) respectively.

south of Channapatna penetrated a region in the Biligirirangan (BR) Hills massif where some, but not all, of the charnockites show pronounced LIL depletion. Depletion in Rb appears sporadically in a few rocks at about the latitude where charnockite becomes the dominant rock. The cause of severe depletion is not yet known, nor whether removal of a granitic melt was essential to the process.

High-grade massifs

Much of the region to the south of the transition zone is occupied by the charnockite-khondalite massifs entirely of the granulite facies, as indicated by the ubiquitous presence of orthopyroxene in appropriate bulk chemistries. The northern Nilgiri Hills (Viswanathiah & Tareen 1970), the Palni Hills, Biligirirangan Hills and Shevaroy

Hills are dominantly made of charnockites. A belt of the southern flanks of the northernmost massif hill ranges contains abundant quartzites, biotite gneisses, mafic layered complexes, cordierite gneisses, and some calc-silicates and marbles. This belt, the Satyamangalam Supergroup (Gopalakrishnan *et al.* 1975), is discontinuous, and complicated by local retrogression and structural interruptions. Another major supracrustal area is the eastern flank of the Anamalai-Palni Hills uplift (Drury *et al.* 1984). The largest supracrustal area in southern India is the Khondalite Belt, a region of mountain slopes and foothills, underlain largely by khondalites, ortho- and para-charnockites, marbles and quartzites, in a NW-SE-trending swath in southernmost Kerala and Tamil Nadu. The Sri Lanka Highlands are rich in supracrustals, with many massive limestones and quartzites among the dominant

FIG. 3. Partially charnockitic Peninsular Gneiss at Kabbaldurga, south of Channapatna, Karnataka (77°15′E). The charnockite (dark) was emplaced in shears and other deformation features in the migmatitic gneiss.

charnockites (Cooray 1962). The southwest portion of the Sri Lanka granulite terrain is particularly rich in layered rocks.

Topographic high areas are occupied by the high-grade massifs with dominant charnockite. The regional Bouguer map of southern India (Subrahmanyam 1983) shows -100 mg anomalies generally coincident with the highest elevations; therefore, the crust is thicker under the high-grade uplift areas. The Sri Lanka region Bouguer map (Hatherton *et al*. 1975) shows a small departure from isostatic compensation of the Central Highlands, but the crust is somewhat thicker there than under the adjoining lowlands areas.

Recumbent folds are the largest-scale structures of the northern massif areas (Drury *et al*. 1984), and of the Sri Lanka Central Highlands (Cooray 1962). The South Indian high-grade terrain is similar in this respect to many or most other granulite provinces. Superposed on the flat-lying structures are the open to tight folds with steep axial planes which give rise to the regional grain. Post-Archaean linear shear belts bound highland segments (Drury & Holt 1980). The eastern boundary of the Sri Lanka Central Highlands may be overthrust onto the Vijayan Complex (Hatherton *et al*. 1975), while the western boundary between the two metamorphic provinces is gradational.

Table 1 gives representative analyses of massif charnockites. Compositions are calc-alkaline and acid to intermediate. Many, but not all, are lower in the LIL elements at a given SiO_2 content than the charnockites of the Transition Zone. Charnockites of the Sri Lanka Highlands are generally

undepleted (Table 1). Some of the most depleted rocks are from the North Slopes of the Nilgiri Hills. They are intermediate charnockites, some with primary hornblende, which is generally lacking from the more acid charnockites.

Retrogression of charnockites is widespread in the massif areas. Minor retrogression in shear veins is often associated with small amounts of acid pegmatites (Janardhan *et al*. 1982; Allen *et al*. 1985). Orthopyroxene was converted to amphibole and biotite and charnockites were bleached. Low levels of Rb and U were often not restored. More extensive retrogression occurs in the northern Anamalai Hills (Gopalakrishnan *et al*. 1975), where charnockites are converted to banded gneisses about invading pink granites. Cooray (1962) regarded such conversion as a general relationship between the Highland Series and the Vijayan Complex. In some areas it may be difficult to tell whether relations between charnockite and migmatitic gneiss are prograde or retrogressive; this is one of the outstanding local problems of southern India.

The chemistry of metasedimentary rocks is very diverse, and major- and minor-element analyses of Indian rocks which have been mineralogically well-characterized are rare. Table 1 gives the analysis of a graphitic paracharnockite from the Khondalite Belt. Analyses of rocks called khondalite by Weaver (1980) from the Madras area probably include several metasedimentary types. It seems to be generally true that quartzofeldspathic metasediments from the high-grade terrain are not particularly depleted in LIL elements, as shown by the relatively low K/Rb ratios.

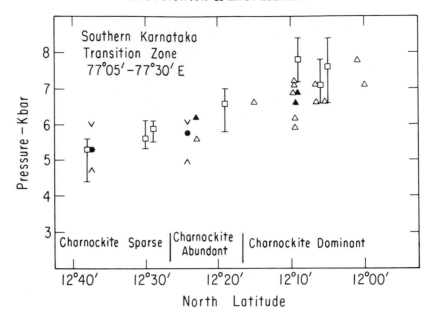

FIG. 4. Geobarometry of the Transition Zone south of Channapatna, Karnataka. The palaeopressure increases from about 5 kbar where charnockite first appears to over 7 kbar in the Biligirirangan Hills massif at 12°00′N. All barometry at 750°C.
Filled circles: Garnet-sillimanite-plagioclase-quartz (Harris & Jayaram 1982). Opposed arrowheads: Garnet-cordierite-sillimanite-quartz, $P(H_2O) = P(total)$ (upper) and $P(H_2O) = O$(lower), from Harris and Jayaram (1982). Filled triangles: Garnet-clinopyroxene-plagioclase-quartz, method of Perkins and Newton (1981), from Hansen et al. (1984).
Open triangles: Garnet-orthopyroxene-plagioclase-quartz, method of Perkins and Newton (1981), data from Hansen et al. (1984).
Squares with error brackets: CO_2 inclusion filling pressures at 750°C, based on method of Touret and Bottinga (1979) (Hansen et al. 1984).

Graphitic biotite gneiss, often garnetiferous, has been found by the present authors in a state of arrested conversion to coarse, dark green charnockite in a number of localities in Tamil Nadu, Kerala and Sri Lanka, including the recently-described quarry at Ponmudi, Kerala (Ravindra Kumar et al. 1985). Charnockite-filled deformation zones with warping of gneissic foliation resemble those of Kabbaldurga. The reaction arrested in progress is manifestly that of biotite-+quartz to orthopyroxene and K-feldspar. The metamorphism is virtually isochemical, as at Kabbaldurga (9, 10, Table 1). The main difference is that orthopyroxene is from hornblende breakdown at Kabbaldurga, and biotite and quartz are stable together, whereas biotite and quartz are unstable in the Ponmudi charnockites and at similar localities. These relations must be due to higher temperatures, lower pressures, or lower $P(H_2O)$ in the high-grade zone than in the transitional zone. The mechanism of CO_2-rich solution flux controlled by deformation was

operative at Ponmudi in virtually the same manner as at Kabbaldurga.
Several mineralogical studies (Harris et al. 1982; Raith et al. 1983; Hansen et al. 1984) concluded that metamorphism in the highest-grade area took place at 680°–820°C. The widespread, nearly uniform temperature distribution suggested to Harris et al. (1982) that heat was transferred pervasively throughout the late Archaean crustal section of southern India by fluids of mantle origin, which they inferred to be dominantly CO_2. A similar 'hot spot' distribution of palaeotemperatures was deduced by Schreurs (1985) for the West Uusimaa granulite terrain of southern Finland.
Figure 5 shows the distribution of palaeopressures calculated from the nearly temperature-independent charnockite assemblage garnet-orthopyroxene-plagioclase-quartz by several authors. Harris et al. (1982) found different pressure ranges in different massif sectors. For the Nilgiri Hills, they found 8–9 kbar in the northern

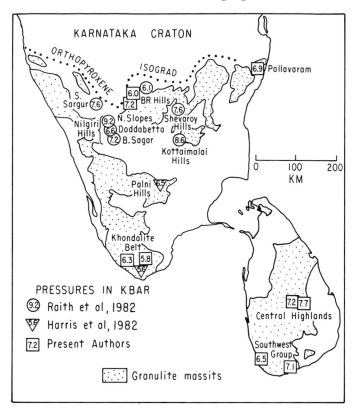

FIG. 5. Palaeopressures in charnockite-khondalite massif areas, by the garnet-orthopyroxene-plagioclase-quartz method. Assumed temperatures 700–750°C.

part and 6–7 kbar in the central part. Raith *et al.* (1983) found over 9 kbar for the Nilgiri North Slopes, which corresponds to depths near the base of a continent of normal thickness. The southernmost massif blocks gave lower pressures.

There appears to be some correspondence of palaeopressures with lithology and degree of depletion. The highest pressures, over 9 kbar, are recorded in the depleted zone of intermediate charnockites in the Nilgiri Hill North Slopes. No rocks of evident supracrustal origin have been found in preliminary surveys there. Other high-pressure areas are the Shevaroy and Kollaimalai Hills, with massive intermediate to acid charnockite sections (Raith *et al.* 1982). Areas of acid charnockites, some undepleted, with some supracrustals, such as the Nilgiri summit, the eastern Sri Lanka Highlands, and the Biligirirangan Hills, show pressures in the range 7–8 kbar. The supracrustal-dominated regions, such as the Khondalite Belt, show pressures of 5.5–7 kbar. The lower-grade southern Karnataka gneiss-greenstone craton falls off steadily in pressure

northward to about 3 kbar in the greenschist areas (Harris & Jayaram, 1982). Thus, there is evidence of a gross vertical stratification of the late Archaean crust which is exposed in southern India.

Peninsular Gneiss—Vijayan Complex

Large areas of Tamil Nadu are designated Peninsular Gneiss on the state geological map of Tamil Nadu (Karunakaran 1974) and are thus tacitly held to be Archaean. The amphibolite-biotite gneisses are cut by discrete, relatively young granite plutons. The major differences between the Tamil Nadu Peninsular Gneiss and the Karnataka Peninsular Gneiss are that low-grade greenstone supracrustal belts are absent from the Tamil Nadu gneisses and that they are interspersed among charnockitic areas. The Tamil Nadu Peninsular Gneiss is somewhat enigmatic in that, if it is actually Archaean, it somehow escaped the charnockitic metamorphism which affected most of the region.

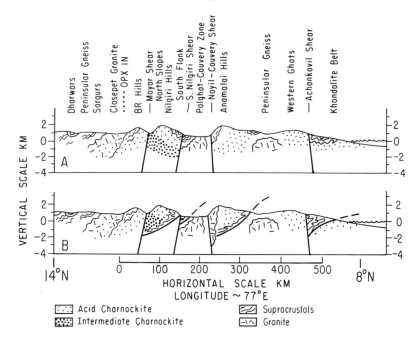

FIG. 6. Generalized N–S cross-sections of southern India. Section A: Assumes that supracrustal-rich assemblages (wavy line) are deformed cover sequences. Peninsular Gneiss relations south of the orthopyroxene isograd are suggested to be retrogressive after charnockite about Proterozoic granites, while Closepet Granite in northern Peninsular Gneiss is Archaean and is partly charnockitized. The Nilgiri Hills may show an Archaean depth profile from supracrustal (south flank) through acid undepleted charnockites down to intermediate high pressure depleted charnockites in the North Slopes.
Section B: Assumes that supracrustal-rich sequences are loci of underthrusting, following Drury *et al.* (1984). Other features are the same as in Section A.

The Sri Lanka analogue of the Tamil Nadu Peninsular Gneiss is the Vijayan Complex, which occupies two coastal bands on either side of the Central Highlands. Migmatitic biotite-amphibole gneisses are cut by numerous granite bodies. The terrain shows a dome and basin structural style lacking in consistent directionality, in contrast to the well-defined trends of the layered Highlands Series. Small bodies of marble and quartzite are found sporadically within the Vijayan Complex. According to Katz (1971), the Highland Series is a younger cover sequence unconformable upon an older Vijayan Complex basement. Cooray (1962) and Holland (1983), following Crawford and Oliver (1969), advocated that the Vijayan Complex is a remobilized terrain invaded by younger granites which consists of reconstituted relics of earlier Highland Series rocks. Munasinghe and Dissanayake (1980) considered that the Highland Series and Vijayan Complex are coeval and cometamorphic.

Available radiometric data support the second of these hypotheses. Crawford and Oliver (1969) obtained a minimum Rb-Sr whole rock age of about 2100 Ma on the Highland series, and an isochron of 1250 Ma for assorted rocks from the Vijayan Complex. Holland (1983) reinterpreted the earlier isochron data to get 2450 Ma for the Highland Series.

These age relations may have implications for much of the Peninsular Gneiss region of Tamil Nadu adjacent to the northwest Vijayan Complex of Sri Lanka. It seems possible by analogy that extensive Proterozoic retrogression affected the southern Peninsular region, displacing charnockites.

Summary and interpretations

Figure 6 gives an interpretive generalized N–S cross-section of the southern Indian peninsula which emphasizes the correlation of palaeopressure with petrologic type. In Figure 6A, the various crustal blocks are shown as bounded by vertical faults. The Karnataka Craton block is slightly tilted, exposing a beveled section of increasing depth to the south, reaching 7.5 kbar equivalent in the southern BR Hills. Here the rocks are acid and intermediate charnockites, variably depleted, with some supracrustals. The greatest uplift is exposed in the Nilgiri Hills North Slopes area, dominated by intermediate depleted charnockites, some hornblende-bearing. If the Nilgiri Hills section is nearly coherent, a 9-kbar palaeopressure surface must dip as shown in Fig. 6A, in order for acid charnockites of palaeopressure 6.5 kbar to be exposed in the Nilgiri summit region. The supracrustal sequence on the Nilgiri Hills southern flank is interpreted as a cover sequence over charnockitic basement. The Palghat-Cauvery gap is intruded by younger granites, with probable retrogression to biotite gneisses about them. Biotite gneiss and granite in the Anamalai summit area are interpreted as retrogressive after charnockite. This explanation is preferred for the relationship between the charnockitic Highlands Series and the granite-gneiss Vijayan Complex of Sri Lanka. The supracrustal sequence of the Anamalai-Palni southeastern flank is interpreted as an Archaean cover sequence, as is the Khondalite Belt.

An alternative reconstruction, advanced by Drury *et al.* (1984), emphasized that thrusting in the Palni-Kodaikanal section seems to be localized in certain supracrustal horizons, notably limestones. The supracrustal sections of the southern flanks of the charnockitic uplifts are thus zones bounding large thrust blocks. Crustal shortening during late Archaean orogenesis was accomplished by oblique stacking of such crustal segments, each floored by a lubricated supracrustal sequence. Under this reconstruction, the metamorphic grade and degree of depletion need not increase monotonically downward. There may, for instance, be a supracrustal-acid charnockite sequence under the Nilgiri Hills North Slopes (Fig. 6B). Such a mid-crustal horizon of deeply-subducted metasediments containing pore fluids has been suggested as a possible explanation of a seismic reflector horizon and high electrical conductivity channel under the Adirondack Highlands granulite province (Brown *et al.* 1983).

Neither hypothetical cross-section of Fig. 6 presumes to suggest what sorts of rocks underlie mid-crustal levels. In particular, it is not certain if the Karnataka gneiss-greenstone terrain is underlain by granulites or whether the granulite-facies rocks are confined to the southern margin of the Karnataka craton.

The major features of the South Indian shield which bear on a possible crustal model are that a gross stratification of exposed crust exists in the high-grade area with supracrustal-rich sections underlain by undepleted or variably depleted acid charnockites, underlain in turn by depleted intermediate charnockites. The supracrustal sequences show palaeopressures of 6 or more kbar, and they were therefore buried to at least 20 km depth and then excavated. The South Indian crust is of normal 30–40 km thickness (Kaila *et al.* 1979). Therefore, something replaced uplifted and eroded crust at its roots.

If the principal mechanism of late Archaean continental accretion was coalescence of island arcs, as envisioned by Langford and Morin (1976) for the Superior Province, the lowermost crust is likely to consist largely of a high-Al_2O_3 silica-oversatured basaltic composition, similar to that inferred for island-arc roots (Anderson 1982). The deformed middle and upper levels of the accreted volcanic arc and adjacent sedimentary basin were exposed at the surface after buoyant rebound and deep erosion. A plausible source of carbonic fluids in this model would be subducted marginal basin sediments. Streaming of CO_2-rich volatiles may have accounted for extreme Rb depletion of deeper levels. A duplex or even more complex crust with repeated sedimentary and acid, intermediate and basic plutonics sections could have resulted from multiple underthrusting.

The voluminous Tertiary Queensland, Australia basalts were considered by Ewart *et al.* (1980) to be a significant addition to the continental crust. This method of accretion is akin to the 'hotspot' model of early continental growth (Fyfe 1978). Voluminous magmas which could have accumulated at the base of the crust were necessarily mafic, from density considerations (Herzberg *et al.* 1983). Such a crust-building model could apply to areas like Enderby Land, Antarctica, with palaeotemperatures over 900°C regionally distributed (Sheraton *et al.* 1980). The crust may have been thickened both by overthrusting and, subsequently, by flooding at the base by mafic magmas, as pictured by Reymer *et al.* (1984) for the Pan-African orogenic province in the southeastern Sinai. Magmatic underplating could provide a mechanism for the excavation of high pressure terrains (O'Hara 1977).

In both the orogenic ('plate-tectonic') and anorogenic ('hot-spot') models, the lowermost crust which is generated is dominantly mafic. Therefore, it seems unlikely that the Precambrian plutonic-sedimentary granulite-facies complexes are representative of all or most of the present-day lowermost continental crust. A basaltic lower crust combined with depleted intermediate granulites at the mid-crustal level may explain the very low heat flow through the shield regions. The exact mechanisms of LIL depletion, particularly of Rb, are still unknown, although it is likely that some process of equilibration of Rb between metamorphic vapours, biotite and K-feldspar, and removal of Rb in a vapour phase during granulite facies metamorphism was instrumental in the depletion in southern India.

ACKNOWLEDGEMENTS: This research was supported by National Science Foundation grants EAR 82-19248 and INT 82-19140.

The authors benefited enormously from field guidance and discussion by many co-workers, including Clark Friend, D. Gopalakrishna, A.S. Janardhan, W.K.B.N. Prame, and G.R. Rafindra Kumar. The logistical services of the Sri Lanka Geological Survey Department and the Centre for Earth Science Studies, Trivandrum, are gratefully acknowledged.

Valuable discussions with Steve Drury, Nigel Harris, Richard Holt, Frank Richter, and Roger Wightman are acknowledged.

References

ALLEN, P., CONDIE, K.C. & NARAYANA, B.L. 1985. The geochemistry of prograde and retrograde charnockite-gneiss relations in southern India. *Geochim et Cosmochim. Acta*, **49**, 323–336.

ANDERSON, A.T. 1982. Parental basalts in subduction zones: implications for continental evolution. *J. Geophys. Res.*, **87**, 7047–7060.

BECKINSALE, R.D., DRURY, S.A. & HOLT, R.W. 1980. 3,360-Myr old gneisses from the South Indian Craton. *Nature*, **283**, 469–470.

BROWN, L., ANDO, C., *et al.* 1983. Adirondack-Appalachian crustal structure. The COCORP Northeast Traverse. *Geol. Soc. Amer. Bull.*, **84**, 1173–1184.

CONDIE, K.C., ALLEN, P. & NARAYANA, B.L. 1982. Geochemistry of the Archean low- to high-grade transition zone, southern India. *Contr. Min. Pet.* **81**, 157–167.

COORAY, P.G. 1962. Charnockites and their associated gneisses in the Pre-cambrian of Ceylon. *Quart. J. Geol. Soc. Lond.* **118**, 239–266.

CRAWFORD, A.R. 1969. Reconnaissance Rb-Sr dating of the Precambrian rocks of southern Peninsular India. *J. Geol. Soc. India.* **10**, 117–166.

—— & OLIVER, R.L. 1969. The Precambrian geochronology of Ceylon. *Geol. Soc. Australia Spec. Publ.* **2**, 283–306.

DRURY, S.A., HARRIS, N.B.W., HOLT, R.W., REEVES-SMITH, G.J. & WIGHTMAN, R.T. 1984. Precambrian tectonics and crustal evolution in South India. *J. Geol.*, **92**, 1–20.

—— & HOLT, R.W. 1980. The tectonic framework of the South Indian Craton: a reconnaissance involving LANDSAT imagery. *Tectonophys.* **65**, T1-T15.

EWART, A., BAXTER, K. & ROSS, J.A. 1980. The petrology and petrogenesis of the tertiary anorogenic mafic lavas of Southern and Central Queensland, Australia—possible implications for crustal thickening. *Contr. Min. Pet.*, **75**, 129–152.

FRIEND, C.R.L. 1981. The timing of charnockite and granite formation in relation to influx of CO_2 at Kabbaldurga, Karnataka, South India. *Nature*, **294**, 550–552.

FYFE, W.S. 1978. The evolution of the earth's crust: modern plate tectonics to ancient hot spot tectonics? *Chem. Geol.*, **23**, 89–114.

GOPALAKRISHNAN, K., SUGAVANAM, E.B. & VENKATA RAO, V. 1975. Are there rocks older than Dharwars?

A reference to rocks in Tamil Nadu. *Ind. Mineral.* **16**, 26–34.

HANSEN, E.C., NEWTON, R.C. & JANARDHAN, A.S. 1984. Pressures, temperatures and metamorphic fluids across an unbroken amphibolite-facies to granulite-facies transition in southern Karnataka, India. *In:* A. KRONER, A.M. GOODWIN & G.N. HANSON (eds)., *Archaean Geochemistry*.

HARRIS, N.B.W., HOLT, R.W. & DRURY, S.A. 1982. Geobarometry, geothermometry, and late Archean geotherms from the granulite facies terrain of South India. *J. Geol.*, **90**, 509–528.

HATHERTON, T., PATTIARATCHI, D.B. & RANASINGHE, V.V.C. 1975. Gravity Map of Sri Lanka. *Sri Lanka Geol. Survey Dept. Prof. Pap.* **3**, 1–39.

HERZBERG, C.T., FYFE, W.S. & CARR, M.H. 1983. Density constraints on the formation of the continental Moho and crust. *Contr. Min. Pet.* **84**, 1–5.

HOLLAND, J.G. 1983. The geochemistry of the Highland Series and Vijayan Series rocks of Sri Lanka. *Abstracts, Symposium on the Geology of Sri Lanka, Perideniya, Sept. 1983*, p. 16.

JANARDHAN, A.S., NEWTON, R.C. & HANSEN, E.C. 1982. The transformation of amphibolite facies gneiss to charnockite in southern Karnataka and northern Tamil Nadu, India. *Contr. Min. Pet.* **79**, 130–149.

KAILA, K.L., ROY CHOWDURY, K., *et al.* 1979. Crustal structure along Kavali-Udipi profile in the Indian peninsular shield from deep seismic sounding. *J. Geol. Soc. India*, **20**, 307–333.

KARUNAKARAN, C. 1974. Geology and mineral resources of the states of India. Pt. VI: Tamil Nadu and Pondicherry. *Geol. Surv. of India. Misc. Publ.* **30**, 28 pp.

KATZ, M.B. 1971. The Precambrian metamorphic rocks of Ceylon. *Geol. Rundsch.* **60**, 1523–1549.

LANGFORD, F.F. & MORIN, J.A. 1976. The development of the superior province of northwestern Ontario by merging island arcs. *Amer. J. Sci.* **276**, 1023–1034.

MUNASINGHE, T. & DISSANAYAKE, C.B. 1980. Pink granites in the Highland Series of Sri Lanka—a case study. *J. Geol. Soc. India*, **21**, 446–452.

O'HARA, M.J. 1977. Thermal history of excavation of Archaean gneisses from the base of the continental crust. *J. Geol. Soc. Lond.*, **134**, 185–200.

PERKINS, D. & NEWTON, R.C. 1981. Charnockite

geobarometers based on coexisting garnet-pyroxene-plagioclase-quartz. *Nature*, **292**, 144–146.

PICHAMUTHU, C.S. 1965. Regional metamorphism and charnockitization in Mysore State, India. *Ind. Mineral.* **6**, 119–126.

RAITH, M., RAASE, P., ACKERMAND, D. & LAL, R.K. 1982. Regional geothermobarometry in the granulite facies terrain of South India. *Trans. Roy. Soc. Edinburgh, Earth Sci.* **73**, 221–244.

RAMIENGAR, A.S., RAMAKRISHNAN, M. & VISWANATHA, M.N. 1978. Charnockite-gneiss-complex relationship in southern Karnataka. *J. Geol. Soc. of India*, **19**, 411–419.

RAVINDRA KUMAR, G.R., SRIKANTAPPA, C. & HANSEN, E.C. 1985. Charnockite formation at Ponmudi in Southern India. *Nature*, **313**, 207–209.

REYMER, A.P.S., MATTHEWS, A. & NAVON, O. 1984. Pressure-temperature conditions in the Wadi Kid metamorphic complex: implications for the pan-African event in SE Sinai. *Contr. Min. Pet.*, **85**, 336–345.

SCHREURS, J. 1985. The amphibolite-granulite facies transition in West-Uusimaa, S.W. Finland: a fluid inclusion study. *J. Meta. Geol.*, **2**, 327–342.

SHERATON, J.W., OFFE, L.A., TINGEY, R.J. & ELLIS, D.J. 1980. Enderby Land, Antarctica—an unusual Precambrian high-grade metamorphic terrain. *J. Geol. Soc. Australia*, **27**, 1–16.

SUBRAHMANYAM, C. 1983. An overview of gravity anomalies, Precambrian metamorphic terrains and their boundary relationships in the southern Indian Shield. *In*: S.M. NAQVI & J.J.W. ROGERS (eds) *Precambrian of South India*. Geol. Soc. India Memoir, **4**, 553–556.

SUBRAMANIAM, A.P. 1967. Charnockites and granulites of southern India: a review. *Dan. Geol. Foren.* **17**, 473–493.

SWAMI NATH, J., RAMAKRISHNAN, M. & VISWANATHA, M.N. 1976. Dharwar stratigraphic model and Karnataka craton evolution. *Rec. Geol. Surv. India*, **107**, 149–175.

TOURET, J. & BOTTINGA, U. 1979. Equation d'etat pour le CO_2: application aux inclusions carboniques. *Bull. Mineral.*, **102**, 577–583.

VARADAN, V.K.S. 1975. Geology and mineral resources of the states of India. Pt. IX-Kerala. *Geol. Surv. of India Misc. Publ.*, **30**, 34 pp.

VISWANATHIAH, M.N. & TAREEN J.A.K. 1970. Petrography of the Nilgiri charnockites. *Ind. Mineral.* **11**, 1–2, 78–86.

WEAVER, B.L. 1980. Rare-earth element geochemistry of Madras granulites. *Contr. Min. Pet.* **71**, 271–279.

R.C. NEWTON, Department of the Geophysical Sciences, University of Chicago, Chicago, Illinois 60637, USA.

E.C. HANSEN, Department of Geology, Hope College, Holland, Michigan 49423, USA.

Geochemistry of granulite-facies lower crustal xenoliths: implications for the geological history of the lower continental crust below the Eifel, West Germany

H.-G. Stosch, G.W. Lugmair & H.A. Seck

SUMMARY: Mafic meta-igneous granulite facies xenoliths from Engeln/Eifel display petrographic, chemical and isotopic features which are not easily understood in terms of igneous fractionation or granulite-facies metamorphism. These features are best explained through a metasomatic modification of many of the xenoliths, characterized by the formation of amphibole and increasing Rb/Sr and Nd/Sm ratios as well as shifts in $^{143}Nd/^{144}Nd$ towards values similar to those of enriched mantle under the Eifel. Granulites which essentially escaped metasomatism have a Sm-Nd depleted mantle model age of ~ 1.6 Ga whereas metasomatically altered granulites have younger and highly variable model ages. A Sm-Nd isochron for minerals from one granulite resulted in an unexpectedly young age of 172 ± 5 Ma which can be regarded as the maximum age of metasomatism. This age may be indicative of a thermal event having affected the lower crust under the Eifel during the Mesozoic.

Introduction

Mafic granulite-facies lower crustal xenoliths are found in Pleistocene alkalic basaltic tuffs near Engeln/Eifel (West Germany). The Eifel is part of the Rhenish Massif which has been undergoing uplift since the late Tertiary (Fuchs et al. 1983). Granulites from this suite have previously been investigated petrographically and, in part, geochemically by Okrusch et al. (1979) and Voll (1983). These investigators inferred equilibration temperatures between 700°C and 850°C and pressures of c. 6.5–12 bar for the xenoliths and ascribed to the granulite-facies event a Caledonian (Okrusch et al. 1979) or even older age (Voll 1983). Caledonian metamorphic ages obtained from Rb-Sr isochrons (or 'errorchrons') of bulk rocks have been suggested for various exposed felsic granulite terrains in Central Europe (Jäger 1969; Arnold & Scharbert 1973). It now appears, however, that the granulite-facies metamorphism in at least some of these terrains is Hercynian (van Breemen et al. 1982; Todt, pers. communication 1984). It was speculated by Okrusch et al. (1979) and Voll (1983) that the granulite xenoliths are fragments of the present-day lowermost crust or crust-mantle transition at about 28 km to 30 km depth. If these interpretations are correct, the granulites from Engeln may preserve information not only on crust-mantle differentiation episode(s) but also on subsequent modification of the lower crust. In this paper we will examine the evidence for such subsequent modification.

Petrography

A comprehensive petrographic description of the granulite suite from Engeln has been given by Okrusch et al. (1979). According to these authors all granulites have granoblastic textures and grain sizes between 0.2 and 2 mm. Foliation or mineral orientation typical of gneiss are only very rarely observed. The xenoliths investigated by us consist of plagioclase (plag) + clinopyroxene (cpx) + amphibole (amph) ± garnet (gnt) + orthopyroxene (opx) + opaques ± (rare) scapolite. The amount of amph is highly variable in these nominally 'dry' rocks, ranging from almost zero to about 50%. According to their modes of appearance three types of amph may be distinguished: (a) coarse grained amph which, by textural criteria, is primary; (b) Secondary amph marginally replacing cpx; this type of amph also occurs in grain fractures of cpx and as streaks of fine lamellae in cpx; (c) small idiomorphic amph in reaction coronas around gnt. There are only minor differences in major element composition between these types of amph. The garnets in many of the granulites are frequently surrounded by reaction coronas of plag + opx + Ti-magnetite + amph (Okrusch et al. 1979). The garnets preserved in these granulites are homogeneous with respect to their major element composition. In contrast, opx and cpx frequently reveal marginal zoning patterns (< 300 μm in width) with Al and (for opx only) Ca concentrations increasing towards the rims. This zoning may reflect a later heating event

From: DAWSON, J.B., CARSWELL, D.A., HALL, J. & WEDEPOHL, K.H. (eds) 1986, *The Nature of the Lower Continental Crust*, Geological Society Special Publication No. 24, pp. 309–317.

TABLE 1. *Major and trace element concentrations in selected granulite xenoliths from the Eifel*

	S11	S16	S37	S61	S35	S30	S43	S24*
SiO_2 (wt. %)	49.5	49.6	50.0	45.3	47.3	46.0	43.9	43.2
TiO_2 (wt. %)	0.20	0.32	0.54	1.40	1.95	0.88	2.00	0.03
Al_2O_3 (wt. %)	18.7	15.5	17.1	13.2	13.6	18.6	18.7	1.44
Fe_2O_3 (wt. %)	0.2	1.7	0.6	2·9	4·7	2.9	6.1	1.0
FeO (wt. %)	2.8	6.4	7.0	11.3	9.7	5.0	6.7	7.4
MnO (wt. %)	0.08	0.17	—	0.24	0.22	0.18	0.28	0.14
MgO (wt. %)	8.6	11.6	10.6	9.7	7.2	7.4	4.9	43.0
CaO (wt. %)	17.1	12.2	11.5	12.8	12.3	14.2	10.4	1.67
K_2O (wt. %)	0.10	0.10	0.06	0.20	0.22	0.64	0.67	0.04
Na_2O (wt. %)	1.43	1.44	1.49	1.68	2.12	2.48	3.91	0.27
P_2O_5 (wt. %)	0.12	0·11	0·09	0.21	0.27	0.32	0.72	0.01
Mg/(Mg+ΣFe)	0.84	0.72	0.71	0.55	0.48	0.63	0.42	0.90
La (wt. ppm)	6.3 ±0.2	3.05	2.55	5.9	13.9	26.4	47.9 ±0.8	4.40 ±0.15
Ce (wt. ppm)	12.2 ±0.8	—	5.8	16.8	39	63	107 ±5	9.1 ±0.6
Nd (wt. ppm)	5.9 ±1.0	5.2	4.7	14.7	26	33	51 ±6	2.4 ±0.8
Sm (wt. ppm)	1.36±0.04	1.25	1.46	3.54	6.25	6.05	9.2 ±0.1	0.27 ±0.03
Eu (wt. ppm)	0.56±0.06	0.69	0.66	1.22	1.69	1.57	2.34±0.09	0.075±0.010
Tb (wt. ppm)	0.19±0.04	0.24	0.23	0.61	1.01	0.66	1.17±0.13	0.06 ±0.03
Yb (wt. ppm)	0.54±0.15	0.67	0.72	2.11	2.68	1.91	3.40±0.15	0.14 ±0.07
Lu (wt. ppm)	0.12±0.03	0.11	0.094	0.38	0.42	0.33	0.51±0.04	0.04 ±0.02
Rb (wt. ppm)	0.73	0.33	0.210	0.40	1.25	3.4	7.1	0.051
Sr (wt. ppm)	577.0	407.6	406.2	157.8	266.1	916.0	759.0	61.0
Sc (wt. ppm)	29.0 ±0.4	29.6	30.1	44.2	39.5	35.6	34.8 ±0.5	11.2 ±0.3
Cr (wt. ppm)	331 ±6	335	341	205	157	82	15 ±3	2400 ±50
Co (wt. ppm)	20.9 ±0.4	42.3	—	56.9	47.7	30.8	28.3 ±0.5	101 ±1
Hf (wt. ppm)	0.48±0.15	0.41	0.43	1.57	2.8	1.72	6.5 ±0.3	<0.1
Ta (wt. ppm)	0.09±0.05	<0.2	—	0.20	0.40	0.23	0.82±0.07	<0.1
Th (wt. ppm)	<0.3	<0.3	<0.2	<0.3	0.41	0.80	2.30±0.20	<0.3

* mantle spinel peridotite nodule from Engeln/Eifel.
Major element concentrations have been obtained by X-ray fluorescence spectrometry and wet chemical techniques. Rb and Sr concentrations are from mass spectrometric isotope dilution (Stosch & Lugmair 1984), all other trace element data from instrumental neutron activation analysis.
Errors quoted for trace element concentrations of samples S11, S43 and S24 are 2σ-counting statistics of γ-spectrometry. Total errors for La, Sm, Sc, Cr and Co are estimated to be ±3–5% in cases where 2σ errors are smaller than about 3%. For trace element concentrations with 2σ errors > 5% the listed errors of counting statistics should be close to the total errors. Analytical errors for trace element analyses of the other granulites may be estimated by interpolation between the concentration levels of S11 and S43. Uncertainties for concentrations of Rb are ±0.5%, for Sr ±0.2%.

up to temperatures exceeding those of granulite metamorphism. If recent measurements of diffusion coefficients are applicable (Sneeringer *et al.* 1984 or compilation of Freer 1981) the width of this zoning indicates a prolonged period of heating and cannot be ascribed to contact heating in the host magma which carried the granulites to the surface. Trails of fluid inclusions are ubiquitous in the granulites but are especially abundant in some amph rich members of this suite. Where amph appears to be texturally primary these trails sometimes cross-cut plag-amph and cpx-amph grain boundaries indicating that the formation of at least part of the fluid inclusion trails is a very late feature in these rocks.

Chemistry

Major and trace element data of representative granulite xenoliths are given in Table 1. These results confirm the observation of Okrusch *et al.* (1979) and Voll (1983) that the bulk composition of the granulites is invariably mafic. This is frequently observed in other lower crustal xenolith populations as well (Kay & Kay 1981) and contrasts with granulite-facies rocks exposed at the earth's surface, which are predominantly felsic. Granulite-facies xenoliths from the Hessian Depression east of the Rhenish Massif display a similar mafic bulk composition. In contrast to the suite from Engeln these granulites

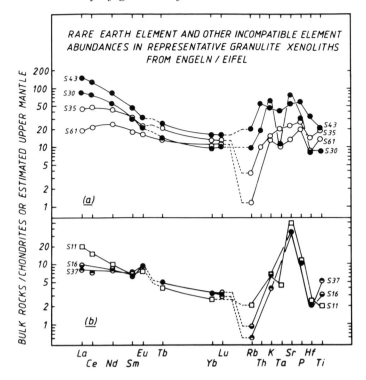

FIG. 1. Normalized trace element abundances in selected granulite xenoliths from Engeln. REE abundances are normalized to chondrite values recommended by Boynton (1984). For all other elements the estimated primitive mantle abundances of Thompson *et al.* (1982) were used. (a) rocks dominated by modal cpx and/or amph, (b) rocks dominated by modal plag.

lack gnt and may have formed at similar temperatures but lower pressures than the suite from the Eifel (Mengel & Wedepohl 1983). A remarkable chemical feature of the lower crustal suite from the Eifel is relatively low, though highly variable, K_2O contents with higher concentrations correlating with high modal amph. A second characteristic is that granulites with the highest Mg-numbers are among the least mafic members within this series. Such rocks also have the highest abundances of compatible trace elements like Cr and the lowest abundances of incompatible 'immobile' trace elements like Ta or Hf (Table 1). Conversely, samples with low Mg-numbers are low and high, respectively, in such trace element concentrations.

Fig. 1 shows normalized paterns of the REE and of some other trace elements of the granulites listed in Table 1. All samples display REE patterns with relative light over heavy REE enrichment. The overall lowest REE abundances and light/heavy REE fractionations are observed for granulites with the highest Mg-numbers (like S16 or S37, Fig. 1b). Their mineralogy is dominated by plag. Only such samples display positive Eu anomalies. Fig. 1a shows data for cpx and/or amph rich granulites. Typically they have intermediate to low Mg-values and higher REE contents as well as light/heavy REE fractionations than granulites rich in plag. The majority of these rocks has negative Eu anomalies and none shows a positive one. Similar correlations between mineralogy (abundance of plag and cpx), Mg-values and REE characteristics have been noted previously for other lower crustal xenoliths suites, e.g. from Lesotho (Rogers 1980; Rogers & Hawkesworth 1982), South Australia (McCulloch *et al.* 1982) or, less pronounced, the French Massif Central (Dostal *et al.* 1980). A sample of amph bearing mantle spinel peridotite from Engeln (S24, data given in Table 1) has low heavy REE abundances but an extremely high light over middle REE enrichment about twice as high as found for alkalic basaltic volcanics from across the Eifel. The relative light REE enrichment of this spinel peridotite is similar to that measured

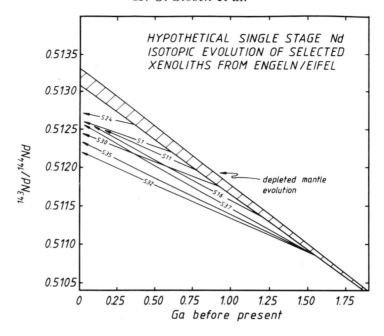

FIG. 2. Hypothetical Nd isotopic evolution diagram for selected granulites and one amph-bearing spinel peridotite (S24) from Engeln. Differentiation of the igneous precursors of these rocks from depleted mantle is assumed. The lower line of the depleted mantle evolution field corresponds to the approximation of DePaolo (1980), the upper line to the model of Zindler (1982).

for the most extreme members of the amph bearing spinel peridotite suite from Dreiser Weiher (Stosch & Seck 1980).

The elements Rb to Ti, also plotted in Fig. 1 are arranged from left to right approximately in the order of increasing compatibility for a mafic mineral assemblage. The distinct Sr peaks for the plag rich granulites in Fig. 1b reflect the compatibility of this element for the plagioclase structure and incompatibility of the other elements. An important observation from this plot is the low Rb values relative to those of K or Sr of most granulites. The highest concentrations of Rb and K have been measured in the rocks with the highest contents of amph. Also, Th is relatively low in these high grade rocks from the lower crust with the exception of one sample (S1, data not given here) which is also aberrant with respect to various other chemical data. These features resemble those observed in some Archaean high grade terrains where lower concentrations of heat producing elements (K, Rb, Th, U) relative to amphibolite-facies metamorphics of similar bulk composition (Heier 1973; Weaver & Tarney 1980) have been measured.

Discussion

A meta-igneous rather than meta-sedimentary origin of the granulite protoliths is indicated by their invariably mafic bulk compositions. Observations supporting such an interpretation are (a) positive Eu anomalies in some and the lack of Eu anomalies in some other granulites; (b) fairly smooth variations are observed if the concentrations of a compatible trace element like Cr or of slightly incompatible trace elements like Yb or Hf are plotted against the Mg-numbers. This suggests that at least many of the granulite-facies xenoliths from Engeln investigated by us may be regarded as a cogenetic igneous suite ranging from early cumulates (samples with high Mg-numbers, low REE abundances and positive Eu anomalies) to residual liquids (samples with low Mg-values, high light REE abundances, negative Eu anomalies). The memory of this fractionation event may be hoped to have been recorded by the Sm-Nd isotope system.

In Fig. 2 we have plotted the hypothetical single stage and isotopic evolution of selected granulites and a sample of mantle peridotite from

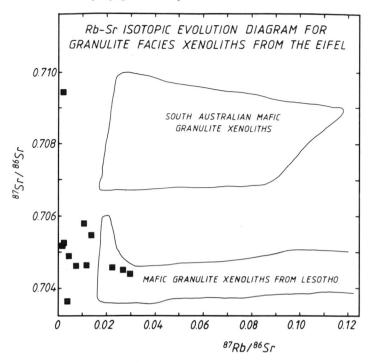

FIG. 3. Rb-Sr isotopic composition of the granulite-facies xenoliths from the Eifel, compared to compositionally similar suites from Lesotho (Rogers & Hawkesworth 1982) and South Australia (McCulloch *et al.* 1982).

Engeln as a function of time using the data of Stosch & Lugmair (1984). Also, we have plotted the depleted mantle evolution path with the parameters suggested by DePaolo (1980) and Zindler (1982). Assuming the granulite igneous precursors differentiated from depleted mantle, crustal residence ages varying from ∼0.65 Ga (S1) to ∼1.6 Ga are indicated by this plot. This suggests that the Sm-Nd isotope system has been disturbed after igneous crystallization of these rocks. In principle, it would be possible to ascribe this dispersion of model ages to partial redistribution of Sm and Nd over significantly larger than xenolith size during granulite-facies metamorphism. In this case the igneous age of the granulite precursors should fall somewhere in between 0.65 and 1.6 Ga. From a granulite terrain in Enderby Land/Eastern Antarctica there appears to be some evidence for such metamorphic repartitioning (DePaolo *et al.* 1982; McCulloch & Black 1984). However, for the granulite xenoliths from Engeln we consider such an interpretation not applicable. Note, that the model ages obtained for the majority of the granulites from Engeln are roughly related to their amph contents: S32, S35 and S37, which bear subordinate amounts of

amph only, have almost identical depleted mantle model ages of ∼1.6 Ga (Fig. 2). In a Sm-Nd isochron diagram the data for these rocks plot along a straight line which corresponds to an age of ∼1.5 Ga (Stosch & Lugmair 1984). All other granulites have higher to much higher modal amph and are displaced from this tentative isochron towards values for an amph-bearing spinel peridotite xenolith from Engeln (S24), that is towards lower Sm/Nd and/or higher ^{143}Nd/^{144}Nd. As a consequence such rocks like S30 or S11 have model ages younger than ∼1.6 Ga (Fig. 2). We suggest this trend reflects variable degrees of metasomatic alteration, caused by infiltration of mantle derived fluids which transported incompatible elements into the lower crust, thus causing Sm-Nd isotopic rejuvenation of many granulite xenoliths.

This metasomatic alteration should be reflected by the REE patterns of the granulites, too. In Fig. 1b, we have plotted the REE data for three rocks whose protoliths are interpreted to be early cumulates. All three have almost identical concentrations of compatible (Cr) to slightly incompatible 'immobile' trace elements (Sc, Hf). The sample with the lowest Yb concentration (S11)

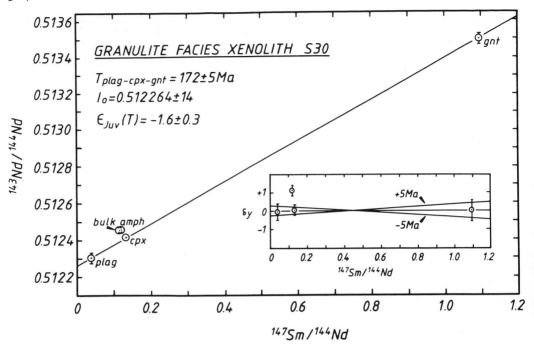

FIG. 4. Sm-Nd isochron diagram for bulk rock and minerals separated from granulite S30. Inset shows the relative deviation $\delta\gamma$ (in parts of 10,000) of the data from the best fit-line through plag, cpx and gnt. Element concentrations and isotopic compositions measured are: bulk Sm 5.82 ppm; Nd 30.7 ppm; $^{147}Sm/^{144}Nd$ 0.1145; $^{143}Nd/^{144}Nd$ 0.512453 $\pm$ 14; amph 11.81; 59.0; 0.1211; 0.512456 $\pm$ 14; cpx 4.97; 22.65; 0.1326; 0.512415 $\pm$ 14; gnt 1.824; 1.007; 1.095; 0.513499 $\pm$ 30; plag 0.1831; 2.804; 0.03946; 0.512306 $\pm$ 24.

has the highest Mg-number whereas the two other granulites have identical $Mg/(Mg + \Sigma Fe)$. Granulite S37 shows incipient formation of amph around some cpxs only. This rock has the lowest La concentration as well as La/Yb ratio. Significantly more amph is present in granulite S16. Apart from its presence as reaction rims around cpx the amph here occurs as abundant small grains in reaction coronas surrounding gnt and, to a smaller extent, as primary grains. As expected, S16 is higher in La and La/Yb than S37. In contrast to S37 the data for S16 plot to the left of the tentative 1.5 Ga Sm-Nd isochron (Stosch & Lugmair 1984). S11 has the highest La concentration and also shows the highest La/Yb ratio among these three granulites even though it has the highest Mg-value. Here amph formation has progressed to form more texturally 'primary' grains but amph rims round cpx are abundant still. The data for S11 plot even more to the left of the tentative 1.5 Ga isochron of Stosch & Lugmair (1984) than for S16. Therefore, we interpret this sequence of increasing light REE abundances and light/heavy REE enrichments as being caused by increasing degrees of metasomatism.

A metasomatic model is also broadly consistent with the Rb-Sr isotopic systematics which have been plotted in Fig. 3 in a $^{87}Sr/^{86}Sr$ vs. $^{87}Rb/^{86}Sr$ diagram (data from Stosch & Lugmair 1984). Compared to granulite facies lower crustal xenoliths from Lesotho (Rogers & Hawkesworth 1982) and South Australia (McCulloch et al. 1982) the suite from Engeln is characterized by extremely low Rb/Sr ratios which render their $^{87}Sr/^{86}Sr$ essentially frozen. The present-day variation in $^{87}Sr/^{86}Sr$ cannot, therefore, be explained on the basis of the measured Rb/Sr ratios. We point out that, apart from the sample with the lowest $^{87}Sr/^{86}Sr$ (S11), higher Rb/Sr ratios and higher abundances of amph in these granulites roughly correlate with lower $^{87}Sr/^{86}Sr$ compositions, that is they are displaced towards the value measured for amphibole bearing mantle peridotite S24 ($\sim$0.7040). The total Sr isotopic variation, however, is unlikely to be explained by metasomatic alteration alone because high Sr concentrations of many granulites (especially those with high Sr/Nd ratios like S16, Table 1) should tend to buffer $^{87}Sr/^{86}Sr$ against modification through infiltration of fluids. Pre-metasomatic (and, probably, pre-granulite facies metamorphic) Sr isotopic variations are likely to have existed as the time integrated response to variable Rb/Sr ratios of the igneous granulite protoliths.

Additional support for our suggestion of metasomatic alteration after granulite facies metamorphism is derived from isotope analyses of minerals separated from granulite S30. The Sm-Nd data have been published previously (Stosch & Lugmair 1984) and are given in Fig. 4 in an isochron diagram. Plag, cpx and gnt plot along a line the slope of which corresponds to an age of 172 ± 5 Ma. The data for amph from this granulite as well as the bulk rock, however, plot significantly above the isochron at slightly lower Sm/Nd ratios than measured for cpx, that is they are displaced towards values for amph bearing mantle peridotite S24 (^{147}Sm/^{144}Nd = 0.066, ^{143}Nd/^{144}Nd = 0.512714 $\pm$ 16, Stosch & Lugmair 1984). The data points for amph and bulk rock are almost identical because by mass balance the amph (modal abundance 40–45%) accounts for 3/4 of this granulite's total Nd. This indicates the presence of a Nd isotopic component in the amph introduced from outside the rock at some time less than 172 ± 5 Ma ago.

In contrast to the Sm-Nd system, plag, cpx and gnt separated from S30 show no isochron relationships in the Rb-Sr isochron plot. Cpx and plag are identical in ^{87}Sr/^{86}Sr (~ 0.7061) which is not surprising in view of their extremely small ^{87}Rb/^{86}Sr ratios (0.0021 and 0.00050, respectively). Amph, the mineral with the highest ^{87}Rb/^{86}Sr ratio in this rock (~ 0.10), however, is by far the lowest in ^{87}Sr/^{86}Sr (~ 0.7054). This clearly indicates open system behaviour of this rock at some time during its evolution, but less than 172 ± 5 Ma ago.

For us the most likely explanation is that the amph contains a Sr as well as Nd isotopic component of the same origin as seen in enriched (amphibole bearing as well as anhydrous) mantle xenoliths from the Eifel (Stosch *et al.* 1980). It should be noted here that texturally the amph from xenolith S30 appears to be 'primary'. Two explanations can be offered to account for this observation: (a) recent metasomatism in the lower crust was accompanied by recrystallization; or (b) the amph indeed formed during the granulite facies event or during a retrograde amphibolite facies overprint. Less than 172 ± 5 Ma ago amph reacted (or continued to react) with infiltrating fluids whereas plag, cpx and gnt behaved as closed systems with respect to Sm and Nd exchange ever since. In either case our data require metasomatic Sr and Nd isotopic modification for several of the granulites from the Eifel lower crust. Only granulites almost free of amphi-

bole escaped this alteration. It is interesting to note that for granulite S35 which shows incipient formation of amph around part of the cpx only, but no 'primary' amph, the bulk falls on an Sm-Nd isochron together with plag, cpx and gnt (Stosch *et al.* 1982).

The young age of this metasomatic event is difficult to reconcile with current ideas on the age of metamorphism of the granulite xenoliths. The geological model of Okrusch *et al.* (1979) or Voll (1983) suggests granulite facies metamorphism during the Caledonian orogeny or even earlier and an amphibolite facies overprint during the Hercynian orogeny. With the recent evidence for granulite-facies metamorphism in Central Europe about 330–335 Ma ago (van Breemen *et al.* 1982) Hercynian ages of high grade metamorphism in the lower crust under the Eifel should also be considered. The mineral isochron from S30, however, gives a Jurassic age which is not easily incorporated into these models. Before the potential time significance of this mineral isochron can be evaluated, it will be essential to constrain the thermal history of these granulites which is not well understood at present. Okrusch *et al.* (1979) and Voll (1983) attributed the estimated equilibration temperatures of between 700° and 850°C to the granulite facies metamorphism, presumably of Palaeozoic or even older age. On the other hand, studies of mantle xenoliths from the Westeifel indicated equilibration temperatures of 900°–1000°C for a group of tabular recrystallized amph bearing spinel peridotites from Dreiser Weiher (Sachtleben 1980; Sachtleben & Seck 1981). Temperatures even down to about 800°C have been inferred for a group of porphyroclastic spinel peridotites from Schönfeld/Westeifel (Witt & Seck, in prep.). In both cases these temperatures are viewed as the ambient temperatures prior to volcanism, i.e. as Quaternary, and have led to models of diapiric upwelling of asthenospheric mantle material by these authors. Such an interpretation gains some support from seismic investigations in the Rhenish Massif which indicate the presence of a low velocity body under the Eifel at depths between some 50 and 200 km and could be explained by the presence of a 1–2% partial melt (Raikes & Bonjer 1983). If the uppermost present-day mantle under the Eifel indeed has anomalously high temperatures of between 800° and 1000°C then one might speculate whether this young (late Tertiary to recent) thermal perturbation inferred from mantle xenoliths has reached the lowermost crust already. As a result of this heating event, associated by an introduction of fluids and recrystallization, lower crustal temperatures may have been elevated above the 'blocking temperature' where Nd (and Sr) isotopic disequilibrium between plag, cpx and gar cannot be retained. In this case a former isochron having recorded the (Hercynian, Caledonian or older?) granulite facies event or a retrograde amphibolite facies event (during the Variscan orogeny?) would be on its way to destruction and complete Nd isotopic homogenization. Such a case of complete Sr and Nd isotopic equilibration has been reported by Richardson *et al.* (1980) between coexisting plag and cpx from a granulite facies orthogneiss from Kilbourne Hole/New Mexico. These authors suggest this equilibrium is a consequence of recent heating of the Proterozoic lower crust under the Rio Grande Rift. Under this aspect we cannot rule out completely that the mineral isochron for granulite S30 has no time significance. However, since different minerals are likely to have different blocking temperatures for isotopic exchange the Sm-Nd isotope data of plag, cpx and gnt are not expected to remain on one line during equilibration with the amph. Therefore, we also consider unlikely the possibility that the temperatures estimated from various geothermometers for the granulites from Engeln reflect a recent heating event in the lower crust. Based on an evaluation of the speed of cation and rare gas diffusion in minerals Harte *et al.* (1981) have argued previously that the temperatures estimated from mineral equilibria in medium to coarse-grained lower crustal xenoliths from stable tectonic regions will not represent the ambient temperatures at the time of eruption of the host basalt or host kimberlite, at least for rocks lacking a fluid phase. Hence, we favour the interpretation that the age of the mineral isochron from granulite S30 relates to an event during the Mesozoic. In this case, the Sm-Nd isotopic system in plag, cpx and gnt of this rock became closed 172 ± 5 Ma ago. Amph, which could have

been formed during such an event, continued to react with available fluids for some time and presumably equilibrated the amph isotopically and chemically over the scale much larger than xenolith size in this granulite complex. Although there is no evidence for a major tectonic event in the Eifel during the Mesozoic, the widespread occurrence of hydrothermal ore deposits of the general age 200 to 100 Ma ago in Central and Western Europe as well as in Britain has been related to tectonic processes during the opening of the North Atlantic (Mitchell & Halliday 1976).

Many of the granulite xenoliths from Engeln investigated by us have been affected by a metasomatic overprint less than 172 ± 5 Ma ago. It is linked to pargasitic amph in these rocks (this paper) and also in a group of peridotites from the subcrustal lithosphere under the Eifel (Stosch & Seck 1980). The scale of this metasomatism cannot be derived from the study of xenoliths. We note, however, that we have no evidence to believe that it may be restricted to zones around basaltic dykes or veins. For the mantle a maximum age of 200 Ma has been estimated previously for this episode of metasomatic modification (Stosch *et al.* 1980). It appears likely, therefore, that metasomatism in upper mantle and lower crust are related in time and by origin.

ACKNOWLEDGEMENTS: H.-G. Stosch thanks Drs W. Herr and U. Herpers for access to the neutron activation analysis facilities of the Institut für Kernchemie, Univ. Köln. C. MacIsaac is thanked for his assistance in the UCSD isotope laboratories. We appreciate constructive reviews of a previous version of this paper by Drs K.H. Wedepohl, A.W. Hofmann, R.L. Rudnick, C.J. Hawkesworth and N.W. Rogers. This research was supported by grants from the DFG to H.A. Seck and by NSF EAR-8213670 to J.D. Macdougall and G.W. Lugmair. SIO-IG^{2}L contribution No. 26.

References

ARNOLD, A. & SCHARBERT, H.G. 1973. Rb-Sr-Altersbestimmungen an Granuliten der südlichen Böhmischen Masse in Österreich. *Schweiz. Mineral. Petrogr. Mitt.* **53**, 61–78.

BOYNTON, W.V. 1984. Cosmochemistry of the rare earth elements: meteorite studies. *In*: HENDERSON, P. (ed.) *Rare Earth Element Geochemistry*. Elsevier, 63–114.

DEPAOLO, D.J. 1980. Neodymium isotopes in the Colorado Front Range and crust-mantle evolution in the Proterozoic. *Nature*, **291**, 193–196.

——, MANTON, W.I., GREW, E.S. & HALPERN, M. 1982. Sm-Nd, Rb-Sr and U-Th-Pb systematics of granulite facies rocks from Fyfe Hills, Enderby Land, Antarctica. *Nature*, **298**, 614–618.

DOSTAL, J., DUPUY, C. & LEYRELOUP, A. 1980. Geochemistry and petrology of meta-igneous granulitic xenoliths in Neogene volcanic rocks of the Massif Central, France—implications for the lower crust. *Earth Planet. Sci. Lett.*, **50**, 31–40.

FREER, R. 1981. Diffusion in silicate minerals and glasses: a data digest and guide to the literature. *Contrib. Mineral. Petrol.*, **76**, 440–454.

FUCHS, K., VON GEHLEN, K., MÄLZER, H., MURAWSKI, H. & SEMMEL, A. (eds) 1983. Plateau Uplift, the Rhenish Shield—a case history. Springer-Verlag Berlin, 411 pp.

HARTE, B., JACKSON, P.M. & MACINTYRE, R.M. 1981. Age of mineral equilibria in granulite facies nodules from kimberlites. *Nature*, **291**, 147–148.

HEIER, K.S. 1976. Geochemistry of granulite facies rocks and problems of their origin. *Phil. Trans. R. Soc. London*, **273A**, 429–442.

JÄGER, E. & WATZNAUER, A. 1969. Einige Rb/Sr Datierungen an Granuliten des Sächsischen Granulitgebirges. *Monatsber. Dtsch. Akad. Wiss. Berlin*, **11**, 420–426.

KAY, R.W. & KAY, S.M. 1981. The nature of the lower continental crust: inferences from geophysics, surface geology, and crustal xenoliths. *Rev. Geophys. Space Phys.*, **19**, 271–297.

McCULLOCH, M.T., ARCULUS, R.J., CHAPPELL, B.W. & FERGUSON, J. 1982. Isotopic and geochemical studies of nodules in kimberlite have implications for the lower continental crust. *Nature*, **300**, 166–169.

—— & BLACK, L.P. 1984. Sm-Nd isotopic systematics of Enderby Land granulites and evidence for the redistribution of Sm and Nd during metamorphism. *Earth Planet. Sci. Lett.*, **71**, 46–58.

MENGEL, K. & WEDEPOHL, K.H. 1983. Crustal xenoliths in Tertiary volcanics from the northern Hessian Depression. *In*: FUCHS, K. *et al.* (eds) *Plateau Uplift, the Rhenish Shield—a Case History*. Springer-Verlag Berlin, 332–335.

MITCHELL, J.G. & HALLIDAY, A.N. 1976. Extent of Triassic–Jurassic hydrothermal ore deposits on the North Atlantic margins. *Trans. Instn. Min. Metall.*, **B85**, 159–161.

OKRUSCH, M., SCHRÖDER, B. & SCHNÜTGEN, A. 1979. Granulite-facies metabasite ejecta in the Laacher See area, Eifel, West Germany. *Lithos*, **12**, 251–270.

RAIKES, S. & BONJER, K.-P. 1983. Large-scale mantle heterogeneity beneath the Rhenish Massif and its vicinity from teleseismic P-residuals measurements. *In*: FUCHS, K. *et al.* (eds) *Plateau Uplift, the Rhenish Shield—a Case History*. Springer-Verlag Berlin, 315–331.

RICHARDSON, S.H., PADOVANI, E.R. & HART, S.R. 1980. The gneiss syndrome: Nd and Sr isotopic relationships in lower crustal granulite xenoliths, Kilbourne Hole, New Mexico. *EOS*, **61**, 338 (abstract).

ROGERS, N.W. 1977. Granulite xenoliths from Lesotho kimberlites and the composition of the lower crust. *Nature*, **270**, 681–684.

—— & HAWKESWORTH, C.J. 1982. Proterozoic age and cumulate origin for granulite xenoliths, Lesotho. *Nature*, **299**, 409–413.

SACHTLEBEN, T. 1980. *Petrologie ultrabasischer Aus-*

würflinge aus der Westeifel. Ph.D. thesis, Univ. Köln.

—— & SECK, H.A. 1981. Chemical control of Al solubility in orthopyroxene and its implications on pyroxene geothermometry. *Contrib. Mineral. Petrol.*, **78**, 157–165

SNEERINGER, M., HART, S.R. & SHIMIZU, N. 1984. Strontium and samarium diffusion in diopside. *Geochim. Cosmochim. Acta*, **48**, 1589–1608.

STOSCH, H.-G., CARLSON, R.W. & LUGMAIR, G.W. 1980. Episodic mantle differentiation: Nd and Sr isotopic evidence. *Earth Planet. Sci. Lett.*, **47**, 263–271.

—— & LUGMAIR, G.W. 1984. Evolution of the lower continental crust: granulite facies xenoliths from the Eifel, West Germany. *Nature*, **311**, 368–370.

——, LUGMAIR, G.W. & SECK, H.A. 1982. Granulite facies xenoliths from the Eifel region, W. Germany: Geochemistry and evolution. *EOS*, **63**, 1151 (abstract).

—— & SECK, H.A. 1980. Geochemistry and mineralogy of two spinel peridotite suites from Dreiser Weiher, West Germany. *Geochim. Cosmochim. Acta*, **44**, 457–470.

THOMPSON, R.N., DICKIN, A.P., GIBSON, I.L. & MORRISON, M.A. 1982. Elemental fingerprints of isotopic contamination of Hebridean Paleocene mantle-derived magmas by Archaean sial. *Contrib. Mineral. Petrol.*, **79**, 159–168.

VAN BREEMEN, O., AFTALION, M., BOWES, D.R., DUDEK, A., MÍSAŘ, Z., POVONDRA, P. & VRÁNA, S. 1982. Geochronological studies of the Bohemian massif, Czechoslovakia, and their significance in the evolution of Central Europe. *Transactions R. Soc. Edinburgh, Earth Sci.*, **73**, 89–108.

VOLL, G. 1983. Crustal xenoliths and their evidence for crustal structure underneath the Eifel volcanic district. *In*: FUCHS, K. *et al.* (eds) *Plateau Uplift, the Rhenish Shield—a Case History*. Springer-Verlag, Berlin, 336–342.

WEAVER, B.L. & TARNEY, J. 1980. Rare-earth geochemistry of Lewisian granulite-facies gneisses, NW Scotland: implications for the petrogenesis of the Archaean lower continental crust. *Earth Planet. Sci. Lett.*, **55**, 279–296.

ZINDLER, A. 1982. Nd and Sr isotope ratios in komatiites and related rocks: implications for petrogenesis and mantle evolution. *In*: ARNDT, N.T. & NESBIT, E. (eds) *Komatiites*. Cambridge University Press, 399–420.

H.-G. STOSCH & H.A. SECK, Mineralogisch-Petrographisches Institut, Universität zu Köln, Zülpicher Str. 49, 5000 Köln 1, West Germany.
G.W. LUGMAIR, Geological Research Division A-012, Scripps Institution of Oceanography, University of California at San Diego, La Jolla, Ca. 92093, USA.

Granulitic xenoliths from the French Massif Central— petrology, Sr and Nd isotope systematics and model age estimates

H. Downes & A. Leyreloup

SUMMARY: The protoliths of granulitic xenoliths found in Tertiary/Quaternary alkali volcanics from the French Massif Central are shown to have been principally immature sediments and calc-alkaline intrusives. The association may indicate formation of the crust in an island-arc or continental margin environment. Pressure and temperature estimates indicate that they have undergone intermediate to high pressure granulite-facies metamorphism (9–15 kbar) followed by uplift to lower pressures (4–9 kbar). Rb-Sr and Sm-Nd isotopic data and REE abundances are presented for both meta-igneous and meta-sedimentary suites. εSr present-day values for the granulitic xenoliths range from $+2.7$ to $+325$, with εNd values between $+0.2$ and -12.0. There are no isochron relationships within the Sr and Nd data, but possible mixing lines in the Rb-Sr system may indicate sedimentation at 700–1000 Myr. T_{DM}^{Nd} model ages indicate a Proterozoic (800–2100 Myr) origin for the protoliths of the meta-igneous samples and a similar provenance age for the meta-sediments.

There are several sources of information about the petrology, mineralogy and chemistry of the lower crust, notably: 1) direct sampling by volcanic activity (xenoliths), 2) analysis of crustal-derived melts (granitoids) and 3) thrust slices in orogenic belts (root zones). This study is concerned with the nature of granulitic xenoliths from the French Massif Central, their Sr and Nd isotope systematics and the age of formation of their protoliths.

The French Massif Central is one of the largest outcrops of the Variscan orogenic belt in Western Europe. The metamorphic grade of the upper crust reaches upper amphibolite facies, with some retrogressed granulites (leptyno-amphibolite group) for which ages of 480–500 Myr have been obtained (Pin & Lancelot 1982). Pin and Vielzeuf (1983) have discussed the differences between these granulitic lenses brought up in thrust slices during the Variscan orogeny and the granulitic xenoliths found in the Tertiary/Quaternary volcanics of the Massif Central. The latter can be compared with granulites from Alpine thrust sheets, e.g. the Ivrea Zone. Such granulitic xenoliths are abundant in the Bournac tuff pipe in Velay (Leyreloup 1973; Leyreloup *et al.* 1977) and the Roche Pointue breccia in Cantal (Cornen 1972). Extensive chemical and petrographic studies have been made of the xenoliths (Dupuy *et al.* 1977; Dostal *et al.* 1980; Leyreloup, in prep.), from which meta-igneous (60%) and meta-sedimentary (40%) suites have been identified.

Xenoliths enclosed in tuffs and volcanoclastic breccias are less affected by pyrometamorphism than those in lava flows or dykes. Only the outermost 5 mm are affected chemically or isotopically by the host basalt (Leyreloup *et al.* 1977; Allègre *et al.* 1975) and only the cores of xenoliths have been analysed in this study.

Petrology of the xenoliths

The xenoliths are angular, 5–40 cm in diameter, fine to medium grained with either granoblastic or granuloblastic textures; some samples are banded and show foliation. Two main xenolith suites are present (Leyreloup *et al.* 1977): 1) a mafic to acidic meta-igneous suite, resembling a recrystallized gabbro-norite series; 2) a felsic meta-sedimentary sequence. Xenoliths showing contracts between meta-igneous and meta-sedimentary lithologies are occasionally found.

In the meta-igneous granulites the main mineral phases are a) in *ultramafic xenoliths* (pyroxenites, hornblendites): opx, cpx, spinel, brown hornblende, phlogopite-biotite, rutile and garnet; b) in *mafic xenoliths* (pyriclasites, pyribolites, pyrigarnites, amphiclasites, coronitic metagabbros): calcic plagioclase, opx, cpx ± garnet ± spinel ± quartz ± brown hornblende ± biotite, ferrian ilmenite, rutile, apatite, very rare zircon and Ti-magnetite; c) in *intermediate and acidic xenoliths* (charno-enderbitic to charnockitic gneisses): quartz, oligoclase or andesine ± K-feldspar ± garnet, opx, cpx ± biotite, rutile, ferrian ilmenite, rare Ti-magnetite, apatite, zircon, monazite. CO_2-rich fluid inclusions are abundant in the meta-igneous xenoliths (Bilal & Touret 1977), yielding pressure estimates which are compatible with granulite-facies metamorphism at 10–12 kbar.

From: DAWSON, J.B., CARSWELL, D.A., HALL, J. & WEDEPOHL, K.H. (eds) 1986, *The Nature of the Lower Continental Crust*, Geological Society Special Publication No. 24, pp. 319–330.

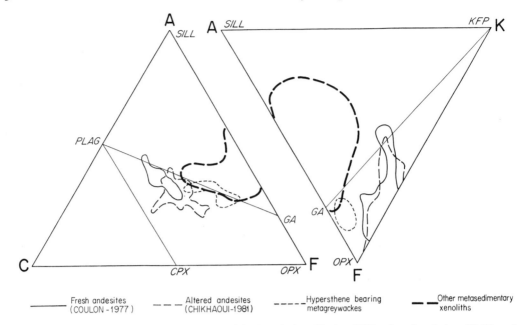

FIG. 1. AKF and ACF diagrams showing field of fresh andesites (Coulon 1977), altered andesites (Chikhaoui 1981), hypersthene metagreywackes and other metasedimentary granulitic xenoliths from Bournac and Roche Pointue. The field of average andesites of Gill (1981) is very similar to the fresh andesite field shown here.

Meta-sedimentary xenoliths have metatexitic or diatexitic textures and contain leucocratic and Al-rich minerals: quartz, K-feldspar, oligoclase or andesine, ±sillimanite, garnet±biotite, with accessory corundum, rutile±spinel (often Zn-rich), ilmenite, Ti-magnetite, apatite, zircon, monazite and graphite. No CO_2-rich fluid inclusions are found in these xenoliths (Touret, pers. comm. 1985). Three main rock-types have been distinguished: a) *khondalito-kinzigitic granulites* (pelitic s.l.) characterized by quartz, albite-oligoclase, K-feldspar, garnet, sillimanite and accessory minerals; b) *garnet granulites* without sillimanite, richer in quartz and feldspar and derived from arenites s.l.; c) hypersthene metagreywackes, probably volcanoclastic in origin, containing quartz, opx, garnet, oligoclase-andesine and K-feldspar, but without sillimanite.

Chemical affinities of the xenoliths

Chemical differences between the meta-igneous and meta-sedimentary suites (e.g. absolute concentrations, relative abundances, inter-element relationships and multivariate analysis) are discussed by Leyreloup *et. al.* (1977) and Leyreloup (in prep. 1985). The average major and trace element contents of the meta-sedimentary series

are comparable with greywacko-pelitic rocks (Curtis 1969; Nance & Taylor 1977; Maccarrone *et al.* 1983), although their LILE content has been modified by high-grade metamorphism (Leyreloup *et al.* 1977). In an ACF diagram (Fig. 1) meta-sedimentary rocks plot in the sillimanite-plagioclase-garnet field while andesitic igneous rocks (both fresh and altered) plot in the gt-plag-opx or opx-cpx-plag sub-triangles. The hypersthene metagreywackes straddle the gt-plag join and are less easily distinguished from meta-igneous rocks, reflecting their volcanoclastic origin. The AKF diagram (Fig. 1) shows that the chemistry of the metasedimentary xenoliths cannot be attributed to K_2O-depletion of parental andesitic material by granulite metamorphism.

Because of the abundance of cumulate rocks in the meta-igneous series, xenoliths with liquid compositions are rare. Leyreloup *et al.* (1977) concluded that meta-igneous xenoliths form a cogenetic saturated rock suite with ultramafic, mafic, intermediate and acidic members. Xenoliths in this suite fall on calc-alkaline differentiation trends on major and trace element variation diagrams (Fig. 2A) and the averaged compositions of important rock-types are similar to calc-alkaline equivalents; the mafic rocks with liquid compositions resemble high-Al basalts (Dostal *et al.* 1980). In trace element discrimination dia-

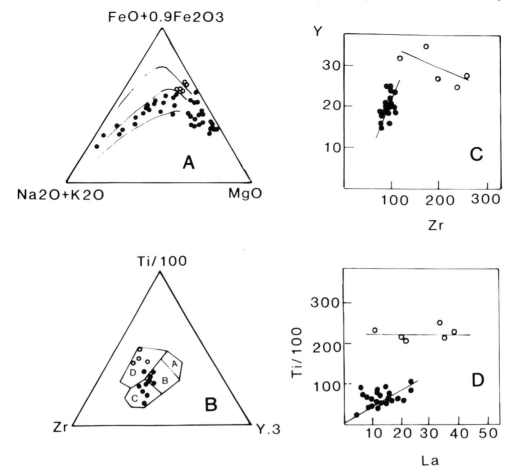

FIG. 2. (A) AFM diagram for meta-igneous xenoliths. Open circles: within plate basalt samples; solid circles: calc-alkaline. Cumulate samples, basic liquids and more differentiated types are distinguishable. (B) Ti-Zr-Y discrimination diagram for meta-igneous xenoliths. Symbols as above. Fields: A low-K tholeiites; B+C calc-alkali basalts; D within-plate basalts. (C) Y vs Zr variation diagram for meta-igneous xenoliths, showing different inter-element trends for within-plate and calc-alkali types. (D) Ti/100 vs La variation diagram for meta-igneous xenoliths as above.

grams (Pearce & Cann 1973) they plot in the calc-alkaline field (Fig. 2B).

Some meta-igneous xenoliths (e.g. BAL36) form a separate group, characterized by higher Ti, Fe, P, Zr, Nb and LREE contents, with low K, Ni and Mg-numbers (Dostal *et al.* 1980). On the Ti-Zr-Y diagram they plot in the within-plate basalt field (Fig. 2B) and have quite different inter-element relationships compared with the calc-alkaline xenoliths (Figs 2C, 2D). These rocks may represent slightly differentiated products of continental tholeiites, with the high Ti, Fe and low Ni contents reflecting Fe-Ti oxide accumulation.

The REE patterns of granulite xenoliths from Bournac have been discussed by Dostal *et al.* (1980), who showed that the meta-igneous granulites with liquid compositions have fractionated REE patterns similar to high-Al basalts and mafic andesites or continental tholeiites. Data for 10 samples from Roche Pointue are presented here (Fig. 3). Meta-igneous samples closely resemble analyses of island-arc calc-alkaline lavas from the Scotia arc (Tarney *et al.* 1982). They are moderately LREE-enriched $(La/Yb)_N = 2.36$–5.46 with variable absolute abundances $(Sm_N = 11.3$–$47.0)$, (Fig. 3A, B). The slope of the HREE is flatter than the slope of the LREE, as

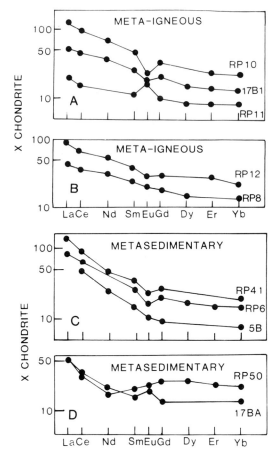

FIG. 3. REE patterns for meta-igneous (A, B) and meta-sedimentary (C, D) granulitic xenoliths from Roche Pointue. Analyses by isotope dilution method of Thirlwall (1982). Chondrite normalization values of Nakamura (1974).

Soudan slate (Wildeman & Haskin 1973). RP50 has a very unusual pattern for a meta-sediment, showing a pronounced trough at Nd but being enriched in both the HREE and in La and Ce. The resulting Sm/Nd ratio does not support the Nd isotopic composition (see below). Two explanations are possible: 1) the pattern is primary and is due to sedimentary processes which gave rise to an immature sediment such as a beach sand, consisting of LREE-enriched minerals (feldspars) and HREE-enriched minerals (garnet, hornblende, pyroxene, zircon) or 2) RP50 is a residue after partial fusion and removal of a granitic melt, with abundant residual plagioclase and garnet.

Comparison with other granulite samples

Petrographically both meta-igneous and meta-sedimentary xenoliths are similar to other xenoliths from Western Europe, e.g. Eifel (Okrusch *et al.* 1980), Cabo de Gata, Spain and to granulites of the Ivrea Zone (Mehnert 1975) and Calabria (Maccarrone *et al.* 1983). From the interpretation of petrographic data (Pin & Vielzeuf 1983); Leyreloup (in prep. 1985), all have undergone a similar P.T. history. However they are quite different from granulites of the Variscan collision belt, e.g. in Haut Allier (Lasnier 1977; Marchand 1974) and Moldanubia (Matejovska 1967; Scharbert 1971). The P.T. conditions recorded by the Variscan granulites are different, as shown by the presence of kyanite coexisting with K-feldspar in the meta-sedimentary rocks and the presence of eclogites in the mafic ones. The same contrast is seen in the geochemical data, since the Variscan granulites resemble olivine tholeiites (Giraud *et al.* 1984) with only rare samples showing calc-alkaline affinities.

Pressure—temperature estimates

Detailed petrological studies (Leyreloup, in prep. 1985) show that most of the granulitic xenoliths, particularly the meta-igneous ones, display complex re-equilibrated mineralogical assemblages, with at least two stages being inferred in their P-T evolution (Fig. 4).
1   Relict parageneses observed in meta-igneous rocks, (a) in Mg-rich, silica-poor basic and ultrabasic samples (ga, Al-opx, Al-cpx, Al-spinel-$\pm$ plag and plag, Al-opx, Al-cpx, Al-spinel), (b) in normal quartz-saturated or olivine tholeiitic compositions (ga, cpx, plag $\pm$ hb and plag, ga, cpx, opx) suggest pressures in excess of 11 kbar and between 8 and 11 kbar respectively, at 900°C.

commonly seen in calc-alkaline volcanics. Negative Eu anomalies are common and are more pronounced in the more fractionated sample RP10. A positive Eu anomaly is seen in RP11, which has low overall abundances and a flat pattern, indicating plagioclase accumulation.

Meta-sedimentary samples have a more complex origin for their REE patterns, involving weathering and erosion of a variety of sources. Such mixing gives rise to a uniform pattern of shales (Nance & Taylor 1976) which three of the samples (Fig. 3C) resemble. All three are strongly LREE-enriched (La/Yb)$_N$ = 5.82–7.1; Sm$_N$ – 15–35, and have negative Eu anomalies. Two other meta-sediments do not follow the shale pattern (Fig. 3D); 17BA is less strongly LREE-enriched, has a positive Eu anomaly and resembles the

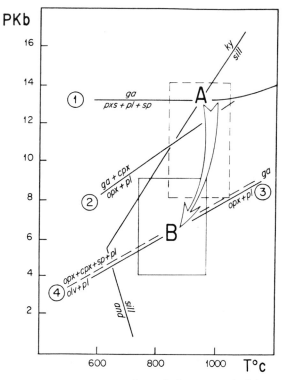

FIG. 4. Pressure–temperature schematic grid showing the evolutionary trend of the granulitic xenoliths from the Massif Central. Field A: probable equilibrium conditions of the lower crust during crustal thickening; Field B: calculated conditions of equilibrium 'primary' assemblages. Details of geothermometry/barometry available from A.L. (1)=(ga⁻)ₐ reaction (Herzberg, 1975). (2)=(ga⁻)ᵦ in quartz tholeiite No 4 (Green & Ringwood 1967, 1972). (3)=(ga⁻)ᵦ reaction in olivine tholeiite NM5 (Ito & Kennedy 1971). (4)=upper limit of the (ol+plag) stability field: (ol⁺) reaction (Green & Ringwood 1967, 1972). The arrow represents the P,T path of the granulitic lower crust during uplift.

2 In such rocks, static reaction textures indicating different 'garnet-out' (ga⁻) reactions occur frequently (i.e. (ga⁻)ₐ: ga→opx+sp+plag+cpx; (ga⁻)ᵦ: ga+cpx→opx+plag±hb±bi. Furthermore the 'olivine in' reactions (ol⁺) have been reached in some rare coronitic-gabbro xenoliths, i.e. (ol⁺)₁: plag+cpx+spinel→opx+plag+spinel+ol and (ol⁺)₂: cpx+opx+spinel+plag→ol+plag. Reaction (ga⁻)ₐ is temperature independent and occurs between 10 and 14 kbar in the CMAS system (Herzberg 1975; Obata 1976; Gasparik 1984). On the other hand, reactions (ga⁻)ᵦ and (ol⁺)₁,₂ are temperature and bulk rock chemistry dependent. They occur between 7 and 12 kbar (Green & Ringwood 1967, 1972; Ito & Kennedy 1971) (Fig. 4). Such reactions are retrograde (P decreasing, T constant or slightly decreasing). These textures are very similar to those described from the Lewisian (Savage 1979; Savage & Ellis 1980; O'Hara & Yarwood 1978) and in Eifel xenoliths (Okrusch *et al.* 1979).

The P-T evolution, which suggests depths as great as 45 km for parts of the lower crust, is interpreted in terms of crustal thickening (field A), probably due to continental collision during the Hercynian orogeny, 380–280 Myr ago (Fig. 4). Petrographic evidence of such a P,T evolution in the meta-sedimentary xenoliths are rare at Bournac or Roche Pointue, but are widespread at a third xenolith locality, La Denise. Reaction textures indicate ga+sill→spinel+plag, spinel+cordierite, or cordierite. However these meta-sediments yield slightly lower P,T estimates than the meta-igneous samples discussed above, suggesting a large-scale zonation of the thickened granulitic lower crust.

In fact, several meta-igneous samples often with simple mineralogy (e.g. plag, opx; plag, opx, cpx±qtz; plag, opx, spinel) showing apparent 'primary' equilibrium textures, without a trace of reaction, have been encountered. Such rocks, and all the meta-sedimentary samples, yield P-T esti-

TABLE 1. *Sr and Nd isotope data, T_{CHUR}, T_{DM}, and mineralogy of granulitic xenoliths. Trace element concentrations in ppm; model ages in Myr. Isotope ratios determined by automated Isomass 54E and manual Micromass MM30 at University of Leeds and at SURRC East Kilbride. Sr blanks: 5ng. NBS987 = 0.71026 ± 0.00005 (N = 13). $^{87}Sr/^{86}Sr$ normalized to $^{86}Sr/^{88}Sr = 0.1194$. Nd blanks: 2ng (Leeds), 1ng (SURRC). $^{143}Nd/^{144}Nd$ normalized to $^{146}Nd/^{144}Nd = 0.7219$. T_{CHUR} calculated from present-day bulk earth $^{143}Nd/^{144}Nd = 0.51264$, $^{147}Sm/^{144}Nd = 0.1966$. T_{DM} calculated by formula of Ben Othman et al. (1984). Sample 5B is from La Griffoul, 7 km S of Roche Pointue. *denotes data from Ben Othman et al. (1984), used for comparison*

Sample	Mineralogy	Rb	Sr	$^{87}Sr/^{86}Sr$	εSr	Sm	Nd	$^{143}Nd/^{144}Nd$	εNd	T_{CHUR}^{Nd}	T_{DM}^{Nd}	T_{DM}^{Sr}
Meta-sedimentary granulitic xenoliths (Roche Pointue)												
RP41	qtz,ksp,gt,sill,plag,rut,ilm,zir,gph,ap	72	254	.71876±3	202.4	7.20	30.10	.512083±24	−10.9	1627	2110	1427
5B	qtz,plag,ilm,gt,zir	4	705	.70545±2	13.5	3.04	16.15	.512104±48	−10.4	985	1440	1376
RP50	qtz,gt,plag,ksp,rut,bi,ap,zir	6	334	.71116±2	94.5	4.30	10.93	.512410±18	−4.5	—	—	—
RP6	gt,plag,ksp,qtz,rut,ilm,ap,zir,bi	47	270	.71425±6	138.4	5.55	26.51	.512070±14	−11.1	1238	1710	1716
17BA	plag,qtz,gt,ksp,rut,ap,ilm,bi	5.3	321.5	.71152±3	99.6	3.33	13.47	.512206±26	−8.5	1398	2000	—
Meta-sedimentary xenoliths (Bournac)												
830	plag,qtz,gt,ksp,sill,ilm,bi,rut	91.8	250.9	.71859±4	200.0	10.10	68.31	.512048±21	−11.5	841	1230	1078
831	qtz,gt,plag,ksp,ilm,rut	58.8	211.6	.71794±4	190.8	7.31	47.67	.512137±65	−9.8	743	1145	1376
832	gt,qtz,ksp,plag,sill,ilm	24.4	292.0	.71449±8	141.8	6.65	35.38	.512052±25	−11.4	1079	1535	1519
8310	qtz,ksp,plag,sill,ilm	54.7	217.9	.71782±7	188.6	6.24	31.54	.512094±45	−10.6	1079	1535	1519
8320	plag,ksp,gt,qtz,ilm,rut	7.7	290.0	.70984±5	75.8	4.31	20.76	.512215±22	−8.3	910	1460	—
8330	qtz,ksp,gt,sill,plag,ilm,bi,rut	60.2	149.7	.72736±5	331.1	8.15	40.72	.512038±21	−11.7	1211	1660	1524
8340	qtz,ksp,gt,plag,sill,ilm	51.3	136.3	.71995±5	218.6	6.78	30.95	.512274±33	−7.1	869	1435	1137
801*	(Ben Othman et al. 1984)	Rb/Sr = .195		.7131	122	7.01	36.69	.51210	−10.5	1015	1471	1360
Meta-igneous xenoliths (Roche Pointue)												
17B1	plag,opx,cpx,parg,sp,ilm	3	187	.70549±2	14.0	5.48	23.59	.512433±20	−4.0	561	1285	—
RP11	plag,cpx,opx,ilm,ap,rut	28	714	.70740±2	41.2	2.29	8.23	.512611±24	−0.6	155	1485	—
RP12	plag,cpx,opx,ilm,ap	16	262	.70583±4	18.9	4.78	19.60	.512451±13	−3.7	578	1360	1464
RP10	plag,gt,qtz,opx,rut,ksp,ilm,bl,zir	15	262	.71077±3	89.0	9.55	44.12	.512159±12	−9.4	1114	1650	—
RP8	alt,plag,ksp,opx,cpx,gt,ilm,ap,qtz,zir	116	645	.70719±3	38.2	7.83	34.25	.512380±22	−5.1	674	1365	625
Meta-igneous granulites (Bournac)												
BAL84	opx,cpx,plag,bi,ilm,mgt	4.2	267.7	.70485±5	5.0	3.40	11.96	.512480±33	−3.1	977	2072	—
BAL904X	qtz,ksp,plag,opx,rut,ilm	118.4	334.3	.71412±8	136.6	2.39	12.04	.512215±24	−8.3	847	1356	804
BAL1038	plag,qtz,opx,ilm,zir	46	920	.70871±4	59.8	3.02	14.02	.512224±35	−8.1	955	1502	—
BAL36	qtz,plag,opx,gt,rut,ilm	5.6	223.1	.70475±5	3.5	9.86	43.30	.512651±49	+0.2	—	805	—
BAL51	plag,opx,cpx,ilm,ap	3.4	184.8	.70469±5	2.7	2.21	9.07	.512508±81	−2.6	487	1194	—
BAL1204	qtz,plag,ksp,opx,cpx,gt,rut,ilm,ap,zir	12.0	363.3	.70802±5	50.0	4.14	25.05	.512027±28	−12.0	966	1367	—
BAL142*		Rb/Sr = −.563		.7274	325	9.90	52.99	.51223	−8.0	746	1236	1081

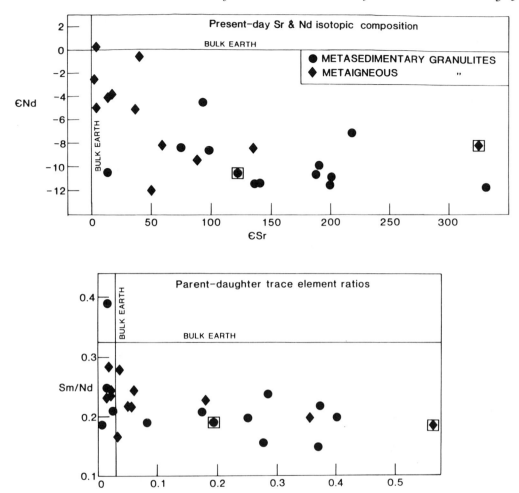

FIG. 5. Upper diagram: Present-day εSr and εNd isotope compositions of xenoliths from Massif Central, calculated from present-day bulk earth $^{87}Sr/^{86}Sr = 0.7045$, $^{143}Nd/^{144}Nd = 0.51264$. Samples from Ben Othman *et al.* (1984) enclosed in squares.
Lower diagram: Rb/Sr and Sm/Nd ratios of xenoliths, showing that many meta-igneous and meta-sedimentary samples have Rb/Sr ratios < bulk earth, and RP50 has a high Sm/Nd > bulk earth.

mates ranging from 4 to 9 kbar. If such 'primary' assemblages are considered to indicate the position of the xenoliths within the crust, these conditions imply a depth of extraction of between 13 and 27 km. Such depths are in good agreement with geophysical data concerning the crustal structure of the Massif Central (Perrier & Ruegg 1973; Hirn 1976) particularly the estimates of the depth of the Moho (27 km). The present-day thin crust of the Massif Central is due firstly to late Variscan post-collisional uplift and secondly to the Tertiary uplift associated with alkali volcanism.

Sr and Nd isotope analyses

The present-day $^{87}Sr/^{86}Sr$ and $^{143}Nd/^{144}Nd$ isotope ratios, Rb, Sr, Sm and Nd contents of Bournac and Roche Pointue granulitic xenoliths are presented in Table 1 and Fig. 5. Two Bournac xenoliths analysed by Ben Othman *et al.* (1984) are also included for comparison. From the εSr–εNd diagram (Fig. 5) it is clear that the meta-igneous samples in general have relatively low εSr and high εNd and are concentrated towards the value of the bulk earth, while the meta-sediments have higher εSr and lower εNd values. However,

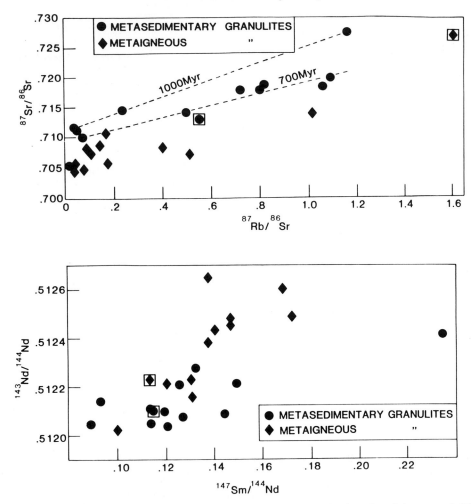

FIG. 6. Sr and Nd isochron diagrams for Massif Central xenoliths. Samples from Ben Othman *et al.* (1984) in squares. No isochron relationship can be seen in the Nd system although positive correlations are present. In the Sr system reference isochrons (dashed lines) at 1000 Myr and 700 Myr are shown for the metasedimentary xenoliths.

exceptions to this observation occur, such as the fractionated meta-igneous samples BAL904X and BAL142 which have high εSr, and also BAL1204 which has a low εNd value.

Comparison of εSr and εNd with the Rb/Sr and Sm/Nd ratios (Fig. 5) shows that many xenoliths have Rb/Sr ratios which are too low to support the observed εSr, indicating fractionation of Rb and Sr during the history of the granulites. RP50 has both a low Rb/Sr ratio and exceptionally high Sm/Nd, neither of which support the isotope ratios. The isotopic and trace element data for the meta-igneous xenoliths are very similar to those of 1400 Myr meta-igneous xenoliths of Lesotho

which are of cumulate origin (Rogers & Hawkesworth 1982) and 2400 Myr xenoliths from Australian kimberlites (McCulloch *et al.* 1982). However they are much less depleted in Rb than the metabasaltic granulitic xenoliths from the Eifel (Stosch & Lugmair 1984). Many of the meta-igneous samples which have low Rb/Sr ratios may be cumulates, particularly those which have mineralogies dominated by opx, cpx and plag. Accumulation of low Rb/Sr phases from a magma with high $^{87}Sr/^{86}Sr$ could explain the discrepancy between the isotope and trace element data. More differentiated rocks with low Rb/Sr ratios may have lost Rb preferentially to Sr during granulite facies metamorphism. Meta-sedimentary xenoliths tend to have higher Rb/Sr and slightly lower Sm/Nd ratios than the meta-igneous samples (Fig. 5). The abundance of 'undepleted' granulites with Rb/Sr > bulk earth is notable and demonstrates that granulite facies metamorphism is not always linked to depletion in Rb. The wide range of Rb/Sr ratios (0.006–0.402) and the restricted range of Sm/Nd reflect the chemical fractionation processes which occur in the sedimentary environment. Rb and Sr have different geochemical affinities and will be present in different minerals of contrasting stability during weathering. Thus a wide variation in Rb/Sr in sediments can be generated by the processes of erosion and deposition. The existence of low Rb/Sr, high $^{87}Sr/^{86}Sr$ sediments could be explained by the weathering of older terrains or volcanics with high Sr isotope ratios. In contrast Sm and Nd behave much more coherently since they have very similar distribution coefficients in the magmatic environment and hence will be found together in the same detrital minerals in immature sediments. The exception to this is RP50 with its high Sm/Nd ratio, in which the two rare earths have been fractionated from each other and do not support the Nd isotope ratio.

Age of the granulitic xenoliths

The Sr and Nd isochron diagrams (Fig. 6) show broad positive correlations for both meta-sedimentary and meta-igneous xenoliths, but in contrast to other studies (Rogers & Hawkesworth 1982; McCulloch *et al.* 1982) no apparent isochron relationships are seen. This is due in part to the great variety of lithologies and the separation between the two localities. There is no reason why the meta-sedimentary xenoliths, being by nature mixtures of pre-existing rocks, should have an isochron relationship with each other. There is, however, a strong suggestion of mixing lines between the meta-sediments (dashed lines, Fig. 6) if the quartzite 5B is ignored. Ages calculated for these lines are 700–1000 Myr when $^{87}Sr/^{86}Sr$ value was homogenized but the resulting Rb/Sr values were variable. Such a mixing process could have occurred between 700 and 1000 Myr ago.

Meta-igneous samples have been shown to have been derived from different sources (Fig. 2) and it is not surprising that no isochron relationship exists in either system. The majority of meta-igneous granulites cluster near the low Rb/Sr end of the Sr isochron diagram and only highly evolved samples have higher Rb/Sr and $^{87}Sr/^{86}Sr$ values. In the Nd isochron diagram there is a poor positive correlation among the meta-igneous

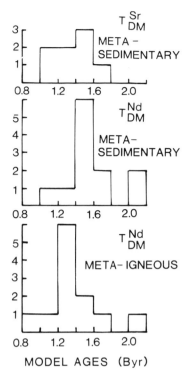

FIG. 7. Frequencies of model ages calculated for a depleted mantle source. Sr model ages also shown to demonstrate similarity in range for meta-sediments.

samples. The lack of an isochron is due to differences in lithology, source and probably age.

Since Sm and Nd are considered to be generally unaffected by metamorphism (Jacobsen & Wasserburg 1978; DePaolo *et al.* 1982; Jahn & Zhang 1984), Nd model ages may be a good guide to the age of the protoliths of the granulites. Model ages relating to CHUR and depleted mantle (DM) have been calculated (Table 1, Fig. 7). The evidence for the existence of the depleted mantle sources in the Proterozoic (DePaolo 1981) suggests that the T_{DM} ages may be more geologically significant than the T_{CHUR} ages. The T_{DM}^{Nd} model ages range from 804 to 2072 Myr for the meta-igneous samples, while meta-sediments have T_{DM}^{Nd} ages ranging from 1145 to 2110 Myr. For the meta-igneous xenoliths these model ages could represent the time of formation by partial melting from a depleted mantle but since present-day calc-alkaline volcanics can be derived from mantle sources which are intermediate between depleted MORB and bulk earth, or even enriched, these T_{DM}^{Nd} model ages can only be an indication of the time of derivation of the volcanics. The ages could be between the calculated T_{DM}^{Nd} values and the T_{CHUR} values. For the metasediments the interpretation of the model ages is complicated by the mixing of different sources. The ages are therefore provenance ages, yet the agreement with the meta-igneous model ages is remarkable (Fig. 7). The peak of the T_{DM}^{Nd} ages for the meta-sediments is slightly older than for the meta-igneous xenoliths; hence the sediments may be formed by a mixture of volcanic detritus associated with the calc-alkaline intrusives and a component derived from older terrains. T_{DM}^{Sr} model ages, where they can be calculated, are in good agreement with the Nd model ages of the meta-sediments.

All the T_{DM} and most of the T_{CHUR} model ages are Proterozoic and confirm previous estimates of the age of the lower crust of the Massif Central (> 1000 Myr, Allègre *et al.* 1975; Hamet *et al.* 1978) but they are much younger than the > 3000 Myr age deduced by Pb isotope analyses (Michard Vitrac *et al.* 1981). The estimates also agree with other lower crustal age estimates for Western Europe: Pyrenees 1200–1900 Myr (Ben Othman *et al.* 1984), Eifel 1500 Myr (Stosch & Lugmair 1984). Relic ages in detrital zircons in the upper crust also reflect crustal formation in the region: 1940–2160 Myr (Gebauer & Grünenfelder 1976), 1700–2000 Myr (Lancelot & Pin 1978).

Conclusions

The variety of petrographic types within the Massif Central xenoliths include immature meta-sediments and calc-alkaline meta-igneous suites, which may indicate formation of the crust in an island-arc or continental margin environment. Incorporation into a large continental landmass occurred during the Variscan orogeny when crustal thickening caused high pressure granulite-facies metamorphism. Later uplift to pressures between 4 and 9 kbar is recorded in the xenoliths. The T_{DM}^{Nd} model ages of the meta-igneous proto-liths are between 800 and 2100 Myr, which is in good agreement with other age estimates for the lower crust of Variscan Europe; provenance model ages for meta-sediments are slightly older. Mixing lines in the Sr isotope system for the meta-sediments may represent the time of deposition (700–1000 Myr).

ACKNOWLEDGEMENTS: The isotopic and REE analyses were obtained while HD was in receipt of a Leeds Postgraduate Studentship and a Post-Doctoral Fellowship from the Natural Environment Research Council. A.N. Halliday, P.J. Hamilton and J. Hutchinson are thanked for technical assistance at SURRC. Isotope research at Leeds University is supported by NERC and by the Royal Society. SURRC is supported by the Scottish Universities and NERC.

References

ALLÈGRE, C.J., HAMET, J. & LEYRELOUP, A. 1975. Etude ^{87}Rb/^{87}Sr des enclaves catazonales remontées par les volcans néogènes du Velay. Présence d'un socle antécadomien sous le Massif Central. *Abst. 3eme R.A.S.T. Montpellier*, p8.

BEN OTHMAN, D., POLVE, M. & ALLÈGRE, C.J. 1984. Nd-Sr isotopic composition of granulites and constraints on the evolution of the lower continental crust. *Nature*, **307**, 510–515.

——, FOURCADE, S. & ALLÈGRE, C.J. 1984. Recycling processes in granite–grandiorite complex genesis: the Querigut case studied by Nd–Sr isotope systematics. *Earth Planet. Sci. Lett.*, **69**, 290–300.

BILAL, A. & TOURET, J. 1976. Les inclusions fluides des enclaves catazonales de Bournac (Massif Central). *Bull. Soc. Fr. Min. Cris.*, **99**, 134–139.

CHIKHAOUI, M. 1981. *Les roches volcaniques du Protérozoique supérieur de la chaîne Pan-Africaine. (Hoggar, Anti Atlas, Adrar des Iforas.) Caractérisation géochimique et minéralogique. Implications géodynamiques.* Thèse d'Etat, Montpellier, 183 pp.

COULON, C. 1977. *Le volcanisme calco-alcalin cénozoique de Sardaigne (Italie). Pétrographie, géochimie et génèse des laves andesitiques et des ignimbrites. Signification géodynamique.* Thèse Doctorat Sciences, Saint Jerome, Marseille, 288 pp.

CORNEN, G. 1972. *Géologie du Bassin des Rhues et Santoire (Massif du Cantal).* Thèse 3ème cycle. Univ. Paris-Sud (Orsay).

CURTIS, C.D. 1969. Trace element distribution in some British Carboniferous sediments. *Geochim. Cosmochim. Acta*, **33**, 519–523.

DEPAOLO, D.J. 1981. Neodymium isotopes in the Colorado Front Range and crust–mantle evolution in the Proterozoic. *Nature*, **291**, 193–196.

——, MANTON, W.I., GREW, E.S. & HALPERN, M. 1982. Sm-Nd, Rb-Sr and U-Th-Pb systematics of granulite facies rocks from Fyfe Hills, Enderby Land, Antarctica. *Nature*, **298**, 614–618.

DOSTAL, J., DUPUY, C. & LEYRELOUP, A. 1980. Geochemistry and petrology of metaigneous granulitic xenoliths in Neogene volcanic rocks of the Massif Central (France). Implications for the lower crust. *Earth Planet. Sci. Lett.*, **50**, 33–40.

DUPUY, C., LEYRELOUP, A. & VERNIERES, J. 1977. The lower continental crust of the Massif Central (Bournac, France)—with special reference to REE, U and Th composition, evolution, heat flow production. *Phys. Chem. Earth*, **11**, 401–416.

GASPARIK, T. 1984. Two pyroxene thermobarometry with new experimental data in the system CaO-MaO-Al$_2$O$_3$-SiO$_2$. *Contrib. Mineral. Petrol.*, **87**, 87–97.

GEBAUER, D. & GRUNENFELDER, M. 1976. U-Pb zircon

and Rb-Sr whole-rock dating of low-grade metasediments. Example: Montagne Noire (Southern France). *Contrib. Mineral. Petrol.*, **59**, 13–32.

GILL, J.B. 1981. Orogenic andesites and plate tectonics. *In: Minerals and Rocks*, **16**. Springer Verlag, Berlin–Heidelberg–New York. 390 pp.

GIRAUD, A., MARCHAND, J., DUPUY, C. & DOSTAL, J., 1984. Trace element geochemistry of Haut Allier leptyno-amphibolitic group (France). *Lithos*, **17**, 203–214.

GREEN, D.H. & RINGWOOD, A.E. 1967. An experimental investigation of the gabbro to eclogite transformation and its petrological application. *Geochim. Cosmochim. Acta.*, **31**, 676–833.

—— & —— 1972. A comparison of recent experimental data on the gabbro-garnet-eclogite transition. *J. Geol.*, **80**, 277–288.

HAMET, J., MICHARD-VITRAC, A. & ALLÈGRE, C.J. 1978. $^{87}Rb/^{86}Sr$ and U-Pb—systematics in granulitic xenoliths in volcanic rocks and limitations on the formation of the lower crust. *Abst. EOS*, **59**, 4, pp 392.

HERZBERG, C.T. 1975. *Phase assemblages of the system $CaO-Na_2O-MgO-Al_2O_3-SiO_2$ in the plagioclase lherzolite and spinel lherzolite mineral facies*. PhD Thesis, Univ. Edinburgh.

HIRN, A. 1976. Sondages seismiques profonds en France. *Bull. Soc. Geol. Fr.* **5**, 1065–1071.

ITO, K. & KENNEDY, G.C. 1971. An experimental study of the basalt garnet granulite-eclogite transition. *In:* HEACOCK, J.G. (Ed.) *The Structure and Physical Properties of the Earth's Crust*. Am. Geophys. Union Monograph. **14**, 303–314.

JACOBSEN, S.B. & WASSERBURG, G.J. 1978. Interpretation of Nd, Sr and Pb isotope data from Archean migmatites in Lofoten-Vesteralen, Norway. *Earth Planet. Sci. Lett.*, **41**, 245–253.

JAHN, B-M. & ZHANG, Z-Q. 1984. Archean granulite gneisses from eastern Hebei Province, China: rare earth geochemistry and tectonic implications. *Contrib. Mineral. Petrol.*, **85**, 224–243.

LANCELOT, J. & PIN, C. 1978. U/Pb evidences of bimodal magmatism of early Paleozoic age in the Massif Central (France): the leptyno-amphibolitic group. *Short paper of the 4th Int. Conf. on Geochron. Cosmochron. Iso. Geol. USGS Open-File Rpt, 78-701*, p. 337.

LASNIER, B. 1977. *Persistence d'une série granulitique au coeur du Massif Central francais, Haut-Allier. Les termes basiques, ultrabasiques et carbonatés.* Thèse d'Etat, Nantes, 300 pp.

LEYRELOUP, A. 1973. *Le socle profond en Velay d'après les enclaves remontées par les volcans néogènes. Son thermometamorphisme et sa lithologie. Granites et serie charnockitique (Massif Central francais).* Thèse 3ème cycle, Nantes 356 pp.

——, DUPUY, C. & ANDRIAMBOLONA, R. 1977. Catazonal xenoliths in French Neogene volcanic rocks; constitution of the lower crust. Part 2: Chemical composition and consequence on the evolution of the French Massif Central Precambrian crust. *Contrib. Mineral. Petrol.*, **62**, 283–300.

MACCARRONE, E., PAGLIONICO, A., PICCARETA, G. &

ROTTURA, A. 1983. Granulite amphibolite metasediments from the Serre (Calabria, southern Italy); their protoliths and the processes controlling their chemistry. *Lithos*, **16**, 95–111.

MARCHAND, J. 1974. *Persistance d'une série granulitique au coeur du Massif Central francais, Haut-Allier. Les termes acides.* Thèse 3ème cycle. Nantes 207 pp.

MATEJOVSKA, A. 1967. Petrogenesis of the Moldanubian granulites near Namest nad Oslavov. *Krystalinikum*, **5**, 85–103.

MICHARD-VITRAC, A., ALBAREDE, F. & ALLÈGRE, C.J. 1981. Lead isotopic composition of Hercynian granitic K-feldspars constrains continental genesis. *Nature*, **291**, 460–464.

McCULLOCH, M.T., ARCULUS, R.J., CHAPPELL, B.W. & FERGUSON, J. 1982. Isotopic and geochemical studies of nodules in kimberlite have implications for the lower continental crust. *Nature*, **300**, 166–169.

NAKAMURA, N. 1974. Determination of REE, Ba, Fe, Mg, Na and K in carbonaceous and ordinary chondrites. *Geochim. Cosmochim. Acta.*, **38**, 575–775.

NANCE, W.B. & TAYLOR, S.R. 1976. Rare earth patterns and crustal evolution. I. Australian post-Archean sedimentary rocks. *Geochim. Cosmochim. Acta.*, **40**, 1529–1551.

—— & —— 1977. Rare earth element patterns and crustal evolution, II. Archean sedimentary rocks from Kalgoorlie, Australia. *Geochim. Cosmochim. Acta.*, **41**, 225–231.

OBATA, M. 1976. The solubility of Al_2O_3 in orthopyroxenes in spinel and plagioclase peridotites and spinel pyroxenites. *Amer. Min.*, **61**, 804–816.

O'HARA, M.J. & YARWOOD, G. 1978. High pressure–temperature point on an Archean geotherm, implied magma genesis by crustal anatexis and consequences for garnet-pyroxene thermometry and barometry. *Phil. Trans. Roy. Soc. Lond.* A **188**, 441–456.

OKRUSCH, M., SCHRODER, B. & SCHNUTGEN, A. 1979. Granulite facies metabasite ejecta in the Laacher See area, Eifel, West Germany. *Lithos*, **12**, 251–270.

PEARCE, J.A. & CAN, J.R. 1973. Tectonic setting of basic volcanic rocks determined using trace element analysis. *Earth Planet. Sci. Lett.*, **19**, 290–300.

PERRIER, G. & RUEGG, J.C. 1973. Structure profonde du Massif Central français. *Ann. Géophysique.*, **29**, (4) 435–502.

PIN, C. & LANCELOT, J.R. 1982. U/Pb dating on an early Paleozoic bimodal magmatism in the French Massif Central and of its further metamorphic evolution. *Contrib. Mineral. Petrol.*, **79**, 1–12.

—— & VIELZEUF, D. 1983. Granulites and related rocks in Variscan median Europe: a dualistic interpretation. *Tectonophysics*, **93**, 47–74.

ROGERS, N.W. & HAWKESWORTH, C.J. 1982. Proterozoic age and cumulate origin for granulitic zenoliths, Lesotho. *Nature*, **299**, 409–413.

SAVAGE, D. 1979. *Geochemistry of layered basic intrusions in the Lewisian complex around Scourie*. PhD Thesis, Imp. Coll. London.

—— & ELLIS, J.D. 1980. High pressure metamorphism

in the Scourian of NW Scotland: Evidence from garnet granulites. *Contrib. Mineral. Petrol.* **74,** 153–163.

SCHARBERT, H.G. 1971. Kyanite and sillimanite in Moldanubischen granulites. *Tschermacks Miner. Petrog. Mitt.,* **16,** 252–267.

STOSCH, H-G. & LUGMAIR, G.W. 1984. Evolution of the lower continental crust: granulite facies xenoliths from the Eifel, W. Germany. *Nature,* **311,** 368–370.

TARNEY, J., WEAVER, S.D., SAUNDERS, A.D., PANKHURST, R.J. & BARKER, P.F. 1982. Volcanic evolution of the Northern Antarctic Peninsula and the Scotia Arc. *In:* THORPE, R.S. (ed.) *Andesites: Orogenic Andesites and Related Rocks.* John Wiley and Sons.

THIRLWALL, M.F. 1982. A triple-filament method for rapid and precise analysis of rare-earth elements by isotope dilution. *Chem. Geol.,* **35,** 155–166.

WILDEMAN, T.R. & HASKIN, L.A. 1973. Rare earths in Precambrian sediments. *Geochim. Cosmochim. Acta,* **37,** 419–438.

H. DOWNES, Grant Institute of Geology, University of Edinburgh, West Mains Road, Edinburgh, EH9 3JW, UK. Present address: Department of Geology, Birkbeck College, University of London, 7–15 Gresse Street, London, W1P 1PA.

A. LEYRELOUP, Laboratoire de Pétrologie, Université des Sciences, Place Eugene Bataillon, 34060 Montpellier Cedex, France.

Mineral reactions in xenoliths from the Colorado Plateau; implications for lower crustal conditions and fluid composition

J.R. Broadhurst

SUMMARY: A study of lower crust and upper mantle xenoliths from Moses Rock Diatreme, Utah, has led to the discovery of evidence for low temperature/high pressure, *in situ* hydration of pyroxene and garnet granulites in isolated parts of the lower crust of the Colorado Plateau; one of the characteristic reactions being:

$$Plagioclase \rightarrow Zoisite + Paragonite + Quartz + Albite$$

This was followed by a second hydration at lower temperature and pressure, under more oxidizing conditions, probably associated with the incorporation of the xenoliths into the diatreme. Under these conditions, pyroxenes were replaced by fibrous amphiboles and the plagioclases were rimmed by chlorite and epidote. Upper mantle xenoliths show similar histories of cooling and hydration. The origin of the cold hydrating fluid is linked to the presence of low T/high P group 'C' eclogites in the suite, xenoliths which may represent relics of a subducted/underplated fragment of the Farollan plate.

Introduction

Moses Rock Diatreme (Fig. 1) is a 6 km long fissure eruption of a serpentinized ultramafic microbreccia on the north-central part of the Colorado Plateau. It runs parallel to the Comb Ridge monocline. The diatreme was mapped in detail and described petrographically and geochemically by McGetchin (1968).

There are two contemporaneous Tertiary volcanic rock types occurring on the Colorado Plateau;

a) Diatreme eruptions of a serpentinized ultramafic microbreccia (SUM); a complex breccia of comminuted wall-rock and ultramafic material, held together by a fine grained serpentinized groundmass containing larger grains of olivine, garnet, chrome-diopside, enstatite, titanoclinohumite, chlorite and apatite, but lacking the typical minerals rich in incompatible elements found in kimberlite such as phlogopite and perovskite.

b) Minette necks and dykes (a potassium rich phlogopite-diopside-sanidine lamprophyre), interpreted as the partial melt from a phlogopite rich garnet lherzolite in a volatile rich upper mantle (Roden 1981; Rogers *et al.* 1982).

The coeval origin of these volcanics is indicated by the cross-cutting relationships of the dykes and the rounded fragments of minette found in the SUM in some of the more complex centres such as Buell Park (Roden 1977) and Moses Rock (this study). The crystallization ages for these rock-types determined by fission track dating of apatites (Naeser 1971) and K/Ar dating of micas

(Roden 1979) indicates a late Oligocene age of eruption, approx. 25–31 Ma. The SUM was emplaced as a solid/fluid mix with no evidence of a melt component whereas the minettes were intruded as melts and have associated lavas and tuff-breccias. Several authors have noted the relationship between minette and SUM, but the exact nature of this relationship still remains obscure. They are the same age, they sometimes use the same vent, and have closely related xenolith suites, yet there are significant differences. It has been suggested (Roden 1981) that there was fluid separation from primitive minette magma in the upper mantle producing a subsolidus eruption of volatiles, but the origin of such a volatile phase is a problem, since vesiculation of the minettes is rare (except at Washington Pass) which would imply that the volatiles were not juvenile (Ehrenberg 1982).

The prolific xenolith suites in both rock types must have some genetic link, but they show significant individual features. Studies of the petrography and geochemistry of xenoliths have been made by Watson (1960), Watson & Morten (1968), O'Hara & Mercy (1966), Helmstaedt *et al.* (1972), Helmstaedt & Shultz (1979), Harley & Green (1981), Smith & Levy (1976), Smith & Zeintek (1979), Hunter & Smith (1981), Ehrenberg (1978), (1982) and Ehrenberg & Griffin (1979). Most of the crystalline rock fragments in the SUM have a crustal origin; granulites, biotite sillimanite gneisses, schists, amphibolites, granites, gabbros and other intrusive rocks. Ultramafic xenoliths include spinel lherzolites, websterites, alkremites, garnetites, serpentinites and

From: DAWSON, J.B., CARSWELL, D.A., HALL, J. & WEDEPOHL, K.H. (eds) 1986, *The Nature of the Lower Continental Crust*, Geological Society Special Publication No. 24, pp. 331–349.

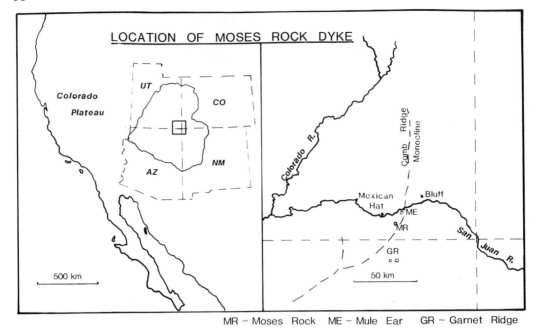

FIG. 1. Location of Moses Rock dyke and neighbouring diatremes on the Colorado Plateau.

serpentine-chlorite schists as well as anomalous low-T/high-P group 'C' eclogites which have been interpreted as pieces of subducted oceanic crust (Helmstaedt & Doig 1975). Many xenoliths are altered, much of the hydration showing a correlation with the micro-cracks which are commonly sealed. Pristine samples have open cracks. In addition, the minettes contain garnet bearing ultramafic xenoliths, but they do not contain the group 'C' eclogites or the very hydrated xenoliths. The hydration of the xenoliths has been the subject of differing models. McGetchin (1968) and McGetchin & Silver (1973) suggest that the entire lower crust has been hydrated whereas other authors (Ehrenberg & Griffin 1979; Padovani et al. 1982) suggest that this hydration is much later related to incorporation in the SUM or even a vent hydration.

Padovani et al. (1982) utilized the seismic velocities measured in altered and unaltered xenoliths, corrected for pressure and temperature from estimates using the anhydrous assemblages, in order to determine which were likely to be representative of the lower crust. From their observations lower crustal velocities were better explained by the unaltered granulite xenoliths and they concluded that the hydrous assemblages must therefore have been produced after incorporation in the diatreme.

The purpose of this paper is to evaluate the causes and conditions of hydration of the xenolith samples in the SUM at Moses Rock.

Petrography and mineral chemistry

Xenoliths litter the surface of Moses Rock. They are commonly rounded and their surfaces are normally weathered or caliche covered. Upper crustal rocks are commonly angular to sub-angular fragments; other xenoliths are rounded to sub-rounded. Most xenoliths have a network of microcracks. It is possible to divide the xenoliths of crystalline rocks into three broad groups:
1 Crustal xenoliths: Granulites, Amphibolites, Biotite sillimanite Gneiss and altered equivalents
2 Eclogite xenoliths: Jadeite/Almandine rocks, Jadeite rich pyroxenites, Garnetites etc.
3 Mantle xenoliths: Spinel Peridotites, Spinel Websterites, Websterites, Alkremites and hydrated equivalents

Brief petrographic descriptions of representative samples are presented in Table 1. Important features of the mineralogy and texture of the crustal xenoliths are discussed here for two of the more important crustal xenolith types.

a) Granulites

These are mainly mafic in bulk composition but

TABLE 1. *Petrographic descriptions*

		Lower crustal xenoliths		
Rock type	Primary mineralogy	Texture	Early hydrous assemblages	Late hydrous assemblages
Basic granulite e.g. MD19 MD116 MRD306	Garnet clinopyroxene quartz plagioclase hornblende ilmenite rutile apatite +/− sillimanite	Large orange/red garnets often as symplectites with plag or as coronas around pyroxene. Plagioclase large and equigranular. Compositional layering and occasional foliation defined by hornblendes.	Plagioclase altered to zoisite paragonite, albite, quartz occasionally some kyanite. Pyroxenes show gradual altn. to amphibole with exsolution of opaques. Ilmenite grains show exsolution of Ti-rich and Ti-poor areas. Cracks in garnet caused by deformation are filled by biotite. Deformation caused granulation of plag and alignment of zoisite and mica.	Plag altered first to clinozoisite and white mica and then to epidote muscovite, chlorite, albite and quartz. Pyroxene replaced by sheaves of pale green amphibole or mosaics of amph, chlorite and quartz. Ilmenite grains replaced by rods of rutile and magnetite. Garnet has coronas of chlorite. Epidotes and some amphiboles which rim the pyroxenes grow across the foliations.
Acid granulite e.g. MRD158	Garnet clinopyroxene quartz plagioclase alkali feldspar ilmenite rutile (kyanite) apatite	Large equant plagioclases with smaller grains of orthoclase. Garnet and pyroxene are minor constituents, no obvious layering.	Early hydration is essentially identical to that of the basic granulites. Development of paragonite is more widespread.	The same as for basic granulites.
Amphibolite e.g. MRD46 MRD47 MRD307	Hornblende plagioclase quartz (clinopyroxene) (garnet)	Amph, plag and quartz as equigranular polygonal grains, strong foliation. Garnets large and inclusion rich.	Amphibolites display only a limited amount of hydration, feldspars display growth of zoisite.	Hornblendes are overgrown by a pale green amphibole. Some chlorite is developed at the rim of garnets and epidote sometime grows at plag-garnet boundaries.
Biotite gneiss e.g. MD10 MD11 MD317	Garnet biotite plagioclase quartz (sillimanite) ilmenite rutile apatite	Comp. layering of felsic and mafic bands. Garnets brown, cracked, inclusion rich; biotite grains are oriented and show exsoln. of needles of Fe oxides. Plag. large, engulfing other mins, no preferred orientn. Sillim. as both fibrolite and euhedral grains, closely assoc. with garnet and biotite.	Cores of plag grains show altn. to a paragonite-muscovite mica, zoisite and some kyanite. Minor amounts of K-feldspar also occur. There appears to be little evidence of early hydration of mafic minerals, except for the growth of rare grains of secondary biotite around some garnets.	Garnets are often rimmed with chlorite. Biotite is also commonly replaced by chorite and specks of oxide. Plag is altered first to clinozoisite and sericite and then to epidote, muscovite, chlorite albite and quartz. Sillimanite is overgrown by talc. Pale green amphiboles commonly engulf areas especially near concentrations of mafic minerals.

TABLE 1 (cont.)

		Eclogites		
Rock type	Primary mineralogy	Texture	Early hydrous assemblages	Late hydrous assemblages
Jadeite-almandine Eclogite e.g. MD4 MD18 MRD9	Almandine-garnet jadeitic pyroxene lawsonite (pseud) (quartz) phengite rutile pyrite	Garnet zoned with Ca-rich rims, commonly as atolls growing from pyroxene, phengite, and lawsonite. Pyroxene oscillatory zoned Jd-rich rims; lawsonite replaced by zoisite needles; strong foliation common.	Lawsonite is invariably altered to mats of zoisite needles. There is only limited development of hydrous assemblages in eclogite xenoliths, although there is ample textural evidence for the presence of fluids.	Late alteration of the eclogites, appears to be limited to surface caliche and infilling of microcracks with carborate, quartz etc.
Jadeitic pyroxenite e.g. MD3 MD101	Jadeitic pyroxene phengite rutile pyrite	Similar texture to above but foliation less common. Rare atoll garnets grown from pyroxenes.	Pyroxenites are very similar to the eclogites.	As above.
Pyrope-omphacite eclogite	Pyropic garnet omphacitic pyroxene chlorite	Coarse pyropes and diopsides with neoblasts of omphacite, more calcic garnet and chlorite.	Diopside-pyrope reacting to give omphacite, garnet and chlorite; pyropes surrounded by coronas of chlorite.	
Garnetite e.g. MRD/GT/1 MRD/GT/2 MRD/GT/5	Garnet ilmenite rutile pyroxene chlorite (glaucophane)	Large orange-brown garnets with minor rutile and ilmenite growing along grain edges sometimes intergrown with chlorite. Rare grains of omphacitic pyroxene, glaucophane as rare inclusions in garnet.	A greater development of chorite As above in garnets nearest to rims.	

Mantle xenoliths

Rock type	Primary mineralogy	Texture	Early hydrous assemblages	Late hydrous assemblages
Spinel lherzolite e.g. MRD26	Olivine orthopyroxene spinel clinopyroxene	Rounded, cracked olivines and large opx grains with fine exsolution of cpx and small platelets of green spinel, smaller grains of cpx; grains of spinel are often small green-brown intergranular.	Pyroxenes altered to Mg-rich amphiboles and chlorites. The chrome-rich spinels are altered to chrome-rich magnesian chlorites and chrome rich oxides. Olivines have been serpentinized.	Some development of iron rich chlorites, amphiboles and also serpentine.
Spinel Websterite e.g. MD131 MD134	Aluminous clinopyroxene porphyroblasts, pleonaste spinel, orthopyroxene, clinopyroxene groundmass	Large multiple twinned porph. of aluminous cpx with coarse lamellae of exsolved opx. and green spinel, set in groundmass of equigranular cpx, opx and spinel neoblasts.	Megacrysts altered to amphiboles mimicking the multiple twins. groundmass overgrown by amph. and chlorite. Some spinels are altered to chlorite, oxides and corundum. Titanoclinohumite is also developed.	As above.
Websterite e.g. MRD36	Orthopyroxene clinopyroxene	Large equant grains of opx showing simple twinning and exsolution of fine lamellae of cpx and also of many small platelets of spinel. Cpx grains are smaller and tend to form small groups.	Widespread growth of tremolite, magnesian chlorite and some serpentine, with small amounts of oxide trapped between grains.	As above.
Alkremite e.g. MRD200	Pyrope garnet pleonaste spinel	Large, equant flesh coloured garnets with intergranular dark red-brown spinel esp. at triple junctions. Rare oriented rods of rutile in the garnets, spinels display complex twinning.	Between grains kornerupine, garnet, corundum and chlorite can be identified.	

TABLE 2. *Hydration reactions observed in granulites*

Early hydration

$$Pyroxene_1 \rightarrow Pyroxene_2 + Opaques$$
$$Pyroxene + Na^+(f) + K^+ + Mg \rightarrow Amphibole + Qz + Ca^{2+}$$
$$Plagioclase + H_2O \rightarrow Zo + Parag + Ab + Qz \pm Ky$$
$$Garnet + K^+ + Al^{3+} + H_2O \rightarrow Biotite + Qz + Ca^{2+}$$
$$Garnet_1 + Plagioclase + K^+ + H_2O \rightarrow Garnet_2 + Biotite + Qz + Ca^{2+} + Na^+$$
$$Garnet_1 + Plag_1 + Pyrox + K^+ + H_2O \rightarrow Garnet_2 + Plag_2 + Amph + Qz$$

Late hydration

$$Plagioclase + Fe^{2+} + Fe^{3+} + K^+ + H_2O \rightarrow clinozoisite + musc + Qz + Albite$$
$$Plag + Pyroxene + K^+ + H_2O + O_2 \rightarrow Albite + Ep + Musc + Qz + Na\text{-}Amph$$
$$Plag + Garnet + K^+ + H_2O + O_2 \rightarrow Albite + Ep + Chlorite + Musc + Qz$$
$$Pyroxene + Plag + K^+ + H_2O + O_2 \rightarrow Actinolite + Qz$$
$$Amphibole + K^+ + Fe^{3+} \rightarrow Hornblende$$
$$Ilmenite \rightarrow Rutile + Magnetite$$
$$Garnet \rightarrow Chlorite + Opaques$$
$$Biotite \rightarrow Chlorite + Opaques$$

some are intermediate and a few are acid. Compositional layering is a common feature. Primary mineralogy is commonly garnet, clinopyroxene, plagioclase, quartz, apatite, ilmenite, rutile, $\pm$ brown amphibole, $\pm$ sillimanite. Most rocks have undergone some degree of retrograde metamorphism with the growth of amphibole, chlorite, epidote, sericite and albite. Such xenoliths generally display reaction or loss of garnet, pyroxene and plagioclase. Observed reactions are summarized in Table 2.

Primary garnets are large calcic almandines, they are cracked and often inclusion rich. In hydrated specimens these inclusions preserve original pyroxene, plagioclases, amphiboles and occasionally some biotite allowing estimation of early temperatures and pressures. There is often a coronitic texture of garnet mantling clinopyroxene in contact with plagioclase. A secondary garnet growing as atolls or symplectites with plagioclase and quartz is not uncommon. These symplectites appear to be associated with an episode of hydration because zoisite and paragonite grains sometimes occur as inclusions and they are not found in unhydrated specimens. They commonly show an association with amphibole-quartz symplectites. A secondary garnet growing with sheaves of biotite which follow small folds or shear zones can be observed in a few specimens. Secondary biotite grows along cracks and where garnet grains have been imbricated or their edges granulated during deformation. When the rock is extensively hydrated, relic garnets are commonly associated with rims or coronas of chlorite.

The original clinopyroxene is a green diopsidic salite with up to 2 wt% Na_2O. The pyroxenes show a progressive alteration associated with a gradual deepening of colour and the exsolution of opaque phases, both in the form of interlocking needles, and also as patches of fine disseminated grains. The alteration, linked to a progressive loss in SiO_2 and an increase in calcium-Tschermak's component, shows a disproportionate increase in Fe due (on charge balance considerations) to an increase in Fe^{3+} and also a rapid loss of sodium. The later pyroxenes grow in association with small sub-grains or lamellae of pargasitic hornblende in apparent optical continuity. These appear to be associated with similar more tschermakitic hornblendes growing with garnet and quartz as symplectites and there is a continuous spectrum of compositions between the two. This alteration is thus interpreted as forming coevally with the garnet-amphibole-plagioclase-quartz intergrowths, that is, in the almandine amphibolite-facies field. There is a second replacement of pyroxene by sheaves or mosaics of an actinolitic amphibole which also grows throughout the groundmass as small stubby crystals. These are interpreted as part of a low temperature, late stage alteration, which appears to have been much more dominated by fluid composition. The actinolite is widely developed throughout the rock with little variation in composition, but there is however a distinct contribution of Na_2O from albite at the edges of the pseudomorphs. Large hornblendes also develop from the pyroxene pseudomorphs, completely overgrowing the sheaves of actinolite and cross-cutting surrounding grain boundaries.

Plagioclase grains were originally An_{35}-An_{38}. They display a complex alteration, which is particularly well developed in felsic bands. The

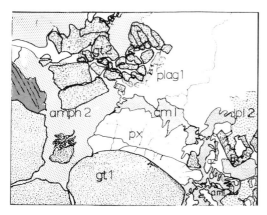

FIG. 2. Textures observed in the mafic band of a hydrated granulite. Pyroxene mantling garnet (gt 1) is reacting to form a pargasitic hornblende (am 1) which shows continuous variation to the more tschermakitic hornblende observed in the symplectite with a second garnet (gt 2), plagioclase (pl 2) and quartz. A second actinolitic amphibole (amph 2) overgrows the primary minerals and the products of the early hydration.

cores of some grains contain zoisite and paragonite as narrow oriented rods or laths within the host grain, which is now albite with dispersed patches of quartz. Occasionally there are small grains of kyanite. Grains of alkali feldspar have also been identified in some samples. In other samples and particularly in association with mafic minerals garnet-quartz-amphibole-oligoclase symplectites pseudomorph the original plagioclase grains. This early hydration is obscured in many samples by the development of later alteration products where the hydrated feldspar has been partly or completely altered to albite, clinozoisite, muscovite and epidote. The rims show a greater development of muscovite, epidote and chlorite especially where in contact with pyroxene or garnet.

Ilmenite grains have a uniform composition in unaltered specimens, but appear to contain a very fine scale exsolution of Ti-rich and Ti-poor areas in samples displaying the early hydration assemblages. In more altered and hydrated xenoliths ilmenite is pseudomorphed by rutile and magnetite rods with some calcite and amphibole.

Some granulite xenoliths display a late stage deformation in which pyroxene and garnet grains show rotation and the edges have been granulated. Some garnet grains have been fractured and imbricated, biotite filling the cracks. In other places sheaves of biotite laths growing with small secondary garnet grains following narrow shear zones. This deformation occurred after the first hydration because, where feldspar grains have been deformed, the grains of zoisite and paragonite are aligned with the foliation. The albite and quartz have regrown as small interlocking polygonal grains. Epidote and amphibole grains overgrow the foliation and both the primary minerals and the products of the early hydration. This is particularly evident where epidotes are growing around pyroxenes, cross-cutting the foliated matrix. Thus the later hydration appears to have occurred after the deformation had ceased.

b) Biotite (sillimanite) gneisses

Garnets are more iron and calcium rich than those in the granulites. They have inclusions of quartz, feldspar and biotite as well as oriented needles of rutile with inclined extinction. They are often rimmed with muscovite or chlorite. In a few samples a secondary garnet appears to have developed with the reaction of biotite, orthoclase and sillimanite to produce garnet and muscovite. Sillimanite is commonly developed both as large grains and as fibrolite. Large grains set in biotite are commonly overgrown by talc and muscovite. Biotite occasionally forms fingerprint symplectites of a second, more iron rich biotite and muscovite but is more commonly replaced by chlorite and oxide. The plagioclase feldspar is more sodic than in the granulites (An_{25}) and shows alteration in the cores to a paragonite-muscovite mica, zoisite and some kyanite. Alkali feldspar in the form of perthitic orthoclase occurs in variable amounts from being the major feldspar in the rock to being present in only trace amounts. The plagioclase feldspar grains show a similar late alteration to that displayed in granulites, with growth of muscovite, epidote, albite and chlorite. Muscovite is particularly well developed near biotite grains which appear to supply K^+ ions. Often some amphibole is also developed.

The early hydration is only obvious in 5% of the granulite and biotite-sillimanite gneiss xenoliths, but this is only to be expected in an area where crystalline basement makes up much of the upper as well as lower crust. Whole rock analyses (Table 3) of crustal xenoliths show that the majority of granulites are basic, similar in composition to continental tholeiites. They do however range to acidic compositions. The early hydration appears to have had little effect on the bulk chemistry of the rocks apart from slight increases in Na and K. Most reactions, however, appear to have been essentially isochemical. Because secondary hydration does seem to have some correlation with microcracks (Padovani *et al.* 1982 and this study), a comparison of whole rock

TABLE 3. Whole rock analyses

Sample number	S.U.M. 1416A	Minette AP4	Sp.Lh. MRD26A	Hyd.Lh. MRD168	Eclog MRD13	Pyrox MD105	Bi.Gn MD10	Gran MD1	Hyd.gran MD1(W)	Gran MD5	Hyd.gran MD5(W)	Gran MD19	Hyd.gran MD19W	Gran MRD306	Hyd.gran MRD306(W)
SiO_2	39.16	48.57	44.32	48.99	51.05	56.05	43.37	54.77	55.44	51.86	51.11	50.14	47.61	49.49	49.86
TiO_2	0.09	1.72	0.11	0.10	1.76	0.06	1.62	0.73	0.74	0.81	0.76	0.77	0.78	0.77	0.71
Al_2O_3	1.58	9.94	3.71	6.53	17.37	16.31	19.39	13.88	14.54	17.58	17.57	17.86	16.10	16.88	17.33
Cr_2O_3		0.05	0.68	1.05	0.03	0.01	0.02	0.02	0.02	0.01	0.01			0.00	0.00
Fe_2O_3	7.05	8.57	8.1	5.23	10.66	6.32	17.32	12.17	11.05	10.53	10.58	7.54	10.96	11.45	10.54
MnO	0.09	0.13	0.17	0.13	0.19	0.10	0.59	0.16	0.16	0.14	0.13	0.08	0.16	0.19	0.18
NiO		0.03	0.05	0.11	0.01	0.01	0.01	0.00	0.00	0.01	0.01			0.00	0.00
MgO	36.97	10.02	40.70	20.64	4.92	3.76	4.98	7.67	6.55	5.22	5.43	4.72	6.10	4.95	4.51
CaO	1.14	9.93	1.65	11.61	6.49	4.67	2.93	6.87	6.62	8.89	8.79	14.22	14.19	8.84	8.91
Na_2O	0.05	2.14	0.17	1.06	7.84	11.30	3.46	3.43	3.65	3.36	3.23	3.65	2.55	3.65	3.92
K_2O	0.06	4.31	0.03	0.08	0.07	0.12	3.87	0.57	0.55	0.44	0.49	0.32	0.30	1.24	1.30
P_2O_5	0.03	0.97			0.02	0.02	0.05	0.09	0.13	0.10	0.09	0.09	0.16	0.35	0.31
SO_3		0.02			0.14	0.60	0.04	0.00	0.01	0.00	0.01				
H_2O	11.20	3.62		5.24	0.2	0.43	2.47	0.04	1.04	0.93	1.67	0.52	1.18	1.82	2.62
Total	99.47	100.04	99.79	99.78	100.51	100.21	100.29	100.38	100.51	99.87	99.88	99.93	100.11	99.64	100.20

MR–1416A-McGetchin (1968)

TABLE 4. *Representative mineral analyses*

Granulite MD19 — *Primary assemblage*: Gt, Pyrox, Plag, ilm; *Minerals from early Hydn*: Amph, Alb, Zo, Parag; *Late hydrous phases*: musc, czo, amph, amph. **Biotite gneiss MRD477**: gt, bi, sill, plag.

Mineral	Gt	Pyrox	Plag	ilm	Amph	Alb	Zo	Parag	musc	czo	amph	amph	gt	bi	sill	plag
SiO_2	38.21	51.37	56.98	0.07	47.96	68.08	39.33	47.2	48.96	37.93	45.10	41.86	39.45	37.80	37.16	60.26
TiO_2	0.71	0.72	0.00	50.30	0.79	0.00			0.39	0.06	0.27	0.09	0.00	1.11	0.00	0.00
Al_2O_3	20.70	5.16	24.26	0.04	9.12	19.48	31.80	28.4	29.01	25.13	7.50	15.55	21.90	16.48	60.74	21.79
Cr_2O_3	0.02	0.03	0.00	0.08	0.05	0.02			0.00	0.00	0.08	0.02	0.00	0.03	0.00	0.02
FeO	23.56	7.28	0.99	49.51	9.71	0.13	1.24		1.77	9.70	23.65	17.87	25.72	12.30	1.37	2.01
MnO	0.57	0.08	0.00	0.29	0.05	0.02			0.00	0.09	0.39	0.23	2.30	0.09	0.00	0.05
MgO	5.54	11.78	0.07	0.43	10.99	0.00			2.26	0.16	7.12	8.08	10.20	16.44		0.02
CaO	10.31	21.21	9.57		16.33	0.46	22.7		0.13	22.65	11.01	7.78	1.51	0.01		7.04
Na_2O		1.57	6.63		2.10	11.42	0.14	11.1	1.05	0.01	1.43	5.04		0.36		8.50
K_2O	0.05	0.05	0.26	0.45				0.4	9.56		0.67	0.73		9.58	0.01	0.05
Total	99.62	99.25	98.76	100.72	97.55	99.63	96.21	93.7	93.13	95.62	97.22	97.25	101.08	94.20	99.04	99.74

Eclogite MD6: Gt (c), Gt (r), px (c), px (r), pheng. **Garnetite MRD/GT/1**: Gt, Pyrox, chl. **Spinel lherzolite MD26** — *Primary minerals*: Ol, Opx, Cpx, Sp; *Secondary minerals*: Amph, Chl, Cor.

Mineral	Gt (c)	Gt (r)	px (c)	px (r)	pheng	Gt	Pyrox	chl	Ol	Opx	Cpx	Sp	Amph	Chl	Cor
SiO_2	38.80	38.33	55.64	57.33	55.78	40.44	57.83	30.63	41.24	56.60	53.60	0.00	45.58	32.74	0.00
TiO_2	0.02	0.02	0.06	0.04	0.13	0.13	0.06	0.00	0.00	0.02	0.09	0.00	0.17	0.00	0.00
Al_2O_3	21.80	21.61	14.26	15.97	18.62	22.45	11.16	18.84	0.00	2.99	4.21	45.62	11.06	19.44	98.86
Cr_2O_3	0.03	0.03	0.00	0.00	0.05	0.03	0.11	0.03	0.00	0.44	1.19	21.63	0.67	0.19	1.37
FeO	29.21	30.62	5.00	6.07	2.57	17.72	2.41	4.54	8.56	5.82	1.76	15.63	3.88	3.55	0.36
MnO	1.43	0.92	0.06	0.16	0.03	0.33	0.08	0.00	0.14	0.16	0.08	0.21	0.05	0.05	0.00
MgO	7.71	4.78	6.56	3.69	6.39	9.14	9.02	31.43	49.53	33.75	15.43	16.34	18.96	28.43	1.62
CaO	2.33	5.21	8.09	4.63	0.01	10.92	13.23	0.00	0.00	0.32	22.17	0.00	12.64	0.16	0.00
Na_2O			9.13	11.73	0.03		7.14		0.00	0.01	1.34	0.00	3.45	0.84	0.00
K_2O					10.43				0.00			0.00	0.46		0.00
Total	101.33	101.52	99.02	99.62	94.04	101.16	100.04	85.26	99.87	100.20	99.92	99.70	97.38	84.60	99.41

TABLE 5. *Pressure-temperature estimates for representative samples of lower crust and upper mantle rock types*

Sample number	Rock type	Temperature (°C)	Pressure (kbar)	Method used
MD1	Granulite	710	7.3	E&G, N&P
MD19	Granulite	650	6.7	E&G, N&P
MD116	Granulite	745	7.5	E&G. N&P
MRD306	Granulite	680	7.1 (6.8) <6.9>	E&G, N&P (N&H) <B>
MD10	Biotite gneiss	590	4.0	F&S (G)
MD11	Biotite gneiss	610	4.3	F&S (G)
MRD317	Biotite gneiss	540	3.8	F&S (G)
MD4	Eclogite	510 (c) 570 (r)		E&G c-core r-rim
MD18	Eclogite	490		E&G
MRD9	Eclogite	530 (510)	15	E&G (G&H)
MRD10	Eclogite	490 (c) 540 (r)		E&G
MRD26	Spinel lherzolite	1040 (990) 1020 (<700)		W (S&S) L (R)
MED36	Websterite	1020 (1030)		W (L)
MD131	Spinel websterite	1150 (1040) 1030 (940)		M (W) L (S&S)
MD134	Spinel websterite	990 (1010) 960		W (L) S&S

Key:

N&P	Newton & Perkins 1982	B	Bohen *et al.* 1983
N&H	Newton & Haselton 1981	E&G	Ellis & Green 1980
F&S	Ferry & Spear 1980	G	Ghent *et al.* 1976
S&S	Sachtleben & Seck 1981	G&H	Green & Hellman 1981
L	Lindsley *et al.* 1982	W	Wells 1977
M	Mercier 1977	R	Roeder *et al.* 1979

analyses for cores and rims of hydrated xenoliths should give some qualitative clue to the chemical effects of the late hydrating fluid. In terms of major elements, rims are enriched in Ca and depleted in Si compared to the cores; Fe^{3+} is enriched in the rims at the expense of Fe^{2+}. As for trace elements, rims are depleted in V and Mn, and enriched in Ba, Sr, Zn and Cu. At least some of these trends must be associated with the growth of calcite in the microcracks, but they also reflect the growth of epidote, micas, chlorite and amphibole poor in Si and rich in Fe^{3+}. Implications are for a cool Ca-rich oxidizing fluid rich in H_2O but also with a substantial contribution of CO_2.

Thermobarometry

For crustal xenoliths, the temperatures of equilibrium formation for the primary mineral assemblages were calculated using Mg-Fe exchange thermometers: garnet-clinopyroxene (Ellis & Green 1979), garnet-biotite (Ferry & Spear 1978), garnet-phengite (Green & Hellman 1982). Pressures were estimated by considering the compositions of coexisting phases in equilibrium: garnet-clinopyroxene-plagioclase-quartz (Newton & Perkins 1982) garnet-clinopyroxene-sillimanite-quartz (Newton & Haselton 1981), garnet-rutile-aluminosilicate-ilmenite (Bohen *et al.* 1983).

Results for representative samples are summarized in Table 5. Temperatures for granulites range from 650–800°C and pressures from 6.5–8 kbar. In general these results were consistent with previous work on similar xenoliths (Newton & Perkins 1982). Biotite-sillimanite gneisses gave pressures and temperatures around 4.5 kbar and 500–600°C.

The temperatures and pressures of the hydrated assemblages were investigated using a variety of methods including consideration of the stability fields for zoisite-paragonite-quartz-albite-plagioclase and zoisite-paragonite-

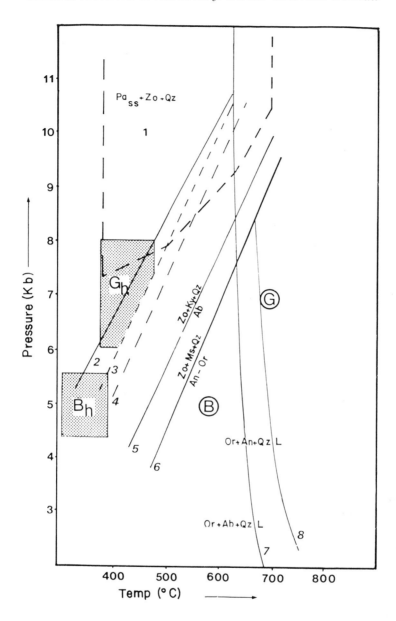

FIG. 3. Stability fields of the hydrated assemblages observed in felsic bands in granulites and biotite sillimanite gneisses.

G—Granulites;

B—Biotite gneisses;

Gh—Granulite hydrated assemblage;

Bh—Biotite gneiss hydrated assemblage.

1 Stability field of zoisite-paragonite-quartz (Franz & Althaus 1977).

2 $Zo + Ky + Qz \rightleftharpoons Plag\ An_{20}$ (Goldsmith 1982).

3 $Zo + Pa + Qz \rightleftharpoons Plag$ (approx.) (Frantz & Althaus 1977).

4 $Zo + Pa\text{-}Ms + Qz \rightleftharpoons Plag\ An_{20}$ (Johannes 1984).

5, 6, 7, 8 Johannes 1980, 1984.

For reactions 2, 3 & 4 Plag appears on the high temperature side of the reaction.

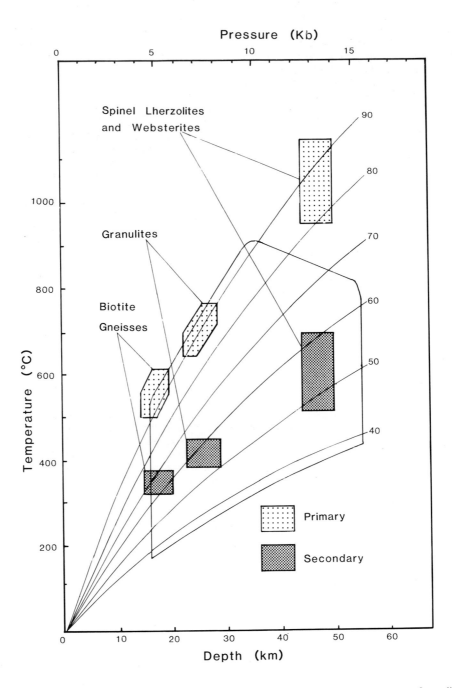

Fig. 4. Pressure and temperature of formation of primary and secondary mineral assemblages of xenoliths from Moses Rock diatreme and their relation to continental geotherms. Heat flow curves after Chapman (this volume).

kyanite-quartz-albite from the work of Gold-smith (1982), Johannes (1984) and Franz & Althaus (1977). Temperature estimates were attempted using the garnet-biotite thermometer where sheaves of biotite and small euhedral garnets cut across old, fractured garnets and altered plagioclases and at the edge of garnet grains where small garnet grains have grown with secondary biotite. Where intergrown garnet and amphibole symplectites occur, Graham & Powell's (1984) garnet-amphibole thermometer was used. In some places K-feldspar has developed and 2-feldspar thermometry was attempted. Analyses were made where the assemblages were least affected by the second hydration.

The latest hydration, accompanied by the growth of sheaves of actinolite was investigated using Spear's (1980) amphibole-feldspar thermometer.

The stability fields of the mineral assemblages associated with the early hydration of the granulites are all high pressure, approx 6–8 kbar at 400–500°C (see Fig. 3), and the temperatures obtained from the thermometers range from 440–470°C. However these are crude estimates as the methods are liable to inaccuracies under these conditions, the feldspar thermometer because of the low temperature and the possibility of variable amounts of potassium and sodium being taken up by the micas and calcium by the zoisites. The garnet-biotite thermometer is liable to inaccuracies because of the low temperature, grossular content of garnet and because of the conditions of growth. These grains are very similar in composition to the larger grains and there is no guarantee that equilibrium was attained. A method of verifying the temperature and pressure estimates, is to consider whether these xenoliths fall on the same low heat flow conductive geotherm (50–60 mWm²) as has been modelled for the cooler values from the mantle xenoliths (Smith & Roden 1979) and xenocrysts (Hunter & Smith 1980). It can be seen from Fig. 4 that, using this geotherm, temperatures of 400–500°C would be entirely compatible with depths equivalent to 6–8 kbar and would also indicate that the xenoliths had been cool for a considerable time, allowing thermal equilibrium to develop. The biotite sillimanite gneisses have similar hydrated assemblages to those found in the granulites, except that the mica contains some potassium. Projected temperatures were in the region of 300–350°C with pressures of 4–5 kbar. These values are also subject to the inaccuracies mentioned above, and they also lie on a low heat flow geotherm (Fig. 4).

The latest hydration occurred under low grade greenschist conditions with temperatures of up to 330°C. To determine whether this was a vent hydration, conodonts were separated from limestone xenoliths and examined for evidence of thermal maturation, i.e for colour and crystallinity variation. No such maturation is observed, and thus the late hydration is taken to have occurred prior to the incorporation of these xenoliths in the SUM.

Temperatures obtained for eclogites support those values calculated by Harley & Green (1981) of between 400–600°C. The pyroxene and garnet zoning cannot be related to any one internal factor although there does seem to be a general decrease in Kgnt−cpx/Fe²⁺−Mg²⁺ from core to rim of mineral grains. However in the case of the garnets this is normally accompanied by an increase in grossular content which makes elucidation of this path difficult as the general increase in temperature indicated by the Fe-Mg partitioning is within the error of the method. Core temperatures are consistently lower by up to 140°C than rim temperatures and this is taken to imply a prograde path, even though the calculated temperatures may well be spurious because it is impossible to determine which part (if any) of the garnet and pyroxene grains were in equilibrium because of the strong and complex zonations. Pressure estimates are very difficult in eclogites. The lack of a buffer for jadeite content in the pyroxenes means that the composition was probably dependent upon circulating fluids and thus it is impossible to use the jadeite content of pyroxenes as a pressure indicator. Another possibility is to use the phengite content of muscovites, which increases with pressure (Velde 1967). Some sort of approximation can be made by extrapolation of thermodynamic data for celadonite calculated by Powell & Evans (1983) and using this for calculations involving Tschermak's type substitutions between pyroxene, garnet and mica. Pressures obtained by this method are in the region of 15 kbar.

Upper mantle xenoliths were investigated using: the 2-pyroxene thermometer of Wells (1977), Lindsley *et al.*'s (1982) calculations on the pyroxene quadrilateral, Al-solubility in orthopyroxene (Sachtleben & Seck 1981) and the olivine-spinel thermometer of Roeder *et al.* (1979). Other authors have also used a 2-spinel thermometer. Spinel lherzolites and websterites show very similar thermal histories. Results support those produced by Smith & Levy (1976) and Smith (1979). Mineral pairs with high blocking temperatures (cpx-opx, Al-opx in equilibrium with homogeneous spinel) give temperatures in the range 920–1030°C, whereas mineral pairs with low blocking temperatures (ol-sp, 2-sp) and zoning studies give temperatures at or below 700°C which indicates considerable cooling within the upper mantle.

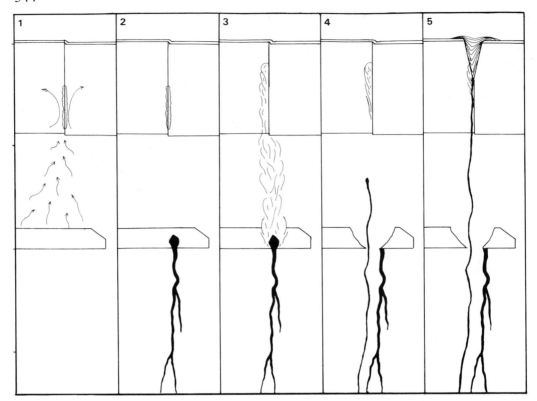

FIG. 5. Model for the development of the lower crust and upper mantle beneath the Colorado Plateau.
1)   An imbricated fragment of subducted oceanic slab beneath the Colorado Plateau, causing hydration and metasomatism of upper mantle and lower crust, as well as cooling the mantle and crust in the area.
2)   Primitive magma intrudes the hydrated, cool imbricated slice. The magma may have been generated by interaction between the mantle and other parts of the subducted slab.
3)   The interaction between the cool, wet slab and the hot magma causes a violent reaction in which the melt is quenched and the subducted slab is disrupted. A solid/fluid mix of hydrated upper mantle material and relics of the imbricated slice is emplaced in the intermediate/upper crust.
4)   The remains of the primary magmatic phase and much of the body of mantle material react with the fluid to produce a matrix of serpentines and amphiboles. The serpentinite thus has a composition closer to hydrated mantle than an alkaline magma.
5)   A second pulse of magma strikes the cool serpentinite body, the consequent explosive reaction causing a diatreme eruption on the surface.

Pressures cannot be estimated for the mineral assemblages studied, but these ultramafic xenoliths are presumed to have originated in the uppermost mantle (45–50 km). Some of the hydrous assemblages found in mantle xenoliths were also stable under upper mantle conditions. The stability fields of amphibole and chlorite bearing ultramafic rocks have been studied by Jenkins (1981) and specific examples from the Alps by Evans & Trommsdorf (1976). The conclusion reached is that magnesian chlorites and amphiboles are stable at mantle-compatible pressures. The association of these minerals with other high-pressure phases such as titanoclinohumite and corundum supports the interpretation that the hydrous assemblages found in lherzolite and websterite xenoliths grew whilst these xenoliths were still in the mantle. The alkremite is similar to that described by Padovani & Tracy (1978) and shows growth of a secondary assemblage of magnesian chorite, garnet, corundum and kornerupine. Later hydration of these xenoliths can be distinguished by the distinctly different (more iron rich) composition of the chlorites which grew cross-cutting the earlier minerals.

Discussion

It is apparent from this study that at least some of the granulite and biotite-sillimanite gneiss xenoliths from Moses Rock Diatreme have undergone two episodes of hydration. The early hydrated assemblages were stable at high pressures and low temperatures. For the granulites these assemblages lie within the almandine amphibolite-facies field. Some xenoliths show evidence for an episode of deformation following this early hydration. This was probably due to the introduction of fluid allowing diffusional mechanisms for deformation to proceed. This early hydration must have occurred whilst the granulites were *in situ* in the lower crust.

The later hydration, affecting a large proportion of all crystalline xenoliths in the diatreme, was at much lower temperatures and pressures under greenschist facies conditions. Opaque phases have been oxidized producing rutile and magnetite instead of ilmenite and in some haematite has developed. This later hydration is not a vent hydration because xenoliths of sedimentary material show maturation due to burial and have not been heated further and the walls of the vent have no aureole. The second hydration must thus represent a reaction of the xenoliths with the fluid component of the SUM and have occurred between entrainment and final diatreme eruption.

Previous models for the nature and history of the lower crust in the Colorado Plateau have ranged from proposing a widely amphibolitized, cool lower crust (McGetchin 1968) to a granulite terrain with hydration and cooling occurring only within the diatreme conduits (Ehrenberg & Griffin 1978). The early hydration reflects the introduction of a fluid into the lower crust. There are a number of possible scenarios to explain the mineral reactions observed and consequently the history of the lower crust in this area, but these can be limited by the consideration of the two major factors. Firstly, whether the hydration is widespread throughout the lower crust of the Colorado Plateau, or localized to the walls of the diatreme vents. Secondly, whether the lower crust was already cool and thus that the granulite-facies assemblages were metastable, only requiring a fluid to initiate the reactions, or whether the fluid caused both the cooling and the hydration of the granulites. Two factors indicate that the hydration was localized to the fractures exploited by the diatremes;

1 The lack of hydrated xenoliths in the contemporaneous minettes which come up through the same lower crust, even though there are abundant anhydrous granulites and also unaltered xenoliths containing hydrous minerals. (Ehrenberg & Griffin 1978, and also this study).

2 The study of the seismic velocities in altered and unaltered xenoliths by Padovani *et al.* (1982) who determined that the lower crustal velocities obtained by Roller (1969) and re-interpreted by Keller *et al.* (1979), were better explained by the unaltered granulite xenoliths.

The remaining problem is to determine whether there was a low heat flow through the area prior to the diatreme eruptions. If this was the case then the granulite xenoliths would have been cool prior to the introduction of the fluid. Heat flow models for the current state of the Colorado Plateau have been refined through the work of Bodel & Chapman (1982). Their measurements show a low heat flow for the interior of the Colorado Plateau (60 mW/m²) with an increased heat flow at the edge (80–90 mW/m²), the variation representing a transient response to increased temperature at depth. Thus the crust of the Colorado Plateau appears to be heating up after being cooler initially. The Mesozoic history of the Colorado Plateau indicates that it was low-lying and stable for a considerable period of time, implying that the crust was cool at that time (Chapman pers. comm.).

Studies of mantle xenoliths and xenocrysts (Smith & Roden 1981; Hunter & Smith 1981; Smith & Ehrenberg 1984) show that the upper mantle beneath the Colorado Plateau had cooled from temperatures between 970 and 1050°C to temperatures below 700°C prior to the eruption of both SUM diatremes and minettes. Hunter & Smith (1981) constructed a conductive geotherm for the mantle beneath the Colorado Plateau using assemblages in xenocrysts to constrain the curve and observed that the cooler assemblages observed in the spinel therzolites also lay on this low heat flow geotherm of approx. 50 mW/m². The stability fields of the early hydration products in the lower crustal xenoliths are consistent with a similar low heat flow conductive geotherm for crustal rocks (Chapman 1985) and thus give some indication of the heat flow through the crust at this time of 50–60 mW/m². For a few xenoliths the anhydrous assemblages of the same rocks give T–P estimates consistent with the higher heat flow for a geotherm on which the higher values for websterites and lherzolites also lie. This does not indicate a coeval origin, merely that such assemblages require a high heat flow to be produced.

The occurrence of the SUM diatremes along monoclines gives some clue to the reasons for the localization. The monoclines are interpreted as the expression of movement along deep, pre-existing crustal faults (Davis 1978) and the SUM may have used the deep faults as conduits (Roden 1979). Previously, a water rich fluid from the mantle may also have used the faults as conduits and may have reacted with rocks in the walls of the conduit. The exploitation of such deep structures by fluids has been suggested by Etheridge *et al.* (1983).

The remarkable similarity of the eclogites in the diatremes to those of the Franciscan formation and to samples from the Alps, and the weight of evidence indicating that oceanic material was being subducted beneath the western United States during and just after the Laramide (Cross & Pilger 1978; Dickenson & Snyder 1978; Keith 1982), has led to the proposal that these xenoliths originated as fragments of subducted oceanic crust (Helmstaedt & Doig 1975). Condie (1982) in a study of Greenstone belts and associated Precambrian rocks has suggested that there was a suture zone in this area at about 1.72 Ga which would provide another possible source for the eclogites as pods within the crust. Their metamorphic origin has already been mentioned but there is still insufficient evidence to completely elucidate the origin of the eclogites. If there were relics of subducted oceanic slab high in the mantle or imbricated within the lower lithosphere, then this would provide a simple source for the eclogites and also a source for the fluids which hydrated both mantle and crustal xenoliths. The microcracks which form a network through the eclogite xenoliths may well indicate that they have travelled some distance within the diatreme unlike the angular fragments of upper crustal material. If the xenoliths had the opportunity to react with the fluid, their lack of hydration implies that either the eclogites were in equilibrium with the hydrating fluid or that they came from so high in the upper crust that they were unaffected by any fluid either percolating up from the mantle or within the diatreme itself. If the eclogites remained in equilibrium with the bulk fluid then most of the fluid must have been derived from the slab and the fluid had to be dominant to prevent any large change in its bulk composition, and thus reaction with the eclogites. This implies external buffering of major variables and when zoning is studied in the eclogites it can be seen that related components have linked zoning patterns in both pyroxenes and garnets (e.g. linked increases and decreases in Fe^{3+} in both minerals which implies an external control on fO_2). Fine scale zoning can be interpreted as an attempt to regain equilibrium with this varying fluid composition. On the break up of the slab, fluids of quite varied composition would be homogenized producing an 'average' fluid. In some of the more fractured specimens, with more open microcracks, zoning in the pyroxenes has been lost by apparent interaction with fluids. Thus the pyroxene composition is more uniform and less jadeitic.

Such homogeneity would be predicted by this model. Preliminary stable isotope studies indicate a close relationship between the fluid in the eclogites and that hydrating the crustal xenoliths (unpub. data).

The remaining problem is to model the conditions of the late hydration of the xenoliths and the emplacement of the SUM in the upper crust. There are two possible methods for the generation of the solid/fluid mix of the serpentinized microbreccia. Both involve a primary (possibly minette) magma migrating into cool, fluid filled areas of the upper continental mantle, the consequent interaction causing the disaggregation of many xenoliths and the hydration of these and many of the other xenoliths. The two models differ in the origin of the fluid.

1 In the first model the fluid is essentially juvenile, from volatiles built up from lower in the mantle, and perhaps from earlier dewatering of subducted slabs and subduction related processes. In this model, the primary magma migrates into these fluid-rich areas and interacts with them, quenching and hydrating the magmatic material.

2 In the preferred second model (Fig. 4) much of the fluid is derived from a relic fragment of Farrollan plate, thus explaining the association of hydrated xenoliths and eclogites in the SUM and the absence of both in the minettes. The primary magma is envisaged to have migrated near to or to have intruded into the relic slab, the consequent violent interaction of hot (minette) magma with the cold fluids in the slab causing fragmentation of the slab and incorporation of eclogite xenoliths in the resulting hydrated material. This does not preclude the earlier build up of fluids beneath the Plateau causing widespread hydration and metasomatism of the upper mantle prior to the diatreme eruption.

This latter model might be expected to produce a perturbed geotherm, with the upper lithosphere being cooled by the slab. The low temperature assemblages observed in various xenolith types appear to display a slight perturbation, but this trend is smaller than would be predicted given the time between emplacement of the subducted slab and eruption of the diatreme. However, if the upper mantle beneath the Colorado Plateau was cold already then the geotherm would show relatively little perturbation.

The eruption of the material can be explained in two ways:

a) A slow uplift in a serpentinite body and then violent eruption from the point at which the fluid went 'critical' or where the water table was reached (Ehrenberg & Griffin 1979).

or b) A two stage eruption; firstly a 'blind'

eruption of a solid/fluid mix to the intermediate/
upper crust where the xenoliths and any remaining magmatic material sat and reacted with the
fluid component, followed by a second eruption
from depth which hit the first phase causing its
eruption to the surface.

The benefits of the second model over the first
are that it would explain the prevalence of
microcracks which are caused by thermal expansion, the presence of rare pristine samples with

clean microcracks and the presence of rare
minette fragments as it is possible that the second
erupting phase was minette.

ACKNOWLEDGMENTS: This work was carried
out whilst the author was in receipt of a NERC research
studentship. I would like to thank D.A. Carswell, D.
Chapman, J.B. Dawson and W. Griffin for helpful
comments. The manuscript benefited greatly from the
reviews of S.L. Harley and D. Smith.

References

BODELL, J.M. & CHAPMAN, D.S. 1982. Heat flow in the
North-Central Colorado Plateau. *J. Geophys. Res.*
87, 2869–2884.

BOHLEN, S.R., WALL, V.J. & BOETTCHER, A.L. 1983.
Geobarometry in granulites. *In: Kinetics and Equi-
librium in Mineral Reactions.* SAXENA, S.K. (eds)
Springer Verlag, New York p. 141–172.

CHAPMAN, D.S. & POLLACK, H.N. 1977. Regional
Geotherms and lithospheric thickness. *Geology*, **5**,
265–268.

COLEMAN, R.G., LEE, D.E., BEATTY, L.B. & BRANNOCK,
W.W. 1965. Eclogites and eclogites: their differences
and similarities. *Bull. geol. Soc. Am.* **76**, 483–508.

CONDIE, K. 1982. A plate tectonics model for Proterozoic Continental accretion in the southwestern
United States. *Geology*, **10**, 37–42.

CONEY, P.J. & REYNOLDS, D. 1980. Cordilleran suspect
terranes. *Nature*, **270**, 403–405.

CROSS, T.A. & PILGER, R.H. 1978. Constraints on
absolute motion and plate interaction inferred from
Cenozoic igneous activity in the Western United
States. *Am. J. Sci.*, **278**, 865–902.

DAVIS, G.H. 1978. Monocline fold pattern on the
Colorado Plateau. *Mem. geol. Soc. Am.*, **151**, 215–
234.

DICKENSON, W.R. & SNYDER, W.S. 1978. Plate tectonics
of the Larimide Orogeny. *Mem. geol. Soc. Am.*, **151**,
355–366.

EHRENBERG, S.N. 1977. The Washington Pass volcanic
centre: evolution and eruption of minette magmas
of the Navajo volcanic field. Extended Absr., *2nd.
Int. Kimberlite Conf. 1977.*

—— 1978. *Petrology of potassic volcanic rocks and
ultramafic xenoliths from the Navajo volcanic field,
New Mexico and Arizona.* Unpubl. Ph.D Thesis,
Univ. Calif. Los Angeles.

—— 1982. Petrogenesis of Garnet Lherzolite and
Megacrystalline nodules from the Thumb, Navajo
volcanic field. *J. Petrol.*, **23**, 507–547.

—— & GRIFFIN, W.L. 1979. Garnet granulite and
associated xenoliths in minette and serpentinite
diatremes of the Colorado Plateau. *Geology*, **7**, 483–
487.

ELLIS, D.J. & GREEN, D.H. 1979. An experimental study
of the effect of Ca upon garnet-clinopyroxene Fe-
Mg exchange equilbria. *Contrib. Mineral. Petrol.*,
71, 13–22.

ESSENE, E. & WARE, N.G. 1970. The low temperature

xenolithic origin of eclogites in diatremes N.E.
Arizona. (abst) *Abst. Prog. geol. Soc. Am.* **2**, 547–
548.

ETHERIDGE, M.A., WALL V.J. & VERNON R.H. 1983.
The role of the fluid phase during regional metamorphism and deformation. *J. Metam. Geol.* **1**, 205–226.

FERRY & SPEAR 1978. Experimental calibration of the
partitioning of Fe and Mg between biotite and
garnet. *Contrib. Mineral. Petrol.* **66**, 113–117.

FRANZ, G. & ALTHAUS, E. 1977. The stability relations
of the paragenesis paragonite-zoisite-quartz. *Neues.
Jahrb. Min. Abhdl.*, **130**, 159–167.

GAVASCI, A.T. & HELMSTAEDT, H. 1969. A pyroxene
rich garnet peridotite inclusion in an ultramafic
breccia dike at Moses Rock, S.E. Utah. *J. Geophys.
Res.*, **74**, 6691–6695.

GHENT, E.D. 1976. Plagioclase-Garnet-Al₂ SiO₅-
Quartz a potential geobarometer/geothermometer.
Am. Mineral., **61**, 1654–1660.

—— & STOUT M.Z. 1981. Geobarometry and geothermometry of Plagioclase-Biotite-Garnet-Muscovite
assemblages. *Contrib. Mineral. Petrol.*, **76**, 92–97.

GOLDSMITH, J.R. 1982. Plagioclase stability at elevated
temperatures and water pressures. *Am. Mineral.*, **67**,
653–675.

GREEN, T.H. & HELLMAN, P.L. 1982. Fe-Mg partitioning between coexisting garnet and phengite at high
pressure, and comments on a garnet-phengite geothermometer. *Lithos*, **15**, 253–266.

HARLEY, S.L. & GREEN, D.H. 1981. Petrogenesis of
Eclogite inclusions in the Moses Dyke Utah, U.S.A.
Tschermak's Min. Petr. Mitt., **28**, 131–155.

HELMSTAEDT, H., ANDERSON, D.L. & GAVASCI, A.T.
1972. Petrofabric Studies of Eclogite, Spinel Websterite and Spinel Lherzolite Xenoliths from Kimberlite-Bearing Breccia pipes in Southeastern Utah
and Northeastern Arizona. *J. Geophys Res.* **77**,
4350–4365.

—— & DOIG, R. 1975. Eclogitic nodules from kimberlite pipes of the Colorado Plateau—Samples of
subducted Franciscan-type oceanic lithosphere.
Phys. Chem. Earth, **9**, 95–111.

—— & SCHULZE, D.J. 1977. Type A–Type C eclogite
transition in a xenolith from the Moses Rock
diatreme—further evidence for the presence of
metamorphosed ophiolites under the Colorado Plateau. Extended Abstr., *2nd. Int Kimberlite Conf.*

—— & —— 1979. Garnet clinopyroxenite-chlorite

eclogite transition in xenoliths from the Moses Rock diatreme—further evidence for the presence of metamorphosed ophiolites under the Colorado Plateau. *In*: BOYD F.R. & MEYER H.O.A. (eds) *The Mantle Sample*; Inclusions in Kimberlites and other Volcanics. Amer. Geophys. Union. Washington p. 357–365.

—— & —— 1982. Eclogite facies ultramafic xenoliths from the Colorado Plateau: Comparision with eclogites in crustal environments and evaluation of the subduction hypothesis. *Terra Cognita*, **2**, 302.

HUNTER, W.C. & SMITH, D. 1981. Garnet Peridotite from Colorado Plateau Diatremes: Hydrates, carbonates and comparative geothermometry. *Contrib. Mineral. Petrol.*, **76**, 312–320.

JOHANNES, W. 1984. Beginning of melting in the Granite system Qz-Or-Ab-An-H_2O. *Contrib. Mineral. Petrol.* **86**, 264–273.

KEITH, S.B. 1982. Paleoconvergence rates determined K_2O/SiO_2 ratios in magmatic rocks and their application to Cretaceous and Tertiary tectonic patterns in S.W. North America. *Bull. geol. Soc. Am.*, **93**, 312–320.

KELLER, G.R., BRAILE, L.W. & MORGAN, P. 1979. Crustal structure, geophysical models and contemporary tectonism of the Colorado Plateau. *Tectonophysics*, **61**, 131–147.

KROGH, E.J. & RAHEIM, A. 1978. Temperature and pressure dependence of Fe-Mg partitioning between garnet and phengite, with particular reference to eclogites. *Contrib. Mineral. Petrol.* **66**, 75–80.

LINDSLEY, D.H., GROVER, J.E. & DAVIDSON, P.M. 1981. The thermodynamics of the $Mg_2Si_2O_6$-$CaMgSi_2O_6$ join: A review and an improved model. *In*: NEWTON, R.C., NAVROTSKY, A. & WOOD, B.J. (eds). *Thermodynamics of Minerals and Melts*. Springer Verlag, New York p. 146–155.

LIPMAN, P.W., PROSTKA, H.J. & CHRISTIANSEN, R.L. 1971. Evolving subduction zones in the western United States, as interpreted from igneous rocks. *Science*, **174**, 821–825.

MCGETCHIN, T.R. 1968. *The Moses Rock dike: geology, petrology and mode of emplacement of a Kimberlite-bearing breccia dike, San Juan Country, Utah*. Unpubl. Ph.D Thesis, Calif inst. Tech. 405 pp.

—— & SILVER, L.T. 1970. Compositional relations in minerals from kimberlite and related rocks from the Moses rock dike, San Juan County, Utah. *Am. Mineral.*, **55**, 1738–1739.

MERCIER, J.C. 1976. Single pyroxene geothermometry and geobarometry. *Am. Mineral.*, **61**, 603–625.

NAESER C.W. 1971. Geochronology of the Navajo-Hopi diatremes, Four Corners Area. *J. Geophys. Res.*, **76**, 4978–4985.

NEWTON, R.C. & HASELTON, H.T. 1981. Thermodynamics of the garnet-plagiocase-Al_2SiO_5-quartz geobarometer. *In*: NEWTON, R.C., NAVROTSKY, A. & WOOD, B.J. (eds). *Thermodynamics of Minerals and Melts*. Springer Verlag, New York p. 129–145.

—— & PERKINS, D. 1982. Thermodynamic calibration of geobarometers based on the assemblages garnet-plagioclase-orthopyroxene (clinopyroxene)-quartz. *Am. Mineral.*, **67**, 203–222.

O'HARA, M.J. & MERCY, E.L.P. 1966. Eclogite, perido-

tite, and pyrope from the Navajo Country, Arizona and New Mexico. *Am. Mineral.*, **51**, 352–366.

PADOVANI, E.R. & CARTER, J.J. 1977. Aspects of the deep crustal evolution beneath south-central New Mexico. *Amer. Geophys. Union Monograph*, **20**, 19–55.

—— & TRACY, R.J. 1981. A pyrope-spinel (alkremite) xenolith from Moses Rock Dike: first known North American occurrence. *Am. Mineral.* **66**, 741–745.

—— HALL, J. & SIMMONS, G. 1982. Constraints on crustal hydration below the Colorado plateau from Vp measurements on crustal xenoliths. *Tectonophysics*, **84**, 313–328.

PERKINS, D., WESTRUM, E.F. & ESSENE, E.J. 1980. The thermodynamic properties and phase relations of some minerals in the system CaO-Al_2O_3-SiO_2-H_2O *Geochim. Cosmochim. Acta*, **44**, 61–84.

POWELL, R. & EVANS, J.A. 1983. A new geobarometer for the assemblage biotite-muscovite-chlorite-quartz. *J. metam. Geol.*, **1**, 331–336.

RÅHEIM, A. & GREEN, D.H. 1974. Experimental determination of the temperature and pressure dependence of the Fe-Mg partition coefficient for coexisting garnet and clinopyroxene. *Contrib. Mineral. Petrol.*, **48**, 179–203.

—— & —— 1975. P,T paths of natural eclogites during metamorphism, a record of subduction. *Lithos*, **8**, 317–328.

RYBURN, R.J., RÅHEIM, A. & GREEN, D.H. 1976. Determination of the P,T paths of natural eclogites during metamorphism—a record of subduction. *Lithos*, **9**, 161–164.

ROEDER, P.L., CAMBELL, I.H. & JAMIESON, H.E. 1979. A re-evaluation of the olivine-spinel geothermometer. *Contrib. Mineral. Petrol.* **68**, 325–334.

RODEN, M.F. 1977. *Field geology and petrology of the minette diatreme at Buell Park, Apache Country, Arizona*. Unpubl. M.A. Thesis University of Texas, Austin.

—— 1981. Origin of coexisting minette and ultramafic breccia; Navajo volcanic field. *Contrib. Mineral. Petrol.*, **77**, 195–206.

—— & SMITH, D. 1979. Field geology and petrology of Buell Park minette diatreme, Apache Country, Arizona. *In*: BOYD, F.R. & MEYER, H.O.A. (eds) *The Mantle sample; Inclusions in Kimberlites and other Volcanics*. Amer. Geophys. Union. Washington p. 400–423.

ROGERS, N.W., BACHINSKI, S.W., HENDERSON, P. & PARRY, S.J. 1982. Origin of potash rich basic Lamprophyres—trace element data for Arizona minettes. *Earth planet. Sci. Lett.*, **57**, 305–312.

SACHTLEBEN, TH. & SECK, H.A. 1981. Chemical control of Al-Solubility in Orthopyroxene and its implications on Pyroxene geothermometry. *Contrib. Mineral. Petrol.*, **78**, 157–165.

SMITH, D. 1979. Hydrous minerals and carbonates in peridotite inclusions from the Green Knobs and Buell Park Kimberlitic diatremes on the Colorado Plateau. *In*: BOYD, F.R. & MEYER, H.O.A. (eds). *The Mantle sample; Inclusions in Kimberlites and other Volcanics*. Amer. Geophys. Union. Washington p. 345–356.

—— & EHRENBERG, S.N. 1984. Zoned minerals in

garnet peridotite nodules from the Colorado Plateau: implications for mantle metasomatism and kinetics. *Contrib. Mineral. Petrol.*, **86**, 264–273.

—— & LEVY, S. 1976. Petrology of the Green Knobs Diatreme and implications for the upper mantle beneath the Colorado Plateau. *Earth Planet. Sci. Lett.*, **29**, 107–125.

—— & RODEN, M.F. 1981. Geothermometry and kinetics in a two-spinel peridotite nodule, Colorado Plateau. *Am. Mineral.*, **66**, 334–345.

—— & ZIENTEK, M. 1979. Mineral chemistry and zoning in eclogite xenoliths from Colorado Plateau diatremes. *Contrib. Mineral. Petrol.*, **69**, 119–131.

SPEAR, F. 1980. NaSi-CaAl exchange equilibrium between plagioclase and amphibole, an empirical model. *Contrib. Mineral. Petrol.*, **72**, 33–41.

THOMPSON, G.A. & ZOBACK, M.L. 1979. Regional geophysics of the Colorado Plateau. *Tectonophysics*, **61**, 149–181.

VELDE, B. 1965. Phengite micas synthesis, stability and natural occurrence. *Am. J. Sci.* **277**, 1152–1167.

—— 1967. Si^{4+} content of natural phengites. *Contrib. Mineral. Petrol.*, **14**, 250–258.

WATSON, K.D. 1960. Eclogite inclusions in serpentinite pipes at Garnet Ridge, Northeastern Arizona. *Bull. geol. Soc. Am.*, **71**, 2082–2083.

—— 1967. Kimberlite pipes of North Eastern Arizona, *In*: WYLLIE, P.J. (ed.) *Ultramafic and Related Rocks*. John Wiley, New York. p. 261–269.

—— & MORTON, D.M. 1968. Eclogite inclusions in kimberlite pipes at Garnet Ridge north-eastern Arizona. *Am. Mineral.* **54**, 267–285.

WELLS, P.R.A. 1977. Pyroxene thermometry in simple and complex systems. *Contrib. Mineral. Petrol.*, **62**, 129–139.

WILSHIRE, H.G. & SHERVAIS, J.W. 1975. Al-Augite and Cr-Diopside ultramafic xenoliths in basaltic rocks from the western United States. *Phys. Chem. Earth*, **9**, 257–272.

J.R. BROADHURST, Department of Geology, University of Sheffield, Mappin Street, Sheffield, S1 3JD UK.

Xenoliths from southern Africa: a perspective on the lower crust

P.W.C. van Calsteren, N.B.W. Harris, C.J. Hawkesworth, M.A. Menzies & N.W. Rogers

SUMMARY: Granulite-facies xenoliths from southern African kimberlites yield temperatures from 780–900°C and pressures from 10–15 kb. They have Nd model ages of 1.4–3.7 Ga and retain Rb-Sr mica and Sm-Nd garnet ages from 0.14–1 Ga, significantly older than kimberlite emplacement ages. It is argued that low abundances of U, K and Rb in lower crustal rocks can be either primary features resulting from plagioclase-dominated cumulate processes, or secondary effects caused by the granulite-facies breakdown of mica in wet intermediate–acid rocks. Combined trace element and isotope arguments suggest that contamination with lower crustal material does not significantly affect the geochemistry of flood basalts from the Karoo, the Parana and the Deccan.

Introduction

The nature of the lower crust may be inferred from its physical properties, but it may only be sampled at surface exposures of rocks which are thought to have crystallized under suitable P–T conditions, or from xenolith suites included in kimberlites and alkali basalts. In this contribution we review the mineralogical, isotope and trace element characteristics of lower crustal xenoliths from southern Africa, and compare them with the granulite terrains of the Lewisian (e.g. Moorbath et al. 1969; Weaver & Tarney 1984) and southern India (Janardhanan et al. 1979; Drury et al. 1983). In particular, we assess the extent to which (i) the distinctive chemical features often attributed to the lower crust are primary or secondary phenomena, and (ii) interaction with the lower crust modifies the composition of continental basalts.

The first models for the lower crust were put forward by petrologists working in high-grade metamorphic terrains (e.g. den Tex 1965) who inferred that in stable cratonic areas the lower crust would consist primarily of mafic to intermediate granulites and eclogites. Since then apparently contradictory arguments have been put forward in support of both granulites and eclogites. Heier (1965) argued that to satisfy the measured heat flow and heat production in upper crustal rocks, the lower crust had to be depleted in U, Th and K, and subsequently Heier and, for example, Lambert & Heier (1968) and Weaver & Tarney (1984) demonstrated that many medium–high pressure granulites are characterized by relatively low U, Th and K abundances. Similarly seismic studies indicate that velocities in the lower crust can be matched with those determined on granulites (Christensen & Fountain 1975).

In contrast, thermal considerations suggest that for likely thermal gradients (Clark & Ringwood 1964) the lower crust in stable continental areas should be in the eclogite- rather than the granulite-facies (see Fig. 3). Granulite-facies implies relatively high thermal gradients, and so the evidence for granulites in the lower crust suggests that much of it formed at times of relatively high heat flow, for example during orogenesis. Then when the thermal profile relaxed the granulite-facies rocks were unable to readjust either mineralogically or compositionally to the new eclogite-facies conditions. Thus much of the lower continental crust may consist of granulites, albeit under eclogite-facies pressures and temperatures.

Regional geology

The geology of southern Africa offers a rare opportunity to study the evolution of the continental lithosphere. The surface outcrop preserves a record of major orogenic and magmatic events ranging in age from 3600–30 Ma (Fig. 1). Archaean rocks (3.6–2.5 Ga) occur in cratonic nucleii surrounded by mobile belts containing varying proportions of new and reworked crustal material. For example, considerable volumes of new crust appear to have been generated in the formation of the Namaqua–Natal belt in the upper Proterozoic (1.4–1.0 Ga, Rogers & Hawkesworth 1982); whereas in the Pan African terrain of northern Namibia many of the igneous rocks were derived from pre-existing Proterozoic crust (Hawkesworth & Marlow 1983). Yet of even greater significance to any potential study of the lithosphere is that after these orogenic episodes and the subsequent stabilization of the crust, both the crust and the underlying subcontinental mantle were sampled by widespread

From: DAWSON, J.B., CARSWELL, D.A., HALL, J. & WEDEPOHL, K.H. (eds) 1986, *The Nature of the Lower Continental Crust*, Geological Society Special Publication No. 24, pp. 351–362.

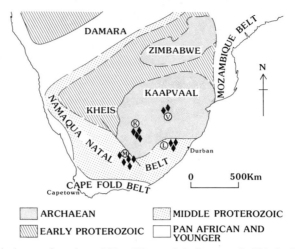

FIG. 1. Schematic geological map of southern Africa. Diamonds indicate studied kimberlite occurrences. V = Voorspoed, K = Kimberley, L = Lesotho, M = Markt.

Karoo magmatism from *c.* 190 Ma, as xenoliths in kimberlites emplaced in the Cretaceous, and by mid-Tertiary alkaline volcanism.

Crustal xenoliths have been reported from many of the kimberlite localities in southern Africa, although for the most part they have received little attention in the scramble for mantle xenoliths and diamonds. They comprise a cosmopolitan selection of the common rock types exposed on the surface, including a variety of granites, granite gneisses, amphibolites and metasediments, together with basalts and sediments of Karoo age. In addition there is an interesting suite of mafic to felsic granulites, which may have surface equivalents in the Natal belt south of Durban (Fig. 1) (McIver & Gevers 1968). Previous studies have indicated that these granulites record temperatures and pressures in the range 550–1000°C and 4.5–20 kb respectively, and that they appear to be restricted to pipes in the younger, Proterozoic belts around the margins of the Archaean cratons (Jackson & Harte 1977; Griffin *et al.* 1979; Robey 1981). To date no comparable high-pressure granulite xenoliths have been reported from the cratonic areas, and the highest pressure rocks in our collection from the Kimberley area still record pressures of less than 10 kb.

Pressure–temperature estimates from crustal xenoliths

Phases from selected equilibrium mineral assemblages in crustal xenoliths from three localities have been analysed (Table 1) to determine the conditions of pressure and temperature at which the assemblages equilibrated.

Garnet granulite xenoliths, from the Markt kimberlite pipe in eastern Namaqualand provide the assemblage garnet-kyanite-clinopyroxene-plagioclase $\pm$ quartz $\pm$ phlogopite, which allows three independent mineral barometers to be used.

1 $3\ CaAl_2Si_2O_8 + 3\ CaMgSi_2O_6 =$
　　Anorthite　　　　Diopside

　　　$2\ Ca_3Al_2Si_3O_{12} + Mg_3Al_2Si_3O_{12} + 3\ SiO_2$
　　　　Grossular　　　　Pyrope　　　　Quartz

2 $3\ CaAl_2Si_2O_8 =$
　　Anorthite

　　　$Ca_3Al_2Si_3O_{12} + 2\ Al_2SiO_5 + SiO_2$
　　　　Grossular　　　Kyanite　　　Quartz

3 $NaAlSi_2O_6 + SiO_2 = NaAlSi_3O_8$
　　Jadeite　　　Quartz　　　Albite

These equilibria have been evaluated by Perkins & Newton 1981 (reaction 1), Newton & Haselton 1981 (reaction 2) and Holland 1980 (reaction 3). a_{anort}^{plag} is taken as 2 and a_{albite}^{plag} as unity after Carpenter & Ferry, 1984. Temperatures have been computed from Fe Mg exchange equilibria using mineral pairs, clinopyroxene–garnet (reaction 4, Ellis & Green 1979) and biotite–garnet (reaction 5, Hodges & Spear 1982). Fe^{3+} was calculated from charge balance and stoichiometry. It was generally minor and did not significantly affect the temperature estimates. Results indicate that the garnet granulite xenolith from Markt (HSA32) equilibrated under conditions of P = 15 ± 2 kbars, T = 780 ± 40°C (Fig. 2a). Pres-

TABLE 1. *Representative mineral analyses in xenoliths*

	HSA 32, Markt				1646, Matsoku, Lesotho				HSA 61, Kimberley		
	Garnet	Clinopyroxene	Plagioclase	Phlogopite	Garnet	Clinopyroxene	Plagioclase	Phlogopite	Clinopyroxene	Orthopyroxene	Plagioclase
SiO_2	41.67	53.34	63.34	40.78	40.21	52.94	64.27	38.30	52.60	53.10	56.04
TiO_2	—	0.45		2.33	—	0.32	—	4.70			
Al_2O_3	23.00	9.82	23.72	16.92	22.97	5.03	22.60	14.84	3.11	2.67	27.94
Cr_2O_3	—	0.30		0.13					0.21		
$FeO*$	12.45	1.88		3.81	21.96	6.30	—	8.73	10.90	22.79	0.42
MnO	0.26	0.12		—	0.40				0.20	0.41	—
MgO	14.79	11.59		20.65	10.80	12.53		17.99	14.22	21.64	
CaO	7.57	18.61	4.39	0.11	5.41	20.00	3.23	0.13	19.42	0.68	10.40
Na_2O	—	3.83	8.62	—		2.32	9.38		0.53		5.42
K_2O	—	—	0.27	10.20	—	—	0.67	9.85			
Total	99.74	99.94	100.34	94.93	101.74	99.44	100.15	94.54	101.19	101.29	100.22
Atoms to:	12(0)	6(0)	8(0)	22(0)	12(0)	6(0)	8(0)	22(0)	6(0)	6(0)	8(0)
Si	3.03	1.91	2.79	5.76	2.99	1.95	2.83	5.60	1.94	1.95	2.52
Al	1.97	0.41	1.23	2.82	2.01	0.22	1.17	2.56	0.14	0.12	1.48
Ti	—	0.01		0.25	—	0.01	—	0.52			
Cr	—	0.01		0.01	—				0.01		
Fe	0.76	0.06		0.45	1.36	0.19	—	1.07	0.34	0.70	0.02
Mn	0.02	—		—	0.03				0.01	0.01	—
Mg	1.61	0.62	—	4.35	1.20	0.69		3.92	0.78	1.19	—
Ca	0.59	0.71	0.21	0.02	0.43	0.79	0.15	0.02	0.77	0.03	0.50
Na	—	0.27	0.74	—		0.17	0.80		0.04		0.47
K	—	—	0.02	1.84	—	—	0.04	1.84			—

* All iron calculated as FeO.

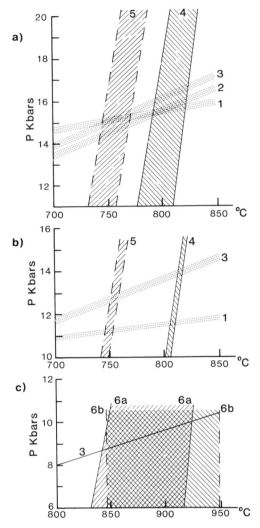

FIG. 2. P-T fields for assemblages from (a) Markt, HSA 32, (b) Lesotho, PHN 1646, (c) Kimberley, HSA 61. Figures refer to reactions discussed in the text. Fields are constructed using averages of several coexisting mineral pairs.

sures may be increased by up to two kilobars if the alumina avoidance model for anorthite activity is used. Robey (1981) reported P-T conditions of 15–20 kbar and 800–1000°C on granulite xenoliths from Markt and neighbouring kimberlite pipes, but pressures would be adjusted to 14 ± 2 kbars applying the barometric equilibria used here.

Northern Lesotho garnet granulite xenoliths provide the assemblage garnet-clinopyroxene-plagioclase $\pm$ quartz $\pm$ phlogopite. These provide two barometers (reactions 1, 3) and two ther-

mometers (reactions 4, 5). Xenolith 1646 equilibrated at 12.5 ± 2 kbar, $780 \pm 40°C$ (Fig. 2b). This is consistent with P–T conditions inferred from the data on several Lesotho xenoliths in Griffin *et al.* (1979) and also with the appearance of sillimanite in those Lesotho xenoliths which contain an aluminosilicate phase.

Granulite xenoliths are exceedingly rare at localities on the Archaean craton, and in contrast to those from the marginal belts the two available samples from Kimberley do not contain garnet. They consist of augite, hypersthene and plagioclase. Maximum pressures (in the absence of quartz) of 10 kbar are inferred from reaction 3 (Fig. 2c), which is consistent with the absence of garnet. Two pyroxene thermometry based on Ca/Mg exchange between clino- and orthopyroxene (reaction 6) provide temperatures of $900 \pm 50°C$ using Wood & Banno 1973 (reaction 6a) and Powell 1978 (reaction 6b). The Wells (1977) calibration of the two pyroxene thermometer yields similar temperature estimates.

These results indicate remarkably consistent P–T fields for xenoliths within a given locality, and while thermometry may yield blocking, rather than peak, temperatures for the mineral pair concerned (Harte & Freer 1982) we are encouraged by the consistency of temperatures inferred from different mineral pairs. As illustrated on Fig. 3, mineral equilibration took place at deeper levels and apparently under lower geothermal gradients, in xenoliths from kimberlite pipes off the craton (Markt and Lesotho) than from those within it (Kimberley); see also Jackson & Harte (1977), Griffin *et al.* (1979) and Robey (1981). The relatively high temperatures from the two analysed samples at Kimberley are thought to be 'real', even though a different thermometer was required, because when the two pyroxene thermometer is applied to the Lesotho data of Griffin *et al.* (1979) it also indicates significantly lower temperatures than those at Kimberley. These P–T fields may not represent equilibration just prior to eruption (Harte *et al.* 1981, and see the following discussion of mineral ages) but they confirm that granulite-facies metamorphic conditions were present at some stage in their pre-emplacement history, and they imply that considerably higher geothermal gradients pertained to the crust beneath Kimberley than that beneath Markt and northern Lesotho, albeit probably at different times.

Sr-, Nd-, and Pb-isotope results

Radiogenic isotope studies have been carried out on whole-rock and a selection of separated

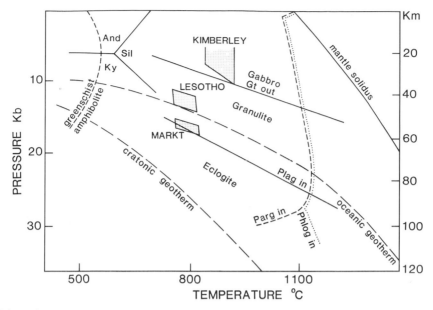

F<small>IG</small>. 3. Schematic P-T diagram with the P-T fields of Fig. 2 indicated.

mineral samples from four areas: northern Lesotho and eastern Namaqualand situated in the marginal Proterozoic belts, and Voorspoed and Kimberley within the Kaapvaal craton (Rogers & Hawkesworth 1982; Hawkesworth *et al.* 1982; van Calsteren *et al.* in prep.). Rock types range from amphibolite-facies granite gneisses to garnet- and two-pyroxene granulites.

Whole rock samples

Rogers & Hawkesworth (1982) reported Nd- and Sr-isotope analyses on garnet granulite xenoliths from northern Lesotho. They obtained an Sm-Nd whole-rock age of 1.4±0.1 Ga, but were concerned that samples with less than 6 ppm Nd, and with lower ^{143}Nd/^{144}Nd ratios than the host kimberlite at the time of emplacement, had had their Nd-isotope composition disturbed by interaction with the kimberlite (see Fig. 4). Thus to investigate the scale of homogeneity within a xenolith, and in particular whether isotope changes accompany those in mineralogy close to the contact with the kimberlite, a large (30 cm across) xenolith was sawn into sub-samples ranging from 12 to 100 g. The rock, a biotite acid gneiss from Letseng (PHN2080, plate 10A in Nixon, 1973), has a well developed foliation and a marked alteration rim *c.* 1 cm across consisting largely of plagioclase and quartz. At the time of kimberlite emplacement ^{87}Sr/^{86}Sr in the four sub-

samples varied from 0.7292–0.7352 and ^{143}Nd/^{144}Nd was indistinguishable analytically at 0.51059±2 (ε_{Nd} = −37.7), even though the host kimberlite had values of 0.7038 and 0.51270 (ε_{Nd} = +3.5) respectively (Kramers *et al.* 1981). Thus entrainment in the kimberlite may have leached material from the outer portions of this xenolith without significantly changing the Sr- and Nd-isotope ratios of even the altered rim. The bulk rock Sm/Nd and ^{143}Nd/^{144}Nd ratios are remarkably uniform and give model T_{Nd}^{CHUR} = 2.7 Ga.

Fig. 4 summarizes the present day ε_{Sr} and ε_{Nd} values of crustal xenoliths from southern Africa and compares them with those of granulite-facies rocks world-wide reported by Ben Othman *et al.* (1984), and the Lewisian (Hamilton *et al.* 1980). Most of the Lesotho and Namaqualand granulite xenoliths are mafic, they tend to have low Rb/Sr but variable Sm/Nd, and hence now form near vertical ε_{Nd}-ε_{Sr} trends. In this they are similar to the granulites of the Lewisian, albeit not as old (mid-Proterozoic rather than Archaean), and they contrast with many of the granulites analysed by Ben Othman *et al.* (1984) which tend to have relatively high ε_{Sr}.

Also illustrated on Fig. 4 is a dashed curve linking the present day ε_{Sr}, ε_{Nd} values of crustal material of different ages with average Sm/Nd = 0.178 and Rb/Sr = 0.12 following Weaver & Tarney (1984). It emphasizes the relatively low

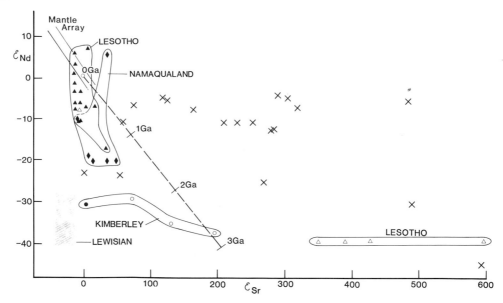

FIG. 4. ε_{Nd} vs ε_{Sr} diagram. Crosses are data from Ben Othman *et al.* 1984, filled symbols are granulite, open symbols are amphibolite-facies rocks.

inferred Rb/Sr ratios of the Lewisian and south African granulites consistent with models in which the lower crust is characterized by relatively low alkali and heat producing element abundances. The amphibolite-facies xenoliths have high ε_{Sr}, and particularly striking is the difference in ε_{Nd} and ε_{Sr} between the amphibolite- and granulite-facies xenoliths from Lesotho. The former yield Archaean model Nd ages, whereas the best estimate of the age of the latter is 1.4 ± 0.1 Ga (Rogers & Hawkesworth 1982).

Reconnaissance Pb-isotope analyses on selected Lesotho mafic granulites indicate that while many have low Rb/Sr ratios, these are not accompanied by strikingly low U/Pb or Th/Pb. Present day Pb isotope ratios plot just below the Pb-ore growth curve (Stacey & Kramers 1975) with $^{206}Pb/^{204}Pb = 17.3$–17.6, implying second stage μ values of ~ 4.6. Thus although the Lewisian and south African granulites arguably have similarly low average Sm/Nd and Rb/Sr ratios, they have significantly different U/Pb and Th/Pb, since the Lewisian is characterized by unusually unradiogenic Pb (inferred $\mu \sim 2.6$, Moorbath *et al.* 1969).

Mineral data

Sm-Nd mineral age studies on lower crustal inclusions in alkalic basalts yielded ages indistinguishable from those of emplacement (van

Breemen & Hawkesworth 1980; Richardson *et al.* 1980). However, a mafic granulite xenolith from Lesotho (Cretaceous kimberlite pipe) which appears to have equilibrated at *c.* 780°C and *c.* 12 kbar preserves a garnet-whole rock Sm-Nd age of 770 Ma. PHN2080 is an amphibolite-facies gneissic xenolith from the same area, and although it has an Archaean model Nd age (see above), its Rb-Sr biotite-whole rock age is 601 ± 18 Ma. Both these results are consistent with cooling after the mid-Proterozoic event in the neighbouring Natal belt (Eglington & Harmer 1984) and they suggest that useful mineral age information can be obtained from at least crustal xenoliths in kimberlite pipes. More perplexing is how biotite and amphibole from similar Lesotho granulite xenoliths also yielded apparently 'reasonable' K/Ar ages of 1005 ± 20 Ma and 714 ± 14 Ma respectively (Harte *et al.* 1981).

Finally, these preliminary mineral data are pertinent to the debate over the nature of the cratonic margin in Lesotho. Rogers & Hawkesworth (1982) postulated that in that area the Proterozoic basic granulites underlay the Archaean amphibolite-facies gneisses at the time of kimberlite emplacement; whereas Nixon *et al.* (1983) suggested that the granulites could have been emplaced tectonically onto the Archaean terrain during the Natal event. The new mineral data imply that any tectonic uplift of the granulites took place at or after 770 Ma, and that the

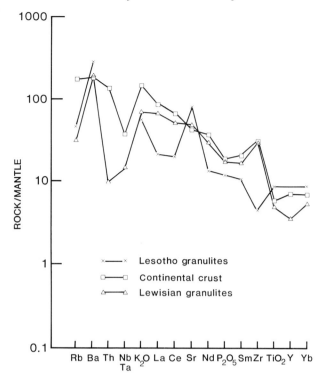

FIG. 5. Trace element abundances normalized versus primordial mantle for the Lesotho granulite xenoliths, Rogers & Hawkesworth 1982. Lewisian granulite and average continental crust data from Weaver & Tarney 1984.

Archaean rocks were then not heated sufficiently to reset the Rb-Sr biotite age. Since major tectonic movements in the Natal belt appear to predate the emplacement of granites at *c.* 1 Ga (Matthews 1981) the granulite emplacement hypothesis may be in conflict with the 770 Ma garnet-whole rock age.

Discussion

Perspectives on the nature of the lower crust differ depending on whether they are based on near surface granulites or the granulite-facies xenolith suites in alkalic basalts and kimberlites (Griffin *et al.* 1979; Wass & Hollis 1983; Weaver & Tarney 1984). The former tend to be of intermediate–acid compositions, whereas the latter are more mafic. Their more significant minor and trace element features are summarized on a mantle-normalized diagram in Fig. 5, which highlights that the continental crust is characterized by relatively low Nb, Ta and Ti contents. Relative to estimates for the average composition of the continental crust, the average Lewisian granulite is depleted

in Rb, Th (U, not shown), and K, but for other elements the two trace element patterns are remarkably similar (Weaver & Tarney 1984). The Lesotho mafic granulites also have low Rb contents, but in addition they have low REE, P and Zr, and a positive Sr anomaly consistent with their relatively high Eu abundances (Rogers 1977; Rogers & Hawkesworth 1982). Thus the Lewisian and Lesotho granulites have similar Rb/Sr ratios (average 0.019 and 0.017 respectively), and although the former tend to have lower Sm/Nd, they are both light REE enriched—and yet the trace element patterns (Fig. 5) suggest that these features reflect different processes.

Not all the rocks from the lower crust appear to have been modified chemically by granulite- and/or eclogite-facies metamorphism, since in some the trace element and isotope ratios are still similar to those of their igneous or sedimentary precursors. On the basis of Rb-Sr and Sm-Nd systematics many of the granulites analysed by Ben Othman *et al.* (1984) would fall into this category. However, most speculation on the lower crust focuses on those aspects which are in some way chemically unusual, the most obvious being the low U/Pb, Th/Pb and hence Pb-isotope ratios that are often regarded as diagnostic of the lower crust (see Moorbath & Taylor, this vol.). One problem is the extent to which these 'unusual' features reflect primary or secondary processes and whether their effects can be distinguished. We consider only igneous and metamorphic, rather than sedimentary, processes since the latter are unlikely to have been responsible for any distinctive aspects of the lower crust.

The Lesotho and Lewisian rocks typify granulites whose distinctive trace element features respectively reflect primary and secondary processes, and in both cases, these features appear to be mineralogically determined. Cox (1980) argued that most continental flood basalts differentiated from picritic magmas and that gabbroic cumulates should therefore be a feature of the crust/mantle boundary in continental areas. Most studies of the Lesotho mafic granulites have concluded that they are of an igneous origin (Rogers 1977; Griffin *et al.* 1979), and that while some represent near-liquid compositions, many are either residua after partial melting or cumulates. The positive Eu and Sr anomalies (Fig. 5) indicate that plagioclase was an important residual mineral, and the variation in the abundance of heavy REE with little change in Sm/Yb suggests that garnet was not. Thus the present garnet-bearing assemblages are metamorphic, and since separated clinopyroxene and garnet also have positive Eu anomalies, the bulk rock REE do not appear to have been disturbed significantly during the granulite-facies metamorphism.

Rogers & Hawkesworth (1982) emphasized that the petrogenesis of the Lesotho granulites therefore had to be argued on the basis of normative rather than modal mineralogy. They identified several reasons, including the larger scatter in compatible rather than incompatible trace elements (Frey & Prinz 1978), against these granulites being residua after partial melting. Instead the xenoliths were successfully modelled as cumulates using whole rock norms calculated solely on the basis of plagioclase, clinopyroxene and olivine, with all other normative components assigned to the intercumulus liquid with $K_D = 1$. Such a model interprets the low average Rb/Sr of 0.017 as a primary feature, and for appropriate U and Pb partition coefficients it predicts slightly low μ values ($\frac{\mu \text{ cumulates}}{\mu \text{ magma}}$ *c.* 0.5) consistent with the observed present day Pb-isotope ratios.

The second way of generating distinctive trace element, and hence isotope ratios in the lower crust is by depletion during granulite-facies metamorphism. Heat flow arguments (Lachenbruch

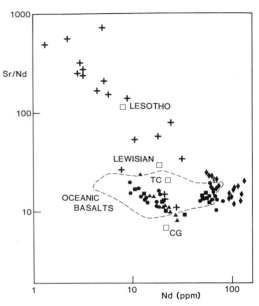

FIG. 6. Sr/Nd vs Nd diagram. Crosses are data points for Lesotho granulites; squares indicate averages, TC = total crust, CG-collision granite, diamonds are kimberlites and lamproites, other symbols as indicated in Fig. 7.

1970) and the many detailed studies on the Lewisian (Weaver & Tarney 1984; Moorbath & Taylor this vol., and references therein) have encouraged speculation that depletion is ubiquitous in the lower crust. However, it is worth reiterating that most of the granulites analysed by Ben Othman *et al.* (1984) do not appear to have been depleted significantly by metamorphic processes, and the same is true of the granulites in the Indian Archaean (Allen *et al.* 1983) and of most mafic granulite xenolith suites (Rogers & Hawkesworth 1982; Dupuy 1979). Why then does granulite metamorphism *sometimes* result in the depletion of elements such as Rb, K, and U? The critical reaction appears to be the breakdown of mica in the presence of quartz to form an aluminium silicate, K feldspar and fluid. Rb, Th and U are released into the fluid, whereas K and Ba are partly retained in the K-feldspar, Sr in plagioclase and any amphibole. This is then accentuated for U with the release of more H_2O on the conversion of amphibole to clinopyroxene. REE are little affected. The implication is that for low Rb/Sr, U/Pb and Th/Pb to be the result of *granulite*-facies metamorphism, the rocks should have contained mica, i.e. they need to have been both H_2O-rich and of intermediate–acid composition. Basic and/or dry rocks are unlikely to have these trace element ratios changed significantly during granulite-facies metamorphism.

Contamination with lower crustal material

Much has been written on the extent to which crustal contamination processes are responsible for the isotope and trace element ratios of continental basalts (Taylor 1980; Carlson *et al.* 1981; De Paolo 1981; James 1981; Cox & Hawkesworth 1984). Xenoliths of the sub-continental mantle can have radiogenic isotope ratios similar to those in the overlying crust (Menzies & Murthy 1980; Erlank *et al.* 1982; Cohen *et al.* 1982), and so, whatever the conclusions, crustal contamination cannot be discussed on the basis of isotopes alone. This is illustrated by arguments over whether contamination of mantle-derived magmas takes place in the lower crust.

The first point is that contamination by material that has been identified as 'lower crustal' need not actually take place in the lower crust. Such potential contaminants may, for example, have been transported tectonically to higher levels in the crust in some previous event.

Secondly, the lower crust is only invoked as a contaminant when the observed isotope ratios are in some way unusual. The temptation has been to presume that, like the Lewisian, the lower crust is distinctive because it has been depleted in Rb, Th and U during granulite-facies metamorphism (see Moorbath & Taylor, this vol.). However, it is becoming increasingly clear that such features are far from ubiquitous among granulites, and it may be that depletion only takes place if the pre-metamorphic rocks were both wet and intermediate–acid in composition. Slightly low Rb/Sr and U/Pb ratios are also a feature of plagioclase-rich, typically gabbroic, cumulates (see also Cox 1980), but it remains difficult to assess whether 'depleted' granulites and cumulates are actually common in the lower crust.

Distinctive isotope ratios in the lower crust require distinctive trace element ratios and, as argued here, these may result from either primary or secondary processes. Fig. 6 illustrates how the low Rb/Sr mafic granulites of Lesotho are characterized by relatively high Sr contents and often extremely high Sr/Nd. These elements appear to be little affected by granulite-facies metamorphism, and hence the high Sr/Nd ratios are a primary feature interpreted as of cumulate origin (Rogers & Hawkesworth 1982). The rocks also have high Ba and Ba/Nd, and so they cannot be the source of much of the Sr and Nd in continental basalts without significantly changing both

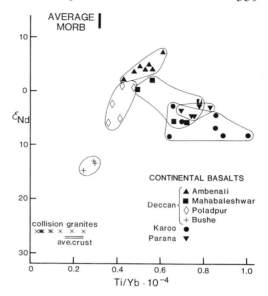

FIG. 7. ε_{Nd} versus Ti/Yb diagram.

Ba/Nd and Sr/Nd. As shown in Fig. 6, continental basalts tend to have similar Sr/Nd ratios to those from oceanic areas.

Much of the crustal contamination debate has relied on the distribution of elements such as Rb, U, Th, K and the LREE. But because these elements are concentrated by enrichment processes in both the mantle and the crust, they may not discrimate well between the two environments. More useful is an element such as Ti, which is both incompatible and significantly more abundant in basalt than in crustal rocks or crustal melts. Although relatively low Ti contents in mantle derived magmas may result from fractionation of oxide phases or crustal contamination, *high* Ti abundances are unlikely to be due to interaction with the continental crust, and presumably reflect enrichment processes within the upper mantle. An ε_{Nd} vs Ti/Yb diagram (Fig. 7) for selected Deccan, Karoo, and Parana basalts (Cox & Hawkesworth 1984, in press; Hawkesworth *et al.* in press; Hawkesworth & Mantovani unpubl.) exhibits both negative and positive trends. The rocks from the Poladpur and Bushe formations in the Deccan have been interpreted in terms of Ambenali magma that has undergone both gabbro fractionation and crustal contamination (Cox & Hawkesworth 1984; in press). Their trend to low Ti/Yb and low ε_{Nd} are clearly consistent with such a model although, because low Ti styles of trace element enrichment have also been recognized in mantle xenoliths (Hawkesworth *et al.*, 1984) it is premature to imply that low Ti/Yb values are *diagnostic* of crustal contamination. In contrast, the Mahabaleshwar rocks of the Deccan and the high Ti rocks of the Karoo and Parana, have higher Ti/Yb ratios than MORB with concomitantly lower ε_{Nd} values. These basalts plot down, or to the left of the mantle ε_{Nd}-ε_{Sr} array, and it is this trend to higher Ti with less radiogenic Nd that is so very difficult to attribute to crustal contamination, and is therefore thought to reflect upper mantle processes (see also Leeman & Hawkesworth, in press).

Conclusions

1 Xenolith suites of granulite-facies rocks tend to be more mafic and to yield higher pressure estimates than most surface granulite terrains, 5–20 kbar (Griffin *et al.* 1979; Robey 1981; Wass & Hollis 1983) and < 10 kbar (Perkins & Newton 1981; Harley & Green 1982; Harris *et al.* 1982) respectively.
2 Crustal xenoliths from southern Africa retain Sm-Nd whole rock, and both Rb-Sr mica and Sm-Nd garnet ages which are older than the age of entrainment in the kimberlite and consistent with the available information on near surface rocks. The kimberlite event does not appear to have affected either the mineral ages or, by implication, the mineral equilibria on which P-T estimates are based.
3 The relatively low abundances of U, K and Rb in lower crustal rocks can be either a primary

or a secondary feature. Many mafic granulites show no evidence of significant depletion during metamorphism, and their low Rb/Sr with low Sm/Nd ratios reflect both the trace element contents of the parent magmas and plagioclase-dominated cumulate processes in the lower crust. In contrast, several granulite-facies surface terrains, such as the Lewisian, show good evidence of having been depleted in U, K and Rb during metamorphism, probably on the breakdown of mica and then amphibole. The implication is that metamorphic (i.e. secondary) depletion of these elements may only be important when the pre-metamorphic rocks were both wet and of intermediate-acid composition. Both magmatic and metamorphic processes can therefore generate low Rb/Sr and low Sm/Nd, resulting with time in low $^{87}Sr/^{86}Sr$ and low $^{143}Nd/^{144}Nd$, but present evidence suggests depletion during metamorphism results in lower U/Pb ratios.
4 Contamination of magmas with lower crustal material is often invoked to explain unradiogenic Pb-isotopes and near vertical trends on ε_{Nd}-ε_{Sr} diagrams. However, combined trace element and isotope arguments suggest that such isotope characteristics can also be a feature of mantle-derived magmas.

ACKNOWLEDGEMENTS: We thank De Beers Consolidated Mines Ltd for hospitality and assistance in the field, and Andrew Gledhill for assistance in the isotope laboratory. Comments by D.A. Carswell, J. Harmer and two anonymous reviewers were appreciated. This work was supported in part by NERC Grant GR3/4330. The manuscript was typed by Janet Dryden.

References

ALLEN, P., CONDIE, K.C. & NARAYANA, B.L. 1983. The Archaean low- to high-grade transition in the Krishnagisi-Dhawnapuri area, Tamil Nadu, southern India. *In*: NAQVI, S.M. & ROGERS, J.J.W. (eds) *Precambrian of south India*, Memoir 9, Geol. Soc. India, 450–461.
BEN OTHMAN, D., POLVÉ, M. & ALLÈGRE, C.J. 1984. Nd-Sr isotopic composition of granulites and constraints on the evolution of the lower continental crust. *Nature*, **307**, 510–515.
VAN BREEMEN, O. & HAWKESWORTH, C.J. 1980. Sm-Nd isotopic study of garnet and their metamorphic host rocks. *Trans. R. Soc. Edin.*, **71**, 97–102.
CARLSON, R.W., LUGMAIR, G.W. & MacDOUGALL, J.D. 1981. Columbia River volcanism: the question of mantle heterogeneity or crustal contamination. *Geochim. Cosmochim. Acta*, **45**, 2483–2499.
CARPENTER, M.A. & FERRY, J.M. 1984. Constraints of the thermodynamic mixing properties of plagioclase feldspars. *Contrib. Mineral. Petrol.*, **87**, 139–148.
CLARK, S.P. JR. & RINGWOOD, A.E. 1964. Density

distribution and constitution of the mantle. *Revs. Geophys.*, **2**, 35–88.
CHRISTENSEN, N.I. & FOUNTAIN, D.M. 1975. Constitution of the lower continental crust based on experimental studies of seismic velocities in granulite. *Geol. Soc. Am. Bull.*, **86**, 227–236.
COHEN, R.S., O'NIONS, R.K. & DAWSON, J.B. 1982. Pb, Nd and Sr isotopes in ultramafic xenoliths: evidence for ancient sub-continental mantle? *EOS*, **63**, 460.
COX, K.G. 1980. A model for flood basalt vulcanism. *J. Petrol.*, **21**, 629–650.
—— & HAWKESWORTH, C.J. 1984. Relative contribution of crust and mantle to flood basalt magmatism, Mahabaleshwar area, Deccan Traps. *Phil. Trans. R. Soc. Lond. A* **310**, 627–641.
—— & —— in press. Geochemical stratigraphy of the Deccan Traps at Mahabaleshwar, Western Ghats, India, with implications for open system magmatic processes. *J. Petrol.*
DEN TEX, E. 1965. Metamorphic lineages of orogenic plutonism. *Geol. en Mijnb.*, **44**, 105–132.

DE PAOLO, D.J. 1981. Trace element and isotopic effects of continued wallrock assimilation and fractional crystallisation. *Earth Planet. Sci. Lett.*, **53**, 189–202.

DRURY, S.A., HARRIS, N.B.W., HOLT, R.W., REEVES SMITH, G.J. & WIGHTMAN, R.T. 1983. Precambrian tectonics and crustal evolution in south India. *J. Geol.*, **92**, 3–20.

DUPUY, C., LEYRELOUP, A. & VERNIÈRES, J. 1979. The lower continental crust of the Massif Central (Bournac, France)—with special reference to REE, U and Th composition, evolution and heat-flow production. *Phys. Chem. Earth*, **11**, 410–415.

EGLINGTON, B.A. & HARMER, R.E. 1984. Age and geochemical character of granitoids from southern Natal. *PRU, abstr.*, 14–15.

ELLIS, D.J. & GREEN, D.H. 1979. An experimental study of the effect of Ca upon garnet-clinopyroxenes Fe-Mg exchange equilibria. *Contrib. Mineral. Petrol.*, **71**, 13–22.

ERLANK, A.J., ALLSOPP, H.L., HAWKESWORTH, C.J. & MENZIES, M.A. 1982. Chemical and isotopic characterisation of upper mantle metasomatism in peridotite nodules from the Bultfontein kimberlite. *Terra Cognita.*, **2**, 261–263.

FREY, F.A. & PRINZ, M. 1978. Ultramafic inclusions from San Carlos, Arizona: petrological and geochemical data bearing on their petrogenesis. *Earth Planet. Sci. Lett.* **38**, 129–176.

GRIFFIN, W.L., CARSWELL, D.A. & NIXON, P.H. 1979. Lower crustal granulites and eclogites from Lesotho and South Africa. *In:* BOYD, F.R. & MEYER, H.O.A. (eds) *The Mantle Sample*, A.G.U. Washington, D.C., 177–204.

HAMILTON, J., EVENSON, N.M., O'NIONS, R.K. & TARNEY, J. 1979. Sm-Nd systematics of Lewisian gneisses: implications for the origin of granulites. *Nature*, **277**, 25–28.

HARRIS, N.B.W., HOLT, R.W. & DRURY, S.A. 1982. Geobarometry, geothermometry, and late Archaean geotherms from the granulite facies terrain of south India. *J. Geol.*, **90**, 509–527.

HARLEY, S.L. & GREEN, D.H. 1982. Garnet-orthopyroxene barometry for granulites and peridotites. *Nature*, **300**, 697–701.

HARTE, B. & FREER, R. 1982. Diffusion data and their bearing on the interpretation of mantle nodules and the evolution of the mantle lithosphere. *Terra Cognita*, **2**, 273–275.

——, JACKSON, P.M. & MACINTYRE, R.M. 1981. Age of mineral equilibria in granulite facies nodules from kimberlites. *Nature*, **291**, 147–148.

HAWKESWORTH, C.J., ROGERS, N.W., VAN CALSTEREN, P., MENZIES, M.A. & REID, D.L. 1982. Nd- and Sr-isotope studies on crustal xenoliths from southern Africa. *Terra Cognita Abstr.*, **2**, 236.

——, ——, —— & —— 1984. Mantle enrichment processes. *Nature*, **311**, 331–335.

—— & MARLOW A. 1985. Isotope evolution of the Damara orogenic belt. *In:* MILLER, R. McG. (ed.) *Evolution of the Damara Orogen, Southwest Africa/ Namibia*, Geol. Soc. S. Afr. Spec. Publ. **11**, 397–407.

HEIER, K.S. 1965. Radioactive elements in the continental crust. *Nature*, **208**, 479–480.

HODGES, K.V. & SPEAR, F.S. 1982. Geothermometry,

geobarometry and the Al_2SiO_5 triple point of Mt. Moosilauki, New Hampshire. *Am. Min.* **67**, 1118–1135.

HOLLAND, T.J.B. 1980. The reactions albite = jadeite + quartz determined experimentally in the range 600–1200°C. *Amer. Mineral.* **65**, 129–134.

JACKSON, P.M. & HARTE, B. 1977. The nature and conditions of formation of granulite facies xenoliths from the Matsoku kimberlite pipe. *2nd Int. Kimberlite Conf.*, Extended Abstr.

JAMES, D.E. 1981. The combined use of oxygen and radiogenic isotopes as indicators of crustal contamination. *Ann. Rev. Earth. Planet. Sci.*, **9**, 311–344.

JANARDHAN, A.S., NEWTON, R.C. & SMITH, J.V. 1979. Ancient crustal metamorphism at low P_{H_2O}: Charnockite formation at Kabbaldurgn, South India. *Nature*, **278**, 511–514.

KRAMERS, J.D., SMITH, C.B., LOCK, N.P., HARMON, R.S. & BOYD, F.R. 1981. Can kimberlites be generated from an ordinary mantle? *Nature*, **291**, 53–55.

LACHENBRUCH, A.H. 1970. Crustal temperatures and heat production: implications of the linear heat flow relation. *J. Geophys. Res.* **75**, 3291–3300.

LAMBERT, I.B. & HEIER, K.S. 1968. Estimates of the crustal abundances of thorium, uranium and potassium. *Chem. Geol.*, **3**, 233–238.

MATTHEWS, P.E. 1981. Eastern or Natal sector of the Namaqua-Natal mobile belt in southern Africa. *In:* GUNTER, D.R. (ed.) *Precambrian of the Southern Hemisphere*, Elsevier, Amsterdam, 705–715.

McIVER, J.R. & GEVERS, T.W. 1968. Charnockites and associated hypersthene-bearing rocks in southern Natal, South Africa. *Int. Geol. Congr. 22nd India*, **13**, 151–168.

MENZIES, M.A. & MURTHY, V.R. 1980. Enriched mantle: Nd and Sr isotopes in diopsides from kimberlite nodules. *Nature*, **283**, 634–636.

MOORBATH, S., WELKE, J. & GALE, N.H. 1969. The significance of lead isotope studies in ancient, high-grade metamorphic basement complexes, as exemplified by the Lewisian rocks of northwest Scotland. *Earth Planet. Sci. Lett.*, **6**, 245–256.

NEWTON, R.C. & HASELTON, H.T. 1981. Thermodynamics of the garnet-plagioclase-Al_2SiO_5-quartz geobarometer. *In:* NEWTON, R.C. & WOOD, B.J. (eds) *Thermodynamics of Minerals and Melts*, Springer-Verlag, New York, 129–145.

NIXON, P.H. 1973. Lesotho Kimberlites. *Lesotho Nat. Dev. Corp.* Maseru.

——, ROGERS, N.W., GIBSON, I.L. & GREY, A. 1981. Depleted and fertile mantle xenoliths from southern African kimberlites. *Ann. Rev. Earth Planet. Sci.*, **9**, 285–309.

PERKINS, D. & NEWTON, R.C. 1981. Charnockite geobarometers based on coexisting garnet-pyroxene-plagioclase-quartz. *Nature*, **292**, 144–146.

POWELL, R. 1978. The thermodynamics of pyroxene geotherms. *Phil. Trans. R. Soc. London*, **288**, 457–469.

ROBEY, J.vA. 1981. *Kimberlites of the central Cape Province, R.S.A.* Unpubl. PhD thesis, Univ. Cape Town.

RICHARDSON, S.H., PADOVANI, E.R. & HART, S.R. 1980. The gneiss syndrome: Nd- and Sr-isotopic relationships in lower crustal xenoliths, Kilbourne Hole, New Mexico. *EOS*, **61**, 388.

ROGERS, N.W. 1977. Granulite xenoliths from Lesotho kimberlites and the lower continental crust. *Nature*, **270**, 681–684.

—— & HAWKESWORTH, C.J. 1982. Proterozoic age and cumulate origin for granulite xenoliths, Lesotho. *Nature*, **299**, 409–413.

STACEY, J.S. & KRAMERS, J.D. 1975. Approximation of terrestrial lead isotope evolution by a two-stage model. *Earth Planet. Sci. Lett.*, **26**, 201–221.

TAYLOR, H.P. 1980. The effects of assimilation of country rock by magmas on $^{18}O/^{16}O$ and $^{87}Sr/^{86}Sr$ systematics in igneous rocks. *Earth Planet. Sci. Lett.*, **47**, 243–254.

WASS, S.Y. & HOLLIS, J.D. 1983. Crustal growth in south-eastern Australia—evidence from lower crustal eclogitic and granulitic xenoliths. *J. Metamorphic Geol.*, **1**, 25–45.

WEAVER, B.L. & TARNEY, J. 1984. Empirical approach to estimating the composition of the continental crust. *Nature*, **310**, 575–577.

WELLS, P.R.A. 1977. Pyroxenes thermometry in simple and complex systems. *Contrib. Mineral. Petrol.*, **62**, 129–139.

WOOD, B.J. & BANNON, S. 1973. Garnet-orthopyroxene and orthpyroxene-clinopyroxene relationships in simple and complex systems. *Contrib. Mineral. Petrol.* **42**, 109–124.

P.W.C. VAN CALSTEREN, N.B.W. HARRIS, C.J. HAWKESWORTH, M.A. MENZIES & N.W. ROGERS, Department of Earth Sciences, The Open University, Milton Keynes, UK.

The lower crust in eastern Australia: xenolith evidence

W.L. Griffin & S.Y. O'Reilly

SUMMARY: Hundreds of Mesozoic to Recent xenolith-bearing basalt flows, diatremes, cinder cones and maars have erupted through the low-grade Palaeozoic metamorphic rocks of the Tasman Fold Belt, which makes up the eastern third of the Australian continent. Granulite-facies xenoliths, interpreted as lower-crustal material, have been found at more than 40 of these localities. A palaeogeotherm, based on garnet pyroxenite xenoliths from Victoria, also fits available data from New South Wales and Queensland. A crustal thickness of 25–30 km, and lower-crustal temperatures of 700–850°C, have been derived by referring T estimates for spinel lherzolite xenoliths to this geotherm. The lower-crustal xenolith suites are dominated by basic pyroxene granulites and garnet granulites; more silicic xenoliths are very rarely reported. The granulites cover a range in composition typical of intraplate basaltic magmas and cumulates. Granoblastic microstructures are typical, but common relict igneous features suggest that all the mafic granulites originated as igneous rocks. The lower crust in this area probably formed through multiple intrusion of basic magmas near the crust/mantle interface over a long time span. The strongly layered 'lower crust' seen on seismic reflection studies in eastern Australia probably represents this thick crust-mantle transition zone. The seismic 'Moho' reflects the depth at which the proportion of basic to ultrabasic rocks drops below a critical value, and probably lies well below the *petrographically* defined base of the crust.

Introduction

The Australian continent can be divided into two broad tectonic (and hence crustal) units: the western two-thirds of the continental area consists dominantly of Proterozoic and Archaean cratonic terrains; the eastern third consists of the complex Paleozoic Tasman Fold Belt (Fig. 1). Information on the nature of the lower crust in this latter region can only be gained indirectly, by geophysical methods, or from lower crustal xenoliths entrained in basaltic host rocks.

A chain of dominantly Cenozoic basaltic provinces parallels the eastern continental margin of Australia for over 2500 km. Many localities, randomly spaced throughout these basaltic provinces, yield abundant xenoliths of high pressure origin, some of which appear to represent lower crustal lithologies.

Recognition of lower crustal lithologies

Numerous studies of xenolith localities in Australia and other continents have identified a 'lower-crustal' suite consisting of 'granulites'. In most cases the term is used loosely to cover metamorphic rocks with pyroxene + plagioclase or garnet + pyroxene + plagioclase assemblages; such suites may or may not contain eclogites or other garnet pyroxenites. The identification of these xenoliths as lower-crustal lithologies is based on the widespread recognition, using on experimen-

tal studies and geophysical data, that the deep crust should consist largely of granulite-facies rocks. Many of these studies ignore two essential points. One is that the *surface outcrop* of country rock in some areas of xenolith-bearing basaltic rocks consists of old granulites; the xenoliths may be derived from these, and thus give no information on the deeper layers of the present-day crust (e.g. Bournac, France; cf. Leyreloup *et al.* 1977; Dostal *et al.* 1980). The other is that many of the observed mineral assemblages are stable to depths well below the normal continental Moho, and there is no guarantee that these granulites come from within the crust. However, the judicious application of geothermometry/geobarometry based on mineral analyses can give a guide to the depth of equilibration of these assemblages.

A Cenozoic geotherm has been constructed for southeastern Australia, using well-equilibrated garnet websterite xenoliths, and is in good agreement with available geophysical data (Fig. 2; O'Reilly & Griffin 1985). New data for similar rocks from northern New South Wales (Stolz 1984) and Queensland (authors' unpubl. data) also fit this geotherm, suggesting that it adequately describes the thermal regime beneath eastern Australia during the eruption of each of the sampled localities; these range in age from c. 10^4–c. 10^8 years. The low-T end of this geotherm, below c. 900°C, is poorly constrained, but fits the available data. The existence of composite garnet websterite-spinel lherzolite xenoliths allows correlation of geothermometers for these two rock types, and thus a rough pressure estimate for the

From: DAWSON, J.B., CARSWELL, D.A., HALL, J. & WEDEPOHL, K.H. (eds) 1986, *The Nature of the Lower Continental Crust*, Geological Society Special Publication No. 24, pp. 363–374.

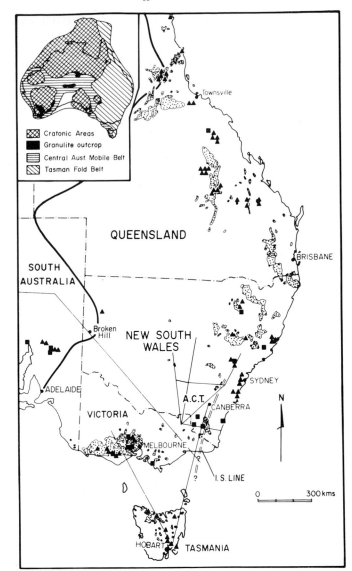

FIG. 1. Distribution of gabbroic and granulite xenoliths from basaltic rocks in eastern Australia (based on Sutherland and Hollis (1981)). Square symbols denote garnet-bearing, triangular symbols garnet-free, assemblages. Stippled areas enclose dominantly Cenozoic basaltic outcrops. Straight lines represent seismic traverses (Finlayson *et al.* 1980). Dashed lines show the I–S line which is considered to represent the eastern edge of a thickened Devonian basement against a dominantly meta-igneous and probably younger basement as deduced from granitoid types (White *et al.* 1976; White *et al.* 1977; Shaw *et al.* 1982). Inset shows the distribution of tectonic terrains in Australia based on Plumb's (1979) map. Areas of Precambrian granulite outcrop are from Wilson (1978).

volumetrically dominant lherzolites (Griffin *et al.* 1984; O'Reilly & Griffin 1985). Temperatures for the lherzolites range down to *c.* 850°C for most of eastern Australia, suggesting that the base of the crust at the time of eruption lay no deeper than *c.* 30 km. We therefore suggest that xenoliths from E. Australia which give temperatures <850°C by two-pyroxene (specifically, the thermometer of Wood & Banno 1973) or garnet-cpx (Ellis & Green 1979) thermometry can be regarded as crustal. Temperatures lower than 850°C are recorded in some lherzolite xenoliths from Victoria, but these have been analysed by others, and we cannot vouch for the quality of the mineral analyses (see O'Reilly & Griffin 1985).

Geothermobarometry must be used with caution, especially at the lower temperatures relevant to crustal rocks. Lack of equilibration during cooling after a thermal episode could lead to 'freezing-in' of spurious temperatures, well above the ambient values (Harte & Freer 1982). These effects should, however, be easily detected by detailed searches for mineral zoning (Wilson & Smith 1984). A further problem arises from the common overestimation of Fe^{3+} in cpx, which will give spuriously low temperatures by gnt + cpx thermometry (Carswell & Griffin 1981). This problem may be countered by the use of very precise analytical techniques, coupled with structural refinements of clinopyroxenes by XRD (Griffin, Oberti, Mellini & Rossi, 1985).

On the basis of geothermometry/geobarometry, and the existence of composite xenoliths showing contacts between spinel lherzolite and granulites or pyroxenites, we suggest that several well-studied suites probably are derived from the upper mantle, rather than representing 'lower crust'. These include:

1 most of the garnet granulites and garnet pyroxenites from Delegate, N.S.W. (Lovering & White 1969). Microstructures show that many of the 'garnet granulites' have formed by exsolution of plagioclase, as well as garnet, from aluminous pyroxenes. Geothermobarometry indicates that this exsolution occurred at mantle pressures and temperatures. The granoblastic two-pyroxene granulites from this locality do appear, on the basis of geothermometry, to be from the lower crust.

2 garnet pyroxenites ± plagioclase from Bullenmerri and Gnotuk maars, Victoria (Griffin *et al.* 1984), and the Walcha area, N.S.W. (Stolz 1984).

3 some of the garnet granulite-eclogite suite from Anakie, Victoria (Wass & Hollis 1983).

4 the abundant cumulates (cpx ± oliv ± opx ± spin ± amph) seen in many xenolith suites from southeastern Australia (Irving 1974). These commonly occur in composite xenoliths with spinel periodite, and Wass (1978) has observed that pyroxenes similar to those of the cumulates form overgrowths on spinel-lherzolite xenoliths. The rapid eruption necessary to entrain such xenoliths (Spera 1980) implies that such rims have formed at mantle depths. However, some of these cumulates may have formed within the crust; Wilkinson (1975a) has described a series which seems to be related to a coexisting suite of pyroxene-spinel granulites.

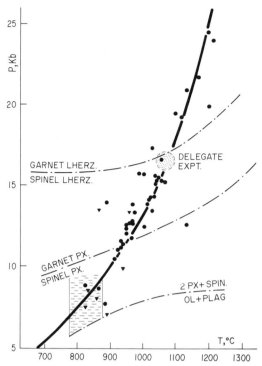

FIG. 2. Xenolith-derived geotherm for southeastern Australia, after O'Reilly & Griffin (1985). Individual data points are based on cpx + opx + gnt assemblages; the shaded field includes all temperatures derived from two-pyroxene granulites of presumed crustal derivation. Black dots are garnet websterites and garnet (± spinel) lherzolites. Black triangles are Sapphire Hill granulites (Kay & Kay 1983). The spinel lherzolite to garnet lherzolite transition is from O'Neill (1981): other experimental boundaries are from Herzberg (1978).

Nature of lower crustal xenoliths

Xenolith localities, mineral assemblages and microstructures are summarized in Table 1. The xenoliths listed here are almost entirely mafic granulites (*sensu lato*). While most have granoblastic microstructures (with or without foliations), many suites show igneous microstructures in various stages of microstructural and mineralogical re-equilibration. Some xenoliths, which may have formed contemporaneously with a relatively young intrusive episode, have the petrographic characteristics of igneous cumulates. Others (for example, from Anakie, Victoria) have doleritic or gabbroic microstructures and mineralogies (Figs. 3 & 4 of Wass & Hollis 1983) showing partial development of granoblastic grain boundaries and arrested development of garnet by

TABLE 1. *Summary of critical mineral assemblages and microstructures in lower-crust xenoliths from eastern Australia*

	Ol +Cpx +Plag ±Spin	Ol +2px +Plag ±Spin	2px +Plag	2px +Plag +Spin	2px +Plag +Gnt ±Spin	Cpx +Plag	Cpx +Plag +Spin	Cpx +Plag +Gnt	Plag +Gnt	Cpx +Gnt ±Opx ±Spin	References
Tasmania											
Pencil Point	—	I	G,I	—	—	G	—	—	—	—	Sutherland 1974
Corra Lin Bridge	G	—	G	G	—	—	—	—	—	—	Sutherland 1969, 1971
Table Cape	—	—	G	G	—	—	—	—	—	—	Wass & Irving 1976
Victoria											
Mt. Franklin	—	—	G	G	—	—	M,G	—	—	—	This work
Mt. Shadwell	—	—	G	—	—	—	—	—	—	—	This work
Anakie	—	—	G	—	G	—	—	—	—	G	Wass & Hollis 1983
New South Wales											
Delegate	—	—	G	—	G	—	—	G	—	G	Lovering & White 1969
											O'Reilly & Griffin 1984
Tumut-Eucumbene	—	—	—	—	G	—	—	G	—	G	This work
Jugiong	—	—	G	G	G	—	—	—	—	—	Arculus et al. 1984
Kelly's Point	I,G,M	—	—	G	G	—	G	—	G	—	Wass & Irving 1976
N. Kiama	S	—	—	—	—	G	G	—	—	—	This work
S. Bulli	—	—	—	—	—	G	G,S	—	—	—	This work
Stanwell Park	—	—	—	—	—	—	G,S	—	—	—	This work
Grabben Gullen	—	—	—	—	—	I	I	—	—	—	This work
S. Highlands	G	—	—	G	—	G	G	—	—	—	This work; Wass 1973
Erskine Park	—	—	—	S	—	—	—	—	—	—	This work
Dundas	—	—	—	S	—	—	—	—	—	—	This work; Benson 1910
											Wilshire & Binns 1961

Locality									Reference
The Basin	S	—	—	—	—	—	—	—	Osborne 1920
Blue Mtns.	—	—	—	—	—	—	I	—	This work
Glen Alice	G	—	—	—	—	G	G	—	Wass & Irving 1976
Kandos	I,G	—	I,M	—	—	I	I	—	This work
Dubbo	G	G	G	—	—	—	G	—	Wass & Irving 1976
Wallabadah	G	—	G	—	—	G	—	—	Wass & Irving; This work
Barrington	—	G	—	—	—	—	—	—	This work
Lawlers Creek	—	—	G	—	—	—	—	—	Wilkinson 1975a
Boomi Creek	G	—	G	—	—	—	—	—	Wilkinson 1975b
Gloucester	—	G	—	G	—	G	—	—	Wilkinson 1974
Comboyne prov.	G	—	—	—	—	G	G	—	Wass & Irving 1976, Knutson & Green 1975
Ruby Hill	—	—	—	G	—	—	—	G	This work; O'Reilly & Griffin 1984
	—	—	—	—	—	—	—	—	Lovering 1964
Kayrunnera	—	—	—	G($\pm$ky $\pm$qtz)	—	—	—	G($\pm$qtz)	Edwards et al. 1979
S. Australia									
El Alamein	G	G	G	G(+Q)	G	—	—	G	Arculus et al. 1984; McCulloch et al. 1982
Calcutteroo	—	—	G	—	G	—	—	—	Arculus et al. 1984; McCulloch et al. 1982
Queensland									
Mt. St. Martin	—	—	G	—	G	—	—	—	Ewart 1982; Sutherland & Hollis 1982
Sapphire Hill	M,S	G,M	G,M	—	G,M	—	—	—	This work; Kay & Kay 1983
Hill 32	G,I,M	G,M	—	—	G,M	—	G	—	This work; Kay & Kay 1983
Batchelor Crater	G,M	—	—	G,M	—	G,M	—	—	This work; Kay & Kay 1983
Hoy Province	G	G	G	—	—	—	—	—	Wass & Irving 1976
Atherton Prov.	G	G	G	—	—	—	—	—	Wass & Irving 1976

Symbols: G, granoblastic; I, igneous; M, recrystallized, relict igneous; S, symplectitic.

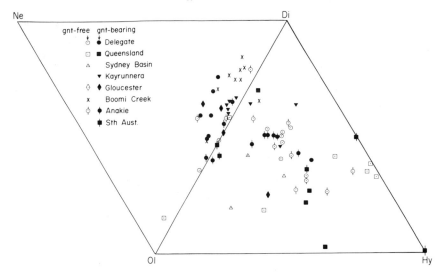

FIG. 3. Normative mineral composition of analyzed granulite-facies xenoliths from eastern Australia. Data sources given in Table 1; representative new analyses are given in Table 2.

reaction of plagioclase and pyroxene. The development of symplectites and corona structures also demonstrates partial mineralogical re-equilibration in response to declining temperatures. Many of the partially re-equilibrated rocks contain zoned plagioclase and garnet (e.g. Wallabadah, Sapphire Hill, Batchelor's Crater (authors' unpubl. data; Kay & Kay 1983)).

Assemblages containing olivine + plagioclase are obviously of low-P/high-T origin (Herzberg 1978), and many of these, in particular, show evidence of reaction during cooling to give the lower-T/higher-P assemblage 2 pyroxenes + plagioclase + spinel. In the xenoliths from Erskine Park and Dundas the latter assemblage is developed in spectacular coarse-grained symplectites. Most localities are dominated by intermediate-pressure two-pyroxene + plagioclase assemblages. The rarer garnet-granulite assemblages *may* represent higher P/lower T, but these rocks in general do not contain quartz, and thus their pressure of formation cannot be accurately assessed. However, Rudnick *et al.* (1984) have recently reported metasedimentary, quartz-bearing garnet granulites from Hill 32 (Queensland). The lower crust of eastern Australia therefore appears to be mainly in the intermediate-pressure to high-pressure granulite facies, and has been so for at least the last 80–100 Ma (O'Reilly & Griffin 1985).

By contrast a few localities on the edge of the craton in South Australia and western N.S.W. (Kayrunnera), contain quartz-bearing ($\pm$ kya-

nite) eclogites and high-P granulites (Arculus *et al.* 1984; Edwards *et al.* 1979; McCulloch *et al.* 1982). Published P-T estimates for these suites scatter widely, but all lie to the low-T side of the geotherm shown in Fig. 2. They may reflect real differences in the thermal regime beneath the Precambrian craton and the Tasman fold belt at the time of eruption, but more and better analytical data are obviously necessary to resolve this question.

Garnet pyroxenites are abundant at Ruby Hill, N.S.W. (O'Reilly & Griffin 1985) and Sapphire Hill, Queensland (Kay & Kay 1983; authors' unpubl. data), and are associated with apparently crustal garnet granulites in both cases. The pyroxenites yield cpx + gnt temperatures in the 875–950°C range after correction for Fe^{3+} in cpx. Comparison with the geotherm (Fig. 2) suggests that these rocks were derived from near the base of the crust.

Xenolith geochemistry

About 50 major-element analyses of granulite xenoliths were retrieved from the literature and from our unpublished data. Almost half of these are from Delegate and Anakie, and many of the garnet-bearing xenoliths of these suites probably are derived from sub-crustal depths. Representative new analyses are given in Table 2 and the available data are summarized in Fig. 3. All of the analyses are basaltic in the broad sense; about

TABLE 2. *Representative analyses of lower-crustal xenoliths from southeastern Australia*

	Delegate, N.S.W.				Sydney Basin			Gloucester, N.S.W.	
	19654	104340A	10430/1	10430/3	EP4	8004	69340	9118	9119
	GG	GG	PG	PG	PG(S)	PG(S)	PG(S)	GG	GG
SiO_2	48.48	46.21	46.85	49.96	48.96	43.97	48.32	45.97	46.23
TiO_2	0.44	0.62	1.72	0.77	0.39	0.84	0.84	0.25	0.93
Al_2O_3	17.03	17.08	16.45	17.45	19.91	12.21	19.05	17.71	15.36
Fe_2O_3	1.28	1.54	3.52	1.66	0.88	2.02	1.09	1.57	1.87
FeO	6.38	7.71	8.66*	6.15*	4.41	10.10	5.47	7.86	9.35
MnO	0.16	0.16	0.13	0.15	0.02	nd	nd	0.16	0.18
MgO	10.25	9.71	7.87	9.08	9.20	18.77	8.58	10.03	9.75
CaO	11.16	11.48	11.04	12.28	12.34	8.37	10.62	13.80	12.03
Na_2O	2.80	2.39	2.78	2.69	2.47	1.17	2.44	1.49	3.05
K_2O	0.67	0.48	0.28	0.12	0.15	0.04	0.47	0.09	0.45
P_2O_5	0.07	0.07	0.07	0.27	0.05	0.03	0.07	0.02	0.02
H_2O^+	1.93	1.91	1.21	1.11	1.86	nd	nd	0.83	0.90
CO_2	0.17	0.28	0.14	0.08	nd	nd	nd	nd	nd
Σ	100.82	99.64	100.92	101.60	100.64	97.34	96.95	99.78	100.12
D	3.04	3.10	3.01	2.94	nd	nd	nd	3.33	3.24
CIPW NORM (volatile-free)									
or	3.96	2.84	1.65	0.71	0.89	0.24	2.78	0.53	2.66
ab	20.26	16.83	23.43	22.76	20.90	9.90	20.65	12.42	12.76
an	31.92	34.46	31.58	35.18	42.80	27.95	39.64	41.37	26.89
ne	1.86	1.86	0.05	—	—	—	—	0.10	7.07
di+hd	18.50	17.85	17.54	20.21	18.46	10.72	10.11	21.86	26.58
en+fs	—	—	—	5.92	4.51	10.33	10.59	—	—
fo+fa	19.36	20.08	18.32	11.72	13.10	33.61	9.84	19.87	18.75
il	0.84	1.18	3.32	1.46	0.74	1.60	1.06	0.47	1.77
mt	1.86	2.23	2.90	1.78	1.28	2.93	1.58	2.28	2.71
ap	0.17	0.17	0.64	0.24	0.12	0.07	0.17	0.05	0.05

* Titrimetric analysis. All others calculated to $Fe_2O_3/FeO = 0.2$ from XRF analysis
GG = garnet granulite; PG = pyroxene granulite.

half are Ne-normative, and a few are true basanites (> 5% normative Ne, plag ⩾ An_{50}). The rest are Hy-normative, and a few are nearly quartz-normative. Most of the Hy-normative rocks contain 40–60% normative plagioclase with compositions ranging from An_{50} to An_{80}. The compositional range (Fig. 3) is similar to that of Cenozoic basaltic lavas from eastern Australia. Most of the xenoliths appear to have compositions typical of intraplate basaltic magmas and cumulates; TiO_2 generally ranges from 0.5–1.5 wt% and MgO from 7–10 wt%. However, K_2O values (⩽ 0.3 wt%) are very low compared to intraplate basalts, and suggest that many of the rocks are cumulates with little trapped liquid, or have lost K_2O during metamorphism.

Most of the Ne-normative rocks are garnet granulites, while nearly all pyroxene granulites are Hy-normative. This suggests that the appearance of garnet in most suites is related primarily to bulk composition, and secondarily to differ-

ences in pressure and temperature of equilibration. However, in the Anakie suite (Wass & Hollis 1983), rock composition seems to exert no control on the presence or absence of garnet and the xenoliths were probably sampled over a relatively large depth range. The two pyroxene + spinel granulites from Boomi Creek (Wilkinson 1975b) appear to have crystallized at lower P than garnet granulites of similar composition.

No obvious differences in composition are seen between the 'crustal' granulites and others, such as some samples from Anakie, which appear to come from greater depth. There is also no clear compositional distinction between the intermediate-pressure granulites found in most localities, and the eclogites and high-pressure granulites from Kayrunnera and South Australia. The available chemical data come from a limited number of localities, and samples from several of these (Delegate pyroxene granulites, Boomi Creek, Gloucester) group rather tightly in different fields on Fig. 3. This emphasizes the lateral and/or vertical heterogeneity which must be present in the lower crust over this large area, and underlines the need for more studies of xenolith suites from other localities.

The 29 granulite xenoliths for which data are available (not including Anakie) have an average density of $3.13 \pm .16$ (1σ) g/cm^3; this is lower than the values for 4 eclogites from Calcutteroo ($3.31 \pm .07$ g/cm^3; Jackson & Arculus 1984). The average mean atomic weight [M] of these granulites and eclogites is 21.8 ± 0.3, and the average [CaO] is 0.12.

Discussion

The validity of reconstructing the nature of the lower crust from xenolith evidence is limited by two intrinsic sampling problems—volcanic and human. The volcanic sampling uncertainty is well-known (e.g. Jackson 1969). Basaltic and kimberlitic magmas sample high-pressure lithologies very selectively, as evidenced by different xenolith populations at adjacent localities. One such example (out of many) is from western Victoria, where Bullenmerri and Gnotuk maars (Griffin *et al.* 1984) yield abundant garnet pyroxenites, amphibole-rich wehrlites and anhydrous and hydrous lherzolites. Mt. Leura cinder-cone, less than 1 km away, contains dominantly anhydrous lherzolites (Irving 1974; authors' unpubl. data) with no recorded garnet pyroxenites or amphibole-rich wehrlites. The reasons for such selective entrainment are not understood and may depend on such factors as the rheological characteristics and the gas contents both of the country rock and of the host magma.

Human sampling also may be very selective, either purposely or inadvertently, unless a sound statistical approach is used (e.g. Jackson 1968; Bloomer & Nixon 1973). Examples include:
(i) Bullenmerri/Gnotuk maars (Victoria, Australia); this locality is very rich in garnet pyroxenites (Griffin *et al.* 1984), but earlier work (Wass & Irving 1976; Ellis 1976) did not record any garnet-bearing assemblages.
(ii) Sapphire Hill (Queensland, Australia); Kay & Kay (1983) document only the occurrence of garnet granulites, but a subsequent collection (J. Knutson & M. Duggan, pers. comm. 1982) revealed the presence of abundant garnet pyroxenites.
(iii) Lesotho (southern Africa); Bloomer & Nixon (1973), using a systematic counting procedure, demonstrated that felsic granulite xenoliths, because they are extensively altered, were

significantly under-represented in most collections.

However, the available material from eastern Australia covers more than 40 localities, extending over a zone 2500 km long (Fig. 1), and many of these have been studied systematically and in detail by several workers. The lithologies inferred to represent lower crustal material are overwhelmingly basic. Thus it appears that the lower crust in eastern Australia probably is in fact dominated by basic granulites of basaltic affinity.

At $T = 700$–$900°C$, these granulites give calculated seismic velocities (V_P) of 6.9–7.0 km/sec, while the garnet pyroxenites give calculated $V_P = 7.8$–7.9 km/sec (O'Reilly & Griffin 1985). These values are similar to those measured in laboratory experiments on Calcutteroo (South Australia) granulites and eclogites (when corrected for temperature; Jackson & Arculus 1984). Measured values of V_P in the depth range 20–30 km beneath southeastern Australia are 6.4–7.2 km/sec (Finlayson *et al.* 1979). These values *may* indicate the presence of more silicic rocks, interlayered with the more abundant mafic granulites. However, so few analytical data exist on silicic granulite xenoliths that any attempt to refine the crustal model further would be premature.

Seismic refraction data have been used to estimate the depth to the 'Moho' (defined seismically as $V_P > 8.0$ km/sec) at several localities in eastern Australia. These estimates include 50–55 km in the southern Lachlan Fold Belt (Finlayson *et al.* 1979), 40–45 km in the northern Lachlan Fold Belt (Finlayson & McCracken 1981), 35–40 km in the Eromanga Basin, and 45–50 km in the Bowen Basin (Finlayson 1983). In all cases, this 'Moho' is not a sharp discontinuity, but is overlain by a zone (from 25–35 km depth) through which V_P increases from *c.* 6.5 to 7.5 km/sec.

Recently published seismic reflection data (Mathur 1983a,b,c) give a more detailed picture of the crust in these areas. The deep crust beneath the Lachlan Fold Belt is characterized by bands of subhorizontal reflectors from 21–27 km and 30–32 km depth. In the Bowen Basin, similar bands are seen between 25 and 36 km. The best picture comes from the Eromanga basin, where a middle-crust zone from 6–24 km shows essentially no reflectors, and is interpreted as consisting mainly of strongly folded, weakly metamorphosed Palaeozoic supracrustal rocks. Beneath a sharp discontinuity, corresponding to a jump in V_P from < 6.5 to *c.* 7 km/sec (Finlayson 1983), a zone with abundant subhorizontal reflectors (many up to 3 km long) extends down to *c.* 36 km. V_P increases to *c.* 7.7 km/sec through this zone, before jumping to 8.0 km/sec at the top of a zone

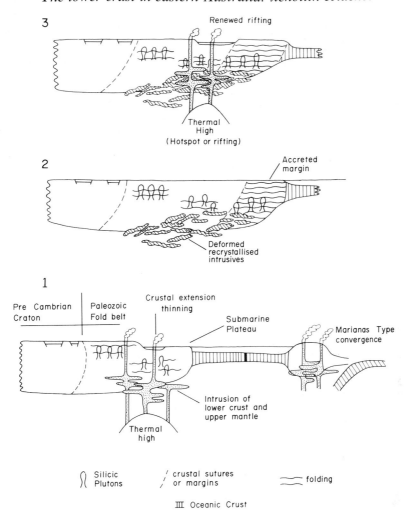

FIG. 4. Crustal development in eastern Australia, modified from Powell (1983). Parts 1, 2 and 3 represent successive stages, which may have been repeated. Intrusion of basaltic rocks in the lower crust and uppermost mantle is superimposed on other processes, including the tectonic accretion of island arcs, submarine plateaus and ocean floor material. The process produces a complex transition zone from the crust to the upper mantle, and results in a dominantly mafic lower crust.

with no reflectors. This latter discontinuity is interpreted as Moho, and the base of the crust.

Care should be taken to avoid circular arguments when interpreting these seismic data in terms of lithology. The equation of layering with 'crust', in particular, may be misleading. The evidence of geothermobarometry, and of composite xenoliths, shows that spinel peridotites, usually regarded as mantle rocks, commonly are interlayered with granulites and pyroxenites at depth beneath eastern Australia. The V_P of these peridotites at the temperatures suggested by

geothermometry will be $\leqslant 8.0$ km/sec (O'Reilly & Griffin 1985). Interlayering of peridotite with subordinate volumes of less dense rocks (such as mafic granulites) will give a mean $V_P < 8.0$ km/sec, so that these parts of the upper mantle would be interpreted as 'crust' from seismic data. Thus the seismic definition of 'Moho', from refraction and reflection data, may not correspond to the top of the mantle which would be defined petrographically by a preponderance of ultramafic rocks.

The seismic data strongly support the idea that the lower crust is layered, probably on a scale of tens to hundreds of metres (Fuchs 1969; Hale & Thompson 1982; Meissner 1973). The crust–mantle transition itself is probably also layered, on a similar scale. The velocity gradients observed in the depth range from 20–45 km in various parts of eastern Australia probably, in our view, reflect an interlayering of basic granulites and pyroxenites with more silicic material in the 'crustal' portions, and with ultramafic material in the 'mantle', with the greatest concentration of basic rocks near the crust–mantle 'boundary'. Our studies of the xenoliths strongly suggest that igneous processes have been responsible for this structure.

As described above, several of the lower crustal xenolith suites show a range in microstructures from igneous types, through partially re-equilibrated types, to granoblastic microstructures. This textural readjustment is accompanied by mineralogical changes which convert igneous mineral assemblages to others, stable under low-T conditions. These observations suggest, by analogy, that most of the mafic granulites formed originally by intrusive igneous processes.

We therefore suggest that the present lower crust beneath eastern Australia has formed mainly by the episodic intrusion of basaltic material over a long time span (Fig. 4). Crook (1980) has suggested that a mafic lower crust

beneath eastern Australia might have formed by stabilization of an oceanic slab, underthrust during subduction and continental accretion. However, the available data on the compositions of mafic xenoliths suggest that most, if not all, of these are of intraplate, rather than oceanic, origin. Tectonic accretion, associated with subduction, may have been important in producing new continental crust during the early Palaeozoic formation of the Tasman Fold Belt. Since that time, addition of basaltic material may have significantly modified the composition of the *lower* crust. Much of this igneous activity also may be related to postorogenic extension, basin formation, and the breakup of Gondwanaland over the last 150 Ma (Falvey & Mutter 1981). Intrusion of basaltic magma into the subjacent spinel lherzolite of the mantle can explain the diffuse 'transitional Moho' seen on seismic interpretations from eastern Australia and other non-cratonic continental areas. Isotopic study of the lower crustal xenoliths, to define their formation ages and source regions, may help to define both the mechanisms and the timing of these events.

ACKNOWLEDGEMENTS: We thank the following colleagues for donating samples in this study: G. Halford (for ANU), J. D. Hollis, A. J. Irving, F. L. Sutherland (for the Australian Museum) and J. G. F. Wilkinson.

References

ARCULUS, R.J., FERGUSON, J. *et al.* in press. Eclogites and granulites in the lower continental crust: examples from eastern Australia and southwestern USA. *Proc. 1st Eclogite Conference*, Clermont Ferrand 1982.

BENSON, W.N. 1910. The volcanic necks of Hornsby and Dundas near Sydney. *Journ. Proc. Roy. Soc. N.S.W.* **44**, 495–555.

BLOOMER, A.G. & NIXON, P.H. 1973. The geology of the Letseng-la-terae kimberlite pipes. *In:* NIXON, P.H. (ed.) *Lesotho Kimberlites*. Lesotho National Development Corporation Maseru, 20–32.

CARSWELL, D.A. & GRIFFIN, W.L 1981. Calculation of equilibration conditions for garnet granulite and garnet websterite nodules in African kimberlite pipes. *Tschermak's Min. Petr. Mitt.*, **28**, 229–244.

CROOK, K.A.W. 1980. Fore-arc evolution and continental growth: a general model. *Jour. Struct. Geol.*, **2**, 289–303.

DOSTAL, J., DUPUY, C. & LEYRELOUP, A. 1980. Geochemistry and petrology of meta-igneous granulite xenoliths in neogene volcanic rocks of the Massif Central, France—implications for the lower crust. *Earth planet. Sci. Lett.*, **50**, 31–40.

EDWARDS, A.B., LOVERING, J.F. & FERGUSON, J. 1979. High pressure basic inclusions from the Kayrunnera kimberlitic diatreme in New South Wales, Australia. *Contr. Min. Petrol.*, **69**, 185–192.

ELLIS, D.J. 1976. High-pressure cognate inclusions in the Newer Volcanics of Victoria. *Contr. Min. Petrol.*, **58**, 149–180.

—— & GREEN, D.H. 1979. An experimental study of the effect of Ca upon garnet-clinopyroxene Fe-Mg exchange equilibria. *Contr. Min. Petrol.*, **71**, 13–22.

EWART, A. 1982. Petrogenesis of the Tertiary anorogenic volcanic series of Southern Queensland, Australia, in the light of trace element geochemistry and O, Sr and Pb isotopes, *J. Petrology*, **23**, 344–382.

FALVEY, D.A. & MUTTER, J.C. 1981. Regional plate tectonics and the evolution of Australia's passive continental margins. *BMR Jour. Aust. Geol. Geophys.*, **6**, 1–29.

FINLAYSON, D.M. 1983. The mid-crustal horizon under the Eromanga Basin, Eastern Australia. *Tectonophysics*, **100**, 199–214.

—— & McCRACKEN, H.M. 1981. Crustal structure studies under the Sydney Basin and Lachlan Fold Belt, determined from explosion seismic studies. *Jour. Geol. Soc. Austr.*, **28**, 177–190.

——, PRODEHL, C. & COLLINS, C.D.N. 1979. Explosion seismic profiles and implications for crustal evolution in southeastern Australia. *BMR Journ. Aust. Geol. Geophys.*, **4**, 243–252.

FUCHS, K. 1969. On the properties of deep crustal reflectors. *Zeitschr. für Geophysik*, **35**, 133–149.

GRIFFIN, W.L., MELLINI, M., OBERTI, R. & ROSSI, G.

1985. Evolution of coronas in Norwegian anorthosites: re-evaluation based on crystal chemistry and microstructures. *Contr. Min. Petrol.*, **91**, 330–339.

——, Wass, S.Y. & Hollis, J.D. 1984. Ultramafic xenoliths from Bullenmerri and Gnotuk maars, Victoria, Australia: petrology of a subcontinental crust–mantle transition. *J. Petrology*, **25**, 53–87.

Hale, L.D. & Thompson, G.A. 1982. The seismic reflection character of the continental Mohorovicic Discontinuity. *Jour. Geophys. Res.*, **87**, 4625–4635.

Harte, B. & Freer, R. 1982. Diffusion data and their bearing on the interpretation of mantle modules and the evolution of the mantle lithosphere. *Terra Cognita*, **2**, 273–275.

Herzberg, C.T. 1978. Pyroxene geothermometry and geobarometry: experimental and thermodynamic evaluation of some subsolidus phase relations involving clinopyroxenes in the system CaO-MgO-Al_2O_3-SiO_2. *Geochim. Cosmochim. Acta*, **42**, 945–957.

Irving, A.J. 1974. Pyroxene-rich ultramafic xenoliths in the Newer Basalts of Victoria, Australia. *N. Jb. Miner. Abh.*, **120**, 147–167.

Jackson, E.D. 1968. The character of the lower crust and upper mantle beneath the Hawaiian Islands. *23rd. Int. geol. Congr. Prague, Proc.* **I**, 135–150.

—— 1969. Discussion on the paper 'The origin of ultramafic and ultrabasic rocks' by P.J. Wyllie. *Tectonophysics*, **7**, 517–518.

Jackson, I. & Arculus, R.J. 1984. Laboratory wave velocity measurements on lower crustal xenoliths from Calcutteroo, South Australia. *Tectonophysics*, **101**, 185–197.

Kay, S.M. & Kay, R.W. 1983. Thermal history of the deep crust inferred from granulite xenoliths, Queensland, Australia. *Am. J. Sci.*, **283-A**, 486–513.

Knutson, J. & Green, T.H. 1975. Experimental duplication of a high pressure megacryst/cumulate assemblage in a near-saturated hawaiite. *Contr. Min. Petrol.*, 121–132.

Leyreloup, A., Dupuy, C. & Andriamboola, R. 1977. Catazonal xenoliths in French neogene volcanic rocks: constitution of the lower crust. *Contr. Min. Petrol.*, **62**, 283–300.

Lovering, J.F. 1964. The eclogite-bearing basic igneous pipe at Ruby Hill, near Bingara, N.S.W. *J. Proc. Roy. Soc. N.S.W.*, **97**, 73–79.

—— & White, A.J.R. 1969. Granulitic and eclogitic inclusions from basic pipes at Delegate, Australia. *Contr. Min. Petrol.*, **21**, 9–52.

Mathur, S.P. 1983a. Deep reflection experiments in north-eastern Australia, 1976–1978. *Geophysics*, **48**, 1588–1597.

—— 1983b. Deep reflection probes in eastern Australia reveal differences in the nature of the crust. *First Break*, July, 9–16.

—— 1983c. Deep crustal reflection results from the central Eromanga Basin, Australia. *Tectonophysics*, **100**, 163–173.

McCulloch, M.T., Arculus, R.J., Chappell, K.W. & Ferguson, J. 1982. Lower continental crust: inferences from isotopic and geochemical studies of nodules in kimberlite. *Nature*, **300**, 166–169.

Meissner, R. 1973. The 'Moho' as a transition zone. *Geophys. Surv.*, **1**, 195–216.

O'Neill, H. St. C. 1981. The transition between spinel lherzolite and garnet lherzolite, and its use as a geobarometer. *Contr. Min. Petrol.*, **77**, 185–194.

O'Reilly, S.Y. & Griffin, W.L. 1985. A xenolith-derived geotherm for southeastern Australia and its geophysical implications. *Tectonophysics*, **111**, 41–63.

Osborne, G.D. 1920. The volcanic neck at the Basin, Nepean River. *J. Proc. Roy. Soc. N.S.W.*, **54**, 113–145.

Plumb, K.A. 1979. The tectonic evolution of Australia. *Earth Sci. Rev.* **14**, 205–249.

Powell, C. McA. 1983. Tectonic relationship between the Ordovician and Late Silurian palaeogeographics of southeastern Australia. *J. Geol. Soc. Aust.*, **30**, 353–373.

Rudnick, R.C., Taylor, S.R. & Jackson, I. 1984. Geochemistry and seismic velocities of the lower crust in northeast Queensland: evidence from granulite facies nodules in recent basalt. *Geol. Soc. Aust.* (in press).

Shaw, S.E., Flood, R.H. & Riley, G.H. 1982. The Wologorong Batholith, New South Wales, and the extension of the I-S line of the Siluro-Devonian granitoids. *J. Geol. Soc. Aust.*, **29**, 41–48.

Spera, F.J. 1980. Aspects of magma transport. *In*: Hargraves, R.B. (ed.) *Physics of Magmatic Processes*. Princeton University Press, pp. 265–323.

Stolz, A.J. 1984. Garnet websterites and associated ultramafic inclusions from a nepheline mugearite in the Walcha area, New South Wales, Australia. *Mineralog. Mag.*, **48**, 167–180.

Sutherland, F.L. 1969. A review of the Cainozoic volcanic province of Tasmania. *Spec. Pub. Geol. Soc. Aust.* **2**, 133–144.

—— 1971. The geology and petrology of the Tertiary volcanic rocks of the Tamar Trough, northern Tasmania. *Rec. Queen Vic. Mus.* 36.

—— 1974. High pressure inclusions in tholeiitic basalt and the range of lherzolite-bearing magmas in the Tasmanian volcanic province. *Earth planet. Sci. Lett.*, **24**, 317–324.

—— & Hollis, J.D. 1982. Mantle–lower crust petrology from inclusions in basaltic rocks in eastern Australia—an outline. *J. Volc. Geotherm. Res.*, **14**, 1–29.

Wass, S.Y. 1973. Plagioclase-spinel intergrowths in alkali basaltic rocks from the Southern Highlands, New South Wales. *Contr. Min. Petrol.*, **38**, 167–175.

—— & Hollis, J.D. 1983. Crustal growth in southeastern Australia—evidence from lower crustal eclogitic and granulitic xenoliths. *J. Metam. Geol.*, **1**, 25–45.

—— & Irving, A.J. 1976. *XENMEG: A Catalogue of Occurrences of Xenoliths and Megacrysts in Volcanic Rocks of Eastern Australia*. Australian Museum, Sydney, 441 pp.

White, A.J.R., Williams, I.S. & Chappell, B.W. 1976. The Jindabyne thrust and its tectonic, physiographic and petrogenetic significance. *J. Geol. Soc. Aust.*, **23**, 105–112.

——, —— & —— 1977. *The geology of the Berridale 1:100,000 Sheet, 8625,* Geological Survey of N.S.W., Sydney.

WILKINSON, J.F.G. 1974. Garnet clinopyroxenite inclusions from diatremes in the Gloucester area, New South Wales. *Contr. Min. Petrol.,* **46,** 275–299.

—— 1975a. Ultramafic inclusions and high pressure megacrysts from a nephelinite sill, Nandewar mountains, north-eastern New South Wales, and their bearing on the origin of certain ultramafic inclusions in alkaline volcanic rocks. *Contr. Min. Petrol.,* **51,** 235–262.

—— 1975b. An Al-spinel ultramafic-mafic inclusion suite and high pressure megacrysts in an analcimite and their bearing on basaltic fractionation at elevated pressures. *Contr. Min. Petrol.* **54,** 71–104.

WILSHIRE, H.G. & BINNS, R.A. 1961. Basic and ultrabasic xenoliths from volcanic rocks of New South Wales. *J. Petrology,* **2,** 185–208.

WILSON, A.F. 1978. Comparison of some of the geochemical features and tectonic setting of Archaean and Proterozoic granulites, with particular reference to Australia. *In:* WINDLEY, B.F. & NAQVI, S.M. (eds) *Archaean Geochemistry* pp. 241–268, Elsevier, New York.

WILSON, C.R. & SMITH, D. 1984. Cooling rate estimates from mineral zonation: resolving power and applications. *In:* KORNPROBST, J. (ed.). *Kimberlites II: The mantle and crust–mantle relationships.* Elsevier, Amsterdam, 265–276.

WOOD, B.J. & BANNO, S. 1973. Garnet-orthopyroxene and orthopyroxene-clinopyroxene relationships in simple and complex systems. *Contr. Min. Petrol.,* **42,** 109–121.

W.L. GRIFFIN, Geologisk Museum, Sars gate 1, 0562 Oslo 5, Norway.
S.Y. O'REILLY, School of Earth Sciences, Macquarie University, North Ryde, N.S.W. 2113, Australia.

Subject Index

Note: page references in italic refer to figures, those in bold to tables.